**Books are to be returned on or before
the last date below.**

Nutritional Genomics

Edited by
Regina Brigelius-Flohé and
Hans-Georg Joost

Nutritional Genomics

Impact on Health and Disease

Edited by Regina Brigelius-Flohé and Hans-Georg Joost

WILEY-VCH

WILEY-VCH Verlag GmbH & Co. KGaA

The Editors

Prof. Dr. Regina Brigelius-Flohé
Prof. Dr. Dr. Hans-Georg-Joost
German Inst. of Human Nutrition,
Potsdam-Rehbruecke
A.-Scheunert-Allee 114-116
14558 Nuthetal
Germany

1st Edition 2006
1st Reprint 2007

Library of Congress Card No.: applied for
British Library Cataloging-in-Publication Data:
A catalogue record for this book is available from the British Library.

Bibliographic information published by Die Deutsche Bibliothek
Die Deutsche Bibliothek lists this publication in the Deutsche Nationalbibliografie; detailed bibliographic data is available in the Internet at ⟨http://dnb.ddb.de⟩.

© 2006 WILEY-VCH Verlag GmbH & Co. KGaA, Weinheim

Printed in the Federal Republic of Germany.
Printed on acid-free paper.

Typesetting   Asco Typesetters, Hong Kong
Printing   Strauss GmbH, Mörlenbach
Binding   Litges & Dopf Buchbinderei GmbH, Heppenheim

ISBN-13 978-3-527-31294-8
ISBN-10 3-527-31294-3

Foreword

The idea of creating a book on nutritional genomics was born at the 55th Mosbach Colloquium in 2004, which covered the topic "How nutrients influence gene activity." Regina Brigelius-Flohé and myself had organized this symposium together with the "Gesellschaft für Biochemie und Molekularbiologie" (GBM) in Germany with the clear intention of reanimating nutritional sciences and bringing them back onto the German research and funding agenda.

This discipline, like Sleeping Beauty, appeared to have fallen into a kind of dormancy after a flourishing period in the first half of the last century, when the basic metabolic pathways had been worked out and most vitamins and trace elements had been identified as precursors of coenzymes, cofactors, or pro-hormones. With the rapid advance of molecular biology during the last decade, however, new aspects of nutrition appeared on the horizon and the Prince was ready to kiss awake his Sleeping Beauty. Of primary importance was the discovery that dietary components interacted with various transcription factors of the nuclear receptor family that had previously been known to be affected only by hormones or drugs. Moreover, several "orphan" receptors now have been adopted by their natural ligands, which constitute nutritive components or are derived therefrom by intermediary metabolism in the organism. Dietary agents were also found to interfere with intracellular central signaling cascades and to induce gene expression, and the variability in such food-responsive systems appeared to provide a key to understanding the differential susceptibility of genetically diverse individuals to food-related diseases.

The Editors, Regina Brigelius-Flohé and Hans-Georg Joost, have succeeded in bringing together many renowned international experts to present their work in this new frontier of nutrition research and have managed to cover all the currently emerging trends in nutritional genomics, ranging from nutrigenomics to nutrigenetics and to basic receptor research, physiology, and pathophysiology. The book thus provides a topical state-of-the-art compilation of recent developments in this exponentially developing field. It certainly deserves a broad readership in the disciplines of nutrition, biology, medicine, and life sciences.

Nutritional Genomics. Edited by Regina Brigelius-Flohé and Hans-Georg Joost
Copyright © 2006 WILEY-VCH Verlag GmbH & Co. KGaA, Weinheim
ISBN: 3-527-31294-3

The newcomer to the field will enjoy the stimulatory introductory reviews and may dig deeper into specialized chapters, and the specialists can update and enlarge their horizons while reading the latest news from their expert colleagues.

Josef Köhrle
November 2005 Charité University Medicine Berlin

Contents

Nutritional Genomics. Edited by Regina Brigelius-Flohé and Hans-Georg Joost
Copyright © 2006 WILEY-VCH Verlag GmbH & Co. KGaA, Weinheim
ISBN: 3-527-31294-3

Part III
Nutrigenetics (Nutrient–Genotype Interactions)

Preface

Nutrition is a key factor in the development of chronic diseases. Nutritional research has therefore evolved from a discipline that determines the required daily intake of calories and essential macro- and micronutrients into a biomedical science with a high potential for disease prevention. The food industry has responded to this awareness by designing foods that not only attempt to satisfy the hedonic preferences of the consumer but also promise additional health benefits. The chemical complexity of diets and the equally complex responses of individuals to a defined diet imply scientific challenges that are easily met neither by the stringent methodology of analytical biochemistry nor by descriptive epidemiology. Instead, the concerted approaches of functional genomics promise to provide reliable and more useful answers within a reasonable time frame.

Within nutritional science, functional genomics comprises two interrelated areas: the influence of nutrients on the transcriptional activity of genes and the heterogeneous response of gene variants to nutrients. The former is usually referred to as "nutrigenomics" and is generally studied with technologies of systems biology such as transcriptomics, proteomics, and metabolomics. The latter area is sometimes referred to as "nutrigenetics"; its findings are to establish a scientific basis for the concept of a genotype-based, personalized nutrition. The title "Nutritional Genomics" is used to indicate that this book intends to cover progress made in both areas.

The first part of this book is devoted to the impact of particular food components on the machinery of gene expression. One of the introductory chapters elaborates on the potential blessings and limitations of current technologies in nutritional science. Another summarizes the state of the art in nuclear receptor research, which has for long remained a domain of endocrinology and pharmacology but is gaining increasing interest in nutritional science. In fact, many of these receptors, in particular the so-called orphan ones, turn out to respond specifically to dietary components with transcriptional activation or repression. In the following chapters particular aspects of food-responsive gene activities are presented by renowned experts, each chapter addressing, as far as possible, the molecular mechanisms, micro- or macronutrients involved, and the potential or established relevance to human health.

Nutritional Genomics. Edited by Regina Brigelius-Flohé and Hans-Georg Joost
Copyright © 2006 WILEY-VCH Verlag GmbH & Co. KGaA, Weinheim
ISBN: 3-527-31294-3

In Part III the problems of predispositions to food-related diseases are addressed. Clearly, the etiology and course of several complex diseases are associated with nutritional habits. For example, hypertension may depend on sodium intake and other dietary parameters, diabetes is a consequence of abdominal obesity, and obesity is to a large part due to a hypercaloric, energy-dense nutrition (the so-called cafeteria diet). There are also clear associations between dietary parameters in certain types of cancer as well as in inflammatory diseases. Yet individuals by no means respond identically to diets that are generally considered unhealthy nor is a "healthy diet" equally tolerated by everybody. This is because the genetic basis of related diseases is heterogeneous and, accordingly, the functional response to nutrients differs between individuals. Current research, therefore, aims to identify the variability of food-responsive genes that modulate the quality of the response and thus determine disease risks or other nutrition-dependent outcomes. The current knowledge on the genetics of some nutrition-associated diseases that is compiled here will reveal that the concept of a personalized nutrition, although being tested for particular conditions, still remains over the horizon.

The Editors wish to thank all the authors who have put considerable effort into their contributions, thereby providing a detailed overview on this exciting and rapidly advancing area of research. They also hope that this compilation of research frontiers will help readers to discriminate between fact and fiction and to plan their future with more certainty, whether in terms of nutritional research or personal lifestyle.

Nuthetal, November 2005

Regina Brigelius-Flohé
Hans-Georg Joost

Contributors

Ana Aranda
Instituto de Investigaciones Biomédicas
Consejo Superior de Investigaciones
Científicas and Universidad Autónoma de
Madrid
Arturo Duperier 4
28029 Madrid
Spain

Andreas G. Bader
The Scripps Research Institute
Department of Molecular and Experimental
Medicine
Institute of Oncovirology
10550 North Torrey Pines Road, BCC239
La Jolla
California 92037
USA

Inês Barroso
Wellcome Trust Sanger Institute
Metabolic Disease Group
Wellcome Trust Genome Campus
Hinxton
Cambridge CB10 1SA
UK

Günter Brönner
Rheinische Kliniken Essen
Department of Child and Adolescent
Psychiatry
University of Duisburg-Essen
Virchowstr. 174
45133 Essen
Germany

Alain Bruhat
Unité de Nutrition et Métabolisme Protéique
Institut National de la Recherche
Agronomique de Theix
63122 Saint Genès Champanelle
France

Bernd Bufe
German Institute of Human Nutrition
Potsdam-Rehbruecke
Department of Molecular Genetics
Arthur-Scheunert-Allee 114-116,
14558 Nuthetal
Germany

Yoan Cherasse
Unité de Nutrition et Métabolisme Protéique
Institut National de la Recherche
Agronomique de Theix
63122 Saint Genès Champanelle
France

Hannelore Daniel
Molecular Nutrition Unit
Department of Food and Nutrition Sciences
Technische Universität München
Am Forum 5
85354 Freising
Germany

Béatrice Desvergne
University of Lausanne
Center for Integrative Genomics
Genopode
CH-1015 Lausanne
Switzerland

Hélène Duez
UR.545 INSERM
Département d'Athérosclérose
Institut Pasteur de Lille and Université de
Lille 2
1, rue du Professeur Calmette
F-59019 Lille Cedex
France

Pierre Fafournoux
Unité de Nutrition et Métabolisme Protéique
Institut National de la Recherche
Agronomique de Theix
63122 Saint Genès Champanelle
France

Fabienne Foufelle
INSERM U671
Centre de Recherches Biomédicales des
Cordeliers
Université Pierre et Marie Curie
15, rue de l'école de médecine
75270 Paris
France

S. Friedel
Rheinische Kliniken Essen
Department of Child and Adolescent
Psychiatry
University of Duisburg-Essen
Virchowstr. 174
45133 Essen
Germany

Jean-Charles Fruchart
UR.545 INSERM
Département d'Athérosclérose
Institut Pasteur de Lille and Université de
Lille 2
1, rue du Professeur Calmette
59019 Lille Cedet
France

Emmanuelle Germain
Department of Cell Biology and Signal
Transduction
Institut de Génétique et de Biologie
Moléculaire et Cellulaire (IGBMC)/CNRS/
INSERM/ULP
1, rue Laurent Fries
67404 Illkirch Cedex
CU de Strasbourg
France

Maolian Gong
Max-Delbrück-Center for Molecular Medicine
(MDC)
Robert-Rössle-Str. 10
13125 Berlin
Germany

Robert F. Grimble
Institute of Human Nutrition
School of Medicine
University of Southampton
Southampton SO16 7PX
UK

Hinrich Gronemeyer
Department of Cell Biology and Signal
Transduction
Institut de Génétique et de Biologie
Moléculaire et Cellulaire (IGBMC)/CNRS/
INSERM/ULP
1, rue Laurent Fries
67404 Illkirch Cedex
CU de Strasbourg
France

J. Hebebrand
Rheinische Kliniken Essen
Department of Child and Adolescent
Psychiatry
University of Duisburg-Essen
Virchowstr. 174
45133 Essen
Germany

John E. Hesketh
Institute for Cell and Molecular Biosciences
The Medical School
Newcastle University
Framlington Place
Newcastle-upon-Tyne NE2 4HH
UK

A. Hinney
Rheinische Kliniken Essen
Department of Child and Adolescent
Psychiatry
University of Duisburg-Essen
Virchowstr. 174
45133 Essen
Germany

Norbert Hübner
Max-Delbrück-Center for Molecular Medicine
(MDC)
Robert-Rössle-Str. 10
13125 Berlin
Germany

Céline Jousse
Unité de Nutrition et Métabolisme Protéique
Institut National de la Recherche
Agronomique de Theix
63122 Saint Genès Champanelle
France

Fred F. Kadlubar
Division of Pharmacogenomics and Molecular
Epidemiology
National Center for Toxicological Research
Jefferson, AR 72079
USA

Sander Kersten
Nutrition, Metabolism and Genomics Group
PO Box 8129
Wageningen University
6700 EV Wageningen
The Netherlands

Markku Laakso
Department of Medicine
University of Kuopio
70210 Kuopio
Finland

Julian Little
Canada Research Chair in Human Genome
Epidemiology
Department of Epidemiology and Community
Medicine
University of Ottawa
451 Smyth Road
Ottawa
Ontario K1H 8M5
Canada

Stéphane Mandard
Nutrition, Metabolism and Genomics Group
PO Box 8129
Wageningen University
6700 EV Wageningen
The Netherlands

Anne-Catherine Maurin
Unité de Nutrition et Métabolisme Protéique
Institut National de la Recherche
Agronomique de Theix
63122 Saint Genès Champanelle
France

Catherine Méplan
Institute for Cell and Molecular Biosciences
The Medical School
Newcastle University
Framlington Place
Newcastle-upon-Tyne NE2 4HH
UK

Wolfgang Meyerhof
German Institute of Human Nutrition
Potsdam-Rehbruecke
Department of Molecular Genetics
Arthur-Scheunert-Allee 114-116,
14558 Nuthetal
Germany

Liliane Michalik
University of Lausanne
Center for Integrative Genomics
Genopode
1015 Lausanne
Switzerland

Susan Nowell
Department of Environmental and
Occupational Health
College of Public Health
University of Arhansas for Medical Sciences
Little Rock, AR 72205
USA

Vasileios Pagmantidis
Institute for Cell and Molecular Biosciences
The Medical School
Newcastle University
Framlington Place
Newcastle-upon-Tyne NE2 4HH
UK

Angel Pascual
Instituto de Investigaciones Biomédicas
Consejo Superior de Investigaciones
Científicas and Universidad Autónoma de
Madrid
Arturo Duperier 4
28029 Madrid
Spain

Jean-Paul Pégorier
Département d'Endocrinologie
Institut Cochin
INSERM U567, CNRS UMR
8104
Université Paris 5 René Descartes
24, rue du Faubourg Saint Jacques
75014 Paris
France

Catherine Postic
Institut Cochin
INSERM U567 CNRS UMR8104
Université René Descartes
Département d'Endocrinologie
24, rue du Faubourg Saint Jacques
Paris
France

K. Reichwald
Rheinische Kliniken Essen
Department of Child and Adolescent
Psychiatry
University of Duisburg-Essen
Virchowstr. 174
45133 Essen
Germany

A. Scherag
Institute of Medical Biometry and
Epidemiology
Philipps-University of Marburg
Germany

Linda Sharp
National Cancer Registry Ireland
Elm Court
Boreenmanna Road
Cork
Ireland

Bart Staels
UR.545 INSERM
Département d'Athérosclérose
Institut Pasteur de Lille and Université de
Lille 2
1, rue du Professeur Calmette
59019 Lille Cedex
France

Catherine Stevenson
MRC Epidemiology Unit
Strangeways Research Laboratory
Worts Causeway
Cambridge CB1 8RN
UK

Frédéric Varnat
University of Lausanne
Center for Integrative Genomics
Genopode
1015 Lausanne
Switzerland

Peter K. Vogt
The Scripps Research Institute
Department of Molecular and Experimental
Medicine
Institute of Oncovirology
10550 North Torrey Pines Road, BCC239
La Jolla
California 92037
USA

Emilie Voltz
Department of Cell Biology and Signal
Transduction
Institut de Génétique et de Biologie
Moléculaire et Cellulaire (IGBMC)/CNRS/
INSERM/ULP
1, rue Laurent Fries
67404 Illkirch Cedex
CU de Strasbourg
France

Walter Wahli
University of Lausanne
Center for Integrative Genomics
Genopode
1015 Lausanne
Switzerland

Nicholas Wareham
MRC Epidemiology Unit
Strangeways Research Laboratory
Worts Causeway
Cambridge CB1 8RN
UK

Uwe Wenzel
Hannelore Daniel
Molecular Nutrition Unit
Department of Food and Nutrition Sciences
Technische Universität München
Am Forum 5
85354 Freising
Germany

A.-K. Wermter
Department of Child and Adolescent
Psychiatry
Philipps-University of Marburg
Germany

List of Abbreviations

AARE	amino acid response element
ABCA1	ATP-binding cassette transporter A1
ACBP	Acyl-CoA binding protein
ACE	angiotensin converting enzyme
ACL	ATP citrate lyase
ACS	acyl-CoA synthetase
ADH	alcohol dehydrogenase
ADRP	adipose differentiation-related protein
AF-1	activation function -1
AF-2	activation function -2
AgRP	agouti-related protein
AGT	angiotensinogen
ALDH	aldehyde dehydrogenase
AML	acute myeloid leukemia
aP2	adipocyte fatty acid-binding protein
Apaf-1	apoptosis protease-activating factor-1
APL	akute promyelocytic leukemia
ApoA1	apolipoprotein A1
ApoE	apolipoprotein E
AR	Androgen receptor
ARE	antioxidant responsive element
ASNS	asparagine synthetase
ASP	acylation-stimulating protein
AT1R	angiotensin II receptor type 1
AT2R	angiotensin II receptor type 2
ATF	activating transcription factor
ATRA	all trans retinoic acid
AZiP	mouse strain
BAF	BRG-1-associated factor
BaP	benzo[a]pyrene
Bcl-2	B-cell leukemia/lymphoma 2
BDNF	brain-derived neurotrophic factor
bHLH	basic helix-loop-helix

Nutritional Genomics. Edited by Regina Brigelius-Flohé and Hans-Georg Joost
Copyright © 2006 WILEY-VCH Verlag GmbH & Co. KGaA, Weinheim
ISBN: 3-527-31294-3

bHLH-LZ	basic helix-loop-helix leucine zipper
BMI	body mass index
BP	blood pressure
BRG-1	ATP-hydrolysing subunit of the chromatin remodeling SWI/SNF complex
C/EBP	CCAAT/enhancer binding protein
CAPN10	calpain 10
CAR	constitutive androstane receptor
CARM	cofactor-associated arginine methyltransferase
CAT	cationic amino acid transporter
CBP/P300	CREB-binding protein/p300(=adenovirus E1A-associated 300 kDa protein)
CD36	oxLDL scavenger receptor CD36
ChIP	chromatin immunoprecipitation
CHOP	C/EBP homologous protein
ChoRE (ChRE)	carbohydrate-responsive element
ChREBP	carbohydrate-responsive element-binding protein
CoRNR box	co-repressor nuclear receptor interaction box
COUP-TF	Chicken ovalbumin upstream promoter transcription factor
COX	cyclooxygenase
CPT-I	carnitin-palmitoyl transferase-I
CRE	cAMP responsive element
CREB	cAMP responsive element binding (protein)
CRP	C-reactive protein
CVD	cardiovascular disease
CYP11B1	11-β-hydroxylase
CYP11B2	aldosterone synthase
CYP7A	Cytochrome P450 7A
DAX-1	dosage-sensitive sex reversal adrenal hypoplasia congenita, critical region on the X chromosome, gene 1
db/db mice	mice lacking a functional leptin receptor
DBD	DNA binding domain
DGAT	diacylglycerol acyltransferase
DHA	docosahexaenoic acid
DNMT	DNA methyltransferase
DR	direct repeat
DRIP	vitamin D receptor-interacting protein
EFSec	selenocysteine-specific elongation factor
EGCG	epigallocatechin gallate
EH	essential hypertension
EH	epoxide hydrolase
eIF	eukaryotic translational initiation factor
EPIC	European Prospective Investigation into Cancer and Nutrition
ER	estrogen receptor
ERR	estrogen related receptor

ERSR	endoplasmic reticulum stress response
FABP	fatty acid-binding protein
FADD	Fas-associated death domain
FAT	fatty acid translocase
FIAF	fasting-induced adipose factor
FIAF/ANGPTL4	fasting-induced adipose factor/angiopoietin-like protein 4
FOXO	forkhead transcription factor
FXR	farnesoid X receptor
GAP	GTPase-activating protein
GCN2	general control non-repressible 2 (an eIF-2a kinase)
GCNF	germ cell nuclear factor
GEF	guanine nucleotide exchange factor
GH	growth hormone
GPAT	glycerol-3-phosphate acyltransferase
GPx	glutathione peroxidase
GR	glucocorticoid receptor
GSK3β	glycogen synthase kinase 3β
GST	glutathione-S-transferase
HAA	heterocyclic aromatic amines
HAT	histone acetyltransferase 37
HBP	high blood pressure
HCA	heterocyclic amine
HCN	hyperpolarization activated cyclic nucleotide-gated potassium channel
HDAC	histone deacetylase
HGP	human genome project
HIF-1	hypoxia-inducible factor-1
HMT	histone methyltransferase
HNF	hepatocyte nuclear factor
HPFS	Health Professional Follow-up-Study
HRE	hormone responsive element
HUGO	human genome organization
IEF	isoelectric focusing
IFNγ	interferon γ
Ihh	indian hedgehog
IKKα	IkB kinase α
ILK	integrin-linked kinase
IMT	intima-media thickness
iNOS	inducible nitric oxide synthase
iNOS	inducible NO-synthase
InR	insulin receptor
INSIG	insulin-induced gene
IOI	iodothyronine deiodinase
IRES	internal ribosome entry site
IRS-1	insulin receptor substrate-1

JDP2	jun dimerization protein-2
KAR	3-keto acyl-CoA reductase
LBD	ligand binding domain
LCE	long chain fatty acid elongase
LD	linkage disequilibrium
LDL-R	LDL receptor
LOD	logarithm of odds
LRH1	liver receptor homologous protein 1
LXR	liver X receptor
MALDI-TOF	matrix-assisted laser desorption/ionization time-of flight mass spectrometry
MAPK	mitogen-activated protein kinase
MC4R	melanocortin-4, receptor
MCP-1	monocyte chemoattractant protein 1
ME	malic enzyme
MeIQ	2-amino-3,4-dimethylimidazo [4,5-f]quinoline
MeIQx	2-amino-3,8-dimethylimidazo [4,5-f]quinoxaline
MODY	maturity-onset diabetes of the young
MR	mineralocorticoid receptor;
MRE	metal responsive element
mRNP	messenger ribonucleotide particles
α-MSH	α-melanocyte-stimulating hormone
MTHFR	methylenetetrahydrofolate reductase
NAT	*N*-acetyltransferase
NCoR	nuclear co-repressor
NEFA	non-esterified fatty acids
NFκB	nuclear factor κB
NGFI-B	NGF-induced clone B
NIDDM1	non insulin dependent diabetes mellitus gene1
NPY	neuropeptide Y
NR	nuclear receptor
NRAMP1	natural restance-associated macrophage protein 1
Nrf2	nuclear factor-erythroid 2 p45-related factor 2
NRRE	nuclear receptor responsive element
NSRE	nutrient sensing response element
NSRU	nutrient sensing regulatory unit
ob/ob mice	mice lacking leptin
ORE	oxygen-responsive element
p160	family of nuclear coactivators 37
P3K	a homolog of the p110α submit of PI3K
PABP	poly(A)-tail-binding protein
PAS	period aryl hydrocarbon receptor/single minded homology
PBMC	peripheral blood mononuclear cells
PCAF	p300/CBP-associating factor
PDK1	phosphoinositide-dependent protein kinase-1

PGC-1	PPAR gamma co-activator 1
PhIP	2-amino-1-methyl-6-phenylimidazo [4,5-b]pyridine
PI3K	phosphoinositol-3-kinase
PLZF	promyeloic leukemia zinc finger
PML	promyelocytic leukemia
PNR	photoreceptor-specific nuclear receptor
POMC	pro-opiomelanocortin
PP2A	protein phosphatase 2A
PPAR	peroxisome proliferator-activated receptor
PR	progesterone receptor
PRMT	protein arginine methyltransferase
PTC	phenylthiocarbamide
PTEN	phosphatase and tensin homolog deleted on chromosome 10
PUFA	polyunsaturated fatty acid
PXR	pregnane X receptor
PYY	peptide $YY_{3\text{-}36}$
QTL	quantitative trait loci
QTN	quantitative trait nucleotide
RAAS	renin-angiotensin aldosterone system
RAR	retinoic acid receptor
RevErbA	reverse ErbA
Rheb	ras homolog enriched in brain
RID	nuclear receptor interacting domain
RNA	RNA interference technique
RXR	retinoid X receptor
RZR/ROR	retinoid Z receptor/retinoic acid-related orphan receptor
S6K	p70 S6 kinase
SBP	systolic blood pressure
SBP2	SECIS-binding protein-2
SCAP	SREBP cleavage-activating protein
SCD-1	stearoyl-CoA desaturase-1
SECiS	selenocysteine insertion sequence
SelB	selenocysteine-specific translation elongation factor
SelP	selenoprotein P
SelW	selenoprotein W
SERM	selective estrogen receptor modulator
SF-1/FTZ-F1	steroidogenic factor 1 Fushi Tarazu Factor 1
SHP	small heterodimeric partner
SHR	spontaneously hypertensive rat
Sin3	yeast transcriptional repressor
SLC 6A 14	solute carrier family 6 (neurotransmitter transporter) member 14
SMCC	Srb and mediator protein-containing complex
SMRT	silencing mediator for retinoic and thyroid hormone receptors
SNP	single-nucleotide polymorphism

SPPARM	selective PPAR modulators
SPS complex	an amino acid sensor complex consisting of three proteins: SSY1p, Pt-3p, SSy5p
SPS2	selenophosphate synthetase-2
SRB	suppressor of RNA polymerase B
SRC	steroid receptor co-activator
SREBP	sterol regulatory element-binding protein
ST2	streptozotocin
SULT	sulfotransferase
SUMO	small ubiquitin-related modifier
SWI/SNF	a chromatin remodeling complex
TCE	translation control element
TDT	transmission distortion/disequilibrium test
TER	trans-2,3-enoyl-CoA reductase
TGF	transforming growth factor
TLR	Toll-like receptors
TLX	Tailles-related receptor
TOR	target of rapamycin
TR	thyroid hormone receptor
TR1	thioredoxin reductase-1
TR2	testis receptor
TRAIL	tumor necrosis factor-related apoptosis inducing ligand
TRAP	thyroid hormone receptor (TR)
TRPM5	transient receptor potential cation channel, subfamily M, member 5
trsp	Selenocysteine tRNA$^{(ser)sec}$ gene
TSC	tuberous sclerosis complex
TSC-22	TGFβ-stimulated clone 22
TZD	thiazolidinedione
USF	upstream stimulating factor
3′UTR	3′ untranslated region
VCAM-1	vascular cell adhesion molecule 1
VDR	Vitamin D receptor
WAF1/Cip1(p21)	a general inhibitor of cyclin-dependent kinases
WINAC	WSTF including nucleosome assembly complex
WNK	with no K (lysine) kinase
WSTF	Williams syndrome transcription factor

Part I
Introduction – Definitions

1
Nutritional Genomics: Concepts, Tools and Expectations

Hannelore Daniel and Uwe Wenzel

1.1
Nutrigenomics: Just Another "omic"?

The age of nutrigenomics is already upon us. Various new programs in molecular nutrition research have been launched in Europe, Asia, and the US under the heading of nutrigenomics. We may for this review consider nutrigenomics as the science that seeks to provide a molecular understanding for how diets and common dietary constituents affect mammalian metabolism and health by altering gene/protein expression on basis of an individual's genetic makeup.

Although nutrigenomics represents in the first place just another "omic," it clearly induces a conceptual shift in nutritional sciences by moving the genome into the center of all the processes that essentially determine mammalian metabolism in health and disease. Moreover, for the first time, nutritional science speaks the same language and uses the same tools as the other biomedical sciences and this is going to change the face of nutrition research. Nutritional sciences is functional genomics "par excellence" and will thereby move the discipline into the heart of biological sciences. Unlike other environmental factors, nutrients, non-nutrient components of foods, and xenobiotics in foods have huge variability in dose and time and hit a rather static genome, affecting the function of a large number of proteins encoded by the respective mRNA molecules that are expressed in a certain cell, organ or organism. Alterations of mRNA levels and in turn of the corresponding protein levels are critical parameters in controlling the flux of a nutrient or metabolite through a biochemical pathway. Nutrients and non-nutrient components of foods, diets, and lifestyle can affect essentially every step in the flow of genetic information from gene expression control to protein synthesis, protein degradation, and allosteric control and consequently alter metabolic functions in the most complex ways (Fig. 1.1).

The advent of high-throughput technologies has led to the rapid accumulation of biological data, ranging from complete genomic sequences, transcripts, proteome and metabolome profiles as well as the first protein–protein interaction maps. Referred to as "omics", these parallel approaches are usually classified by the measured target molecules. Transcriptomics determines the transcript levels or pat-

Nutritional Genomics. Edited by Regina Brigelius-Flohé and Hans-Georg Joost
Copyright © 2006 WILEY-VCH Verlag GmbH & Co. KGaA, Weinheim
ISBN: 3-527-31294-3

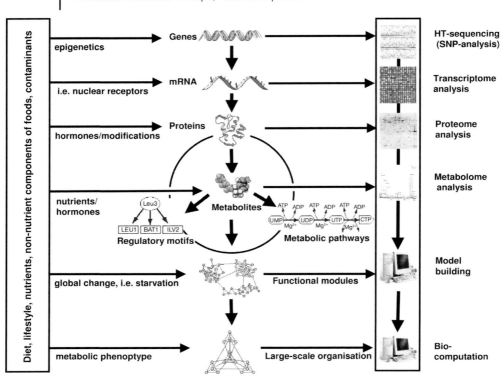

Fig. 1.1. Nutrigenomics as the paradigm for research on environment–genome interactions. Nutritional factors can affect essentially every step from information storage and retrieval, to processing and the execution of biological processes. The emerging new profiling technologies as well as data processing and interpretation tools make the corresponding adaptive changes of mammalian metabolism on a global scale accessible.

terns of subclasses or even of all expressed genes of a given genome. Likewise, proteomics refers to the analysis of the protein complement and metabolomics (also called metabonomics) determines in parallel the accessible metabolites in a cell, tissue, organ, or organism. The data output of these approaches is enormous and often overwhelms our ability to understand the underlying biological processes.

Nutritional science in the past was characterized by well-defined experimental studies based on the experience and knowledge that there is hardly anything else as difficult to standardize as mammalian nutrition. In terms of the biological read-outs of nutritional studies, in most cases only a few parameters could be determined simultaneously. The conceptual shift in biological science towards application of high-throughput profiling technologies poses a particular challenge to nutrition researchers as they now have additionally to handle huge data sets derived from the "omic" approaches. How can we use this information to build metabolic topology maps that are easy to comprehend and to interpret and that allow us to navigate to the specific information that we need? Here nutritional science clearly relies on the new systems biology tools of pathway construction that are based on concepts of control theory, numerical analysis, and stochastic processes. Although systems biology is dependent on "omics" and technologies for data input, it really encompasses the design and use of new analysis tools, and the development of new ways to represent data in a meaningful manner. Nutritional systems biology is the high end of systems biology when it comes to describing the highly diverse changes in metabolism occurring at the same time in different organs or even within an organ in its different cell populations.

1.1.1
What makes Nutrigenomic Research Exceptional?

In contrast to applications of the profiling technologies in drug discovery or toxicology testing, nutrigenomics deals with some exceptional problems. Drug and xenobiotic testing usually determines the consequences of just one compound on the background of a limited number of relevant genes but an otherwise fairly stable environment. Of course, the test compounds may undergo extensive metabolism and the bioactivity could as an integrated read-out result from both the parent compound and the metabolite(s). However, assessing the metabolic response to complex foods is like looking at hundreds of test compounds at the same time and a highly diverse response over time and spatial location (i.e. organ, cell type in an organ).

The human genome and the genetic variation within the human population are the result of high and persistent evolutionary pressures via processes of gene mutation, selection, and random drift. Nutrition has thereby shaped the human genome like no other environmental factor. Individual dietary components can affect gene mutation rates and nutrient availability affects, for example, fetal viability and modifies the penetrance of deleterious genetic lesions [1–5].

As part of the evolutionary pressure it was essential for life that mammals can adapt quickly to changes in their nutritional environment while maintaining metabolism to satisfy the needs of a high rate of ATP production and the production of all building blocks required for cell and tissue renewal and maintenance. Adaptation to food availability in terms of energy as well as individual (essential) nutrients requires very fast but also sustained responses that simultaneously change a huge set of interconnected metabolic processes. This is mainly achieved by hormones that can be classified by their chemical nature (i.e. peptide hormones, amino acid derivatives, or steroids) and/or the mode and time frame of their action. When looking at the effects of a diet on the genetic response, individual nutrients such as carbohydrates, lipids, proteins, or minerals such as calcium directly affect hormone secretion and these hormones adjust cellular functions via specific receptors and a multitude of intracellular signaling events for allowing the required metabolic changes to occur within milliseconds and/or by sustained responses over hours. Moreover, certain nutrients and metabolites directly affect gene expression via interaction with specific cellular targets, including nuclear receptors and response elements, and thereby mediate the integration of extracellular signals (hormones) and signals from the intracellular environment. Allosteric control mechanisms of protein functions are also an integral part of this synchronization of signal inputs from the extracellular and intracellular environment. Figure 1.2 provides a simplified view of the integrative nature of the input signals for adjusting metabolism to alterations in the nutritional environment. They key ques-

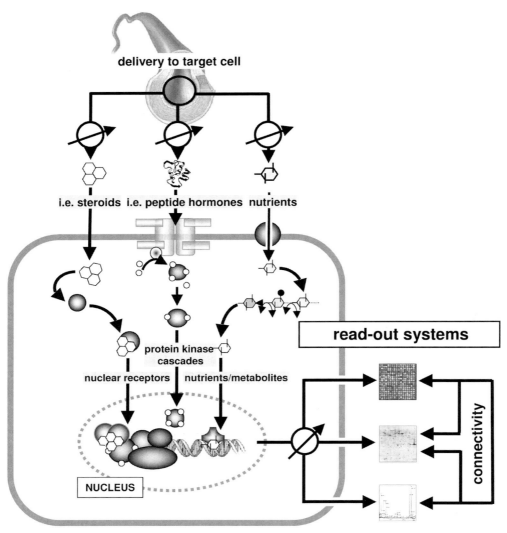

Fig. 1.2. A simplified model depicting the integrative nature of signal processing for transmitting changes in the nutritional environment into the adaptation of the transcriptome, proteome, and metabolome of a cell system.

tion is whether the "omics" technologies combined with advanced data analysis and interpretation tools allow us to reconstruct and understand the underlying sensing and signal integration mechanisms and their multidimensional wiring that in the end permit cells to regulate rates of nutrient transport and storage capacity, to fine-tune the flux of intermediates through metabolic routes and branching points, and to restructure the cellular transcriptome and proteome.

1.1.2
Transcript Profiling in Nutrition Research

For historical reasons, transcript profiling has dominated high-throughput genomic studies in mammalian systems since this technology has been around for quite some time [6–9]. Moreover, various commercial systems for easy-to-handle array-based screening applications are available [6–10]. Transcript-profiling experiments so far have often followed a simple experimental design in which, for example, cells or organisms are exposed to an altered nutritional environment (absence or presence of a particular compound) and are then assayed for changes in gene expression [11–13]. These first-generation experiments led to the general conclusion that the cells often respond to quite different environmental conditions with an overlapping response of a battery of genes, although these outputs most probably originate from multiple signaling pathways.

Most microarray studies in the nutrigenomics area so far have the character of snapshots. Based on the high costs of the arrays, pooled RNA samples and/or only a few arrays have been used for analysis. To come from the snapshot approach to more consistent and reliable data, time-series of changes in gene expression as well as repeated and statistically valid measurements are required. As the costs of commercial arrays are expected to drop considerably in the future and as more small-scale targeted and cheaper arrays become available, better microarray data are expected to be produced. It is also essential that the procedures of how the study was conducted and how the array experiments have been performed are well described and data need to be collected and deposited in a standardized format. ArrayExpress (www.ebi.ac.uk/arrayexpress) [14] is the database for collecting information about microarray experiments and is provided by the European Bioinformatics Institute (EBI). ArrayExpress is the world's first database for storage of microarray information that conforms agreed community standards of MIAME (Minimum Information About a Microarray Experiment) devised by the Microarray Gene Expression Data (MGED) Society (www.mged.org) [15]. Since the nutritional science community has no tradition yet in using transcriptome analysis tools it is advised to adopt these standards quickly. Under the umbrella of the European Nutrigenomics Organization (www.nugo.org) a first nutrigenomics-specified MIAME version has been developed and this should in the future allow via ArrayExpress the sharing of vast amounts of microarray-based data with the global science community. In addition, many journals require or recommend authors of microarray data-based papers to submit their data to a MIAME-compliant database.

In its application as a screening approach to nutrition- or nutrient-dependent gene expression analysis, transcript profiling may lead to numerous newly identified genes/mRNA species that respond – not necessarily as expected – to the particular treatment. Before starting to bring a biological meaning into the observation it is highly recommended to use an independent method such as reverse transcriptase polymerase chain reaction (RT-PCR) or Northern blotting to check the magnitude of the changes in the mRNA level of the identified target gene(s), as the reliability of gene expression changes depends on a variety of parameters and particularly on the applied normalization method. In most cases, array data slightly underestimate the changes in transcript levels but there is also often a considerable number of transcripts that are not confirmed as significantly changed in level when assayed by other methods. Nevertheless, global transcript profiling can be seen as an expedition into the terra incognita of molecular nutrition by identifying novel genes, mechanisms and/or pathways by which a dietary maneuver changes cell physiology. The downside of transcriptomics is that one can get lost in the attempt to understand why the changes happen and although hours of scanning of the relevant literature is a rewarding learning exercise it may not provide the answer.

Although currently mainly used in basic science applications, global gene expression analysis is beginning to move from the laboratories to large-scale clinical trials as a tool in diagnostics [16–18]. In the field of human nutrition, signatures or unique patterns of gene expression profiles are expected to be used to describe a nutritional condition or may even allow disease states – even preclinical ones – to be determined [19]. The potential of this technology to improve diagnosis and tailored treatment of human diseases becomes obvious in the area of cancer diagnosis. Several comprehensive studies have demonstrated the utility of gene expression profiles for the classification of tumors into characteristic and clinically relevant subtypes and the prediction of clinical outcomes [16–18, 20, 21]. Applied to human nutrition, gene expression profiling is of course limited (a) by the available cells that should preferentially be obtained by non-invasive techniques, (b) by the genetic heterogeneity of the human population, and (c) by the highly diverse dietary habits and lifestyles. Nevertheless, transcriptome analysis studies for exploring whether characteristic patterns or signatures reflecting the nutritional status in a human population can be obtained need to be performed to explore the scientific and diagnostic value of this technology.

1.1.3
Proteome Profiling in Nutritional Sciences

The term "proteome" was introduced as the complement of the genome and relates to the goal of determining all transcribed and translated open reading frames from a given genome. Analysis of the proteome is beginning to emerge as a second high-throughput tool for nutrition research. The revival of two-dimensional gel electrophoresis (2D-PAGE) but with high resolution, the advanced instrumentation and elegant software tools now available for gel analysis, and the enormous advancements in mass spectrometry have made proteomics applications a practical alternative screening method in the nutrigenomics tool box.

2D-PAGE separates proteins according to charge (isoelectric point: pI) by isoelectric focusing (IEF) in the first dimension and according to size (molecular mass) by sodium dodecyl sulfate PAGE (SDS-PAGE) in the second dimension. It therefore has a unique capacity for the resolution of complex mixtures of proteins, permitting the simultaneous analysis of hundreds or even thousands of gene products [22]. However, not all proteins are resolved and separated equally well by 2D-PAGE. Very alkaline, hydrophobic, and integral membrane proteins as well as high molecular weight proteins are still a problem. In some cases, a prefractionation according to cellular compartment (membranes, microsomes, cytosol, mitochondria) or according to protein solubility by classical means may be necessary [23, 24]. In addition, proteins of low cellular abundance, which may be particularly important in view of their cellular functions for example in signaling pathways, are still very difficult to be resolved in the presence of large quantities of housekeeping proteins [25]. However, new concepts are constantly being developed that employ for example tagging techniques [26] or enrich the minor proteins prior to separation in 2D gels.

The most common procedure for the identification of a protein spot in a gel is currently the peptide mapping or "fingerprint" analysis, but other techniques and approaches can also be applied. For peptide mapping, protein-containing spots are excised from the gel before the gel is altered chemically to make the protein accessible for hydrolysis by a protease such as trypsin [27]. Based on this site-specific enzymatic hydrolysis, a distinct and characteristic pattern of peptide fragments of a given protein serves as the peptide mass fingerprint. The mixture of peptides isolated by digestion with the protease is usually submitted to matrix-assisted laser desorption/ionization time-of-flight mass spectrometry (MALDI-TOF-MS) analysis to determine the corresponding peptide masses that are characteristic for a given protein. The mass spectrum obtained is submitted to computer programs that apply various algorithms for interpretation of the peptide pattern and to predict the protein based on a comparison with masses predicted by "virtual digestion" of identified open reading frames in a given genome [28]. Post-translational modifications of proteins such as addition of phosphate groups, hydroxylations at lysine or proline residues, glycosylations, or addition of fatty acids may also be identified by fragment analysis and structural TOF (FAST) or other techniques. Deviations of measured from predicted masses may be due to polymorphisms in coding sequences with subtle amino acid substitutions or even more pronounced with deletions or insertions. Such changes in the primary sequence can be resolved but strongly depend on the type of substitution and may require internal peptide Edman sequencing or more advanced mass analysis by electrospray ionization mass spectroscopy.

It needs to be emphasized here that proteome analysis is straightforward if one assesses the effects of a treatment for example in cultured cells or cell lines since here a homogeneous population of cells is analyzed. When tissue samples are utilized that contain different cell populations with different expression profiles, proteome analysis becomes a difficult task and may require the separation of the different cell populations by means of cell-specific surface markers and immuno-affinity techniques or by laser-driven microdissection approaches [29, 30]. The future of more simple proteome analysis tools may be the use of antibody libraries that contain specific antibodies raised against any expressed open reading frame and taking proteome analysis onto the format of high-throughput microplate assays that allow essentially every protein to be identified and quantified easily [31].

There are only very few examples of proteome analysis studies in nutrition research. However, proteome analysis is an interesting tool that assesses changes of the steady-state protein levels as the prime functional units without the need for proof and the worry that changes in the transcript level may not translate into corresponding changes in the level of the encoded protein. Combining transcriptome and proteome techniques in analysis of the same sample has the charm of assessing both layers of information flow in adaptation of metabolism and to separate true co-regulation processes and seemingly uncoupled changes in mRNA and corresponding protein level simultaneously.

1.1.4
Metabolite Profiling in Nutritional Science

Various new approaches to assess globally the pattern and concentrations of a vast spectrum of metabolites in biological samples are currently under development. Metabolomics or metabolite profiling techniques are mainly based on gas chromatography combined with mass spectrometry (GC/MS) or liquid chromatography in combination either with electrospray ionization/mass spectrometry (LC/ESI/MS) or with nuclear magnetic resonance spectroscopy (LC/NMR). In contrast to genome, transcriptome, and proteome profiling technologies, which monitor target molecules of similar chemical nature such as DNA, RNA, and proteins, metabolite profiling has to deal with metabolites that vary considerably in chemical nature, molecular weight, and physical properties. This is a real challenge for analytical techniques and consequently there is no single analytical platform that allows the multiparallel analysis of the complete metabolome. However, GC/MS- and LC/NMR-based technologies have proven to be valid in producing robust metabolite profiles from biological samples. Similar to data from transcriptome and proteome analysis, metabolite fingerprints may be used to generate and refine metabolic pathway maps and to identify co-regulation phenomena of whole metabolic networks or functional modules. A variety of statistical methods and visualization tools, such as the principal component analysis, can be used to describe the mostly pleiotropic changes in metabolite spectra.

It needs to be emphasized that at the current state of technology only around 20% of the metabolites present in the mammalian cell can be identified and reliably annotated. To determine the nature of the fast growing number of the yet unknown analytes requires a huge international effort to turn metabolomics into a more powerful tool. The number of metabolites in a typical eukaryotic organism is predicted to range from 4000 to 20 000 individual compounds [32]. Although this number is impressively high and may be frightening, metabolite spectra reduce the number of components to deal with compared with the much higher number of mRNA and protein entities dealt with when performing transcript and proteome profiling.

Similar to the other high-throughput approaches, data obtained from metabolomics need highly standardized formats for disposition and their linkage to interpretation tools. A consortium recently has outlined these requirements and has given a framework by using examples from plant metabolomics [33].

As in most areas of post-genomic profiling technologies, nutritional science is way behind other fields such as microbiology, plant sciences, or drug and environmental toxicology in applying metabolomics approaches. It is obvious that body fluids such as serum or urine that can be obtained easily are primarily used in both animal and human studies to assess the signatures of the contained metabolites. In most cases ^{1}H nuclear magnetic resonance (NMR) spectroscopic analysis is applied to the body fluid samples from animal or human studies [34, 35]. The obtained complex metabolic profiles are usually submitted to multivariate statistical analysis to obtain patterns. Such a pattern-recognition analysis of NMR spectroscopic data can be performed without the need to assign all of the spectral peaks to specific metabolites before analysis and even provides time-related metabolic changes. Various examples applied in toxicology research – mainly in rodent models with urine sample NMR analysis – demonstrate the power of the techniques. The most impressive metabonomics study in humans was performed with serum samples of normal volunteers and patients at various stages of coronary heart disease. Based on pattern-recognition analysis not only the presence but also the severity of the disease could be determined based on the NMR spectra [36]. It can be envisaged that the technology will soon be taken into population screening.

A first urine sample NMR-based screen of 150 volunteers from Britain and Sweden has recently been reported [37]. The urine samples analysed via principal component analysis displayed characteristic differences related to dietary and cultural habits between the subjects of both countries. Various centers dedicated to nutritional sciences in Europe and the US have identified metabolomics as their strategic field in the nutrigenomics area and we expect to see a rapid increase in the number of studies in this area. It is anticipated that metabolite screens will be used to identify signatures that resemble certain dietary habits, that define the intake of particular food components, or that classify disease states linked to nutrition and nutrient intake.

1.1.4.1 **Metabonomics Goes Dynamic**

The next level of metabolite analysis uses advanced mass spectrometry and NMR to assess the route of individual metabolites by isotope tracer techniques. The labeling of a compound *in vivo* with stable isotopes enables the biosynthesis of differentially mass-labeled metabolite mixtures, which then can be detected by mass isotopomer ratio analysis to follow the flow of atoms through metabolites and pathways and help to identify the molecular switches that guide the compounds through metabolic chains [38]. The beauty of the application of mass isotopomer analysis in combination with powerful calculation algorithms for the quantification of intracellular metabolic fluxes has recently been demonstrated for central carbon metabolism in *Escherichia coli*. The proposed new method proved to be reliable and capable of obtaining information on the biochemical changes involved in the regulation of acetate and glucose metabolism in *E. coli* K12 cells [39]. As nutritional science has expertise in applying stable isotopes in human studies for the characterization of metabolic processes, metabolomics combined with isotopomer ratio analysis to assess metabolic isotope fluxes, for example in genetically well-characterized individuals or subgroups, is the arena for the next generation of nutrition researchers.

1.1.5
Cell Biology and Genetic Tools for Nutrigenomics Research

Conceptually, nutrigenomics research is based on either gene-driven or phenotype-driven approaches. The gene-driven approaches use genomic information for identifying, cloning, expressing, and characterizing genes and their products at the molecular level. As we are still far from understanding the role of every encoded open reading frame in a mammalian genome, animal models are of central importance for assigning genes to functions. Phenotype-driven approaches characterize phenotypes of naturally occurring variants to identify the genes, the relevant single-nucleotide polymorphisms (SNP) or haplotypes that are either responsible for or associated (in statistical terms) with the particular phenotype. In most cases this is done without knowing the exact underlying molecular mechanisms. Of course, the two strategies are highly complementary at virtually all levels of analysis and lead collectively to the correlation of genotypes and phenotypes. Because nature has not provided human inborn errors of metabolism that demonstrate the phenotypical consequences of individual gene or protein malfunction, the role of single genes or groups of genes in the makeup of metabolism needs to be analyzed in more simple models than humans. Targeted gene inactivation ("knock-out") or selective overexpression ("knock-in") models employing experimental animals from fruitflies (*Drosophila melanogaster*) to nematodes (*Caenorhabditis elegans*) or mice and rats or human cell lines will eventually reveal the roles that individual genes play in the orchestrated way metabolism works. These approaches have already produced a large number of animal lines missing one or several genes or overexpressing others. The availability of the large-scale knockout collections will accelerate the wet-laboratory work necessary to provide an understanding of the biological roles of the various players in nutrition-triggered signal transduction and gene regulation processes. Although very elegant as genetic tools to unravel metabolic changes, unfortunately, these maneuvers quite often do not produce the predicted or any distinct phenotype.

The more advanced transgenic technologies in animals through controlled cell- or organ-specific and/or time-dependent gene inactivation or induction of expression allow the analysis of phenotypical consequences in even more elegant ways. They appear also particularly helpful when simple gene disruption is lethal for the developing fetus or newborn. In simple cell models and even complex organisms (the best example being *Caenorhabditis elegans*) RNA interference techniques (RNAi) have made it easier to suppress or at least markedly reduce expression of the protein of interest in order to assess the phenotypic consequences [40, 41].

Assigning gene to function is the most critical part and this relies currently on genetic models. In combining the technologies of targeted gene inactivation and RNAi or selective overexpression with the "omics" technologies, the annotation of gene functions is greatly improved but the redundancy in biological systems also becomes visible.

Understanding the consequences of operational shifts in genetic circuits and cellular systems is and will remain a challenge. In emerging new and sophisticated metabolic network analysis tools, the metabolites are represented by interconnected nodes that show correlative behavioral changes and the actions of these metabolic networks are studied on the basis of the strength of correlations between the metabolites that make up the network [42, 43]. To understand these nodular systems and to determine the connectivity of the layers of the transcriptome, proteome, and metabolome, comprehensive approaches to measure metabolites, proteins, and/or mRNA simultaneously from the same sample are required.

How do we cope with the data generated by the high-throughput data acquisition and systems approaches? Well, the best answer appears to be to have someone on the research team who is expert in the area of analysis of these data and who is willing to learn a bit of nutritional science, with the nutritionists willing to learn a bit of advanced statistics and bioinformatics.

1.2
Nutrigenetics – Examples and Limitations

Nutrigenetics aims to understand the effect of genetic variations on the interaction between the diet and disease or on nutrient requirement. Consequently the major goal is to identify and characterize gene variants associated or responsible for differential responses to nutritional factors. In the final stage, nutrigenetics could provide the rationale for recommendations regarding the risks and benefits of a particular diet or dietary components based on the individual's genetic makeup. The quite impressive variations in the phenotype of "classical" monogenetic diseases such as phenylketonuria or familial hypercholesterolemia, however, tells us what kind of challenge we are facing when nutrigenetic approaches are applied to common multifactorial disorders such as diabetes, cardiovascular disease, or cancer. Although the methods for detecting single-nucleotide polymorphisms (SNPs) or haplotypes are improved constantly and the next generation of microarrays that cover 500 000 SNPs on one chip will be available soon, phenotype analysis and assigning alterations in protein functions to an SNP or haplotype is going to be the pinhole. Although mostly inconclusive, preliminary results involving gene–diet interactions for cardiovascular diseases and cancer suggest that the concept could work and that we will be able to harness the information contained in our genome.

Most of the available data are derived from molecular epidemiology studies. As all multifactorial nutrition-dependent diseases require a long period of exposure to the same or similar dietary patterns to develop a disease phenotype [44] epidemiological studies are the tool of choice to assess genetic variation and disease development or progression.

1.2.1
Genes, Diet, and Cardiovascular Disease

Dyslipidemia is commonly associated with the development of atherosclerosis and can be caused by improper function of a variety of proteins that control lipid homeostasis, such as nuclear factors, binding proteins, apolipoproteins, enzymes, lipoprotein receptors, and hormones (see also chapter 15). Polymorphisms have been identified in most of these components and many of the underlying genes have been explored in terms of diet–gene interactions [45–47]. Amongst these, the *apolipoprotein E* gene (*apoE*) is the most intensively studied with regard to its effects on low-density lipoprotein (LDL)-cholesterol levels in response to dietary interventions. Genetic variation at the *apoE* locus results from three common alleles in the population, *E4*, *E3*, and *E2*. However, other genetic variants at the *apoE* locus have been described as well [48].

Besides the fact that LDL-cholesterol levels were highest in subjects carrying the *apoE4* isoform [49, 50], this association was especially prominent in populations consuming diets rich in saturated fats and cholesterol [51]. These epidemiology data, therefore, indicate that high LDL-cholesterol levels are manifested primarily in the presence of an atherogenic diet but that an individual's response to dietary saturated fat and cholesterol may differ depending on the individual *apoE* alleles. However, it needs to be stressed that especially for *apoE*, investigations of diet–gene interactions have yielded quite diverse outcomes [46, 47]. Significant diet–*apoE* interactions occurred in studies focusing on males, suggesting a significant gene–gender interaction [46, 47]. Baseline lipid levels seem to affect the outcome and significant associations were frequently found only in subjects who were moderately hypercholesterolemic. More consistent effects were reported on the impact of alcohol intake on LDL-cholesterol depending on the *apoE* genotype in men [52]. A negative association between alcohol consumption and LDL-cholesterol was found for carriers of *apoE2*, whereas subjects with *apoE4* displayed a positive correlation. Within these genotype studies *apoA1* has emerged as a primary candidate for genetic variability in high-density lipoprotein (HDL) levels and its gene product plays a crucial role in lipid metabolism and for cardiovascular disease risk [53].

In women it has been found that a G to A transition in the *apoA1* gene is associated with an increase in HDL-cholesterol levels depending on the dietary intake of polyunsaturated fatty acids [54]. Similar to this G/A single-nucleotide polymorphism in *apoA1*, increased HDL levels were found to be associated with a homozygous −514(CC) polymorphism in the *hepatic lipase* gene in response to higher fat contents in the diet [55]. This increase in the level of protective HDL particles was interpreted as a defense mechanism that was not found in subjects carrying the TT genotype. Interestingly the TT genotype is common in certain ethnic groups, such as African-Americans, and might help to explain their limited ability to adapt rapidly to new nutritional environments [56].

1.2.2
Genes, Diet, and Cancer

Similar to cardiovascular diseases, dietary factors were shown to contribute significantly to the development of cancers [57] with the most prominent effects on colon, gastric, and breast cancer (see also chapter 17). Although there are general guidelines to reduce cancer risk at the population level, a specific protective food or food component has not been identified [58, 59]. Numerous studies using quantitative dietary assessments in large cohort studies and assessing genetic variation in the cohorts such as the European Prospective Investigation into Cancer and Nutrition (EPIC) with 519 978 participants in 23 centers in 10 European countries [60] are being conducted to understand and define the role dietary factors play in the causes of cancer development on the basis of genetic variations.

One of the polymorphisms that is significantly associated with cancer risk is the homozygous (TT) form of the *methylenetetrahydrofolate reductase* (*MTHFR*) gene. The cytosine to thymidine substitution, which converts an alanine residue to a valine, is relatively common and results, in its homozygous form, in hyperhomocysteinemia and an increased cardiovascular disease risk but simultaneously reduced cancer risk [61, 62]. The *MTHFR*-TT genotype displays a reduced enzymatic activity; less 5,10-methylenetetrahydrofolate is used for the remethylation of homocysteine to methionine and hence more substrate appears to be available for thymidine synthesis. In contrast, an increased misincorporation of desoxyuridine nucleotides into DNA in folate deficiency was shown to be mutagenic and this could, for example, explain the increased colon cancer risk observed in humans with a low folate status [63]. This example shows the complexity of the problem and the difficulty of transferring these observations to the level of recommendations.

Regarding homocysteine as an atherogenic factor, the recommendation for individuals with the *MTHFR*-TT genotype must be to normalize the enzyme activity and reduce homocysteine levels by higher rates of remethylation to methionine. It has to be suggested, however, that under conditions where MTHFR activity is normalized, one-carbon flux into the thymidylate cycle may be reduced and, thus, the protective function of the TT polymorphism with regard to cancer development may be lost. Although this hypothesis has not yet been proven, an increased risk for the development of colorectal adenomas was shown for the TT genotype associated with a low folate, vitamin B_{12} and B_6 intake [64].

Another gene for which polymorphisms seem to predispose to cancers by exposure to food carcinogens is the *N-acetyltransferase* (*NAT*) gene (see also chapter 19). NAT is a phase 2 enzyme that is found in two isoforms (NAT1 and NAT2) and is involved in the acetylation of heterocyclic aromatic amines (HAA) as found in heated products. Several polymorphisms have been characterized in NAT1 and NAT2 and some of these polymorphisms have been related to NAT activities of so-called "slow", "intermediate," or "fast acetylators." Although the outcome of studies investigating the association between acetylator phenotype and cancer risk are quite controversial, the NAT2 fast acetylator genotype consistently revealed a higher risk of developing colon cancer in people who consumed relatively large quantities of red meat, which may reflect the greater ability of fast acetylators to activate aromatic amines within the colon mucosa [65, 66].

Glutathione-S-transferases (GST) have also been studied in detail with respect to individual cancer risk (see also chapter 19). GSTs are subdivided into the four classes alpha (A), pi (P), mu (M), and theta (T) and for each class various polymorphisms have been described. *GSTM1*- and *GSTT1*-null genotypes appear to confer a high risk for several types of cancer [67]. By their ability for detoxification GSTs play a crucial role in xenobiotic metabolism and respond to a variety of dietary factors with changes in expression level [68]. In humans, it has been suggested that the cancer protective effects (e.g. of cruciferous vegetables) may depend on the ability to induce GSTs and other phase 2 enzymes [69, 70]. On the other hand, besides toxic compounds they also conjugate isothiocyanates, the active ingredients of cruciferae, leading to their excretion. Indeed, a significant protective effect of a high broccoli consumption was only found in subjects with the *GSTM1*-null genotype [71]. The response to isothiocyanates requires the nuclear factor-erythroid 2 p45-related factor 2 (Nrf2). In $Nrf2^{+/+}$ but not $Nrf2^{-/-}$ mice isothiocyanates from broccoli caused a modest increase in GSTM1 and a significant increase in GSTA1/2 and GSTA3 protein [72]. Strain-specific Nrf2 mRNA expression and a T to C transition in the promoter that co-segregates with susceptibility phenotypes in mice [73] and the detection of three SNPs and one triplet repeat in the human Nrf2 promoter [74] makes Nrf2 a candidate itself for individual responses to dietary factors with chemopreventive properties.

Altogether, the examples demonstrate that we are far away from a reasonable dietary advice for an individual or a subpopulation on the basis of the genetic makeup. It is important to keep in mind that gene–nutrient interactions can occur at any time during the process of a disease development and the multistage process of carcinogenesis is a perfect example. Dietary factors can affect essentially every step in cancer initiation and development starting at the level of an initial mutation, by blocking promotion or by stopping progression from the premalignant state to carcinomas or by preventing invasion or metastasis. Only long-term studies with appropriate population sizes, well-reported dietary intakes and reliable genotype and phenotype analysis will help to prove the concept of nutrigenetics, with its final goal of providing a solid scientific basis for individualized "genotype-based" dietary advice.

1.3
Conclusions

Nutritional genomics is still in its infancy, but it is predicted to rapidly move to the systems-based "holistic" level by using high-throughput technologies and advanced data analysis tools. Transcript, proteome, and metabolite profiling technologies are constantly being improved and are becoming more convenient but in the end require a substantial investment in equipment and specialized personnel. Although the new technologies already generate insights into nutrition-dependent signal transduction mechanisms and gene regulation phenomena, it is obvious that at present, these studies more often generate hypotheses than deliver true answers. In spite of the apparent profusion of data that overwhelm us already, magnitudes of new data are needed to reach the goal of a comprehensive understanding of signal transduction and gene regulation phenomena that allow the adaptation of mammalian metabolism for maintaining health or that may eventually lead to disease.

References

1 FENECH M (**2002**) Biomarkers of genetic damage for cancer epidemiology. *Toxicology* 181–182, 411–416.

2 BRAWLEY OW (**2003**) Population categorization and cancer statistics. *Cancer Metastasis Rev.* 22, 11–19.

3 GOLDMAN R, SHIELDS PG (**2003**) Food mutagens. *J. Nutr.* 133, 965S–973S.

4 GLUCKMAN PD, HANSON MA (**2004**) Maternal constraint of fetal growth and its consequences. *Semin. Fetal Neonatal Med.* 9, 419–425.

5 SACKS DA (**2004**) Determinants of fetal growth. *Curr. Diab. Rep.* 4, 281–287.

6 HUGHES TR, SHOEMAKER DD (**2001**) DNA microarrays for expression profiling. *Curr. Opin. Chem. Biol.* 5, 21–25.

7 TUTEJA R, TUTEJA N (**2004**) Serial analysis of gene expression (SAGE): application in cancer research. *Med. Sci. Monit.* 10, 132–140.

8 WELLE S. Gene transcript profiling in aging research (**2002**) *Exp. Gerontol.* 37, 583–590.

9 COOK SA, ROSENZWEIG A (**2002**) DNA microarrays: implications for cardiovascular medicine. *Circ. Res.* 91, 559–564.

10 SHEDDEN K, CHEN W, KUICK R, GHOSH D, MACDONALD J, CHO KR, GIORDANO TJ, GRUBER SB, FEARON ER, TAYLOR JM, HANASH S (**2005**) Comparison of seven methods for producing Affymetrix expression scores based on False Discovery Rates in disease profiling data. *BMC Bioinformatics* 6, 26.

11 DANIEL H, TOM DIECK H (**2004**) Nutrient–gene interactions: a single nutrient and hundreds of target genes. *Biol. Chem.* 385, 571–583.

12 TOM DIECK H, DORING F, FUCHS D, ROTH HP, DANIEL H (**2005**) Transcriptome and proteome analysis identifies the pathways that increase hepatic lipid accumulation in zinc-deficient rats. *J. Nutr.* 135, 199–205.

13 HERZOG A, KINDERMANN B, DORING F, DANIEL H, WENZEL U (**2004**) Pleiotropic molecular effects of the pro-apoptotic dietary constituent flavone in human colon cancer cells identified by protein and mRNA expression profiling. *Proteomics* 4, 2455–2464.

14 BROOKSBANK C, CAMERON G, THORNTON J (**2005**) The European Bioinformatics Institute's data resources: towards systems biology. *Nucleic Acids Res.* 33 Database Issue, D46–53.

15 BRAZMA A. HINGAMP P, QUACKENBUSH J, SHERLOCK G, SPELLMAN P, STOECKERT C, AACH J, ANSORGE W, BALL CA, CAUSTON HC, GAASTERLAND T, GLENISSON P, HOLSTEGE FC, KIM IF, MARKOWITZ V, MATESE JC, PARKINSON H, ROBINSON A, SARKANS U, SCHULZE-KREMER S, STEWART J,

Taylor R, Vilo J, Vingron M (**2001**) Minimum information about a microarray experiment (MIAME)-toward standards for microarray data. *Nat. Genet.* 29, 365–371.

16 Perez EA, Pusztai L, Van de Vijver M (**2004**) Improving patient care through molecular diagnostics. *Semin. Oncol.* 31, 14–20.

17 Jain KK (**2004**) Applications of biochips: from diagnostics to personalized medicine. *Curr. Opin. Drug Discov. Devel.* 7, 285–289.

18 Mantripragada KK, Buckley PG, de Stahl TD, Dumanski JP (**2004**) Genomic microarrays in the spotlight. *Trends Genet.* 20, 87–94.

19 Muller M, Kersten S (**2003**) Nutrigenomics: goals and strategies. *Nat. Rev. Genet.* 4, 315–322.

20 Baker M, Gillanders WE, Mikhitarian K, Mitas M, Cole DJ (**2003**) The molecular detection of micrometastatic breast cancer. *Am. J. Surg.* 186, 351–358.

21 Macgregor PF, Squire JA (**2002**) Application of microarrays to the analysis of gene expression in cancer. *Clin. Chem.* 48, 1170–1177.

22 Gorg A, Obermaier C, Boguth G, Harder A, Scheibe B, Wildgruber R, Weiss W (**2000**) The current state of two-dimensional electrophoresis with immobilized pH gradients. *Electrophoresis* 21, 1037–1053.

23 Blackstock WP, Weir MP (**1999**) Proteomics: quantitative and physical mapping of cellular proteins. *Trends Biotechnol.* 17, 121–127.

24 Cordwell SJ, Nouwens AS, Verrills NM, Basseal DJ, Walsh BJ (**2000**) Subproteomics based upon protein cellular location and relative solubilities in conjunction with composite two-dimensional electrophoresis gels. *Electrophoresis* 21, 1094–1103.

25 Corthals GL, Wasinger VC, Hochstrasser DF, Sanchez JC (**2000**) The dynamic range of protein expression: a challenge for proteomic research. *Electrophoresis* 21, 1104–1115.

26 Adam GC, Cravatt BF, Sorensen EJ (**2001**) Profiling the specific reactivity of the proteome with non-directed activity-based probes. *Chem. Biol.* 8, 81–95.

27 Conrads TP, Anderson GA, Veenstra TD, Pasa-Tolic L, Smith RD (**2000**) Utility of accurate mass tags for proteome-wide protein identification. *Anal. Chem.* 72, 3349–3354.

28 Chamrad DC, Koerting G, Gobom J, Thiele H, Klose J, Meyer HE, Blueggel M (**2003**) Interpretation of mass spectrometry data for high-throughput proteomics. *Anal. Bioanal. Chem.* 376, 1014–1022.

29 Banks RE, Dunn MJ, Forbes MA, Stanley A, Pappin D, Naven T, Gough M, Harnden P, Selby PJ (**1999**) The potential use of laser capture microdissection to selectively obtain distinct populations of cells for proteomic analysis – preliminary findings. *Electrophoresis* 20, 689–700.

30 Simone NL, Paweletz CP, Charboneau L, Petricoin EF III, Liotta LA (**2000**) Laser capture microdissection: beyond functional genomics to proteomics. *Mol. Diagn.* 5, 301–307.

31 LaBaer J, Ramachandran N (**2005**) Protein microarrays as tools for functional proteomics. *Curr. Opin. Chem. Biol.* 9, 14–19.

32 Fiehn O, Kopka J, Dormann P, Altmann T, Trethewey RN, Willmitzer L (**2000**) Metabolite profiling for plant functional genomics. *Nat. Biotechnol.* 18, 1157–1161.

33 Jenkins H, Hardy N, Beckmann M, Draper J, Smith AR, Taylor J, Fiehn O, Goodacre R, Bino RJ, Hall R, Kopka J, Lane GA, Lange BM, Liu JR, Mendes P, Nikolau BJ, Oliver SG, Paton NW, Rhee S, Roessner-Tunali U, Saito K, Smedsgaard J, Sumner LW, Wang T, Walsh S, Wurtele ES, Kell DB (**2004**) A proposed framework for the description of plant metabolomics experiments and their results. *Nat. Biotechnol.* 22, 1601–1606.

34 Williams RE, Jacobsen M, Lock EA (**2003**) ^{1}H NMR pattern recognition and 31P NMR studies with d-Serine in rat urine and kidney, time- and dose-related metabolic effects. *Chem. Res. Toxicol.* 16, 1207–1216.

35 Liu M, Nicholson JK, Lindon JC (**1996**) High-resolution diffusion and relaxation edited one- and two-dimensional ^{1}H NMR spectroscopy of biological fluids. *Anal. Chem.* 68, 3370–3376.

36 Brindle JT, Antti H, Holmes E, Tranter G, Nicholson JK, Bethell HW, Clarke S, Schofield PM, McKilligin E, Mosedale DE, Grainger DJ (**2002**) Rapid and noninvasive diagnosis of the presence and severity of coronary heart disease using 1H-NMR-based metabonomics. *Nat. Med.* 8, 1439–1444.

37 Lenz EM, Bright J, Wilson ID, Hughes A, Morrisson J, Lindberg H, Lockton A (**2004**) Metabonomics, dietary influences and cultural differences: a ^{1}H NMR-based study of urine samples obtained from healthy British and Swedish subjects. *J. Pharm. Biomed. Anal.* 36, 841–849.

38 Birkemeyer C, Luedemann A, Wagner C, Erban A, Kopka J (**2005**) Metabolome analysis: the potential of in vivo labeling with stable isotopes for metabolite profiling. *Trends Biotechnol.* 23, 28–33.

39 Zhao J, Shimizu K (**2003**) Metabolic flux analysis of *Escherichia coli* K12 grown on 13C-labeled acetate and glucose using GC-MS and powerful flux calculation method. *J. Biotechnol.* 101, 101–117.

40 Poulin G, Nandakumar R, Ahringer J (**2004**) Genome-wide RNAi screens in *Caenorhabditis elegans*: impact on cancer research. *Oncogene* 23, 8340–8345.

41 Sugimoto A (**2004**) High-throughput RNAi in *Caenorhabditis elegans*: genome-wide screens and functional genomics. *Differentiation* 72, 81–91.

42 de la Fuente A, Bing N, Hoeschele I, Mendes P (**2004**) Discovery of meaningful associations in genomic data using partial correlation coefficients. *Bioinformatics* 20, 3565–3574.

43 Kell DB (**2004**) Metabolomics and systems biology: making sense of the soup. *Curr. Opin. Microbiol.* 7, 296–307.

44 Leong NM, Mignone LI, Newcomb PA, Titus-Ernstoff L, Baron JA, Trentham-Dietz A, Stampfer MJ, Willett WC, Egan KM (**2003**) Early life risk factors in cancer: the relation of birth weight to adult obesity. *Int. J. Cancer* 103, 789–791.

45 Loktionov A (**2003**) Common gene polymorphisms and nutrition: emerging links with pathogenesis of multifactorial chronic diseases (review). *J. Nutr. Biochem.* 14, 426–451.

46 Masson LF, McNeill G, Avenell A (**2003**) Genetic variation and the lipid response to dietary intervention: a systematic review. *Am. J. Clin. Nutr.* 77, 1098–1111.

47 Rubin J, Berglund L (**2002**) Apolipo-protein E and diets: a case of gene–nutrient interaction? *Curr. Opin. Lipidol.* 13, 25–32.

48 Artiga MJ, Bullido MJ, Sastre I, Recuero M, Garcia MA, Aldudo J, Vazquez J, Valdivieso F (**1998**) Allelic polymorphisms in the transcriptional regulatory region of apolipoprotein E gene. *FEBS Lett.* 421, 105–108.

49 Ordovas JM, Litwack-Klein L, Wilson PW, Schaefer MM, Schaefer EJ (**1987**) Apolipoprotein E isoform phenotyping methodology and population frequency with identification of apoE1 and apoE5 isoforms. *J. Lipid Res.* 28, 371–380.

50 Schaefer EJ, Lamon-Fava S, Johnson S, Ordovas JM, Schaefer MM, Castelli WP, Wilson PW (**1994**) Effects of gender and menopausal status on the association of apolipoprotein E phenotype with plasma lipoprotein levels. Results from the Framingham Offspring Study. *Arterioscler. Thromb.* 14, 1105–1113.

51 Davignon J, Gregg RE, Sing CF (**1988**) Apolipoprotein E polymorphism and atherosclerosis. *Arteriosclerosis* 8, 1–21.

52 Corella D, Tucker K, Lahoz C, Coltell O, Cupples LA, Wilson PW, Schaefer EJ, Ordovas JM (**2001**) Alcohol drinking determines the effect of the APOE locus on LDL-cholesterol concentrations in men: the Framingham Offspring Study. *Am. J. Clin. Nutr.* 73, 736–745.

53 Segrest JP, Li L, Anantharamaiah GM, Harvey SC, Liadaki KN, Zannis V (**2000**) Structure and function of apolipoprotein A-I and high-density lipoprotein. *Curr. Opin. Lipidol.* 11, 105–115.

54 Ordovas JM, Corella D, Cupples LA, Demissie S, Kelleher A, Coltell O, Wilson PW, Schaefer EJ, Tucker K (**2002**) Polyunsaturated fatty acids modulate the effects of the APOA1 G-A polymorphism on HDL-cholesterol concentrations in a sex-specific manner: the Framingham Study. *Am. J. Clin. Nutr.* 75, 38–46.

55 Ordovas JM, Corella D, Demissie S, Cupples LA, Couture P, Coltell O, Wilson PW, Schaefer EJ, Tucker KL (**2002**) Dietary fat intake determines the effect of a common polymorphism in the hepatic lipase gene promoter on high-density lipoprotein metabolism: evidence of a strong dose effect in this gene-nutrient interaction in the Framingham Study. *Circulation* 106, 2315–2321.

56 Tai ES, Corella D, Deurenberg-Yap M, Cutter J, Chew SK, Tan CE, Ordovas JM (**2003**) Dietary fat interacts with the −514C > T polymorphism in the hepatic lipase gene promoter on plasma lipid profiles in a multiethnic Asian population: the 1998 Singapore National Health Survey. *J. Nutr.* 133, 3399–3408.

57 Doll R, Peto R (**1981**) The causes of cancer: quantitative estimates of avoidable risks of cancer in the United States today. *J. Natl Cancer Inst.* 66, 1191–1308.

58 Riboli E, Norat T (**2003**) Epidemiologic evidence of the protective effect of fruit and vegetables on cancer risk. *Am. J. Clin. Nutr.* 78, 559S–569S.

59 Voorrips LE, Goldbohm RA, van Poppel G, Sturmans F, Hermus RJ, van den Brandt PA (**2000**) Vegetable and fruit consumption and risks of colon and rectal cancer in a prospective cohort study: The Netherlands Cohort Study on Diet and Cancer. *Am. J. Epidemiol.* 152, 1081–1092.

60 Riboli E, Hunt KJ, Slimani N, Ferrari P, Norat T, Fahey M, Charrondiere UR, Hemon B, Casagrande C, Vignat J, Overvad K, Tjonneland A, Clavel-Chapelon F, Thiebaut A, Wahrendorf J, Boeing H, Trichopoulos D, Trichopoulou A, Vineis P, Palli D, Bueno-De-Mesquita HB, Peeters PH, Lund E, Engeset D, Gonzalez CA, Barricarte A, Berglund G, Hallmans G, Day NE, Key TJ, Kaaks R, Saracci R (**2002**) European Prospective Investigation into Cancer and Nutrition (EPIC): study populations and data collection. *Public Health Nutr.* 5, 1113–1124.

61 Chen J, Giovannucci E, Kelsey K, Rimm EB, Stampfer MJ, Colditz GA, Spiegelman D, Willett WC, Hunter DJ (**1996**) A methylenetetrahydrofolate reductase polymorphism and the risk of colorectal cancer. *Cancer Res.* 56, 4862–4864.

62 Ma J, Stampfer MJ, Giovannucci E, Artigas C, Hunter DJ, Fuchs C, Willett WC, Selhub J, Hennekens CH, Rozen R (**1997**) Methylene-tetrahydrofolate reductase polymorphism, dietary interactions, and risk of colorectal cancer. *Cancer Res.* 57, 1098–1102.

63 Slattery ML, Potter JD, Samowitz W, Schaffer D, Leppert M (**1999**) Methylenetetrahydrofolate reductase, diet, and risk of colon cancer. *Cancer Epidemiol. Biomarkers Prev.* 8, 513–518.

64 Ulrich CM, Kampman E, Bigler J, Schwartz SM, Chen C, Bostick R, Fosdick L, Beresford SA, Yasui Y, Potter JD (**1999**) Colorectal adenomas and the C677T MTHFR polymorphism: evidence for gene-environment interaction? *Cancer Epidemiol. Biomarkers Prev.* 8, 659–668.

65 CHEN J, STAMPFER MJ, HOUGH HL, GARCIA-CLOSAS M, WILLETT WC, HENNEKENS CH, KELSEY KT, HUNTER DJ (**1998**) A prospective study of N-acetyltransferase genotype, red meat intake, and risk of colorectal cancer. *Cancer Res.* 58, 3307–3311.

66 ROBERTS-THOMSON IC, RYAN P, KHOO KK, HART WJ, MCMICHAEL AJ, BUTLER RN (**1996**) Diet, acetylator phenotype, and risk of colorectal neoplasia. *Lancet* 347, 1372–1374.

67 HABDOUS M, SIEST G, HERBETH B, VINCENT-VIRY M, VISVIKIS S (**2004**) [Glutathione S-transferases genetic polymorphisms and human diseases: overview of epidemiological studies]. *Ann. Biol. Clin.* 62, 15–24.

68 ROCK CL, LAMPE JW, PATTERSON RE (**2000**) Nutrition, genetics, and risks of cancer. *Annu. Rev. Public Health* 21, 47–64.

69 SEOW A, SHI CY, CHUNG FL, JIAO D, HANKIN JH, LEE HP, COETZEE GA, YU MC (**1998**) Urinary total isothiocyanate (ITC) in a population-based sample of middle-aged and older Chinese in Singapore: relationship with dietary total ITC and glutathione S-transferase M1/T1/P1 genotypes. *Cancer Epidemiol. Biomarkers Prev.* 7, 775–781.

70 NIJHOFF WA, GRUBBEN MJ, NAGENGAST FM, JANSEN JB, VERHAGEN H, VAN POPPEL G, PETERS WH (**1995**) Effects of consumption of Brussels sprouts on intestinal and lymphocytic glutathione S-transferases in humans. *Carcinogenesis* 16, 2125–2128.

71 LIN HJ, PROBST-HENSCH NM, LOUIE AD, KAU IH, WITTE JS, INGLES SA, FRANKL HD, LEE ER, HAILE RW (**1998**) Glutathione transferase null genotype, broccoli, and lower prevalence of colorectal adenomas. *Cancer Epidemiol. Biomarkers Prev.* 7, 647–652.

72 MCWALTER GK, HIGGINS LG, MCLELLAN LI, HENDERSON CJ, SONG L, THORNALLEY PJ, ITOH K, YAMAMOTO M, HAYES JD (**2004**) Transcription factor Nrf2 is essential for induction of NAD(P)H:quinone oxidoreductase 1, glutathione S-transferases, and glutamate cysteine ligase by broccoli seeds and isothiocyanates. *J. Nutr.* 134, 3499S–3506S.

73 CHO H-Y, JEDLICKA AE, REDDY SP, ZHANG L-Y, KENSLER TW, KLEEBERGER SR (**2002**) Linkage analysis of susceptibility to hyperoxia. Nrf2 is a candidate gene. *Am. J. Respir. Cell Mol. Biol.* 26, 42–51.

74 YAMAMOTO T, YOH K, KOBAYASHI A, ISHII Y, KURE S, KOYAMA A, SAKAMOTO T, SEKIZAWA K, MOTOHASHI H, YAMAMOTO M (**2004**) Identification of polymorphisms in the promoter region of the human NRF2 gene. *Biochem. Biophys. Res. Commun.* 321, 72–79.

Part II
Nutrigenomics (Nutrient–Gene Interactions)

2
Nuclear Receptors: An Overview

Ana Aranda and Angel Pascual

2.1
The Nuclear Receptor Superfamily

Small lipophilic molecules, such as fat-soluble hormones, vitamins, and intermediary metabolites, play an important role in the growth, differentiation, metabolism, reproduction, and morphogenesis of higher organisms and humans. Unlike polypeptide hormones that act on membrane-bound receptors and activate signaling pathways that lead to gene regulation, most cellular actions of the lipophilic hormones are mediated through direct binding to nuclear receptors, which act as ligand-inducible transcription factors to activate or repress many target genes [1].

The superfamily of nuclear receptors includes the receptors for the lipophilic hormones, for the active forms of vitamin A (the retinoids) and vitamin D (1,25-dihydroxyvitamin D$_3$), as well as "orphan" receptors with unknown ligands [2]. That orphan receptors play key roles in development, homeostasis, and disease has been proven by targeted deletion in mice and by their association with different diseases including atherosclerosis, cancer, diabetes, or lipid disorders. Some orphan receptors may have a still unidentified ligand, but others may act in a constitutive manner or could be activated by other means, i.e. phosphorylation. In recent years several orphan receptors have been "adopted." Some of the novel ligands are products of lipid metabolism such as fatty acids, leukotrienes, prostaglandin and cholesterol derivatives, bile acids, pregnanes, or even benzoate derivatives [3]. Therefore, as opposed to classic hormones, other ligands originate intracellularly as metabolic products, which may explain why their role as regulators of nuclear receptors was not previously identified by physiological experimentation. Also, unlike classical endocrine receptors that are activated by high-affinity ligands with dissociation constants in the nanomolar range, these "metabolic receptors" are activated by abundant but low-affinity ligands with dissociation constants in the micromolar range.

A list of classical, metabolic, and orphan hormone receptors and their ligands is shown in Table 2.1. Evolutionary analysis of the receptors has led to a subdivision into six different subfamilies [4]. One large family is formed by thyroid hormone receptors (TRs), retinoic acid receptors (RARs), vitamin D receptors (VDRs), and

Nutritional Genomics. Edited by Regina Brigelius-Flohé and Hans-Georg Joost
Copyright © 2006 WILEY-VCH Verlag GmbH & Co. KGaA, Weinheim
ISBN: 3-527-31294-3

Table 2.1. Subfamilies of nuclear receptors.

	Receptor	Subtype	Nomenclature	Denomination	Ligand	Response element	Monomer (M) Homodimer (D) Heterodimer (H)
Class I	TR	α β	NR1A1 NR1A2	Thyroid hormone receptor	Thyroid hormone (T3)	Pal, DR-4, IP	H
	RAR	α β γ	NR1B1 NR1B2 NR1B3	Retinoic acid receptor	Retinoic acid	DR-2, DR-5 Pal, IP	H
	VDR		NR1I1	Vitamin D receptor	1,25(OH)$_2$ vitamin D$_3$, litocholic acid	DR-3, IP-9	H
	PPAR	α β γ	NR1C1 NR1C2 NR1C3	Peroxisome proliferator-activated receptor	Benzotriene B4. Wy 14.643 Eicosanoids Thiazolidinediones (TZD$_s$) 15-deoxy-12,14-prostaglandin J2 Polyunsaturated fatty acids	DR-1	H
	PXR		NR1I2	Pregnane X receptor	Pregnanes. C21 steroids, xenobiotics, PCN	DR-3	H
	CAR	α	NR1I3	Constitutive androstane receptor	Androstanes. 1,4-bis[2-(3,5-dichloropyridyloxy)benzene, xenobiotics, phenobarbital	DR-5	H
	LXR	α β	NR1H3 NR1H2	Liver X receptor	Oxysterols	DR-4	H

FXR		NR1H4	Farnesoid X receptor	Bile acids, lanosterol	DR-4, IR-1	H
RevErbA	α	NR1D1	Reverse ErbA	Unknown	DR-2, Hemisite	M, D
	β	NR1D2				
RZR/ROR	α	NR1F1	Retinoid Z receptor/Retinoic acid-related orphan receptor	Retinoic acid	Hemisite	M
	β	NR1F2				
	γ	NR1F3				
Class II						
RXR	α	NR2B1	Retinoid X receptor	9-*cis* Retinoic acid, docosahexaenoic acid	Pal, DR-1	D
	β	NR2B2				
	γ	NR2B3				
COUP-TF	α	NR2F1	Chicken ovalbumin upstream promoter transcription factor	Unknown	Pal, DR-5	D, H
	β	NR2F2				
	γ	NR2F6				
HNF-4	α	NR2A1	Hepatocyte nuclear factor 4	Unknown	DR-1, DR-2	D
	γ	NR2A2				
TLX		NR2E2	Tailles-related receptor	Unknown	DR-1, Hemisite	M, D
PNR		NR2E3	Photoreceptor-specific nuclear receptor	Unknown	DR-1, Hemisite	M, D
TR2	α	NR2C1	Testis receptor	Unknown	DR-1 to DR5	D, H
	β	NR2C2				
Class III						
GR		NR3C1	Glucocorticoid receptor	Glucocorticoids	Pal	D
AR		NR3C4	Androgen receptor	Androgens	Pal	D
PR		NR3C3	Progesterone receptor	Progestins	Pal	D
ER	α	NR3A1	Estrogen receptor	Estradiol	Pal	D
	β	NR3A2				

Table 2.1 (*continued*)

	Receptor	Subtype	Nomenclature	Denomination	Ligand	Response element	Monomer (M) Homodimer (D) Heterodimer (H)
	ERR	α β γ	NR3B1 NR3B2 NR3B3	Estrogen-related receptor	Unknown	Pal, Hemisite	M, D
Class IV	NGFI-B	α β γ	NR4A1 NR4A2 NR4A3	NGF-induced clone B	Unknown	Pal, DR-5	M, D, H
Class V	SF-1/FTZ-F1		NR5A1	Steroidogenic factor 1 Fushi Tarazu factor 1	Unknown	Hemisite	M
	LRH1		NR5A2	Liver receptor homologous protein 1	Unknown	Hemisite	M
Class VI	GCNF		NR6A1	Germ cell nuclear factor	Unknown	DR-0	D
Class 0	SHP		NR0B2	Small heterodimeric partner	Unknown		H
	DAX-1		NR0B1	Dosage-sensitive sex reversal	Unknown		

peroxisome proliferator-activated receptors (PPARs), as well as different orphan receptors. Some of these orphan receptors have been recently "adopted" (see Table 2.1). The second subfamily contains the retinoid X receptors (RXRs), together with chicken ovalbumin upstream stimulators (COUPs), hepatocyte nuclear factor 4 (HNF4), testis receptors (TR2), and receptors involved in eye development (TLX and PNR). RXRs play an important role in nuclear receptor signaling, as they are partners for different receptors that bind as heterodimers to DNA. 9-*cis*-Retinoic acid binds with high affinity to RXR and it has been widely used to study RXR signaling. The fatty acid docosahexaenoic acid (DHA) binds with a lower affinity but could be an endogenous RXR ligand. Ligands for other receptors have not yet been identified. The third family is formed by the steroid receptors and the highly related orphan receptors ERRs (estrogen-related receptors). The fourth, fifth, and sixth subfamilies contain the orphan receptors NGFI-B, FTZ-1/SF-1, and GCNF, respectively. Most subfamilies appear to be ancient since they have an arthropod homolog, with the exception of steroid receptors that have no known homologs.

2.2
Mechanism of Action

Some steroid receptors, such as the glucocortioid receptor (GR), are sequestered in the cytoplasm by association with a large multiprotein complex of chaperones, including Hsp90 and Hsp56. Ligand binding induces dissociation of the complex and nuclear translocation. More recent data have demonstrated the presence of unbound receptors for other steroid hormones in the nucleus, with equilibrium between both cell compartments. Many unliganded non-steroid receptors are located in the nucleus and can interact directly with chromatin (Fig. 2.1). Once in the nucleus the receptors regulate transcription by binding, generally as dimers, to DNA sequences termed positive or negative hormone response elements (HREs), normally located in regulatory regions of target genes [5].

The effects of nuclear receptors on transcription are mediated through recruitment of co-regulators. A subset of receptors binds co-repressors and actively represses target gene expression in the absence of ligand. Upon ligand binding the receptors undergo a conformational change that causes co-repressor release and the recruitment of co-activator complexes and transcriptional activation of genes containing HREs. Nuclear receptors can also regulate expression of genes that do not contain HREs by modulation of the activity of signaling pathways and transcription factors that bind to the target promoter.

Alternative ligand-independent pathways for activation of nuclear receptors exist (Fig. 2.1). For example, some receptors can be activated by phosphorylation mediated by hormones and growth factors that stimulate diverse signal transduction pathways [6]. These signaling pathways can also affect hormone-mediated transcription by modification of co-activators and co-repressors. Receptors and co-regulators are also targets for other modifications such as methylation, acetylation,

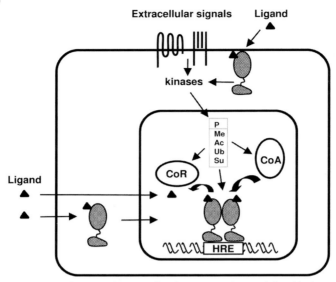

Fig. 2.1. Mechanism of action of nuclear receptors (overview). The unliganded receptor may have a nuclear location. However, some steroid receptors are cytoplasmic in the absence of ligand and ligand binding induces nuclear translocation. Once in the nucleus the receptors regulate transcription by binding, generally as dimers, to hormone response elements (HREs) located in regulatory regions of target genes. Activity is regulated by an exchange of co-repressor (CoR) and co-activator (CoA) complexes. Receptor activity is also modulated by hormones and growth factors that stimulate diverse signal transduction pathways. Both receptors and coregulators are targets for phosphorylation (P) as well as for modifications such as acetylation (Ac), methylation (Me), ubiquitination (Ub), or sumolyation (Su), that regulate their activity, levels or localization. Ligand binding to nuclear receptors located at the plasma membrane can also elicit rapid "non-genomic" effects that lead to stimulation of kinase pathways.

ubiquitination, sumoylation, etc., that can regulate their activity, levels, and localization [7].

Lastly, nuclear receptor ligands can also elicit rapid responses also called "non-genomic" or "non-genotropic" responses, which are not blocked by inhibitors of transcription or translation. These rapid actions could be mediated by a fraction of membrane-associated nuclear receptors, or by occupancy of a putative membrane receptor coupled through appropriate second-messenger systems to the generation of the biological response. Through these "non-genomic" mechanisms steroid hormones can increase intracellular calcium or activate mitogen-activated protein kinases (MAPKs) or phosphoinositol-3-kinase (PI3K) to provoke cellular effects [8].

The ligand can be generated in three different ways: an active ligand or hormone can be synthesized in a classical endocrine organ, the ligand may be generated from a precursor or prohormone within the target cell, and the ligand may be an endogenous cell metabolite.

a)

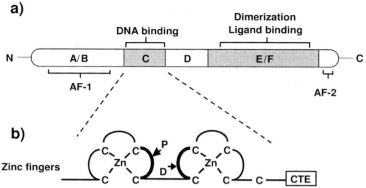

b)

Fig. 2.2. (a) Schematic representation of a nuclear receptor showing the functional domains. The A/B region contains the ligand-independent AF-1 transactivation domain. The DNA-binding domain (DBD) or region C, is responsible for the recognition of specific DNA sequences. A variable linker region D connects the DBD to the conserved E/F region that contains the ligand-binding domain (LBD) as well as the dimerization surface and the ligand-dependent AF-2 transactivation domain. (b) A diagram of the two zinc fingers and the C-terminal extension (CTE) of the DBD. In the zinc fingers four conserved cysteines coordinate a zinc ion. P box residues are involved in the discrimination of the response element. Residues in the second zinc finger labeled as D box form a dimerization interface. The CTE contains the T and A boxes critical for monomeric DNA binding.

2.3
Structure of Nuclear Receptors

Nuclear receptors exhibit a modular structure with different regions corresponding to autonomous functional domains that can be interchanged between related receptors without loss of function. A typical nuclear receptor consists of a variable N-terminal region (A/B), a conserved DNA-binding domain (DBD) or region C, a linker region D, and a conserved E region that contains the ligand-binding domain (LBD). Some receptors possess also a C-terminal region (F) of unknown function. A scheme of a nuclear receptor is shown in Fig. 2.2a. The receptors also have regions required for transcriptional activation: a ligand-independent transcriptional activation function (AF-1) located in the hypervariable A/F region, and a conserved ligand-dependent transcriptional activation domain, termed AF-2, located in the C-terminus of the LBD.

2.3.1
The A/B Region

This modulatory region is the most variable both in size and sequence and in many cases contains an AF-1 domain. Multiple receptor isoforms generated from a single gene by alternative splicing or by the use of alternative promoters, diverge in their A/B regions in most cases. On the other hand, the modulatory domain is the target for phosphorylation mediated by different signaling pathways and this modification can significantly affect transcriptional activity.

2.3.2
The DNA-binding Domain

This is the most conserved domain of nuclear receptors, and confers the ability to recognize specific DNA target sequences (Fig. 2.2b). The DBD contains nine cysteines, as well as other residues that are conserved across the nuclear receptor superfamily and are required for high-affinity DNA binding. This domain comprises two "zinc fingers" which span approximately 60–70 amino acids and a C-terminal extension (CTE), which contains the so-called T and A boxes. In each zinc finger, four of the invariable cysteines coordinate tetrahedrically one zinc ion. Amino acids required for discrimination of core DNA recognition motifs are present at the base of the first finger in a region termed the "P box," and other residues of the second zinc finger that form the so called "D box" are involved in dimerization. Structural studies have shown that the core DBD is composed of two α-helices. The first one, the recognition helix, that begins at the third conserved cysteine residue and binds the major groove of DNA, contains the P box that makes contacts with specific DNA bases. The second helix contains the D box and forms a right angle with the recognition helix [9].

2.3.3
The Hinge Region

The D domain is not well conserved among the different receptors and serves as a hinge between the DBD and the LBD, allowing rotation of the DBD. The D domain in many cases harbors nuclear localization signals and also contains residues whose mutation abolishes interaction with nuclear receptor co-repressors. In addition, this region associates strongly with the LBD but only in the presence of ligand or co-repressors, exerting a stabilizing effect on the overall structure of the receptor.

2.3.4
The Ligand-binding Domain

The LBD is a multifunctional domain that, besides the binding of ligand, mediates homo- and heterodimerization, interaction with heat-shock proteins, ligand-dependent transcriptional activity and, in some cases, hormone reversible transcriptional repression. The LBDs contain two well-conserved regions: a "signature motif" or Ti and the C-terminal AF-2 motif responsible for ligand-dependent transcriptional activation.

Although a three-dimensional structure of an entire nuclear receptor has not yet been obtained, the crystal structures of the LBDs of multiple nuclear receptors have been solved [10] (see also chapter 3). As illustrated in Fig. 2.3, the LBDs are

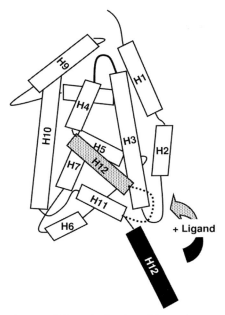

Fig. 2.3. Schematic drawing of the nuclear receptor ligand-binding domain (LBD). Cylinders represent α-helices that are numbered from 1 to 12. Note the different position of the C-terminal H12 that contains the core AF-2 domain in the absence and presence of ligand.

generally formed by 12 conserved α-helical regions numbered from H1 to H12. The ligand-binding pocket, which accommodates the ligand, is mainly made up of non-polar amino acids and is buried within the bottom half of the LBD. The size of the ligand-binding pocket varies among the different receptors, being small in the classical receptors which bind ligand with high affinity and very large in "metabolic" receptors which can bind very differently sized ligands with a lower affinity. By contrast, the ligand-binding pocket of the orphan receptor NURR is filled with bulky hydrophobic side-chains, suggesting that it could be a true orphan receptor.

2.3.5
The AF-2 domain

The AF-2 domain, contained in H12, is required for ligand-dependent transactivation. This domain possesses a high homology over a very short region that adopts an amphipathic α-helical conformation with a consensus motif $\phi\phi X E \phi\phi$ (ϕ being a hydrophobic amino acid). Although H12 of the LBD contains the core AF-2 activity, this domain comprises other dispersed elements brought together upon ligand binding. One of such elements is the "signature motif," encompassing the C-terminal half of helix 3 and helix 4. Mutations in this region affect neither ligand binding nor dimerization, but impair ligand-dependent transactivation. Specifically, a highly conserved lysine in the C-terminus of helix 3 is important for transcriptional activity of several receptors [5].

Several differences are evident when comparing unliganded and ligand-bound receptors. The liganded structures are more compact than the unliganded ones, demonstrating that upon ligand binding the receptors undergo a clear conformational change. The most striking difference observed in the receptors upon ligand binding is the position of H12 [10, 11]. This helix projects away from the body of the LBD in unliganded RXR. However, in liganded receptors H12 moves in a "mouse-trap" model, being tightly packed against helix 3 or 4 and making direct contacts with the ligand (see Fig. 2.3). This change generates a hydrophobic cleft responsible for co-activator interactions. It has been demonstrated that in the estrogen receptor (ER) LBD bound to antagonists the position of H12 is different from that shown by the agonist-bound LBD. In the antagonist-bound receptor, H12 lies in a position that overlaps with the surface of co-activator interaction, thus precluding co-activator binding and consequently transcriptional activity.

2.4
Binding of Monomers, Homodimers and Heterodimers to Hormone Response Elements

Nuclear receptors regulate transcription by binding to specific DNA sequences in regulatory regions of target genes known as hormone response elements or HREs (Fig. 2.4). Steroid hormone receptors typically bind as homodimers to palindromes of the AGAACA sequence spaced by three nucleotides, with the exception of the ERs that possess a different P box and recognize a consensus AGG/TTCA motif with the same configuration. Two steroid hormone receptor monomers bind co-operatively to their response elements, and dimerization interfaces have been identified both in the LBD and in the DBD. The dimerization interface in the DBD involves the D box that contacts the same box of the partner receptor. The dimerization in the LBD is mediated through a hydrophobic sequence in H10 that forms a coiled-coil structure, or by formation of an intermolecular β-sheet in the case of the glucocorticoid receptor [12].

In contrast with steroid receptors that almost exclusively recognize palindromic elements, non-steroidal receptors can bind as homodimers or heterodimers to HREs composed of two copies of the AGG/TTCA motif configured as palindromes (Pal), inverted palindromes (IPs) or direct repeats (DRs) (Fig. 2.4). In fact, the most potent HREs for non-steroid receptors appear to be configured as DRs in which the length of the spacer region is an important determinant of the specificity of hormonal responses. Thus, DRs separated by 3, 4, and 5 base pairs (i.e. DR3, DR4, and DR5) are HREs for VDR, TR, and RAR, respectively. A DR1 serves as the preferred HRE for RXR or PPAR and RAR can also activate transcription through a DR2. The configuration of the preferred HREs for different classical, metabolic, and orphan receptors has been included in Table 2.1.

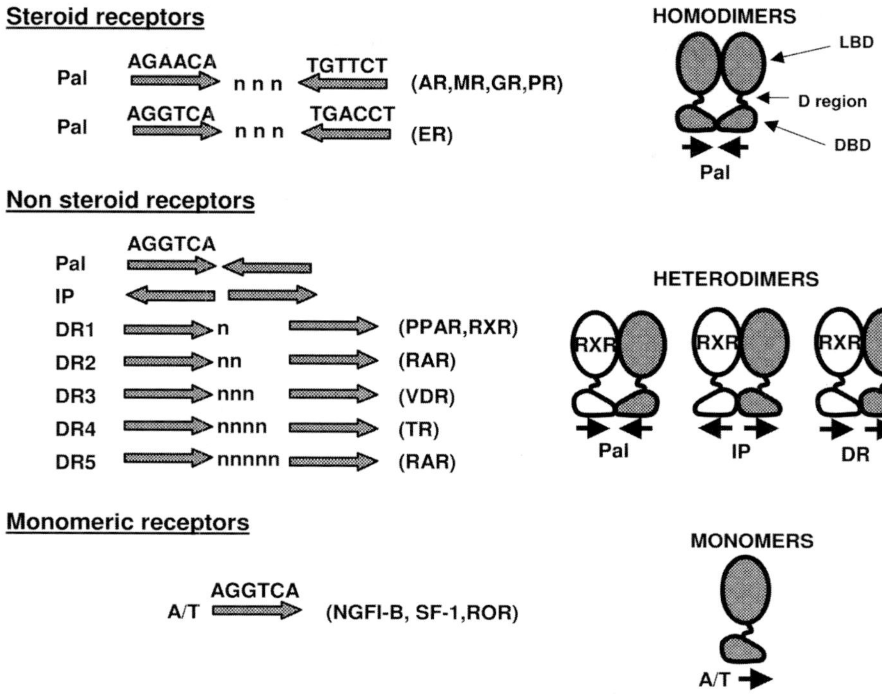

Fig. 2.4. Steroid receptors bind as homodimers to palindromic elements spaced by three nucleotides (n) in a symmetrical way. Non-steroid receptors can bind to the HREs as monomers, homodimers, or RXR heterodimers and can recognize diverse HREs in which half-core motifs can be arranged as palindromes (Pal), inverted palindromes (IP), or direct repeats (DR) with different spacing. Some receptors bind as monomers to DNA. Monomeric binding requires the half-core motif preceded by a 5′-flanking A/T rich sequence (see also Table 2.1).

In the case of heterodimers, the retinoid X receptor RXR is the promiscuous partner for different receptors [13]. Although typical heterodimeric receptors can bind to their response elements as homodimers, heterodimerization with RXR strongly increases the affinity for DNA and transcriptional activity. Therefore, RXR plays a dual role in nuclear receptors signaling. On one hand, this receptor binds to a DR1 as a homodimer and activates transcription in response to 9-*cis*-retinoic acid, and on the other hand serves as a heterodimer partner for other nuclear receptors. Since DRs are inherently asymmetric, heterodimeric complexes may bind to them with two distinct polarities. Indeed it has been established that on DR3, DR4, and DR5, RXR occupies the upstream half-site and the heterodimeric partner (e.g. VDR, TR, or RAR) occupies the downstream motif. In contrast, RAR/RXR heterodimers bind with a reversed polarity (with RXR occupying the 3′ half-site) on DR1 elements, switching the activity of the heterodimer from an activator to a repressor of retinoic acid responsive genes.

The ability of heterodimeric receptors to bind to palindromes, inverted palindromes, and DR elements implies that the DBDs must be rotationally flexible with respect to the LBD dimerization interface (see Fig. 2.4), and on a DR the receptors must use a different region of the DBD of each receptor to create the dimerization interface.

Several orphan nuclear receptors can bind DNA with high affinity as monomers [14] (Fig. 2.4). For monomeric HREs, a single AGG/TTCA half-site is preceded by a 5′-flanking A/T rich sequence. In this case, the CTE of the DBD can make extensive contacts with the minor groove of DNA and effectively extends the surface contact of the receptor DBD to beyond the consensus half-site recognition sequence, providing additional receptor–DNA contacts in monomeric sites necessary for specific and high affinity binding. In Table 2.1, the receptors that bind as monomers, homodimers, and heterodimers to their HREs are indicated.

The existence of two types of nuclear receptor heterodimers, non-permissive and permissive, has been described [5]. Permissive heterodimers, such as PPAR/RXR, FXR/RXR, LXR/RXR or NGFI-B/RXR, can be indistinctly activated by ligands of either RXR or its partner receptor, and are synergistically activated in the presence of both ligands. However, in non-permissive heterodimers the ligand-induced transcriptional activities of RXR are suppressed, and it was believed that formation of the heterodimer actually precluded binding of ligand to RXR. Thus, in these complexes, RXR is said to be a "silent partner." TRs as well as the receptors for vitamin D (VDRs) or for retinoic acid (RARs) were considered to be non-permissive. However, recent data indicate that RXR can bind ligand and recruit co-activators as a heterodimer with RAR or TR. Lack of autonomous transcription on binding of the RXR agonist would be due to the fact that in the usual cellular environment co-repressors do not dissociate from the receptors and they prohibit co-activator access because co-regulator binding is mutually exclusive [4]. This model predicts that transcription by RXR agonists (rexinoids) could be obtained under some conditions, for instance in cells expressing high co-activator levels.

2.5
Ligand-dependent Activation. Nuclear Receptor Co-activators

Formation of the transcriptional initiation complex in RNA polymerase II-dependent promoters requires binding of the general transcription factors (GTFs). Performed complexes, composed of the RNA polymerase II, GTFs, SRBs (suppressor of RNA polymerase B), and several other proteins, termed the "holoenzyme," can be directly recruited to the promoter by sequence-specific transcription factors. The current hypothesis is that transcription factors will finally cause their effect on gene expression by influencing the rate of assembly of these complexes to the regulated promoter. One aspect of the mechanisms by which nuclear receptors affect the rate of RNA polymerase II-directed transcription likely involves the interaction of receptors with components of the transcription preinitiation complex. Thus, different nuclear receptors are able to interact directly with GTFs including

TBP (TATA box-binding protein), certain TAFs (TBP-associated factors), TFIIB, or TFIIH. However, modulation of the assembly of preinitiation complexes at the target promoter by transcriptional activators involves not only direct actions on components of the basal transcriptional machinery but also indirect actions mediated by the recruitment of co-regulators (co-activators and co-repressors).

Co-activators, also termed transcription intermediary factors, are bridging molecules that mediate the interactions of transcription factors with the basal transcriptional machinery. The packing of DNA in nucleosomes provides a major impediment to transcription. Two major mechanisms alleviate the block of transcription caused by the nucleosomal structure: histones can be post-translationally modified to destabilize chromatin, and nucleosomes can be disrupted through the activity of ATP-driven machines [15]. Not surprisingly, some receptor co-activators are ATP-dependent chromatin-remodeling factors or possess acetylase, methylase, or ubiquitin ligase activity, whereas others may interact directly with the basic transcriptional machinery and help to recruit the RNA polymerase II holoenzyme. Recruitment of co-activator complexes to the target promoter causes chromatin decompactation and transcriptional activation. Conversely, co-repressors can bind transcriptional activators and inhibit the formation of transcriptionally active complexes. Co-repressors are found within multicomponent complexes, which can contain histone deacetylase activity.

2.5.1
Histone Acetyltransferases

Histones are subjected to a great variety of post-translational modifications, including acetylation, methylation, phosphorylation, ubiquitination, sumoylation, and ADP-ribosylation. These modifications normally occur at the N-terminal and C-terminal histone "tail" domains, which play an essential role in controlling the folding of nucleosomal arrays into higher order structures. Histone modifications can cause nucleosome unfolding and increase access of transcription factor to the promoter. On the other hand, modifications could create novel recognition surfaces, thus promoting the association of positive regulators. Indeed, bromodomains can recognize acetylated lysines, and chromodomains bind to methylated lysines. This has led to the "histone code" hypothesis [16], in which specific combinatorial sets of histone modification signals can dictate transcriptional activation or repression.

A well-known family of co-activators that are recruited to the nuclear receptors in a ligand-dependent manner is the SRC/p160 family (steroid receptor co-activator) with three related members: SRC-1/NCoA-1, SRC/2TIF-2/GRIP-1/NCoA-2, and p/CIP/ACTR/AIB1/TRAM1/RAC3. These proteins act as primary co-activators, interacting with different nuclear receptors including classical, metabolic and orphan receptors in an agonist and AF-2-dependent manner. They also serve as platforms for the recruitment of secondary co-activators [17].

The three members of the p160 family of co-activators show a conserved structure, with a nuclear receptor-interacting domain (RID) in their central region

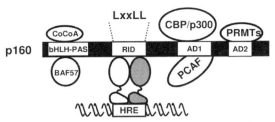

Fig. 2.5. p160 receptor co-activator complexes. p160 co-activators contain a bHLH (basic helix-loop-helix) motif and a PAS (Per-Arnt-Sim) homology region at the N-terminus which mediates interaction with the secondary co-activators CoCoA and BAF57 (BRG1-associated factor, whereby BRG-1 is the ATPase subunit of the chromatin remodeling complex SWI/SNF). The nuclear receptor-interacting domain (RID) contains three LxxLL motifs. Two activation domains AD1 and AD2 are located at the C-terminus. AD1 and AD2 are regions of interaction with CBP/p300 and protein arginine methyltransferases (PRMTs), respectively. p160 co-activators also interact with the p300/CBP-associating factor (PCAF).

(Fig. 2.5). Conservation is maximal in their N-terminal domains, which contain the nuclear localization signal and bHLH (basic helix-loop-helix) and PAS (period/aryl hydrocarbon receptor/single minded homology) domains. These domains mediate interactions with several transcriptional activators, such as BAF57 (see 2.5.4), and a coiled-coil co-activator (CoCoA) of still an unknown activity. Two activation domains with intrinsic transcriptional activity are also well conserved in p160 co-activators. The stronger transactivation domain (AD1) is the region of interaction with the histone acetyltransferase (HAT) CBP/p300, which serves co-activator roles for many different types of transcription factors, acting as co-integrator of extracellular and intracellular signaling pathways. The interaction of CBP with the nuclear receptors is also ligand- and AF-2-dependent, and CBP/p300 appears to function as an essential co-activator for the receptors. A weaker transactivation domain (AD2) located in the far C-terminus of p160 co-activators has been shown to interact with secondary co-activators with histone arginine methyltransferase activity such as CARM1 and PRMT (see Section 2.5.2) [18].

The RID of the p160 co-activators contains three highly conserved LxxLL motifs, where L is leucine and x is any amino acid, necessary and sufficient to mediate association of co-activators to ligand-bound receptors. LxxLL motifs form amphipathic α-helices with the leucines forming a hydrophobic surface on one face of the helix. The co-crystal structure of the LBD of several receptors and a p160 fragment containing two LxxLL motifs indicate that the conserved glutamic acid in H12 and lysine in H3 of the receptor make hydrogen bonds to leucines 1 and 5 of the co-activator RID. These contacts form a charge clamp that orients and positions the co-activator RID into the hydrophobic groove formed in the LBD after the conformational change elicited by ligand binding. Two LxxLL motifs from a single co-activator molecule interact with the AF-2 domains of both dimer partners and each member of the homo or heterodimer can cooperatively recruit one molecule of co-activator [10].

It has been shown that p160 co-activators also interact with the AF-1 domain of several nuclear receptors. Binding of co-activators to the AF-1 domain does not involve LxxLL motifs, but rather a glutamine-rich region of the co-activator [19]. CBP/p300 association to the p160 co-activators C-terminal region appears to be primarily responsible for the recruitment to the receptors, but there is also a direct interaction between CBP and the receptors that maps to the CBP N-terminus that contains one LxxLL-like motif. As different regions of CBP are involved in interaction with receptors and p160 co-activators, they can form a functional ternary complex [17]. A domain exhibiting intrinsic HAT activity is present in CBP/p300 and removal or mutation of domain results in loss of function for many transcription factors, indicating the importance of this activity. The p160 co-activators also possess HAT activity that maps to the C-terminal region and can acetylate both free histones and nucleosomal histones *in vitro*.

p160 co-activators also associate with PCAF, which is an ortholog of yeast GCN5, and the first identified mammalian HAT. PCAF (p300/CREB-binding protein-associated factor), additionally interacts directly and independently with the receptors and CBP/p300 through different regions and acts as a nuclear receptor co-activator. Thus, p160 co-activators might serve as a docking platform to bridge protein complexes with HAT activity to DNA-bound nuclear receptors. One of the purified, large multiprotein complexes that interact with liganded nuclear receptors contains PCAF, the c-Myc-interacting protein TRRAP, and TAFll30, which are common factors shared with the TFTC complex (TBP-free TAFII-HAT co-activator complexes). Three LxxLL motifs in TRRAP protein are responsible for the direct and ligand-dependent interactions with the receptors. Therefore, TFTC-type HAT complexes also appear to act as a novel class of co-activators for nuclear receptor function [20]. This suggests that the assembly of large, modular transcriptional complexes with HAT activity is involved in transcriptional regulation by activated nuclear receptors.

It has been demonstrated that *in vivo* histone acetylation levels of nuclear receptor target genes are strongly induced upon treatment with the corresponding ligand. Hyperacetylation is only triggered by agonists and is AF-2-dependent, confirming that histone acetylation is a critical step in nuclear receptor-mediated hormone signaling. Unexpectedly, hormone-induced histone hyperacetylation at the target promoter is transient and cyclic. The underlying mechanism for this observation appears to be that p160 co-activators can be acetylated by CBP/p300 and acetylation neutralizes the positive charges of two lysine residues adjacent to the core LxxLL motif and disrupts the association of HAT co-activator complexes with promoter-bound receptors [21, 22].

2.5.2
Protein Arginine Methyltransferases (PRMTs)

Methylation of histones by PRMTs has recently shown to be linked to gene activation. PRMT1 and CARM1 (cofactor-associated arginine methyltransferase 1, also known as PRMT4) bind to the AD2 region of the p160 co-activators and act as secondary co-activators in nuclear hormone receptor-regulated gene expression

(Fig. 2.5). Their co-activator potential is dependent on an intact histone methyl-transferase (HMT) domain, and the HMT, p160 binding, and homo-oligomerization activities reside in the central region. Further, CARM1, p160, and CBP/p300 act synergistically to enhance ligand-dependent transcriptional activation by nuclear receptors. While the N-terminal region has no known activity at present, the C-terminal part of CARM1 contains an autonomous activation domain, suggesting that it interacts with other proteins that help to mediate CARM1 co-activator function [18].

Recent studies have demonstrated the existence of an interplay between lysine acetylation and arginine methylation. By following the *in vivo* ligand-dependent pattern of modifications on histone H3 in a target promoter, it has been shown that arginine methylation follows prior acetylation of H3. A mechanism for the observed cooperation between acetylation and arginine methylation comes from the finding that acetylation tethers CARM1 to the H3 tail and allows it to act as a more efficient methyltransferase. Furthermore, acetylation of histone H4 inhibits methylation by PRMT1, whereas methylation of arginine 3 of H4 facilitates its subsequent acetylation by CBP/p300. In addition to histones, other chromatin proteins are methylated by PRMTs. Thus, CARM1 methylation of CBP/p300 was also reported to contribute to receptor-mediated transcriptional activation. In contrast, this methylation inhibits binding to the transcription factor CREB, causing the loss of CREB-dependent transcription [7, 18, 22].

2.5.3
Ubiquitination and Sumoylation

Ligand-dependent degradation of nuclear receptors occurs via the 26S proteasome, that degrades poly-ubiquitinated target proteins. This process involves the ubiquitination of the receptors and the recruitment of the proteasome at the AF-2 domain through SUG-1, a proteasome component. Remarkably, blocking of the proteasome abrogates not only receptor degradation but also ligand-dependent transactivation by different receptors. The paradoxical mechanism for this phenomenon is currently unknown, but several proteins involved in proteolysis, such as ubiquitin ligases and proteasome components, have been suggested to act as nuclear receptor co-regulators and are recruited *in vivo* to the receptor regulated promoters. The dual role of the ubiquitin-proteasome machinery may play a role in dynamic assembly/disassembly of the receptors to the promoter of target genes, which has been demonstrated to occur in the cyclic manner [23, 24]. In addition, both co-activator and co-repressor complexes can recruit ubiquitination complexes and some of them have been demonstrated to be targets of the proteasome. Recent studies have also shown that histones are ubiquitinated and that this modification regulates histone H3 methylation, linking the proteasome to epigenetic gene regulation [22].

The small ubiquitin-related modifier (SUMO) also modifies a number of nuclear receptors and co-regulators with different transcriptional outcomes. SUMO-conjugating enzyme, Ucb9, interacts with the glucocorticoid receptor (GR), enhancing its transcriptional activity and this protein as well as SUMO E3 ligases interact with the androgen receptor (AR) repressing its activity. Sumoylation does not appear to target proteins for degradation, and is rather involved in protein stabilization and subcellular localization. For instance, sumoylation of p160 proteins can increase interaction with the receptor and prolong retention in the nucleus [7, 21, 22].

2.5.4
ATP-dependent Chromatin Remodeling Complexes

ATP-dependent remodeling factors, such as SWI2/SNF2, ISWI/SNFL2, or WINAC use the energy derived from ATP hydrolysis to catalyze nucleosome mobilization, which is a net change in the position of the histone octamer relative to DNA. This change is believed to facilitate the access and function of key components of the transcriptional apparatus. Chromatin remodeling complexes comprise an ATPase subunit (BRG-1 or hBrahma) with a conserved nucleotide-binding motif along with other polypeptides such as BAFs (BRG-1-associated factor) [25].

It is known that receptors can bind to their response elements packaged into chromatin, and that receptor binding to DNA indeed facilitates binding of other factors whose binding sites were not previously exposed in the nucleosomes in an ATP-dependent manner. Different receptors can interact with the SWI/SNF complex that is recruited to the target promoters in a ligand-dependent manner [26]. The receptors do not interact directly with BRG-1 or hBrahma, but different receptors interact with different BAF subunits. Thus, the glucocorticoid receptor (GR) interacts with BAF250 and BAF60a, PPARγ with BAF60c, VDR/RXR with BAF60a and ERα with BAF57. Interestingly, this protein also contacts the p160 co-activators, linking SWI/SNF with HAT complexes.

SWI/SNF complexes also appear to be required for periodic binding and displacement of receptors during chromatin remodeling. Transient GR binding to a target promoter occurs in concert with nucleosome remodeling, as the process is completely dependent on the presence of SWI/SNF and ATP. During nucleosome remodeling histones H2A and H2B undergo extensive reorganization. When SWI/SNF leaves the promoter, GR is released and chromatin is then ready for the next cycle of GR-directed SWI/SNF action [23].

WINAC represents a new member of the ATP-dependent remodeling complexes. It contains BRG-1 and hBrahma as ATPases, but it also has subunits associated with DNA replication and transcriptional elongation as well as WSTF (Williams syndrome transcription factor). WSTF appears to function as a platform for the assembly of WINAC and interacts directly with the vitamin D receptor (VDR) both in the presence and absence of ligand. WSTF is targeted to vitamin D receptors *in vivo*, and appears to be required for VDR-mediated transcriptional regulation [26].

2.5.5
TRAP/DRIP/Mediator Complexes

Multiprotein complexes denominated TRAP and DRIP that interact with TR or VDR (and other receptors) in a ligand-dependent manner and enhance the ligand-dependent transcriptional activity have been isolated. Both complexes are equivalent to the yeast mediator complex that together with SRB proteins associates with the large subunit of RNA polymerase. It is believed that the TRAP/DRIP complex acts by recruiting the polymerase to the target promoter. The TRAP/DRIP complexes are recruited to the core AF-2 receptor region in response to ligand binding through a single subunit (DRIP205/TRAP220) via an LxxLL motif identical to that found in the p160 co-activators [27]. This subunit anchors the other proteins comprising the TRAP/DRIP complex, which is presumably preformed in the cell.

Ligand-dependent transcriptional activity of nuclear receptors requires recruitment of both TRAP/DRIP and histone acetyltransferase and methyltransferase-containing co-activator complexes. As the receptor binding subunits from both functionally distinct co-activator complexes interact with the same receptor region they compete with each other for binding to the receptor. They may act independently or consecutively. Recent studies, primarily employing chromatin immunoprecipitation (ChIP) assays, have shown that p160 and TRAP/DRIP complexes are recruited to hormone-regulated promoters in a sequential manner with distinct kinetics. p160 complexes act earlier than TRAP/DRIP and this ordered recruitment may proceed for multiple cycles of factor association and dissociation, with multiple rounds of transcription occurring within each cycle. Furthermore, evidence for a model of "facilitated recruitment" in which the prior actions of the p160 complexes facilitate the recruitment and actions of TRAP/DRIP has been obtained [28]. The observation that acetylation of p160 proteins by p300/CBP disrupts the association with the receptors, might provide the opportunity for TRAP/DRIP complexes to associate with the receptor.

In summary, binding of a ligand to the nuclear receptors allows the recruitment of co-activators with unique biochemical activities at temporally appropriate times during the transcription process. The receptors can recruit first co-activators with HAT and HMT activity resulting in histone acetylation and methylation. The p160 complexes then dissociate, subsequent to their acetylation or to their degradation by the proteasome. The ATP-dependent remodeling complexes can then be recruited causing the displacement of nucleosomes. However, it cannot be excluded that remodeling complexes could be recruited before p160 co-activators in a context of highly condensed chromatin. Once chromatin has been descondensed, the receptors via their association with the TRAP/DRIP complex would be able to recruit the transcriptional machinery resulting in stimulation of gene expression (see Fig. 2.7).

2.6
Ligand-independent Repression. Nuclear Receptor Co-repressors

In addition to ligand-dependent gene activation, selected receptors including TR and RAR repress basal transcription in the absence of ligand. Binding of hormonal ligand to the receptor releases the transcriptional silencing and leads to gene acti-

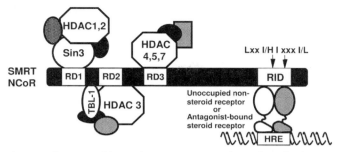

Fig. 2.6. Structure of the nuclear receptor co-repressors SMRT and NCoR. Schematic representation of a co-repressor showing the location of the repressor domains (RD1, RD2 and RD3) and the receptor interacting domain (RID). The RID contains two extended helical motifs Lxx I/H I xxx I/L. The RD1 interacts with mSin3A that in turn recruits class I deacetylases (HDAC1 and 2). Class II deacetylases bind at RD3 without Sin3 as a mediator. HDAC3 has been shown to interact directly with a SANT (according to its presence in Swi3, Ada2, NCoR, and TFIIB) domain located between RD1 and RD2. Co-repressor complexes bind to unliganded non-steroid receptors or to antagonist-bound steroid receptors.

vation. The current model of gene regulation by these receptors assumes that the unliganded receptors are bound to the HRE and that under these conditions are associated with co-repressors responsible for the silencing activity. The conformational changes elicited in the receptors by ligand binding would cause the dissociation of co-repressors and the recruitment of co-activator complexes responsible for transcriptional activation.

The best characterized co-repressors that associate with TR and RAR are 270-kDa cellular proteins named NCoR (nuclear co-repressor) and SMRT (silencing mediator for retinoic and thyroid hormone receptors). Unliganded TRs and RARs interact strongly with NCoR and SMRT and addition of ligand induces dissociation from the co-repressors. NCoR and SMRT are related both structurally and functionally. They contain three autonomous repressor domains (RD) and a receptor interacting domain (RID) located toward the C-terminus (Fig. 2.6). The RID is composed of two motifs (or CoRNR boxes), with the consensus sequence Lxx I/H I xxx I/L, likely to adopt an amphipathic α-helical conformation. However, when compared to the LxxLL motif, the CoRNR motif presents an N-terminal extended helix. This extension appears to be required for effective binding to the unliganded receptor [29, 30].

Although a receptor CoR box, located in H1 of the LBD within the hinge region, is essential for interaction of receptors with the co-repressors, the CoRNR box does not interact directly with residues in this region, but docks to a hydrophobic groove in the surface of the LBD helices 3 and 4. Since this surface is similar to that involved in co-activator interaction, co-activator and co-repressor binding is mutually exclusive. The structure of a ternary complex containing the PPARα LBD bound to an antagonist with a SMRT co-repressor motif has been recently solved. In this structure, the CoRNR motif adopts a three-turn α-helix that indeed binds the co-activator groove, preventing the AF-2 helix from assuming the active conformation.

Ligand binding by itself is not sufficient to induce dissociation of co-repressors. Rather, it appears that the AF-2 region serves to trigger the release of co-repressors from the receptors. H12 is fully inhibitory for co-repressor binding to most nuclear receptors. Nuclear receptors lacking this AF-2 region act as constitutive transcriptional repressors. For instance, RXR does not bind co-repressors, but deletion of this region allows co-repressor interaction and *in vivo* repression. In the case of TRs and RARs, mutation or deletion of the AF-2 domain increases interactions with co-repressors and reduces the release of the co-repressors after ligand binding, indicating again that H12 is inhibitory [31].

Although agonist liganded steroid hormone receptors do not interact effectively with NCoR or SMRT, clear interactions both *in vivo* and *in vitro* are observed with receptor-bound antagonists. It is possible, that the repositioning of H12 by the antagonists permits co-repressor binding into the hydrophobic pocket. This suggests that steroid receptors occupied by antagonists can act as transcriptional silencers by binding of cellular co-repressors.

Transcriptional repression by the co-repressor-bound receptors appears to be mediated by the recruitment of histone deacetylases (HDACs) to the promoter region. HDAC1 and 2 are found in the cells in large multiprotein complexes associated with mSin3 proteins. Sin3 is a large multidomain protein that most likely forms the scaffold upon which the rest of the complex assembles [29]. mSin3/HDAC complexes are abundant and stable and could be available for binding and recruitment by the repressors. mSin3 associates with SMRT and NCoR through the C-terminal repressor domain (RDI). The interaction between unliganded receptors and mSin3 is therefore not direct but is mediated by NCoR and SMRT whose function would be to link the receptors to HDAC complexes.

The co-repressors were thought to act exclusively through the indirect recruitment of HDAC1 or 2 (class I deacetylases), via the adapter mSin3 protein. However, the RD3 has been demonstrated to repress transcription by directly interacting with class II deacetylases (HDACs 4, 5, and 7). Endogenous NCoR and SMRT each associate with class II HDACs in a complex that does not contain mSin3A or HDAC1 [29, 30]. Therefore, a single co-repressor could use distinct domains to engage class I HDAC complexes in a Sin3-dependent manner and class II HDAC complexes in a Sin3-independent manner. Furthermore, a novel SMRT-containing complex has been isolated from HeLa cells. This complex contains HDAC3 and transducin beta-like protein 1 (TBL1), a protein that interacts with histone H3 and is associated with human sensorineural deafness [32]. *In vivo*, TBL1 is bridged to HDAC3 through SMRT and can potentiate repression by the receptors.

Above observations suggest that compactation of chromatin structure due to recruitment of histone deacetylases complexes by the co-repressors is involved in transcriptional silencing by the unliganded non-steroid receptors or antagonist-bound steroid receptors. Agonist binding would allow the release of co-repressors and enable the receptors to recruit co-activators and stimulate transcription (Fig. 2.7).

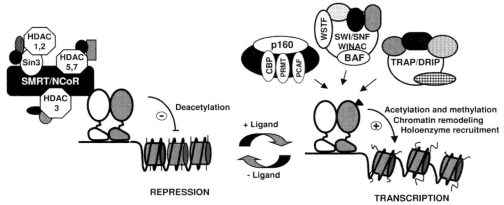

Fig. 2.7. Exchange of co-activator and co-repressor complexes. In the absence of ligand the nuclear hormone receptor heterodimer is associated with co-repressor complexes. The co-repressors (SMRT/NCoR) recruit histone deacetylases (HDACs) either directly or through their interaction with Sin3. Deacetylation of histone tails leads to chromatin compactation and transcriptional repression. Upon ligand binding, the receptors recruit different co-activator complexes. Some complexes possesses histone acetyltransferase and methyltransferase activity (p160), other possess ATP-dependent chromatin remodeling activity (SWI/SWF, WINAC), and finally the TRAP/DRIP complex may recruit the RNA polymerase II (RNAP II) holoenzyme. Recruitment of co-activator complexes causes chromatin decompactation and transcriptional stimulation. WINAC, WSTF including nucleosome assembly complex; WSTF, Williams syndrome transcription factor.

2.7
Ligand-dependent Transcriptional Repression

2.7.1
Negative Response Elements

Although most of the attention has been focused on transcriptional activation by binding of nuclear receptors to positive HREs, nuclear receptors can also repress gene expression in a ligand-dependent manner. In some cases repressive effects may be due to passive inhibition, which can occur due to competition for DNA sites with other transactivators or to formation of transcriptionally inactive heterodimers. However, there are also "negative HREs" which bind the receptors and mediate negative regulation by the ligand. Although at the present time the properties of the negative HREs are not totally understood, location of the element may play a role since negative HREs are generally very close to the transcription initiation sites, and some are positioned downstream of the TATA box or even have an unusual location at the 3′-untranslated region. On the other hand, there is evidence that co-repressors and deacetylase activity could be also involved in ligand-dependent negative regulation by nuclear receptors. In contrast to positively regulated genes, it is known that thyroid hormone receptors increase basal activity of negatively regulated promoters, and addition of ligand reverses this stimulation. It has been reported that co-repressors and HDACs paradoxically enhance rather than suppress basal activity [5].

2.7.2
"Cross-talk" of Nuclear Receptors with Other Signaling Pathways

Nuclear receptors can also modulate gene expression by mechanisms independent of binding to an HRE. Thus, they can alter expression of genes that do not contain a HRE through positive or negative interference with the activity of other transcription factors, a mechanism generally referred to as "transcriptional cross-talk" [5, 33]. Thus, the receptors can negatively regulate target gene promoters that carry AP1, NFκB, or CREB-binding sites, without binding to these DNA elements themselves. The mechanisms responsible are not totally understood and can involve direct protein-to-protein interaction of the receptors with these transcription factors, or with intermediary factors that tether the receptors to the regulated promoter. The receptors can also regulate the activity of the signaling pathways that lead to the activation of these factors by different mechanisms. It is believed that many of the antiproliferative effects and anti-inflammatory actions of ligands of nuclear receptors could be mediated by "cross-talk" mechanisms.

That transrepression plays a very important role *in vivo* has been demonstrated in "knock-in" mice in which wild-type glucocorticoid or thyroid hormone receptors have been replaced by mutants that cannot bind DNA. In addition, it has been possible to generate synthetic ligands that dissociate transactivation from transrepression. These ligands have a large potential as pharmacological tools in the treatment of a variety of diseases including cancer and inflammatory diseases [4].

The cross-talk between nuclear receptors and other signaling pathways is not restricted to the transcriptional antagonism described above. Phosphorylation of nuclear receptors provides an important link between signaling pathways. Depending on the receptor and on the residue involved, in some cases phosphorylation can inhibit ligand-dependent activation by nuclear receptors due to a reduction in ligand binding or in DNA-binding affinity or by promoting receptor degradation. However, in other cases the receptors can be activated in the absence of their cognate ligands by phosphorylation through signals originated in membrane receptors. In certain diseases such as breast or prostate cancer this modification can contribute to disease progression and lead to resistance to antagonist therapy.

Also co-activators and co-repressors can be phosphorylated by different kinases and there is increasing evidence that this modification alters their activity. Thus, phosphorylation of SMRT inhibits its interaction with the receptors and provokes redistribution of the co-repressor from the nucleus to the cytoplasm. In contrast, phosphorylation of different co-activators by a variety of kinases can enhance their enzymatic activity and their affinity for the receptors leading to an increased transcriptional response [7, 22].

References

1 MANGELSDORF, D. J., THUMMEL, C., BEATO, M., HERRLICH, P., SCHUTZ, G., UMESONO, K., BLUMBERG, B., KASTNER, P., MARK, M., CHAMBON, P. et al. (**1995**) *Cell* 83, 835–839.

2 ROBINSON-RECHAVI, M., ESCRIVA GARCIA, H., LAUDET, V. (**2003**) *J Cell Sci* 116, 585–586.

3 MOHAN, R., HEYMAN, R. A. (**2003**) *Curr Top Med Chem* 3, 1637–1647.

4 GRONEMEYER, H., GUSTAFSSON, J. A., LAUDET, V. (**2004**) *Nat Rev Drug Discov* 3, 950–964.

5 ARANDA, A., PASCUAL, A. (**2001**) *Physiol Rev* 81, 1269–1304.

6 ROCHETTE-EGLY, C. (**2003**) *Cell Signal* 15, 355–366.

7 HERMANSON, O., GLASS, C. K., ROSENFELD, M. G. (**2002**) *Trends Endocrinol Metab* 13, 55–60.

8 NORMAN, A. W., MIZWICKI, M. T., NORMAN, D. P. (**2004**) *Nat Rev Drug Discov* 3, 27–41.

9 RENAUD, J. P., MORAS, D. (**2000**) *Cell Mol Life Sci* 57, 1748–1769.

10 GRESCHIK, H., MORAS, D. (**2003**) *Curr Top Med Chem* 3, 1573–1599.

11 NAGY, L., SCHWABE, J. W. (**2004**) *Trends Biochem Sci* 29, 317–324.

12 BLEDSOE, R. K., MONTANA, V. G., STANLEY, T. B., DELVES, C. J., APOLITO, C. J., MCKEE, D. D., CONSLER, T. G., PARKS, D. J., STEWART, E. L., WILLSON, T. M., LAMBERT, M. H., MOORE, J. T., PEARCE, K. H., XU, H. E. (**2002**) *Cell* 110, 93–105.

13 MANGELSDORF, D. J., EVANS, R. M. (**1995**) *Cell* 83, 841–850.

14 GIGUERE, V. (**1999**) *Endocr Rev* 20, 689–725.

15 NARLIKAR, G. J., FAN, H. Y., KINGSTON, R. E. (**2002**) *Cell* 108, 475–487.

16 JENUWEIN, T., ALLIS, C. D. (**2001**) *Science* 293, 1074–1080.

17 MCKENNA, N. J., LANZ, R. B., O'MALLEY, B. W. (**1999**) *Endocr Rev* 20, 321–344.

18 STALLCUP, M. R., KIM, J. H., TEYSSIER, C., LEE, Y. H., MA, H., CHEN, D. (**2003**) *J Steroid Biochem Mol Biol* 85, 139–145.

19 WARNMARK, A., TREUTER, E., WRIGHT, A. P., GUSTAFSSON, J. A. (**2003**) *Mol Endocrinol* 17, 1901–1909.

20 YANAGISAWA, J., KITAGAWA, H., YANAGIDA, M., WADA, O., OGAWA, S., NAKAGOMI, M., OISHI, H., YAMAMOTO, Y., NAGASAWA, H., MCMAHON, S. B., COLE, M. D., TORA, L., TAKAHASHI, N., KATO, S. (**2002**) *Mol Cell* 9, 553–562.

21 SMITH, C. L., O'MALLEY, B. W. (**2004**) *Endocr Rev* 25, 45–71.

22 BAEK, S. H., ROSENFELD, M. G. (**2004**) *Biochem Biophys Res Commun* 319, 707–714.

23 NAGAICH, A. K., WALKER, D. A., WOLFORD, R., HAGER, G. L. (**2004**) *Mol Cell* 14, 163–174.

24 METIVIER, R., PENOT, G., HUBNER, M. R., REID, G., BRAND, H., KOS, M., GANNON, F. (**2003**) *Cell* 115, 751–763.

25 BECKER, P. B., HORZ, W. (**2002**) *Annu Rev Biochem* 71, 247–273.

26 BELANDIA, B., PARKER, M. G. (**2003**) *Cell* 114, 277–280.

27 FREEDMAN, L. P. (**1999**) *Cell* 97, 5–8.

28 ACEVEDO, M. L., LEE, K. C., STENDER, J. D., KATZENELLENBOGEN, B. S., KRAUS, W. L. (**2004**) *Mol Cell* 13, 725–738.

29 JEPSEN, K., ROSENFELD, M. G. (**2002**) *J Cell Sci* 115, 689–698.

30 PRIVALSKY, M. L. (**2004**) *Annu Rev Physiol* 66, 315–360.

31 HU, X., LAZAR, M. A. (**2000**) *Trends Endocrinol Metab* 11, 6–10.

32 GUENTHER, M. G., LANE, W. S., FISCHLE, W., VERDIN, E., LAZAR, M. A., SHIEKHATTAR, R. (**2000**) *Genes Dev* 49, 1048–1057.

33 GOTTLICHER, M., HECK, S., HERRLICH, P. (**1998**) *J Mol Med* 76, 480–489.

3
Mechanism of Action and Cancer Therapeutic Potential of Retinoids

Emilie Voltz, Emmanuelle Germain, and Hinrich Gronemeyer

3.1
Introduction

Our understanding of the molecular mechanisms that underlie retinoid action has increased tremendously in recent years. Retinoids, a collective term for natural and synthetic derivatives of vitamin A, regulate complex biological functions, such as embryonic development, organogenesis, bone formation, metabolism, vision, and reproduction. At the cellular level, these regulations are achieved, among others, by modulation of proliferation, differentiation, and apoptosis. Retinoids are believed to trigger these cellular biological events by acting as transcription modulatory factors [1, 2]. The strong differentiative and apoptosis-inducing capacity of retinoids, originally observed in cellular systems *in vitro*, was the basis for their successful development as anticancer agents. We are currently witnessing an increasing number of studies that reveal the mechanistic basis of this anticancer action (for recent reviews, see Refs [3, 4–7]). To date, the most striking example of so-called cancer differentiation therapy, based on a combination of retinoic acid and chemotherapy, is that used for acute promyelocytic leukemia (APL) in which more than 72% of the patients are cured. In the past few years, paradigms for combotherapies have been reported that comprise treatment with retinoids or rexinoids (RXR-selective ligands) and a diverse set of other compounds such as signaling drugs, kinase activators or inhibitors, or drugs affecting epigenetic enzymes.

3.2
Origin of Retinoids

Retinoids are formed from dietary vitamin A (all-*trans* retinol). The only source of vitamin A is plants carotenoid pigment (β-carotene) but the main vitamin A intake of carnivores comes from the long-chain retinyl esters present in animal products (eggs, milk, butter, fish-liver oil). Vitamin A undergoes a series of metabolic con-

Nutritional Genomics. Edited by Regina Brigelius-Flohé and Hans-Georg Joost
Copyright © 2006 WILEY-VCH Verlag GmbH & Co. KGaA, Weinheim
ISBN: 3-527-31294-3

versions in the intestine. Retinol within the mucosal cell is largely re-esterified with long-chain fatty acids. The retinyl esters are then incorporated, together with other lipids and with apolipoproteins, into chylomicron particles that are secreted from the cell into the lymph. Approximately 75% of chylomicron retinoids are eventually taken up in the liver, the main store of retinoids in the body. The cleavage of retinyl esters mobilizes retinol which is subsequently oxidized to retinal and retinoic acid (RA) [8]. Note that only a very small proportion of plasma and tissue retinol (0.2–5%) is converted to all-*trans*-retinoic acid (ATRA), the main signaling retinoid. Multiple metabolites and retinoids with putative signaling functions have been described (for details, the reader is referred to the corresponding reviews, see Ref. [9]). Interestingly, one metabolite, 4-oxo-RA was reported to act as a positional specifier during frog embryogenesis [10]. Another retinoid, 1,4 hydroxy-4,14-*retro*-retinol regulates T lymphocyte growth and appears to act by directly interacting with protein kinase C [11, 12]. Finally there is considerable debate about the origin and signaling function of the 9-*cis* isomer of ATRA, 9-*cis*-RA, which is particularly interesting due to its high affinity for the second class of retinoic acid receptors (RXRs) (see below). It is important to point out that the production of ATRA in the body needs to be tightly regulated for proper organogenesis, as too little vitamin A or too high RA concentrations causes severe malformation (see Ref. [1], and references therein).

3.3
Mechanistic Basis of RA Signaling

3.3.1
Retinoid Action is Mediated through RAR/RXR Heterodimers

There are two classes of RA receptors: retinoic acid receptors (RARs), which bind ATRA and 9-*cis*-RA with similar affinities, and retinoid X receptors (RXRs), which bind only 9-*cis*-RA [13, 14; for recent review, see Ref. [15]). Each family of these nuclear receptors is composed of three subtypes, referred to as α, β, and γ, which are encoded by different genes. In addition, each subtype can be expressed as a number of isoforms due to differential splicing and alternative promoter usage. It is believed that this diversity of receptors contributes to the cell/development-specific effect of retinoids.

Retinoid receptors are functionally and structurally related members of the nuclear receptor family (see also chapter 2). Molecular cloning and structure/function analyses have revealed that these members share a common architecture with six regions designated A through F (Fig. 3.1). The non-conserved N-terminal region A/B, contains an autonomous transcriptional activation function (AF-1), called the transactivation domain, which is of variable length and provides the sequence differences that define the various isoforms of the receptors. The three RAR types differ in their B regions, and their main isoforms (α1 and α2, β1–β4, γ1, and γ2) have different N-terminal A regions. Similarly, the three RXRs characterized differ in their A/B regions. The N-terminal A/B regions of RAR and RXR possess promoter context-specific transcription activation functions which could synergize with and modulate the autonomous ligand-dependent AF-2 function associated with region E [16].

The C region harbors the well-conserved DNA-binding domain (DBD), composed of two zinc fingers which are crucial for the recognition of the cognate DNA response element. While the first zinc finger mediates DNA interaction, the second zinc finger appears to be involved mainly in receptor dimerization (for original references and reviews, see Refs [15, 17, 18]). It has to be noted that the DBD has the most conserved amino acid sequence within the nuclear receptor superfamily. The hinge domain (D), located between the DBD (C) domain and the ligand-binding (E) domain, allows rotation of the ligand-binding domain (LBD) and is less conserved among the nuclear receptors. The E region is a highly structured domain that encodes several distinct functions, most of which operate in a ligand-dependent manner. The LBD harbors the ligand-dependent activation function (AF-2) and a major dimerization interface. In addition, some nuclear receptors possess a repression function in their LBD which can negatively regulate transcription of target genes in the absence of ligands or in the presence of a certain type of antagonists. The function of the C-terminal F region of the RARs (not found in the RXRs), if any, has remained elusive (for details, reviews and references to the original studies, see Refs [1, 2, 15, 19, 20]).

Like other members of the nuclear receptors, retinoid receptors act as ligand-activated DNA-binding transcription factors which bind as RAR/RXR heterodimers, to *cis*-acting retinoic acid response elements (RAREs/RXREs) present in the target genes (for a list of RAREs, see Refs [15, 19, 20]). These elements consist of two hexameric half-sites with the consensus sequence 5′ PuG(G/T)TCA 3′ (where "Pu" stands for a purine residue) which can be arranged as palindromes or direct repeats (DR) separated by a spacer of generally 0 (palindrome), 1 (DR1), 2 (DR2), or 5 (DR5) nucleotides [21–24]. The selective recognition of DRs based on inter-half-site spacings has been formalized in a 1–5 rule that defines the highest affinity binding sites and determines, at least in part, the receptor specificity for optimal binding and activation [25, 26]. When the RAR/RXR heterodimer binds to an RARE with spacer of 2 or 5 nucleotides (DR2, DR5), RXR binds to the 5′ and RAR to the 3′ of the hexameric motif [24, 27–29]. In contrast, this polarity is reversed on DR1 elements, with the RAR in 5′ and the RXR in 3′ position (see Fig. 3.1) [30]. The crystal structure and mutational analysis of the RAR/RXR and RXR/RXR DNA-binding complexes revealed that the distinct dimerization surfaces within the DBDs of each heterodimerization partner dictates the specificity of response element recognition while the dimerization interface in the LBD increases DNA-binding efficacy but has no role in response element selection [31].

In the absence of an agonist ("apo-heterodimer"), or in presence of some RAR antagonists, the RAR/RXR heterodimer is bound to co-repressors (CoRs) such as nuclear receptor co-repressor (NCoR) and silencing mediator for retinoid and thyroid hormone receptors (SMRT). These CoRs are recruited to the receptor heterodimer as complexes with several other factors, among them the epigenetically

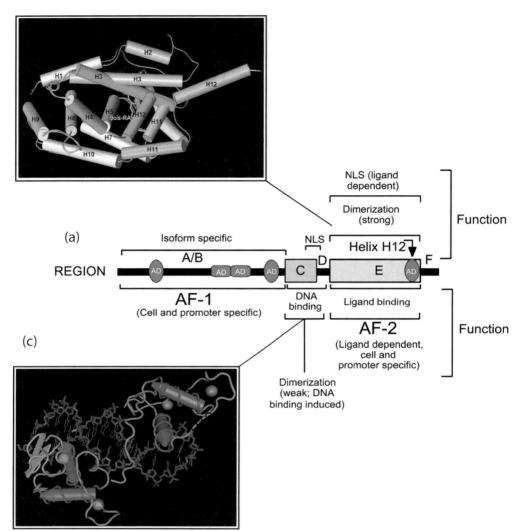

Fig. 3.1. Structural and functional organization of nuclear receptors. (a) Nuclear receptors consist of six domains (A–F) based on regions of conserved sequence and function. The evolutionarily conserved regions C and E are indicated as boxes (yellow and blue, respectively), and a black bar represents the divergent regions A/B, D and F. The N-terminus (A/B region) contains one autonomous transcriptional activation function (AF-1). The highly conserved C region harbors the DNA-binding domain that confers sequence-specific DNA recognition. The ligand-binding domain (E region) is a highly structured domain comprising a ligand-dependent activation function (AF-2). The activation domains (ADs) contain transcriptional activation functions that can activate transcription when fused to a heterologous DNA-binding domain. (b) Ligand

active histone deacetylases (HDACs) (see Ref. [32], and refences therein; also Refs [33–36]). The accumulation of these CoR/HDAC complexes results in local chromatin condensation and consequent gene silencing. Upon agonist binding, a conformational change occurs at the level of the LBD (see below) which generates a novel interaction surface for co-activators (CoAs) and concomitantly destabilizes CoR binding, thus leading to the dissociation of the HDAC-containing complex. The binding of CoAs results in the establishment of at least two types of complexes. The first comprises the histone acetyltransferase (HAT) complex, which contains CBP/p300 (CREB-binding protein) recruited by the CoAs. The local acetylation of the core histones tails leads to chromatin decondensation and derepression of the target gene (reviewed in Refs [37, 38]). This facilitates recruitment of another type of complex, variously termed thyroid hormone receptor (TR)-associated protein (TRAP), vitamin D receptor-interacting protein (DRIP), or Srb and mediator protein-containing complex (SMCC), which binds to the holoreceptor and establishes contacts with the basal transcription machinery [39, 40].

Transactivation may also require the recruitment of other multiprotein complexes such as the ATP-dependent chromatin remodeling machineries (e.g. SWI/SNF) but the kinetics and order of recruitment have not yet been fully established and may be gene-specific [33]. Consequently, transcription can be regarded as two successive phenomena: (1) derepression caused by chromatin decondensation and (2) bona fide transactivation by receptor-dependent increase of the frequency of transcription initiation.

3.3.2
Structural Basis of Retinoid Action

Crystallization studies of several ligand-free (apo) and agonist-occupied (holo) nuclear receptor LBDs alone or in complex with co-activator fragments have provided molecular details of the various allosteric effects and, moreover, have revealed how the chemical information present in the ligand structure is transformed into a defined order of selective protein–protein interactions.

◄───

binding induces a conformational change of the ligand-binding domain structure of nuclear receptors. A comparison of the crystal structures of the apo-RXRα ligand-binding domain (LBD) (PDB: 1LBD) with the holo-RXRα LBD complexed with 9-*cis*-retinoic acid (PDB: 1FBY). The figure reveals the ligand-induced *trans*-conformation that generates the transcriptionally active form of the receptor. The coloured helices H2, H3, H11, and H12 (purple in the apo-form; red in the holo-form) are relocalized during the conformational change. In this model, ligand binding induces a structural transition that triggers a mousetrap-like mechanism. (c) Schematic representation of the structure of the RXR/RAR DNA-binding domain heterodimer in complex with the retinoic acid response element DR1. The sequence of the DBD sequence is comprised of two zinc-nucleated modules and two alpha-helices that fold into a single globular domain. The receptors recognize identical half-sites through extensive base-specific contacts; however, RXR binds exclusively to the 3′ site to form an asymmetric complex with the reverse polarity of other RXR heterodimers (PDB: 1DSZ).

The first crystal structure reported was that of the unliganded LBD of the human RXRα which revealed a structure composed of 12 α-helices (H1 to H12) organized in three layers, referred to as an "antiparallel α-helical sandwich", and one antiparallel β-sheet [41]. The subsequent crystal structure of the human RARγ LBD bound to ATRA revealed the ligand-binding interactions and led to the proposition of the "mousetrap model" [42]. According to this model, ligand binding induces a structural transition, which triggers a mousetrap-like mechanism: pushed by the ligand, H11 is repositioned in the continuity of H10, and the concomitant swinging of H12 unleashes the Ω-loop, which flips over underneath H6, carrying alone the N-terminal part of H3. In its final position, H12 seals the ligand-binding cavity as a "lid" and further stabilizes ligand-binding by contributing to the hydrophobic pocket (see Fig. 3.1). These agonist-induced conformational changes make the helix H12 able to interact with the LxxLL NR box motif of bona fide co-activators [15, 42, 43].

3.3.3
Structural Basis of Ligand Action: Agonist and Antagonist

As already mentioned, non-liganded receptors expose a surface that accommodates co-repressors. There are now multiple examples which demonstrate that the precise chemical nature of a ligand can differently modulate the surfaces for co-activator and co-repressor binding, thereby differently modulating the transcriptional activity of receptors. For example, "inverse agonists" may stabilize the receptor/co-repressor complexes, while other ligands may exert pure or partial antagonistic or agonistic effects. Below, we will review some of the general principles originating from structural studies which demonstrated that it is primarily the AF-2 activation function that is differently affected by various classes of ligands.

A general feature common to all pure AF-2 antagonists is the presence of a bulky side-chain that cannot be accommodated without structural changes within the agonist-binding cavity, occupied by the cognate agonist. As observed in the RARα-BMS 614 crystal structures, the chemical moiety that is responsible for its antagonistic feature, a quinoline group, interferes with the positioning of H12 in the agonist structure and induces the unwinding of the C-terminus of H11, thus enabling H12 to adopt a second low-energy position by binding to the co-activator LxxLL recognition cleft [44]. Similar structural features were also observed for antagonist complexes of other receptors [45]. Therefore, it appears that the action of pure antagonist apparently originates from at least two structural principles (1) the presence of a large antagonistic ligand which sterically prevents proper positioning of H12 and thus does not form an interaction surface for co-activator binding, and (2) the competition between H12 and the nuclear receptor boxes (NR box) of co-activators for a common binding surface.

In addition to pure AF-2 antagonist, partial AF-2 agonist/antagonists have also been crystallized with the corresponding receptors. These data reveal, as exemplified by the RXRα/F318A/oleic acid complex, that H12 adopts the antagonist conformation even though the corresponding ligand elicits *in vivo* a weak but significant transcriptional activity [44]. It is believed that this apparent discrepancy reflects the concomitant presence of both the agonist and antagonist conformations in solution and the preferential crystallization of the antagonist structure. Similar observations were made for genistein and GW0072 in the complexes with estrogen receptor beta (ERβ) or peroxisome proliferator-activated receptor gamma (PPARγ), respectively [46, 47]. In the presence of such mixed agonist/antagonist ligands, the equilibrium between the holo and antagonist positions of H12 very likely depends on the intracellular concentration of co-activators and co-repressors. Therefore, these ligands may act as either AF-2 agonists or antagonists depending on the cellular context. Hence, oleic acid or genistein do not preclude sterically the agonist position of H12 and act in this respect like agonists. On the other hand, these ligands also induce unwinding of helix H11, which permits H12 to access the antagonist groove; in this respect, partial agonists or antagonists act as antagonists. In opposition with such "active antagonism," another mechanism accounting for the particular biological properties of such ligands has been proposed from the structural resolution of ERα and ERβ complexes with THC (5,11-*cis*-diethyl-5,6,11,12-tetrahydrochrysen-2,8-diol). This latter compound is an ERα agonist and an ERβ antagonist. THC, which lacks the bulky side-chain of pure antagonists, antagonizes ERβ by stabilizing the H12 in an "inactive conformation" that prevents co-activator association [48].

Inverse agonists have been observed to increase interaction between SMRT and NCoR with receptors. For example, some RAR antagonists (e.g. BMS 493) reinforce co-repressor binding and consequently enhance silencing [49]. The detailed structural basis of this stabilization remains to be established. However, it is reasonable to assume that H12 adopts an alternative position which favors the interaction with CoRs. Note in this respect that surfaces for the NR box and surfaces for CoR binding (CoRNR box) are in close structural proximity.

3.3.4
RAR Isotype-selective Ligands and Rexinoids

Since the individual RAR subtypes have distinct tissue distribution patterns and appear to regulate different subset of genes, compounds that are selective for each RAR should have more restricted pharmacological activities, limited side effects and better therapeutic indexes in specific disease applications. A structure-based sequence alignment revealed that only three residues have diverged in the LBDs of RARα, β, and γ [42, 43]. This led to the prediction that these divergent residues would be critically involved in the ability of the receptor to differentiate between selective retinoids. Indeed, swapping of the residues confirmed this hypothesis [50]. All crystal structures of RARγ-selective agonists complexes have shown that selectivity is supported through formation of a hydrogen bond between the proximal hydroxyl group of the ligand and the RARγ Met272 [51, 52]. For RARα, the possibility of establishing a hydrogen bond between an amino group present in the linker of the identified RARα-selective agonists and the RARα-specific Ser232 is predicted to favor RARα selectivity. No such bonds can be formed in the RARβ or RARγ ligand-binding pockets, which harbor an alanine residue instead of a serine. Guidelines for the design of RAR isotype-selective ligands, considering also the tumor-suppressor RARβ, have been reported recently [53].

Even though we increasingly understand the structural principles that govern receptor selectivity and agonist/antagonist features, we are still far from a situation where we can design *in silico* ligands with a predetermined biological activity. The reason for this is that within the living cells a number of additional signaling options which are only partly understood influence the final outcome. These effects comprise mechanisms which involve the so-called "crosstalk of signaling" [54], aspects of cell differentiation, promoter context, even, and less well-understood phenomena, such as the "non-genomic action" of nuclear receptor ligands. In conclusion, the crystallization data are highly useful in proposing ligand structures and reducing the number of hypothetical ligand candidates, but will not replace drug screening in properly designed biological systems.

3.3.5
RXR Subordination and Permissive RXR Heterodimers

RXRs are promiscuous dimerization partners for several other nuclear receptors, such as the PPARs, the vitamin D_3 receptor (VDR), the thyroid receptor (TR), the liver X receptor (LXRs) [15, 20, 31]. In contrast to homodimerization, heterodimerization allows, in principle, fine-tuning of nuclear receptor action by using combinatorial sets of ligands, and regulation of alternative target-gene repertoires, and therefore provides interesting pharmacological opportunities. Although RAR agonists can autonomously activate transcription through an RAR/RXR heterodimer, RXR is unable to respond to RXR-selective agonists in the absence of an RAR ligand. This phenomenon, referred to as RXR "subordination" or "silencing," is biologically important as it provides signaling pathways identity to the RXR partner and avoids simultaneous activation of retinoic acid, thyroid hormone, and vitamin D_3 signaling pathways [49, 55–57]. The molecular basis of this subordination is that agonist binding to RXR is unable to induce the dissociation of co-repressor from the RAR/RXR heterodimers, preventing co-activator recruitment. Consequently, in principle, the only way for RXR to modulate transactivation in response to its ligand in RAR/RXR heterodimers is through synergy with RAR ligands leading to increased interaction efficiency of a single co-activator molecule with both holoRAR and holoRXR of the heterodimer. Interestingly, however, there are exceptions from this rule. Recently, a mechanism has been described by which RXR subordination can be modulated as a consequence of kinase activation. Indeed, this de-subordination in the presence of activated protein kinase A not only allows RAR agonist but also RXR agonists to activate the RAR/RXR heterodimer and thus its cognate gene program. In consequence, myeloid cells which are resistant to RA will undergo differentiation and apoptosis upon treatment with rexinoids under conditions where cAMP levels are increased [58].

While RAR cannot form homodimers, RXR homodimers are able to activate transcription from cognate reporter genes. *In vitro* studies have shown that the RXR LBD forms homodimers with relatively low affinity compared with its hetero-dimeric association with RAR, but it has been postulated, albeit not proven, that RXR homodimers may activate certain PPARE-containing genes directly [59]. RXR subordination does not apply to all apo-nuclear receptor partners, as hetero-dimers with nuclear receptors such as PPARs, LXRs, or Nur77 could be induced by RXR ligands [60, 61]. Such "permissive" heterodimers can therefore respond to two distinct ligands. If this is indeed the case *in vivo*, the biological signifi-cance of a double-ligand input into the corresponding gene network remains to be understood.

3.4
Retinoic Acid and Cancer

A strong rationale exists for the use of retinoids in cancer treatment and chemo-prevention based on experimental animal models, preclinical, epidemiological, and early clinical findings [62, 63]. Already 80 years ago, it was demonstrated that vitamin A deficiency in animal models leads to a higher incidence of cancer and increases sensitivity to chemical carcinogens [64]. Retinoids, predominantly ATRA and 13-*cis*-RA, have been used for a long time for the treatment of psoriasis and acne and only later retinoids were shown to be effective for the therapy of precan-cerous lesions like oral leukoplakia, actinic keratosis, and cervical dysplasia and were able to delay the development of skin cancer in individuals with xeroderma pigmentosum, thus demonstrating their chemopreventive potential [3, 6, 63]. Moreover, several malignancies are being treated with retinoid-based therapies, as single agent for pathologies including acute promyelocytic leukemia (APL), Kaposi's sarcoma, cutaneous T-cell lymphoma, juvenile chronic myelogenous leu-kaemia, and squamous cell carcinoma [65, 66], and in combination therapy with interferon-α for treatment of renal cancer [67] (for review see Ref. [68]).

3.4.1
Molecular Basis of Acute Promyelocytic Leukemia and ATRA Action

APL (classified as FAB M3) represents 10–15% of all acute myeloid leukemia (AMLs) and is characterized by a defect in the myeloid progenitor cells pro-gram leading to a differentiation arrest at the promyelocytic stage [69]. APL originates from a chromosomal translocation, in more than 95% of the cases t(15;17)(q22;q21), which fuses the RARα gene on chromosome 17 with the pro-myelocytic leukemia (PML) gene on chromosome 15, and leads to the expression of a PML/RARα fusion protein. In very rare cases of APL patients, RARα is not fused to PML but to the promyelocytic leukemia zinc finger (PLZF), the nuclear mitotic apparatus (NUMA), nucleophosmin (NPM), or the signal transducer and activator of transcription 5B (STAT5B) proteins [70] (for review, see Ref. [71]).

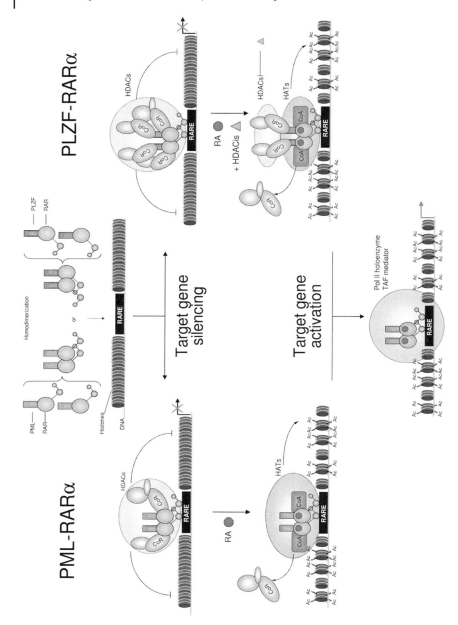

Clues to understand the altered functionality of PML/RARα came from the mechanistic studies of nuclear receptor-mediated gene activation and silencing (in the absence of a ligand), which revealed that gene expression is regulated at the level of chromatin by machineries containing HATs or HDACs. In APL, the fusion protein has gained the ability to form dimers or oligomers, which is both necessary and sufficient for its increased binding efficiency to co-repressors/HDAC complexes [72]. This aberrant recruitment of multisubunit complexes that contain HDAC and DNA methyltransferases (DNMT) by the PML/RARα protein is responsible for the repression of transcription resulting in permanent silencing of the RA-regulated gene programs, such as those triggering cell maturation and death [73]. While physiological ATRA concentration are inefficient, pharmacological doses induce dissociation of PML/RARα/HDAC complex and degradation of the fusion protein, restoring the normal differentiation pathway. The clinical treatment of APL patients with RA is, thus the paradigm for cancer "differentiation therapy" (Fig. 3.2). The situation with PLZF/RARα is different since co-repressors bind to both the apo-RARα and PLZF moieties, so upon ATRA treatment, the co-repressors are released from the RARα parts of the fusion protein but remain bound to PLZF, still inducing transcriptional repression, therefore these patients do not respond to ATRA therapy. However, HDAC inhibitors block the HDAC activity at RA target promoters and allow RA to signal even through PLZF/RARα [74] (Fig. 3.2).

A variety of recent observations indicate that the etiology of APL is more complex and we still do not understand all of the facets. First, PML signaling is altered by

Fig. 3.2. Molecular basis of retinoid responsivity and non-responsivity in APL cells The t(15;17) chromosomal translocation generates the PML/RARα fusion protein. PML/RARα homodimerizes and binds within the regulatory region of target genes through the RAREs. In contrast to RARα, PML/RARα can efficiently dimerize or oligomerize, thereby recruiting multiple transcriptional CoRs (NCoR or SMRT), which then recruit HDACs. This is thought to account for a gene-silencing effect through chromatin condensation. Pharmacological doses of RA are required to disrupt such a repressive complex and to allow recruitment of CoAs/HATs complexes such as CBP and p160. The subsequent acetylation of histone tails leads to chromatin decondensation and gene activation. As the last step, the RNA polymerase II holoenzyme, together with the TATA-binding protein (TBP) and TBP-associated factors (TAFs), and mediator complexes, are recruited, which increases the frequency of transcription initiation. The t(11;17) chromosomal translocation generates the PLZF/RARα fusion protein, which contains two CoR interaction surfaces. The one in the PLZF portion of the molecule cannot be dissociated by RA, accounting for the insensitivity of this acute promyelocytic leukemia (APL) type to retinoid therapy. However, combining RA treatment with an HDAC inhibitor (HDACi) blocks the enzyme activity and so prevents chromatin-mediated repression. Ac, acetyl group; APL, acute promyelocytic leukemia; CoA, co-activator; CoR, co-repressor; HAT, histone acetyltransferase; HDAC, histone deacetylase; HDACi, histone deacetylase inhibitor; NCoR, nuclear receptor co-repressor; PML, promyelocytic leukemia; PLZF, promyelocytic leukemia zinc finger; p160, co-activator (e.g. TIF2/RAC3/SRC-1); RA, retinoic acid; RAR, retinoic acid receptor; RARE, retinoic acid response elements; SMRT, silencing mediator for retinoid and thyroid hormone receptor.

PML/RARα. PML is a multifunctional protein involved in regulation of apoptosis, cell proliferation, and senescence [75] and is typically found as a part of multiprotein structures called nuclear bodies (NBs), which co-localize with more than 30 proteins, including p53 and CBP [76]. In APL, PML/RARα through its interaction with wild-type PML causes the disintegration and relocalization of the NBs, leading to aberrant nuclear structures linked to a loss of several PML functions including growth suppression, transformation suppression, and pro-apoptotic actions [75]. Recently, the role of PML in leukemogenesis has been further clarified: the wild-type PML proteins, associated with PML/RARα, recruit p53 to the fusion protein complexes, causing its deacetylation by HDACs. This leads to p53 degradation by the proteasome and genomic instability, resulting in downregulation of p53-dependent transcription and resistance to apoptosis [77]. The inactivation of p53 favors further mutations which are essential for the development of APL. Besides the alteration of cross-talk to key regulatory factors such as p53, it has also been shown that sumoylation of PML/RARα on lysine 160 recruits another co-repressor complex called Daxx to the PML moiety. This event appears to be critical for the differentiation block *ex vivo* and APL development *in vivo* since transgenic mice bearing a mutation in the sumoylation site of PML/RARα never developed APL but rather myeloproliferations [78].

Taken together, the studies of APL have revealed not only the genetic and epigenetic basis of the disease and of the mechanistic basis of the retinoic acid-based "differentiation therapy" that cures more than 72% of the patients, but also revealed novel treatment paradigms for APL and, moreover, for acute myeloid leukemias (AML) which constitutes a significantly larger patient collective with very heterogeneous genetic background.

3.4.2
Retinoids and Apoptosis: Induction of TRAIL

Studying the cancer preventive potential of retinoids, recent observations have revealed that retinoids (classical or atypical) induce apoptogenic programs involving the signaling pathways activated by the death ligand TRAIL (tumor necrosis factor-related apoptosis-inducing ligand, also named Apo-2L or TNFSF10). Initial evidence was provided by *ex vivo* studies with APL patients' blasts which were shown to die upon exposure to retinoids in a TRAIL-dependent manner [4, 79]. TRAIL is a member of the TNF family and has attracted significant interest for therapy due to its specific induction of apoptosis in malignant cells with normal cells being largely resistant to its effects [4, 80, 81]. TRAIL is a type II membrane protein whose binding to TRAIL receptor 1 (TRAIL-R1/DR5) and/or TRAIL-R2 (DR4) results in apoptosis. No apoptosis is observed with the so-called decoy receptors TRAIL-R3 (Dc-R1) or TRAIL-R4 (Dc-R2), which lack a functional cytoplasmic death domain or are completely devoid of a cytosolic region and, consequently, cannot transmit any apoptosis signal [82–85].

TRAIL ligation to its cognate receptors induces trimerization and recruitment, through the receptor death domains, of specific cytoplasmic proteins resulting in the formation of the so-called death-inducing signaling complex (DISC) [86]. Important components of this complex are the adapter protein Fas-associated death domain (FADD/Mort-1) and the apoptosis initiator caspases 8 or 10. These caspases are autocatalytically cleaved at the DISC, thus initiating a caspase cascade that involves caspases 3, 6, and 7 and executes the death program. This pathway is called the "cell-extrinsic pathway," but in certain types of cells, TRAIL can also couple to the "cell-intrinsic pathway." In this case, caspase 8- or 10-mediated cleavage of the Bcl-2 family member Bid results in its translocation to the mitochondria, causing a loss of the mitochondrial transmembrane potential and release of cytochrome *c* to the cytosol. Cytochrome *c*, together with Apaf-1 and the initiator caspase 9 form the autocatalytically active apoptosome, resulting in caspase 9 activation and initiation of the caspase cascade described above [7, 87, 88].

In APL blasts it has been observed that prior to induction of TRAIL-mediated apoptosis, anti-apoptotic survival programs are activated and it has been speculated that these survival programs set a time-frame during which the cell can exert its differentiated phenotype [89].

The induction of TRAIL has a number of implications for future therapies. For example, TRAIL could be combined with lower doses of ATRA, in order to limit the RA syndrome [90]. In addition, atypical retinoids or chemotherapeutic drugs that induce expression of TRAIL receptors DR4 and DR5 may be used in therapy together with RAR ligands, to activate both receptors and TRAIL expression and improve apoptosis efficacy [91, 92]. It has also been shown that TRAIL is induced in other cell systems than the APL model, by retinoids alone or in synergy with interferons (e.g. in breast cancer or AML cells) [93, 94].

3.4.3
Retinoids and Rexinoids are Cancer-preventive Drugs

Chemoprevention is based on the use of specific natural or pharmacological agents to reverse, suppress, or prevent the progression of precancerous lesions to invasive cancers. Retinoids are prime candidates for cancer chemoprevention since a number of experimental, epidemiological, and clinical studies performed with retinoids have revealed their efficacy in the treatment of three precancerous lesions: leukoplakia, actinic keratosis, and cervical dysplasia, and in the delay of the development of skin cancer in individuals with xeroderma pigmentosum and inherited predisposition to ultraviolet-induced cancers (see Ref. [6] and references therein).

Moreover, retinoids have also been found to be effective in suppressing tumor development in several carcinogenesis models, such as those of the skin, breast, lung, prostate, bladder, pancreas, head and neck, or liver [95]. One model frequently used to assess this antitumor activity is the skin carcinogenesis chemically induced in two steps, first, the initiation induced by a compound such as 7,12-dimethybenz(a)anthracene (DMBA), and a step of promotion often induced by the phorbol ester 12,13-tetradecanoyl phorbol acetate (TPA). *In vivo* studies with such models revealed that retinoids are able to block the promotion step by inhibition of the AP1 activity induced by TPA [96].

However, the use of natural retinoids for therapeutic purposes is often limited by their relatively low specificity that results in toxicity when these ligands are used at pharmacological doses. Another important point is the intrinsic or acquired resistance limiting retinoid clinical activity. Thus, several problems will have to be solved, perhaps by targeted ligand design, to reduce retinoid toxicity for cancer chemoprevention.

3.4.4
RARβ as Tumor Suppressor

Accumulative evidence indicates that the expression of RARβ is critically altered during the pathogenesis of various solid tumors, and favors the emerging concept that RARβ functions as a potential tumor suppressor. Indeed, the loss of RARβ mRNA expression has been associated with human tumor progression in a variety of carcinomas, including non-small-cell lung cancer, squamous cell carcinomas of head and neck, prostate of breast cancers [97–103]. Furthermore, experiments showing first the loss of tumorigenicity in nude mice of lung cancer cells transfected with RARβ, and inversely the gain of tumorigenicity of transgenic mice expressing antisense aimed to downregulate RARβ protein [104] strongly support that deregulation of RARβ expression could be an initial mechanism by which tumor cells escape from normal cellular homeostasis. Due to the identification of the RARE located in the promoter of RARβ, it is speculated that this receptor may be a pivotal factor in the antiproliferative activity of the retinoids. In agreement with this hypothesis, a study with patients with premalignant oral lesions demonstrated a correlation between the ability of RA to elevate RARβ levels and clinical outcome [105].

Frequently, epigenetic changes of the RARβ gene have been demonstrated to contribute to the transcriptional silencing of RARβ despite the availability of physiological levels of RA in tumor cells. Indeed, promoter CpG islands were mapped in the RARβ2 gene and several studies with patients presenting epithelial malignancies highlighted that DNA methylation of this promoter could be a marker for early carcinogenesis [106–108]. On the other hand, a loss of histone H3 acetylation consistently correlates with RA resistance in lung cancer cells lines and loss of RARβ expression, both in presence or absence of hypermethylation [109]. In application to these mechanisms, combinations of RA to HDAC and DNMT inhibitors were found to be effective in overcoming RA resistance or restoring RA signaling [110, 111]. Additional studies are needed to fully decipher the genetic networks mediated by RARβ and definitely address the tumor-suppressor function of this protein. In this respect, the recent elucidation of the crystal structure of LBD of the RARβ and the description of RARβ-selective agonist [112] give the opportunity to assess pharmacologically the tumor-suppressor role of RARβ *in vitro* and in animal models.

3.4.5

Clinical Applications of Retinoids in Cancer: Updated and Future Strategies

The present and future research on retinoid action and the corresponding clinical studies will help to find ways to overcome retinoid resistance in APL patients and decrease the side effects of the currently used ATRA therapy. In addition, such mechanistic studies will reveal the regulations within the molecular pathways that are activated by retinoids and their receptors. For two decades, various strategies, including the development of novel retinoids and combination therapy with other differentiation-inducing, cytotoxic or chromatin-remodeling agents have been proposed, and several strategies are currently undergoing ongoing clinical trials (see Table 3.1) [62, 113, 114].

Atypical retinoids, referring to synthetic analogs of RA that bind and activate particular subsets of nuclear receptors isotypes, such as fenretinide (4-HPR) and CD437, have shown promise as cancer therapeutics owing to their antiproliferative and apoptotic effects *in vitro*. Indeed, fenretinide has shown anticancer potential in several preclinical studies (reviewed in Ref. [113]) and is being used in clinical trials for ovarian cancer (treatment and prevention), prostate cancer, neuroblastoma, glioblastoma, and advanced solid tumors. In addition, atypical retinoids exert their growth-regulatory activity in retinoid-resistant cells [115]. The "heteroarotinoids" is another class of these retinoid-related molecules, which consists of a group of compounds modified on the basis of arotinoid chemistry (containing aromatic ring(s) and at least one heteroatom within the skeletal framework). These have shown marked anticancer activities *in vitro* [116] and much lower toxicities when compared with some clinically used retinoids. One example of a heteroarotinoid was selected by the National Institute of Health (NIH) for preclinical screening (now in progress) for potential use in treating ovarian and cervical cancer.

Another type of retinoid-related molecule, MX781, is an RAR antagonist that induces apoptosis and showed exceptional anticancer activity against estrogen-independent breast cancer cells, although high concentrations of the compound are required *in vitro* to induce apoptosis [117, 118]. Finally, LGD1069 (Targretin) is a prototype of selective ligand classified as RXR receptor agonist. Its antitumor activity is similar to that achieved by tamoxifen in *N*-nitroso-*N*-methylurea animal systems and most notably, the classical signs of retinoid-associated toxicity were limited in chronic therapy [119, 120].

Other specific RXR-selective agonists are being developed as therapeutic agents, such as the bexarotene, currently used as treatment for persistent or refractory cutaneous T-cell lymphoma [121, 122].

Alternatively, combination of existing retinoids and other agents is another potentially strategy to optimize retinoid-based treatment. Pan-agonists for RAR and RXR (e.g. ATRA, 9-*cis*-RA and 13-*cis*-RA) have shown potent antitumor activity when combined with agonists or antagonists for other members of the NR subfamily (e.g. vitamin D, PPARγ ligands and steroids). The anti-estrogen tamoxifen, which has antiproliferative effects by itself, is more effective against breast carci-

Table 3.1. Ongoing clinical trials using retinoids alone or in combination therapies

Compound	Receptor activity	Type of cancer	Combination	Phase
Isotretinoin (13c-RA)	Pan-RAR agonist	Non-small-cell lung cancer	IFNα, paclitaxel	II
		T-cell lymphoma	IFNα	II
		Adult solid tumor, leukemia, lymphoma, small intestine cancer	MS-275	I
Tretinoin (at-RA)	Pan-RAR agonist	Stage I/II/III multiple myeloma	Dexamethasone	II
		Metastatic renal cell cancer	IFN-α2b	II
		Stage IV kidney cancer	IL-2	II
Fenretinide (4-HPR)	RARγ and RARβ agonist, additional unknown activities?	Stage II/III and metastatic/hormone-refractory prostate cancer, stage III/IV recurrent malignant glioma, small-cell lung cancer, metastatic head/neck cancer, peritoneal cavity cancer, recurrent/resistant neuroblastoma, stage III/IV renal cell carcinoma		II
		Refractory solid tumors	Paclitaxel, cisplatin	I
		Head/neck cancer	FTI	I
		Head/neck cancer	lonafarnib	I
		Refractory/relapsed hematologic cancer		I
		Ovarian epithelial cancer		Prevention
Bexarotene	RXR-selective	Breast cancer		Prevention

For further information, see http://www.clinicaltrials.gov.
IFN, interferon; IL, interleukin; FTI, farnesyl transferase inhibitor.

noma or hepatoma cells when used in combination with retinoids [123]. Together with vitamin D_3 analogs, retinoids can effectively reduce breast tumor mass in nude mice [124] and can inhibit cell growth and induce apoptosis in lung, prostate, breast, and ovarian cancer cells. Recently, the combination of dexamethasone and

ATRA has been shown to inhibit cell proliferation and induce differentiation in human osteosarcoma cells [125]. Many associations of classic retinoids or atypical retinoids with chemotherapeutic agents (e.g. cisplatin, etoposide, camptothecin, taxol, vinblastin, gencitabine, and cytosine arabinoside) have a synergistic antiproliferative effect or can sensitize certain types of cancer to chemotherapy-induced apoptosis. In support of this, several clinical trials to test anticancer activities of 4-HPR – used alone or in combination – in neuroblastoma, ovarian cancer, prostate cancer, glioblastoma, or advanced solid tumors are ongoing at present [114].

Finally, the interferon (IFN)/retinoid association is a promising cocktail of such therapeutic approaches and has been extensively studied at the molecular level. It has been found that the mechanism implicated in the synergistic antiproliferative effect induced by the IFN-retinoid association might be related to intrinsic and extrinsic apoptotic pathways [126–128]. In the light of the fact that TRAIL and the death receptor pathway is responsible for retinoid-induced apoptosis in certain systems, IFNs could potentiate the anticancer effect of retinoids by modulating factors that interact with or take part in this pathway, leading to cocktails that are more effective in killing cancer cells. Recently, the combination of IFNα/isotretinoin has shown antitumor effects and was well tolerated in patients with lymphoid malignancies [129].

Despite the success of ATRA-based differential therapy in APL, the broad promise of retinoid in the clinic has not yet been realized. Presented observations argue for the use of retinoid combination therapies that can activate only a subset of the functions induced by the cognate ligand or to act in a cell type-selective manner. A better understanding of the underlying molecular mechanisms promoting and limiting the multiple pathways by which retinoids exert their anticancer action is essential to pave the way towards novel types of ret(x)inoid-based (combo)therapies.

References

1 Sporn, M.B., A.B. Roberts, D.S. Goodman (**1994**) *The Retinoids: Biology, Chemistry and Medicine.* New York: Raven Press.

2 Chambon, P. (**1996**) A decade of molecular biology of retinoic acid receptors. *FASEB J* 10, 940–954.

3 Soprano, D.R., P. Qin, K.J. Soprano (**2004**) Retinoic acid receptors and cancers. *Annu Rev Nutr* 24, 201–221.

4 Altucci, L., H. Gronemeyer (**2004**) Retinoids and TRAIL: Two cooperating actors to fight against cancer. In: *Vitamins and Hormones.* G. Litwack, editor. Oxford: Academic Press, Elsevier, pp. 319–345.

5 Sporn, M.B., N. Suh (**2002**) Opinion: Chemoprevention: an essential approach to controlling cancer. *Nat Rev Cancer* 2, 537–543.

6 Sun, S.Y., R. Lotan (**2002**) Retinoids and their receptors in cancer development and chemoprevention. *Crit Rev Oncol Hematol* 41, 41–55.

7 Deng, Y., Y. Lin, X. Wu (**2002**) TRAIL-induced apoptosis requires Bax-dependent mitochondrial release of Smac/DIABLO. *Genes Dev* 16, 33–45.

8 Chen, H., M. Namkung, M. Juchau (**1995**) Biotransformation of all-trans-retinol and all-trans-retinal to all-trans-retinoic acid in rat conceptal homogenates. *Biochem Pharmacol* 50, 1257–1264.

9 Napoli, J.L. (**1996**) Retinoic acid

biosynthesis and metabolism. *FASEB J* 10, 993–1001.

10 PIJNAPPEL, W., H. HENDRIKS, G. FOLKERS, C. VAN DEN BRINK, E. DEKKER, C. EDELENBOSCH, P. VAN DER SAAG, A. DURSTON (1993) The retinoid ligand 4-oxo-retinoic acid is a highly active modulator of positional specification. *Nature* 366, 340–344.

11 O'CONNELL, M., R. CHUA, B. HOYOS, J. BUCK, Y. CHEN, F. DERGUINI, U. HAMMERLING (1996) Retro-retinoids in regulated cell growth and death. *J Exp Med* 184, 549–555.

12 IMAM, A., B. HOYOS, C. SWENSON, E. LEVI, R. CHUA, E. VIRIYA, U. HAMMERLING (2001) Retinoids as ligands and coactivators of protein kinase C alpha. *FASEB J* 15, 28–30.

13 MANGELSDORF, D.J., E.S. ONG, J.A. DYCK, R.M. EVANS (1990) Nuclear receptor that identifies a novel retinoic acid response pathway. *Nature* 345, 224–229.

14 HEYMAN, R.A., D.J. MANGELSDORF, J.A. DYCK, R.B. STEIN, G. EICHELE, R.M. EVANS, C. THALLER (1992) 9-cis retinoic acid is a high affinity ligand for the retinoid X receptor. *Cell* 68, 397–406.

15 LAUDET, V., H. GRONEMEYER (2002) *The Nuclear Receptor FactsBook*. San Diego: Academic Press.

16 NAGPAL, S., M. SAUNDERS, P. KASTNER, B. DURAND, H. NAKSHATRI, P. CHAMBON (1992) Promoter context- and response element-dependent specificity of the transcriptional activation and modulating functions of retinoic acid receptors. *Cell* 70, 1007–1019.

17 KHORASANIZADEH, S., F. RASTINEJAD (2001) Nuclear-receptor interactions on DNA-response elements. *Trends Biochem Sci* 26, 384–390.

18 CHASSE, S., F. RASTINEJAD (2001) Physical structure of nuclear receptor-DNA complexes. *Methods Mol Biol* 176, 91–103.

19 LEID, M., P. KASTNER, P. CHAMBON (1992) Multiplicity generates diversity in the retinoic acid signalling pathways. *TIBS* 17, 427–433.

20 MANGELSDORF, D.J., R.M. EVANS (1995) The RXR heterodimers and orphan receptors. *Cell* 83, 841–850.

21 GLASS, C.K. (1994) Differential recognition of target genes by nuclear receptor monomers, dimers, and heterodimers. *Endocr Rev* 15, 391–407.

22 DE THE, H., M.M. VIVANCO-RUIZ, P. TIOLLAIS, H. STUNNENBERG, A. DEJEAN (1990) Identification of a retinoic acid responsive element in the retinoic acid receptor beta gene. *Nature* 343, 177–180.

23 SUCOV, H.M., K.K. MURAKAMI, R.M. EVANS (1990) Characterization of an autoregulated response element in the mouse retinoic acid receptor type beta gene. *Proc Natl Acad Sci USA* 87, 5392–5396.

24 DURAND, B., M. SAUNDERS, P. LEROY, M. LEID, P. CHAMBON (1992) All-trans and 9-cis retinoic acid induction of CRABPII transcription is mediated by RAR-RXR heterodimers bound to DR1 and DR2 repeated motifs. *Cell* 71, 73–85.

25 UMESONO, K., K.K. MURAKAMI, C.C. THOMPSON, R.M. EVANS (1991) Direct repeats as selective response elements for the thyroid hormone, retinoic acid, and vitamin D3 receptors. *Cell* 65, 1255–1266.

26 KLIEWER, S.A., K. UMESONO, R.A. HEYMAN, D.J. MANGELSDORF, J.A. DYCK, R.M. EVANS (1992) Retinoid X receptor-COUP-TF interactions modulate retinoic acid signaling. *Proc Natl Acad Sci USA* 89, 1448–1452.

27 RASTINEJAD, F., T. PERLMANN, R.M. EVANS, P.B. SIGLER (1995) Structural determinants of nuclear receptor assembly on DNA direct repeats. *Nature* 375, 203–211.

28 ZECHEL, C., X.Q. SHEN, P. CHAMBON, H. GRONEMEYER (1994) Dimerization interfaces formed between the DNA binding domains determine the cooperative binding of RXR/RAR and RXR/TR heterodimers to DR5 and DR4 elements. *EMBO J* 13, 1414–1424.

29 ZECHEL, C., X.Q. SHEN, J.Y. CHEN, Z.P. CHEN, P. CHAMBON, H. GRONEMEYER (1994) The dimerization interfaces formed between the DNA

binding domains of RXR, RAR and TR determine the binding specificity and polarity of the full-length receptors to direct repeats. *EMBO J* 13, 1425–1433.

30 Rastinejad, F., T. Wagner, Q. Zhao, S. Khorasanizadeh (**2000**) Structure of the RXR-RAR DNA-binding complex on the retinoic acid response element DR1. *EMBO J* 19, 1045–1054.

31 Rastinejad, F. (**2001**) Retinoid X receptor and its partners in the nuclear receptor family. *Curr Opin Struct Biol* 11, 33–38.

32 Nagy, L., H.Y. Kao, D. Chakravarti, R.J. Lin, C.A. Hassig, D.E. Ayer, S.L. Schreiber, R.M. Evans (**1997**) Nuclear receptor repression mediated by a complex containing SMRT, mSin3A, and histone deacetylase. *Cell* 89, 373–380.

33 Glass, C.K., M.G. Rosenfeld (**2000**) The coregulator exchange in transcriptional functions of nuclear receptors. *Genes Dev* 14, 121–141.

34 Aranda, A., A. Pascual (**2001**) Nuclear hormone receptors and gene expression. *Physiol Rev* 81, 1269–1304.

35 Hu, X., M.A. Lazar (**2000**) Transcriptional repression by nuclear hormone receptors. *Trends Endocrinol Metab* 11, 6–10.

36 McKenna, N.J., R.B. Lanz, B.W. O'Malley (**1999**) Nuclear receptor coregulators: cellular and molecular biology. *Endocr Rev* 20, 321–344.

37 Freedman, L.P. (**1999**) Increasing the complexity of coactivation in nuclear receptor signaling. *Cell* 97, 5–8.

38 McKenna, N.J., J. Xu, Z. Nawaz, S.Y. Tsai, M.J. Tsai, B.W. O'Malley (**1999**) Nuclear receptor coactivators: multiple enzymes, multiple complexes, multiple functions. *J Steroid Biochem Mol Biol* 69, 3–12.

39 Ito, M., R.G. Roeder (**2001**) The TRAP/SMCC/Mediator complex and thyroid hormone receptor function. *Trends Endocrinol Metab* 12, 127–134.

40 Rachez, C., M. Gamble, C.P. Chang, G.B. Atkins, M.A. Lazar, L.P. Freedman (**2000**) The DRIP complex and SRC-1/p160 coactivators share

similar nuclear receptor binding determinants but constitute functionally distinct complexes. *Mol Cell Biol* 20, 2718–2726.

41 Bourguet, W., M. Ruff, P. Chambon, H. Gronemeyer, D. Moras (**1995**) Crystal structure of the ligand-binding domain of the human nuclear receptor RXR-alpha. *Nature* 375, 377–382.

42 Renaud, J.P., N. Rochel, M. Ruff, V. Vivat, P. Chambon, H. Gronemeyer, D. Moras (**1995**) Crystal structure of the RAR-gamma ligand-binding domain bound to all-trans retinoic acid. *Nature* 378, 681–689.

43 Bourguet, W., P. Germain, H. Gronemeyer (**2000**) Nuclear receptor ligand-binding domains: three-dimensional structures, molecular interactions and pharmacological implications. *Trends Pharmacol Sci* 21, 381–388.

44 Bourguet, W., V. Vivat, J.M. Wurtz, P. Chambon, H. Gronemeyer, D. Moras (**2000**) Crystal structure of a heterodimeric complex of RAR and RXR ligand-binding domains. *Mol Cell* 5, 289–298.

45 Shiau, A.K., D. Barstad, P.M. Loria, L. Cheng, P.J. Kushner, D.A. Agard, G.L. Greene (**1998**) The structural basis of estrogen receptor/coactivator recognition and the antagonism of this interaction by tamoxifen. *Cell* 95, 927–937.

46 Pike, A.C., A.M. Brzozowski, R.E. Hubbard, T. Bonn, A.G. Thorsell, O. Engstrom, J. Ljunggren, J.A. Gustafsson, M. Carlquist (**1999**) Structure of the ligand-binding domain of oestrogen receptor beta in the presence of a partial agonist and a full antagonist. *EMBO J* 18, 4608–4618.

47 Oberfield, J.L., J.L. Collins, C.P. Holmes, D.M. Goreham, J.P. Cooper, J.E. Cobb, J.M. Lenhard, E.A. Hull-Ryde, C.P. Mohr, S.G. Blanchard, D.J. Parks, L.B. Moore, J.M. Lehmann, K. Plunket, A.B. Miller, M.V. Milburn, S.A. Kliewer, T.M. Willson (**1999**) A peroxisome proliferator-activated receptor gamma ligand inhibits adipocyte differentiation. *Proc Natl Acad Sci USA* 96, 6102–6106.

48 SHIAU, A.K., D. BARSTAD, J.T. RADEK, M.J. MEYERS, K.W. NETTLES, B.S. KATZENELLENBOGEN, J.A. KATZENELLENBOGEN, D.A. AGARD, G.L. GREENE (2002) Structural characterization of a subtype-selective ligand reveals a novel mode of estrogen receptor antagonism. *Nat Struct Biol* 9, 359–364.

49 GERMAIN, P., J. IYER, C. ZECHEL, H. GRONEMEYER (2002) Coregulator recruitment and the mechanism of retinoic acid receptor synergy. *Nature* 415, 187–192.

50 GEHIN, M., V. VIVAT, J.M. WURTZ, R. LOSSON, P. CHAMBON, D. MORAS, H. GRONEMEYER (1999) Structural basis for engineering of retinoic acid receptor isotype-selective agonists and antagonists. *Chem Biol* 6, 519–529.

51 KLAHOLZ, B., A. MITSCHLER, D. MORAS (2000) Structural basis for isotype selectivity of the human retinoic acid nuclear receptor. *J Mol Biol* 302, 155–170.

52 KLAHOLZ, B., A. MITSCHLER, M. BELEMA, C. ZUSI, D. MORAS (2000) Enantiomer discrimination illustrated by high-resolution crystal structures of the human nuclear receptor hRARgamma. *Proc Natl Acad Sci USA* 97, 6322–6327.

53 GERMAIN, P., S. KAMMERER, E. PEREZ, C. PELUSO-ILTIS, D. TORTOLANI, F. ZUSI, J. STARRETT, P. LAPOINTE, J. DARIS, A. MARINIER, A. DE LERA, N. ROCHEL, H. GRONEMEYER (2004) Rational design of RAR-selective ligands revealed by RARbeta crystal stucture. *EMBO Rep* 5, 877–882.

54 HERRLICH, P. (2001) Cross-talk between glucocorticoid receptor and AP-1. *Oncogene* 20, 2465–2475.

55 LEBLANC, B.P., H.G. STUNNENBERG (1995) 9-cis retinoic acid signaling: changing partners causes some excitement. *Genes Dev* 9, 1811–1816.

56 CHEN, J.Y., J. CLIFFORD, C. ZUSI, J. STARRETT, D. TORTOLANI, J. OSTROWSKI, P.R. RECZEK, P. CHAMBON, H. GRONEMEYER (1996) Two distinct actions of retinoid-receptor ligands. *Nature* 382, 819–822.

57 ROY, B., R. TANEJA, P. CHAMBON (1995) Synergistic activation of retinoic acid (RA)-responsive genes and induction of embryonal carcinoma cell differentiation by an RA receptor alpha (RAR alpha)-, RAR beta-, or RAR gamma-selective ligand in combination with a retinoid X receptor-specific ligand. *Mol Cell Biol* 15, 6481–6487.

58 ALTUCCI, L., A. ROSSIN, O. HIRSCH, A. NEBBIOSO, D. VITOUX, E. WILHELM, F. GUIDEZ, D.S.M., E.M. SCHIAVONE, D. GRIMWADE, A. ZELENT, H. DE THE, H. GRONEMEYER (2005) Rexinoid-triggered differentiation and tumor-selective apoptosis of AML by protein kinase A-mediated de-subordination of RXR. *Cancer Res* 65, 8754–8765.

59 IJPENBERG, A., N. TAN, L. GELMAN, S. KERSTEN, J. SEYDOUX, J. XU, D. METZGER, L. CANAPLE, P. CHAMBON, W. WAHLI, B. DESVERGNE (2004) *In vivo* activation of PPAR target genes by RXR homodimers. *EMBO J* 23, 2083–2091.

60 DIRENZO, J., M. SODERSTROM, R. KUROKAWA, M.H. OGLIASTRO, M. RICOTE, S. INGREY, A. HORLEIN, M.G. ROSENFELD, C.K. GLASS (1997) Peroxisome proliferator-activated receptors and retinoic acid receptors differentially control the interactions of retinoid X receptor heterodimers with ligands, coactivators, and corepressors. *Mol Cell Biol* 17, 2166–2176.

61 KLIEWER, S.A., K. UMESONO, D.J. NOONAN, R.A. HEYMAN, R.M. EVANS (1992) Convergence of 9-cis retinoic acid and peroxisome proliferator signalling pathways through heterodimer formation of their receptors. *Nature* 358, 771–774.

62 ALTUCCI, L., H. GRONEMEYER (2001) The promise of retinoids to fight against cancer. *Nat Rev Cancer* 1, 181–193.

63 HONG, W.K., M.B. SPORN (1997) Recent advances in chemoprevention of cancer. *Science* 278, 1073–1077.

64 WOLBACH, S., P. HOWE (1925) Tissue

changes following deprivation of fat-soluble A vitamin. *Nutrition Classics. J Exp Med* 42, 753–777.

65 CHEER, S., R. FOSTER (**2000**) Alitretinoin. *Am J Clin Dermatol.* 1, 307–314; discussion 315–306.

66 BOUWHUIS, S., M. DAVIS, R. EL-AZHARY, M. McEVOY, L. GIBSON, J. KNUDSEN, J. KIST, M. PITTELKOW (**2005**) Bexarotene treatment of late-stage mycosis fungoides and Sezary syndrome: development of extracutaneous lymphoma in 6 patients. *J Am Acad Dermatol* 52, 991–996.

67 BERG, W.J., D.M. NANUS, A. LEUNG, K.T. BROWN, B. HUTCHINSON, M. MAZUMDAR, X.C. XU, R. LOTAN, V.E. REUTER, R.J. MOTZER (**1999**) Up-regulation of retinoic acid receptor beta expression in renal cancers *in vivo* correlates with response to 13-cis-retinoic acid and interferon-alpha-2a. *Clin Cancer Res* 5, 1671–1675.

68 FREEMANTLE, S., M. SPINELLA, E. DMITROVSKY (**2003**) Retinoids in cancer therapy and chemoprevention: promise meets resistance. *Oncogene* 22, 7305–7315.

69 GRIGNANI, F., P.F. FERRUCCI, U. TESTA, G. TALAMO, M. FAGIOLI, M. ALCALAY, A. MENCARELLI, C. PESCHLE, I. NICOLETTI, *et al.* (**1993**) The acute promyelocytic leukemia-specific PML-RAR alpha fusion protein inhibits differentiation and promotes survival of myeloid precursor cells. *Cell* 74, 423–431.

70 PANDOLFI, P.P., M. ALCALAY, M. FAGIOLI, D. ZANGRILLI, A. MENCARELLI, D. DIVERIO, A. BIONDI, F. LO COCO, A. RAMBALDI, F. GRIGNANI, *et al.* (**1992**) Genomic variability and alternative splicing generate multiple PML/RAR alpha transcripts that encode aberrant PML proteins and PML/RAR alpha isoforms in acute promyelocytic leukaemia. *EMBO J* 11, 1397–1407.

71 ZELENT, A., F. GUIDEZ, A. MELNICK, S. WAXMAN, J. LICHT (**2001**) Translocations of the RARalpha gene in acute promyelocytic leukemia. *Oncogene* 20, 7186–7203.

72 MINUCCI, S., M. MACCARANA, M. CIOCE, P. DE LUCA, V. GELMETTI, S. SEGALLA, L. DI CROCE, S. GIAVARA, C. MATTEUCCI, A. GOBBI, A. BIANCHINI, E. COLOMBO, I. SCHIAVONI, G. BADARACCO, X. HU, M.A. LAZAR, N. LANDSBERGER, C. NERVI, P.G. PELICCI (**2000**) Oligomerization of RAR and AML1 transcription factors as a novel mechanism of oncogenic activation. *Mol Cell* 5, 811–820.

73 DI CROCE, L., V.A. RAKER, M. CORSARO, F. FAZI, M. FANELLI, M. FARETTA, F. FUKS, F. LO COCO, T. KOUZARIDES, C. NERVI, S. MINUCCI, P.G. PELICCI (**2002**) Methyltransferase recruitment and DNA hypermethylation of target promoters by an oncogenic transcription factor. *Science* 295, 1079–1082.

74 HE, L.Z., F. GUIDEZ, C. TRIBIOLI, D. PERUZZI, M. RUTHARDT, A. ZELENT, P.P. PANDOLFI (**1998**) Distinct interactions of PML-RARalpha and PLZF-RARalpha with co-repressors determine differential responses to RA in APL. *Nat Genet* 18, 126–135.

75 SALOMONI, P., P.P. PANDOLFI (**2002**) The role of PML in tumor suppression. *Cell* 108, 165–170.

76 JENSEN, K., C. SHIELS, P. FREEMONT (**2001**) PML protein isoforms and the RBCC/TRIM motif. *Oncogene* 20, 7223–7233.

77 INSINGA, A., S. MONESTIROLI, S. RONZONI, R. CARBONE, M. PEARSON, G. PRUNERI, G. VIALE, E. APPELLA, P. PELICCI, S. MINUCCI (**2004**) Impairment of p53 acetylation, stability and function by an oncogenic transcription factor. *Embo J* 23, 1144–1154.

78 ZHU, J., J. ZHOU, L. PERES, F. RIAUCOUX, N. HONORE, S. KOGAN, H. DE THE (**2005**) A sumoylation site in PML/RARA is essential for leukemic transformation. *Cancer Cell* 7, 143–153.

79 ALTUCCI, L., A. ROSSIN, W. RAFFELSBERGER, A. REITMAIR, C. CHOMIENNE, H. GRONEMEYER (**2001**) Retinoic acid-induced apoptosis in leukemia cells is mediated by paracrine action of tumor-selective death ligand TRAIL. *Nat Med* 7, 680–686.

80 ALMASAN, A., A. ASHKENAZI (**2003**) Apo2L/TRAIL: apoptosis signaling, biology, and potential for cancer therapy. *Cytokine Growth Factor Rev* 14, 337–348.

81 LITWACK, G., editor (**2004**) *TRAIL*. San Diego: Academic Press. 448 pp.

82 WILEY, S.R., K. SCHOOLEY, P.J. SMOLAK, W.S. DIN, C.P. HUANG, J.K. NICHOLL, G.R. SUTHERLAND, T.D. SMITH, C. RAUCH, C.A. SMITH, *et al.* (**1995**) Identification and characterization of a new member of the TNF family that induces apoptosis. *Immunity* 3, 673–682.

83 WALCZAK, H., M.A. DEGLI-ESPOSTI, R.S. JOHNSON, P.J. SMOLAK, J.Y. WAUGH, N. BOIANI, M.S. TIMOUR, M.J. GERHART, K.A. SCHOOLEY, C.A. SMITH, R.G. GOODWIN, C.T. RAUCH (**1997**) TRAIL-R2: a novel apoptosis-mediating receptor for TRAIL. *EMBO J* 16, 5386–5397.

84 PAN, G., K. O'ROURKE, A.M. CHINNAIYAN, R. GENTZ, R. EBNER, J. NI, V.M. DIXIT (**1997**) The receptor for the cytotoxic ligand TRAIL. *Science* 276, 111–113.

85 MACFARLANE, M., A.M., S.M. SRINIVASULA, T. FERNANDES-ALNEMRI, G.M. COHEN, E.S. ALNEMRI (**1997**) Identification and molecular cloning of two novel receptors for the cytotoxic ligand TRAIL. *J Biol Chem* 272, 25417–25420.

86 KISCHKEL, F., S. HELLBARDT, I. BEHRMANN, M. GERMER, M. PAWLITA, P. KRAMMER, M. PETER (**1995**) Cytotoxicity-dependent APO-1 (Fas/CD95)-associated proteins form a death-inducing signaling complex (DISC) with the receptor. *EMBO J* 14, 5579–5588.

87 SULIMAN, A., A. LAM, R. DATTA, R.K. SRIVASTAVA (**2001**) Intracellular mechanisms of TRAIL: apoptosis through mitochondrial-dependent and -independent pathways. *Oncogene* 20, 2122–2133.

88 WERNER, A., E. DE VRIES, S. TAIT, I. BONTJER, J. BORST (**2002**) TRAIL receptor and CD95 signal to mitochondria via FADD, caspase-8/10, Bid, and Bax but differentially regulate events downstream from truncated Bid. *J Biol Chem* 277, 40760–40767.

89 YIN, W., W. RAFFELSBERGER, H. GRONEMEYER (**2005**) Retinoic acid determines life span of leukemic cells by inducing antagonistic apoptosis-regulatory programs. *Int J Biochem Cell Biol* 37, 1696–1708.

90 FENAUX, P., C. CHOMIENNE, L. DEGOS (**2001**) All-trans retinoic acid and chemotherapy in the treatment of acute promyelocytic leukemia. *Semin Hematol* 38, 13–25.

91 BONAVIDA, B., C. NG, A. JAZIREHI, G. SCHILLER, Y. MIZUTANI (**1999**) Selectivity of TRAIL-mediated apoptosis of cancer cells and synergy with drugs: the trail to non-toxic cancer therapeutics (review). *Int J Oncol* 15, 793–802.

92 SUN, S.Y., P. YUE, W.K. HONG, R. LOTAN (**2000**) Augmentation of tumor necrosis factor-related apoptosis-inducing ligand (TRAIL)-induced apoptosis by the synthetic retinoid 6-[3-[(1-adamantyl)-4-hydroxyphenyl]-2-naphthalene carboxylic acid (CD437)] through up-regulation of TRAIL receptors in human lung cancer cells. *Cancer Res* 60, 7149–7155.

93 FERRARA, F.F., F. FAZI, A. BIANCHINI, F. PADULA, V. GELMETTI, S. MINUCCI, M. MANCINI, P.G. PELICCI, F. LO COCO, C. NERVI (**2001**) Histone deacetylase-targeted treatment restores retinoic acid signaling and differentiation in acute myeloid leukemia. *Cancer Res* 61, 2–7.

94 CLARKE, N., A.M. JIMENEZ-LARA, E. VOLTZ, H. GRONEMEYER (**2004**) Tumor suppressor IRF-1 mediates retinoid and interferon anticancer signaling to death ligand TRAIL. *EMBO J* 23, 3051–3060.

95 OKUNO, M., S. KOJIMA, R. MATSUSHIMA-NISHIWAKI, H. TSURUMI, Y. MUTO, S. FRIEDMAN, H. MORIWAKI (**2004**) Retinoids in cancer chemoprevention. *Curr Cancer Drug Targets*. 4, 285–298.

96 DONG, Z., M. BIRRER, R. WATTS, L. MATRISIAN, N. COLBURN (**1994**) Blocking of tumor promoter-induced

AP-1 activity inhibits induced transformation in JB6 mouse epidermal cells. *Proc Natl Acad Sci USA* 91, 609–613.

97 YANG, Q., T. SAKURAI, K. KAKUDO (**2002**) Retinoid, retinoic acid receptor beta and breast cancer. *Breast Cancer Res Treat* 76, 167–173.

98 CASTILLO, L., G. MILANO, J. SANTINI, F. DEMARD, V. PIERREFITE (**1997**) Analysis of retinoic acid receptor beta expression in normal and malignant laryngeal mucosa by a sensitive and routine applicable reverse transcription-polymerase chain reaction enzyme-linked immunosorbent assay method. *Clin Cancer Res* 3, 2137–2142.

99 QIU, H., W. ZHANG, A.K. EL-NAGGAR, S.M. LIPPMAN, P. LIN, R. LOTAN, X.C. XU (**1999**) Loss of retinoic acid receptor-beta expression is an early event during esophageal carcinogenesis. *Am J Pathol* 155, 1519–1523.

100 WIDSCHWENDTER, M., J. BERGER, G. DAXENBICHLER, E. MULLER-HOLZNER, A. WIDSCHWENDTER, A. MAYR, C. MARTH, A.G. ZEIMET (**1997**) Loss of retinoic acid receptor beta expression in breast cancer and morphologically normal adjacent tissue but not in the normal breast tissue distant from the cancer. *Cancer Res* 57, 4158–4161.

101 WIDSCHWENDTER, M., J. BERGER, H.M. MULLER, A.G. ZEIMET, C. MARTH (**2001**) Epigenetic downregulation of the retinoic acid receptor-beta2 gene in breast cancer. *J Mammary Gland Biol Neoplasia* 6, 193–201.

102 WIDSCHWENDTER, M., J. BERGER, M. HERMANN, H.M. MULLER, A. AMBERGER, M. ZESCHNIGK, A. WIDSCHWENDTER, B. ABENDSTEIN, A.G. ZEIMET, G. DAXENBICHLER, C. MARTH (**2000**) Methylation and silencing of the retinoic acid receptor-beta2 gene in breast cancer. *J Natl Cancer Inst* 92, 826–832.

103 KHURI, F.R., R. LOTAN, B.L. KEMP, S.M. LIPPMAN, H. WU, L. FENG, J.J. LEE, C.S. COOKSLEY, B. PARR, E. CHANG, G.L. WALSH, J.S. LEE, W.K. HONG, X.C. XU (**2000**) Retinoic acid receptor-beta as a prognostic indicator in stage I non-small-cell lung cancer. *J Clin Oncol* 18, 2798–2804.

104 XU, X.C., J.S. LEE, J.J. LEE, R.C. MORICE, X. LIU, S.M. LIPPMAN, W.K. HONG, R. LOTAN (**1999**) Nuclear retinoid acid receptor beta in bronchial epithelium of smokers before and during chemoprevention. *J Natl Cancer Inst* 91, 1317–1321.

105 LOTAN, R., X.C. XU, S.M. LIPPMAN, J.Y. RO, J.S. LEE, J.J. LEE, W.K. HONG (**1995**) Suppression of retinoic acid receptor-beta in premalignant oral lesions and its up-regulation by isotretinoin. *N Engl J Med* 332, 1405–1410.

106 KWONG, J., K. LO, K. TO, P. TEO, P. JOHNSON, D. HUANG (**2002**) Promoter hypermethylation of multiple genes in nasopharyngeal carcinoma. *Clin Cancer Res* 8, 131–137.

107 IVANOVA, T., A. PETRENKO, T. GRITSKO, S. VINOKOUROVA, E. ESHILEV, V. KOBZEVA, F. KISSELJOV, N. KISSELJOVA (**2002**) Methylation and silencing of the retinoic acid receptor-beta 2 gene in cervical cancer. *BMC Cancer* 2, 4.

108 NAKAYAMA, T., M. WATANABE, M. YAMANAKA, Y. HIROKAWA, H. SUZUKI, H. ITO, R. YATANI, T. SHIRAISHI (**2001**) The role of epigenetic modifications in retinoic acid receptor beta2 gene expression in human prostate cancers. *Lab Invest* 81, 1049–1057.

109 SUH, Y., H. LEE, A. VIRMANI, J. WONG, K. MANN, W.J. MILLER, A. GAZDAR, J. KURIE (**2002**) Loss of retinoic acid receptor beta gene expression is linked to aberrant histone H3 acetylation in lung cancer cell lines. *Cancer Res* 62, 3945–3949.

110 SIRCHIA, S.M., M. REN, R. PILI, E. SIRONI, G. SOMENZI, R. GHIDONI, S. TOMA, G. NICOLO, N. SACCHI (**2002**) Endogenous reactivation of the RARbeta2 tumor suppressor gene epigenetically silenced in breast cancer. *Cancer Res* 62, 2455–2461.

111 FAZI, F., L. TRAVAGLINI, D. CAROTTI, F. PALITTI, D. DIVERIO, M. ALCALAY, S.

McNamara, W. Miller, F. Coco, i.P. Pelicc, C. Nervi (**2005**) Retinoic acid targets DNA-methyltransferases and histone deacetylases during APL blast differentiation *in vitro* and *in vivo*. *Oncogene* 24, 1820–1830.

112 Germain, P., S. Kammerer, E. Pérez, C. Peluso-Iltis, D. Tortolani, F.C. Zusi, J. Starrett, P. Lapointe, J.-P. Daris, A. Marinier, A.R. de Lera, N. Rochel, H. Gronemeyer (**2004**) Rational design of RAR selective ligands revealed by RARb crystal structure. *EMBO Rep* 5, 877–882.

113 Clarke, N., P. Germain, L. Altucci, H. Gronemeyer (**2004**) Retinoids: potential in cancer prevention and therapy. *Expert Rev Mol Med* 6, 1–23.

114 Ortiz, M.A., Y. Bayon, F.J. Lopez-Hernandez, F.J. Piedrafita (**2002**) Retinoids in combination therapies for the treatment of cancer: mechanisms and perspectives. *Drug Resist Updat* 5, 162–175.

115 Brtko, J., J. Thalhamer (**2003**) Renaissance of the biologically active vitamin A derivatives: established and novel directed therapies for cancer and chemoprevention. *Curr Pharm Des* 9, 2067–2077.

116 Chun, K., D. Benbrook, K. Berlin, W. Hong, R. Lotan (**2003**) The synthetic heteroarotinoid SHetA2 induces apoptosis in squamous carcinoma cells through a receptor-independent and mitochondria-dependent pathway. *Cancer Res* 63, 3826–3832.

117 Bayon, Y., M.A. Ortiz, F.J. Lopez-Hernandez, F. Gao, M. Karin, M. Pfahl, F.J. Piedrafita (**2003**) Inhibition of IkappaB kinase by a new class of retinoid-related anticancer agents that induce apoptosis. *Mol Cell Biol* 23, 1061–1074.

118 Fanjul, A.N., F.J. Piedrafita, H. Al-Shamma, M. Pfahl (**1998**) Apoptosis induction and potent antiestrogen receptor-negative breast cancer activity *in vivo* by a retinoid antagonist. *Cancer Res* 58, 4607–4610.

119 Rizvi, N.A., J.L. Marshall, W. Dahut, E. Ness, J.A. Truglia, G. Loewen, G.M. Gill, E.H. Ulm, R. Geiser, D. Jaunakais, M.J. Hawkins

(**1999**) A Phase I study of LGD1069 in adults with advanced cancer. *Clin Cancer Res* 5, 1658–1664.

120 Esteva, F.J., J. Glaspy, S. Baidas, L. Laufman, L. Hutchins, M. Dickler, D. Tripathy, R. Cohen, A. DeMichele, R.C. Yocum, C.K. Osborne, D.F. Hayes, G.N. Hortobagyi, E. Winer, G.D. Demetri (**2003**) Multicenter phase II study of oral bexarotene for patients with metastatic breast cancer. *J Clin Oncol* 21, 999–1006.

121 Hurst, R.E. (**2000**) Bexarotene ligand pharmaceuticals. *Curr Opin Invest Drugs* 1, 514–523.

122 Zhang, C., M. Duvic (**2003**) Retinoids: therapeutic applications and mechanisms of action in cutaneous T-cell lymphoma. *Dermatol Ther* 16, 322–330.

123 Herold, C., M. Ganslmayer, M. Ocker, M. Hermann, E.G. Hahn, D. Schuppan (**2002**) Combined *in vitro* anti-tumoral action of tamoxifen and retinoic acid derivatives in hepatoma cells. *Int J Oncol* 20, 89–96.

124 Koshizuka, K., T. Kubota, J. Said, M. Koike, L. Binderup, M. Uskokovic, H.P. Koeffler (**1999**) Combination therapy of a vitamin D3 analog and all-trans-retinoic acid: effect on human breast cancer in nude mice. *Anticancer Res* 19, 519–524.

125 Wen, J., L. Wang, M. Zhang, D. Xie, L. Sun (**2002**) Repression of telomerase activity during *in vitro* differentiation of osteosarcoma cells. *Cancer Invest* 20, 38–45.

126 Ruiz-Ruiz, C., C. Ruiz de Almodovar, A. Rodriguez, G. Ortiz-Ferron, J.M. Redondo, A. Lopez-Rivas (**2004**) The up-regulation of human caspase-8 by interferon-gamma in breast tumor cells requires the induction and action of the transcription factor interferon regulatory factor-1. *J Biol Chem* 279, 19712–19720.

127 Leaman, D.W., M. Chawla-Sarkar, K. Vyas, M. Reheman, K. Tamai, S. Toji, E.C. Borden (**2002**) Identification of X-linked inhibitor of apoptosis-associated factor-1 as an interferon-stimulated gene that augments TRAIL Apo2L-induced apoptosis. *J Biol Chem* 277, 28504–28511.

128 LISTON, P., W.G. FONG, N.L. KELLY, S. TOJI, T. MIYAZAKI, D. CONTE, K. TAMAI, C.G. CRAIG, M.W. MCBURNEY, R.G. KORNELUK (**2001**) Identification of XAF1 as an antagonist of XIAP

anti-Caspase activity. *Nat Cell Biol* 3, 128–133.

129 TSIMBERIDOU, A.M., F. GILES, J. ROMAGUERA, M. DUVIC, R. KURZROCK (**2004**) Activity of interferon-alpha and isotretinoin in patients with advanced, refractory lymphoid malignancies. *Cancer* 100, 574–580.

4

Nuclear Receptors and the Control of Gene Expression by Fatty Acids

Jean-Paul Pégorier

The last decade has provided evidence that major (glucose, fatty acids, amino acids) and minor (iron, vitamins, etc.) dietary constituents regulate gene expression in a hormonal-independent manner. This review focuses on molecular mechanisms by which fatty acids and/or their metabolites control the expression of genes involved in their own metabolism or in carbohydrate metabolism. These effects are mediated either by direct binding on transcription factors such as PPARs, LXR, HNF-4, RXR (all of which belong to nuclear receptor superfamily), or alternatively through modifications in nuclear abundance and/or activity of numerous transcription factors such as SREBP-1c, ChREBP, PPARs, etc. Knowledge of the mechanisms that govern fatty acid-induced gene expression will provide insight into the role that dietary fat plays in physiology and health, especially in humans.

4.1
Introduction

The high energetic value of fatty acids (9 kcal/g versus 4 kcal/g for glucose) coupled with their low storage bulk (as anhydrous shape in lipid droplets in white adipose tissue, liver and muscles) make them a major source of energy for the organism. Stored fatty acids are either derived from the diet or *de novo* by synthesis from dietary carbohydrates (lipogenesis). Dietary fat is an important macronutrient for the growth and development of all organisms. Excessive levels of dietary fat or imbalance in its composition (saturated versus unsaturated fat) have been related to the onset or development of several chronic diseases such as coronary artery disease [1], obesity and type 2 diabetes [2], and certain forms of cancer [3].

The biological functions of lipids are mainly carried out by fatty acids and/or derived signaling molecules, such as ceramides, diacylglycerol, eicosanoids, and coenzyme A thioesters (acyl-CoA). A large number of cellular systems and functions are affected by these bioactive macromolecules, including regulation of ion channels or pumps, membrane trafficking and composition, protein acylation and

Nutritional Genomics. Edited by Regina Brigelius-Flohé and Hans-Georg Joost
Copyright © 2006 WILEY-VCH Verlag GmbH & Co. KGaA, Weinheim
ISBN: 3-527-31294-3

sorting, control of enzyme activities and immune process and the regulation of gene expression and energy metabolism (reviewed in Ref. [4]).

This review focuses on the transcriptional effect of fatty acids mainly in the liver because this organ plays a central role in whole body lipid metabolism.

4.2
Regulation of Gene Expression by Fatty Acids

4.2.1
Role of Chain Length and Degree of Saturation

When looking at the role of fatty acid chain length in the regulation of gene expression, most of the studies investigated chains with more than 12 carbons. However, a growing number of studies now reported an effect of short-chain fatty acids (4–6 carbons) on the expression of genes. For instance, butyrate affected the expression of a number of genes involved in the regulation of cell proliferation/differentiation/apoptosis in colorectal cancer cell lines or in immortalized colon cells [5, 6]. Similarly, butyrate induced calcitonin in cultured human medullary thyroid carcinoma [7] and the plasminogen activator inhibitor type 1 (PAI-1) in HepG2 cells [8]. Finally, it was suggested that the decreased expression of mitochondrial hydroxy-methylglutaryl-CoA synthase (mHMG-CoA synthase) in the colon of germ-free rats [9] could be due to the absence of butyrate, the most abundant short-chain fatty acid produced by the fermentation of dietary fibers. Indeed, the transcription-stimulating effect of butyrate has recently been demonstrated in human colon cancer cell line where it induced WAF1/Cip1, a protein that inhibits the G_1 to S-phase transition [10]. Stimulation of transcriptional activity by of butyrate seems to be mainly exerted by its inhibitory effect on histone deacetylase activity [11, 12], which alters chromatin structure and transcription rate. To get more insight into the regulation of gene transcription by short-chain fatty acids the author recommends reading recent reviews in this field [12–14].

The rest of the present review will be focused on the transcriptional effect of long-chain fatty acids (more than 12 carbons).

Four classes of long-chain fatty acids are typically encountered in the diet: saturated fatty acids, *n*-9 monounsaturated fatty acids, *n*-3 and *n*-6 polyunsaturated fatty acids (PUFAs). Dietary fish oil PUFAs from the *n*-3 series are considered to have protective effects on cardiovascular diseases, diabetes, cancer, and neurological diseases (reviewed in Ref. [15]). Among the pleiotropic effects responsible for these beneficial actions of PUFAs [15] the decrease in the concentrations of circulating very low-density lipoprotein (VLDL), and chylomicrons plays a central role. This mainly results from a decrease in the activity of hepatic lipogenic enzymes because of an inhibition of gene transcription and/or modifications in mRNA maturation and/or stability (reviewed in Refs [16–18]). Interestingly, it seems that downregulation of gene expression by fatty acids is restricted to fatty acid having more than 18 carbons and at least two double bonds (reviewed in Ref. [19]), whereas the upregulation of gene expression is independent of the degree of saturation of the carbon chain of fatty acids.

4.2.2
Metabolite(s) Responsible for the Effect of Long-chain Fatty Acids

Fatty acids are delivered to cells either as complex lipoproteins (VLDL, chylomicrons) or as non-esterified fatty acids (NEFAs). Triglycerides in chylomicrons and VLDLs are hydrolyzed by the action of a lipoprotein lipase and NEFAs enter cells via fatty acid transporters (reviewed in Ref. [20]). Once in cells, the NEFAs are rapidly converted to fatty acyl-CoA thioesters by acyl-CoA synthetases (ACSs) specific for carbon chain length. At least six different ACSs have been characterized. While each isoform can activate a wide range of fatty acids, they have tissue-specific expression, subcellular distribution, and a specific spectrum of activity. For instance, ACS-1, 4, and 5 are expressed in the liver, ACS-1 activates C12–C20 fatty acids, whereas the activity of ACS-4 is restricted to arachidonic and eicosapentenoic acids [21]. The intracellular location of ACS-1 and 4 inside the endoplasmic reticulum is in agreement with their involvement in triglyceride synthesis. Conversely, ACS-5 is located in outer mitochondrial membrane and thus plays a crucial role in the regulation of β-oxidation [21]. Once activated by ACS, fatty acyl-CoAs are metabolized in many different metabolic pathways (β-oxidation, elongation, desaturation, triglyceride or cholesterol synthesis, prostanoid or leukotriene synthesis, etc.) where each intermediate metabolite or end product can be responsible for the transcriptional effect of long-chain fatty acids (reviewed in Ref. [18]). For instance, it was shown that NEFA itself, long-chain acyl-CoA, lipooxygenase-derived metabolite leukotriene B4, prostacyclins, 15-deoxy-Δ12,14-prostaglandin J2, etc. are potent regulators of transcription depending upon the gene considered (reviewed in Refs [22, 23]). This suggests that fatty acids can control gene transcription by different mechanisms according to the cell-specific context and the target gene.

The regulation of gene transcription by fatty acids seems to be achieved through two different mechanisms. First, the direct binding of the fatty acid or its metabolites onto transcription factors that all, in our actual knowledge, belong to the nuclear receptor superfamily. Second, indirect effects of fatty acids through changes in the activity or abundance of transcription factors such as SREBP (sterol regulatory element-binding protein) or ChREBP (carbohydrate-responsive element-binding protein).

4.3
The Role of Nuclear Receptors in the Regulation of Gene Expression by Fatty Acids

4.3.1
General Structure and Basic Mechanism of Action of Nuclear Receptors

The 48 nuclear receptors that have been identified in the human and mouse genomes share a common structure of four main domains named A/B, C, D, and E/F (Fig. 4.1; reviewed in Refs [24, 25]) (see also chapters 2 and 3). Key functions

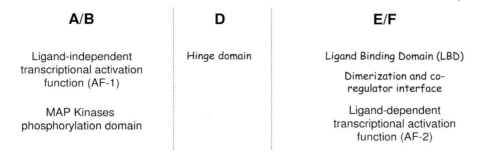

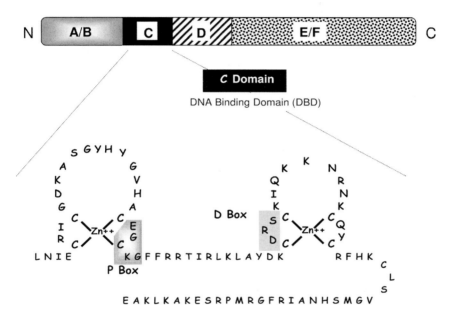

Fig. 4.1. Schematic structure of nuclear receptors. The amino acid sequence of the zinc finger region is from PPARα.

have been assigned to each of these domains. The N-terminal A/B domain harbors a ligand-independent transcriptional activation function (AF-1). The sequence and length of the A/B domain are highly variable between receptors and among receptor subtype (reviewed in Ref. [24]). The C domain, or DNA-binding domain (DBD), is formed by two zinc finger-like motifs folded in a globular structure that can recognize a nuclear response element (NRE) present on target genes (Fig. 4.1).

Nuclear receptors bind to NRE as monomers, homodimers, or heterodimers (generally with RXR, 9-*cis* retinoic acid receptor), depending on the class of receptor (reviewed in Ref. [24]). Three distinct sequences for NRE have been described: (1) direct repeats (DRx: AGGTCA-Nx-AGGTCA) where x represents the number (from 0 to 10) of any nucleotide (N) between the two hexanucleotides; (2) everted repeats (ERx: TGACCT-Nx-AGGTCA), and (3) inverted repeats (IRx: AGGTCA-Nx-

TGACCT). The D or hinge domain permits protein flexibility due to conformational changes induced by ligand binding. This region also contains the C-terminal extension of the DBD domain that seems to be involved in recognition of the 5' extension of the NRE (see Section 4.3.2). Finally, there is a large C-terminal domain, the E/F domain or ligand-binding domain (LBD). The structure of the E/F domain varies substantially between nuclear receptors but they all share a common sequence of 10–13 α-helices organized around the hydrophobic binding pocket. In addition to the ligand binding, this domain encompasses the nuclear receptor's dimerization and co-regulator (co-repressor and -activator of transcription) interfaces and a strong ligand-dependent transcriptional activation function (AF-2; Fig. 4.1). Briefly, upon ligand binding the nuclear receptors undergo a conformational change that coordinately dissociates co-repressors and facilitates recruitment of co-activator proteins to enable transcription of target genes (Fig. 4.2). Finally it must be underlined that in addition to ligand-induced activation of nuclear receptors, some of them including estrogen (ER), androgen (AR), progesterone (PR), vitamin D (VDR), retinoic acid (RAR), and peroxisome proliferator-activated (PPAR) receptors are targets of several kinases which modulate their transcriptional activity through phosphorylation mainly in the A/B domain (reviewed in Refs [26–28]).

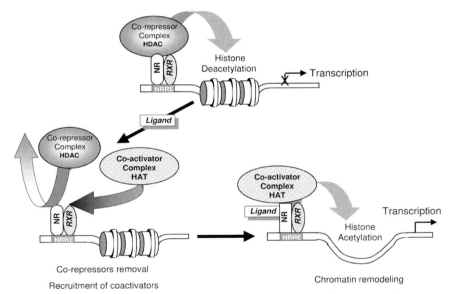

Fig. 4.2. Schematic representation of the mechanism of action of nuclear receptors. In the unliganded state, nuclear receptors (NR) are bound to their specific responsive element (NRRE) generally as a heterodimer with *cis*-retinoic acid receptor (RXR). In this condition, heterodimers are associated with a multiprotein co-repressor complex that contains histone deacetylase activity (HDAC). The deacetylated status of histones keeps the nucleosome in a conformation in which transcription is inhibited. Once a ligand binds to the receptor, the co-repressor complex dissociates and a co-activator complex containing histone acetyltransferase activity (HAT) is recruited to the heterodimer. Acetylation of histone induces chromatin remodeling, a major event in activation of gene transcription.

4.3.2
Peroxisome Proliferator-activated Receptors

Among the fatty acid-regulated nuclear receptors, the PPARs are the most extensively characterized. Three isoforms of PPAR have been cloned: (1) PPARα is mainly expressed in liver, digestive tract, and kidney; (2) PPARβ, -δ or NUC-1 (respectively cloned in *Xenopus*, mouse, and human) are ubiquitously expressed; and (3) PPARγ (γ1, γ2, γ3 arising from an alternative splicing of a single gene) is mainly expressed in adipose tissues and in macrophages (reviewed in Ref. [24]). Initially characterized for their capacity to be activated by peroxisome proliferators (fibrates, xenobiotics, etc.) it was later shown that fatty acids (saturated and PUFA) and some eicosanoids (15-deoxy-Δ12,14-prostaglandin J2, leukotriene B4) are potent ligands of PPARs (reviewed in Ref. [24]). Structural analysis of PPARs reveals that the hydrophobic binding pocket of these receptors is bigger than that of most other members of this family (around 1400Å^3, reviewed in Ref. [29]). Although many fatty acids can bind PPARs *in vitro* they do not have the same potency to activate PPAR. This was particularly well demonstrated in hepatic cells in which oleate (C18:1*n*-9) or eicosapentaenoate (C20:5*n*-3) bind to PPARα (the predominant hepatic isoform) with quite similar affinity, but only eicosapentanoate or docosahexanoate (C22:6*n*-3) activate PPARα [30].

Binding of ligands leads to an active conformation of the receptor through the stabilization of the AF-2 region of the LBD (see Section 4.3.1) [31]. This conformational change leads to the removal of co-repressor complex from the PPAR/RXR heterodimer and the recruitment of the co-activator complex [32, 33] (Fig. 4.2) essential for the interaction with the transcriptional machinery. The recruitment of the transcriptional machinery can occur either directly [34] or in response to the chromatin remodeling (histone acetylation, Fig. 4.2) (reviewed in Ref. [35]). The modulation of gene transcription is due to the binding of the heterodimer PPAR/RXR to a consensus sequence (PPRE) consisting of a direct repeat of a hexamer AGGTCA with an interspacing of 1 base pair (DR1, see Section 4.3.1). Moreover, the 5′-extension (AACT) is essential for the polarity of PPAR/RXR heterodimer binding [36], PPAR interacting with the 5′ repeat and RXR binding to the 3′ motif [37].

Using cDNA microarray technology, it was shown that dietary fish oil PUFAs play a major role in the regulation of an extensive network of genes involved in hepatic fatty acid metabolism [38, 39]. Nevertheless, this does not mean that these effects are dependent upon PPARα activation, most of these genes have a PPRE in their promoter region. Until recently, this observation has led to the established dogma that regulated genes containing one or more PPRE sequences in their promoter respond to fatty acids via PPAR activation. However, a growing number of reports show that the regulation of gene expression by fatty acids is certainly more complex than simple acceptance of this dogma. For instance, apoA-II and FAT-CD36 genes do not respond to fatty acids despite the presence of PPRE sequences in their promoter [40] (reviewed in Ref. [41]). Similarly, adenovirus-mediated overexpression of a dominant negative PPARα in hepatoma cells antagonizes the fibrate-induced liver CPT-I gene expression whereas it does not affect LCFA-induced L CPT-I gene transcription [42]. In the liver of PPARα-null mice

(PPARα$^{-/-}$) the inhibitory effect of PUFAs on the expression of genes encoding regulatory proteins of lipogenesis (acetyl-CoA carboxylase, fatty acid synthase, spot 14) or glycolysis (L-pyruvate kinase) is still present despite the absence of PPARα [43, 44]. PUFA inhibits the transcription of Δ5 and Δ6 desaturase genes, whereas PPAR agonists stimulate the transcription of theses genes [45, 46]. These results suggest that transcription factors different from PPAR are involved in the regulation of gene expression by fatty acids.

4.3.3
Liver X Receptors

Two LXR isoforms have been described: LXRα mainly expressed in liver, kidney, intestine, adipose tissue, and adrenals, and LXRβ which is ubiquitously present [47]. Oxysterols (22-*R*-hydroxycholesterol, 24,25-epoxycholesterol, etc.) are natural ligands of LXR which are commonly known as cholesterol sensors. Recently, it was shown that mono- and polyunsaturated fatty acids bind to LXRα [48, 49] and antagonize the binding of oxysterols leading to an inhibition of LXRα transcriptional activity [48]. While it seems clear that LXRβ is insensitive to fatty acid antagonism [49], it must be underlined that the antagonistic effect of PUFA on LXR-regulated genes is not always found, especially in liver [50]. Structural analysis of the LXRβ-binding domain reveals that the binding pocket is intermediate (830 Å^3) between RXR (489 Å^3) and PPARs (see Section 4.3.2) (reviewed in Ref. [29]). These receptors regulate the expression of genes involved in hepatic bile acid synthesis (7α-hydroxylase CYP7A), cholesterol reverse transport (ATP-binding cassette genes), lipogenesis (see Section 4.4.1) and fatty acid and glucose uptake (reviewed in Ref. [25, 51]) upon binding to the DR4 (see Section 4.3.2) regulatory element (LXRE) as heterodimers with RXR.

In addition to PPAR and LXR receptors as fatty acid sensors, another nuclear receptor has been involved in fatty acid-mediated gene expression. For instance, the gene encoding 7α-hydroxylase (*CYP7A*), the rate-limiting protein in bile acid synthesis, is upregulated by oxysterols (via the activation of LXR) [52] and by fatty acids [53] but downregulated by fibrates (PPARα agonists) [54, 55]. Indeed, analysis of the promoter region of *CYP7A* gene reveals the presence of a DR1 sequence (see Section 4.3.1) that binds specifically HNF-4α receptor but not PPAR/RXR heterodimer [54, 55]. The relative contribution of these two receptors in the control of *CYP7A* gene transcription has been clearly demonstrated by phenotypic analysis of PPARα [55] or HNF-4α [56] knockout mice. The next section on HNF-4α may provide some clues to explain the apparent contradictory effects of fatty acids and fibrates on *CYP7A* gene.

4.3.4
Hepatic Nuclear Factor-4 Alpha

The HNF-4 class of nuclear receptor contains two subtypes in humans, HNF-4α (α1, α2, α4) and HNF-4γ that differ in the A/B and F domains (see Section 4.3.1).

Originally identified in liver, HNF-4α is also present in kidney, intestine, and pancreatic islets [57, 58]. The expression pattern of HNF-4γ is more ubiquitous but high levels are also detected in liver and islets [59]. Unlike PPAR and LXR, HNF-4 is not able to bind native fatty acids but has a great affinity for fatty acyl-CoA [60]. While binding of saturated acyl-CoA (C14:0, C16:0) stimulates the transcriptional activity of HNF-4α, binding of polyunsaturated fatty acyl-CoA (C18:3, C20:5, C22:6) inhibits the effects of HNF-4α on gene transcription [60].

The recent X-ray crystallographic analysis of HNF-4α [61, 62] and HNF-4γ [63] revealed that, although the receptor was crystallized in the absence of ligand, the LBD was occupied by C14–C18 fatty acids interacting with Arg226 through the carboxyl group [62, 63]. These fatty acids do not exchange with exogenously added fatty acids [62, 63], a situation that markedly differs from PPAR and LXR which show a binding and displacement of ligand typical of nuclear receptor. These results suggest that unliganded HNF-4 receptor is not sufficiently stable for crystallization and that bound fatty acids lock the receptor in an active configuration. Similar observations were reported for two other crystallized nuclear receptors, the RORβ containing a stearic acid molecule in the binding pocket [64] and the RXRα containing an oleic acid in ligand-binding pocket [65]. Finally, it was shown that the volume of the ligand-binding pocket of HNF-4α (370 Å^3) is much smaller than the estimated volume for fatty acyl-CoA (around 850 Å^3) [61]. Altogether, these observations underline the need for additional work to understand the role of fatty acids in the function of endogenous HNF-4.

The phenotypic analysis of CRE-Lox conditional HNF-4α-null mice has shown that this nuclear receptor controls either directly or indirectly the expression of several hepatic genes. These include genes encoding proteins of lipoprotein metabolism (apoC-II, C-III, A-II, A-IV, reviewed in Ref. [16]), iron metabolism (transferrin [66]), carbohydrate metabolism (L-pyruvate kinase, glucose-6-phosphatase, phosphoenolpyruvate carboxykinase [44, 67]) and bile acid synthesis (CYP7A [56]). Unlike PPAR or LXR, HNF-4α binds to DR1 sequences as homodimer, making it a competitor of the PPAR/RXR heterodimer for binding to these particular DNA motifs [44, 55, 66]. This has been recently demonstrated for the gene encoding glucose-6-phosphatase; the transcription thereof is inhibited by PUFA-CoA thioesters by preventing HNF-4α binding to the glucose-6-phosphatase promoter [67]. Finally, it was reported that fibrates could be converted to CoA thioesters (as fatty acids) which then bind to HNF-4α leading to an inhibition of its transcriptional activity [68]. This dual level of competition with PPAR receptors (binding to DR1 motifs, fibrate-CoA binding) could explain, in part, the complex regulation of CYP7A gene by fatty acids and fibrates.

4.3.5
Other Fatty Acid-binding Nuclear Receptors

In addition to these three main fatty acid sensors (i.e. PPAR, LXR, and HNF-4), it has been shown that at least three other nuclear receptors are able to bind fatty acids. Firstly, the dimerization partner of most nuclear receptors, RXR, binds mono- and polyunsaturated fatty acids including oleic acid (C18:1), arachidonic acid (C20:4), and DHA (C22:6) as evidenced by both crystal structure [65] and electrospray mass spectrometry [69] experiments. Moreover, these fatty acids have been shown to activate RXR in various cells or organs (brain, heart, testis, colonocytes [70–72]), suggesting that fatty acid ligands have the potential to exert important effects on RXR-mediated gene transcription. As discussed above, fatty acids control the expression of numerous genes through PPAR/RXR or LXR/RXR hetodimers. Recent reports showing that fatty acids are able to bind and activate not only PPAR and LXR but also RXR provide interesting pointers to the possibility that fatty acids act as heteroligands and also clarify the permissive and/or synergistic consequences in the regulation of gene transcription.

The second additional nuclear receptor that binds PUFAs is the bile acid-activated farnesoid X receptor (FXR) [73]. Indeed, it has been shown in hepatoma cells that PUFAs antagonized the agonist-induced FXR activation, leading to specific changes in FXR target genes [73]. Finally, the third additional nuclear receptor that binds PUFA is the retinoic acid-related orphan receptor (RORβ) [64]. This isoform of RORs is exclusively expressed in the central nervous system, whereas two other isoforms, α and γ, are expressed in many tissues [74, 75]. RORα plays a central role in the regulation of lipid metabolism as evidenced by its stimulating effect on intestinal apoC-III, liver apoA-I and skeletal muscle CPT I gene expression [76–78]. Recent studies reveal that cholesterol is a natural ligand of RORα (reviewed in Ref. [75]). Whether fatty acids can bind this isoform and interfere with natural ligand remains to be determined.

A general overview of fatty acid-regulated genes by nuclear receptors is depicted in Fig. 4.3.

In addition to their direct effects as ligands, fatty acids affect the nuclear abundance and/or activity of many transcription factors that control the expression of genes involved in lipid or glucose metabolism. The last part of this chapter will describe two examples of such kind of regulation on well-characterized fatty acid-regulated genes.

4.4
Effect of Fatty Acids on Nuclear Abundance and Activity of Transcription Factors

4.4.1
Sterol Regulatory Element-binding Protein

Three SREBP isoforms have been cloned; SREBP-1a and 1c (derived from alternative transcription start site of a single gene) and SREBP-2 encoded by another gene (reviewed in Refs [79, 80]) (see also chapters 6 and 12). SREBP-1c preferentially enhances transcription of genes involved in fatty acid, triglyceride, and phospholipids synthesis, whereas SREBP-1a and SREBP-2 activate genes involved in cholesterol synthesis (reviewed in Refs [79, 80]). SREBPs belong to the basic helix-loop-helix-leucine zipper (bHLH-Zip) family of transcription factors, but differ from

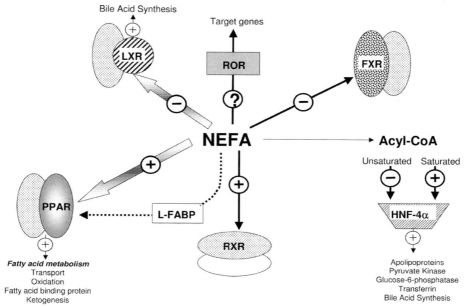

Fig. 4.3. Schematic overview of fatty acid sensor proteins. Non-esterified fatty acids (NEFA) or their respective CoA thioesters regulate the transcription of target genes through direct activation of some nuclear receptors PPAR (peroxisome proliferator-activated receptor), LXR (liver X receptor), HNF-4α (hepatic nuclear factor 4 alpha) or RXR (*cis*-retinoïc acid receptor). In addition, fatty acids were also found as structural ligands in the ligand pocket of FXR (farnesoid X receptor) and ROR (retinoic acid-related orphan receptor) but the contribution of these receptors in the regulation of gene transcription by fatty acids remains to be determined. Finally, ʟ-FABP (fatty acid-binding protein) may be involved in the regulation of gene expression by fatty acids.

other bHLH-Zip proteins in that they are synthesized as inactive precursors bound to the endoplasmic reticulum (reviewed in Refs [79, 80]). In order to reach the nucleus and to act as a transcription factor, the N-terminal domain of SREBP proteins must be cleaved proteolytically from the endoplasmic reticulum membrane (reviewed in Refs [79, 80]).

Briefly, four proteins play a major role in the maturation and translocation of SREBP. In the endoplasmic reticulum, SREBP binds to SREBP-cleavage activating protein (SCAP) that interacts with a third protein, INSIG (insulin-induced gene). When intracellular sterol levels fall, SCAP escorts SREBP to the Golgi whereas INSIG facilitates retention of SREBP in the endoplasmic reticulum [81–83]. In the Golgi, two proteases, S1P and S2P (site 1 and 2 protease), cleave inactive SREBP precursor and release mature transcription factor from the Golgi for its translocation to the nucleus (Fig. 4.4). SREBP binds, as homodimer, to SRE (sterol responsive element) in the promoter of many genes involved in lipid metabolism (reviewed in Refs [18, 84]).

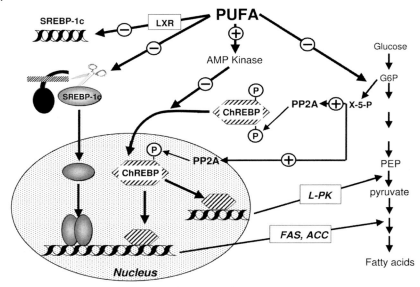

Fig. 4.4. Schematic representation of PUFA-induced repression of glycolytic and lipogenic gene expression. Polyunsaturated fatty acids (PUFAs) inhibit, in an indirect manner, the transcription of genes encoding lipogenic (FAS (fatty acid synthase), ACC (acetyl-CoA carboxylase)) and glycolytic enzymes (L-PK (liver-type pyruvate kinase)) secondary to a reduction in nuclear abundance of stimulatory transcription factors such as SREBP-1c (sterol regulatory element binding protein) or ChREBP (carbohydrate response element binding protein). For SREBP-1c, this mainly results from a decrease in mRNA levels (via an inhibition of LXR-mediated gene transcription) and in the cleavage of the precursor form of SREBP-1c. For ChREBP this is due to a decrease in xylulose-5-phosphate (X-5-P) concentration (secondary to a reduction in glycolytic flux). Under these conditions, ChREBP is not dephosphorylated by PP2A (phosphatase 2A) and thus not translocated in the nucleus. A second putative mechanism is due to an activation of AMP kinase that phosphorylates ChREBP and leads to an inhibition of its translocation in the nucleus.

As mentioned before (Section 4.3.2), PUFAs inhibit the transcription of lipogenic genes (ATP citrate lyase, acetyl-CoA carboxylase, fatty acid synthase, spot 14) in the liver of PPARα-null mice [43]. Indeed, overexpression of mature SREBP-1c in transgenic mice liver [85] or in cultured hepatocytes [86] overrides the PUFA suppression of lipogenic gene expression, supporting a role for SREBP-1c in mediating the negative effects of PUFA.

Although, it has been known for many years that the inhibition of lipogenic enzyme gene expression is restricted to PUFAs [87, 88], recent experiments suggest that saturated fatty acid could have opposite effects (i.e. a stimulation of lipogenic gene expression) [89]. These surprising effects seem to be due to an increased in SREBP-1c gene expression secondary to a strong co-activation by PGC-1β (PPAR gamma co-activator 1β) [89]. Conversely, the inhibitory effects of PUFA through SREBP-1c are complex. At least two mechanisms have been described. First, it was reported that treatment of CHO cells with PUFA activated a sphingomyelinase leading to the redistribution of cholesterol from the plasma membrane to the endoplasmic reticulum [90]. The increased level of cholesterol in endoplasmic reticulum membrane led to the inhibition of proteolytic process and, thus, to a decrease in the abundance of SREBP in the nucleus [90] (Fig. 4.4). This mechanism probably represents the main mechanism by which PUFAs suppress the expression of fatty acid synthesis genes (see below). Secondly, PUFAs markedly reduce hepatic SREBP-1c mRNA levels (Fig. 4.4) by inhibiting its gene transcription [48, 91] and accelerating its mRNA turnover [92]. While PUFA-induced SREBP-1c mRNA degradation is still unresolved, the effect of unsaturated fatty acids on SREBP-1 gene transcription is controversial.

Originally, the effect was attributed to an antagonistic effect of PUFA on LXR activity (see Section 4.3.3), a strong activator of SREBP-1c gene transcription [93, 94]. However, this has been recently questioned. For instance, in rat liver or isolated hepatocytes it was shown that PUFAs did not antagonize the oxysterol/LXR-mediated transcription of target genes such as *CYP7A* or certain ATP-binding cassettes [50]. Moreover, the PUFA-induced suppression of SREBP-1c promoter activity does not require the LXR responsive element (LXRE) [95]. Thus, despite evidence of an inhibitory effect of PUFAs on SREBP-1c gene expression, additional work is required to fully understand the underlying molecular mechanisms.

4.4.2
Carbohydrate Response Element-binding Protein (ChREBP)

As already mentioned (Section 4.3.2), liver-type pyruvate kinase (L-PK) gene transcription is markedly decreased by PUFA in a PPARα-independent mechanism. The PUFA-mediated suppression of L-PK gene expression cannot be attributed to SREBP-1c since the L-PK promoter does not contain an SRE-binding site [86, 96, 97]. Functional mapping analysis of the L-PK promoter has shown that the PUFA response element contains a carbohydrate response element (ChoRE) binding site [98, 99]. Recently, ChREBP was shown to play a crucial role in the induction of glycolytic and lipogenic genes by glucose [100, 101] by its capacity to bind to ChoRE present in promoters of these genes [102–104]. The fact that the ability of ChREBP to bind to the L-PK promoter is decreased in hepatic nuclear extracts from rats fed a high-fat diet suggests that ChREBP may be involved in fatty acid-induced down-regulation of glycolytic genes [102] (see also chapter 6).

At least two different mechanisms have been involved in the negative effects of PUFAs on L-PK gene expression (Fig. 4.4): (1) a decrease in ChREBP gene expression [105, 106] due to an acceleration of ChREBP mRNA decay [107]; and (2) a decrease in the nuclear translocation of ChREBP [105]. Indeed, the translocation of ChREBP from the cytosol to the nucleus is an important process to activate target genes in response to glucose [108]. ChREBP is a phosphoprotein that contains at least three phosphorylation sites important for its activation [109]. The dephosphorylation of Ser196 is stimulated by xylulose-5-phosphate, a metabolite of the pentose phosphate pathway, which activates a phosphatase (PP2A). PP2A dephosphorylates ChREBP leading to its migration into the nucleus [110] (Fig. 4.4). High-fat diets are known to reduce the activity and expression of key enzymes of the pentose phosphate pathway [107, 111–113]. In this way they would reduce the nuclear shuttling of ChREBP. Inside the nucleus, dephosphorylation of Ser568 as well as of Thr666 might be required for ChREBP DNA binding [105, 110] (Fig. 4.4). The dephosphorylation is ensured by a nuclear PP2A, also stimulated by xylulose-5-phosphate [105] (Fig. 4.4). Conversely, Kawaguchi and co-workers reported that fatty acids (whatever the chain length and degree of saturation) induced an increase in intracellular AMP concentration that activated AMP kinase and led to the phosphorylation of ChREBP on Ser568 [105] (Fig. 4.4). However, these results are questionable since it was reported by several groups that dietary PUFAs suppress lipogenic and glycolytic gene transcription without affecting hepatic ATP levels [114, 115].

A possible explanation for these discrepancies is that PUFAs could modulate AMP kinase activity by governing the activity of phosphatases controlling its dephosphorylation [115]. Another explanation is that PUFAs could change the nuclear abundance of ChREBP in an AMP kinase-independent mechanism. Indeed we recently demonstrated that PUFAs (but not saturated or monounsaturated long-chain fatty acids) suppress the nuclear content of ChREBP in mice liver without any activation of AMP kinase both *in vivo* and *in vitro* [107]. Similar observations were found in AMPKα1$^{-/-}$ and AMPKα2$^{-/-}$ mice fed a high-carbohydrate diet supplemented with PUFAs [107]. This impaired translocation of ChREBP inside the nucleus in response to PUFAs is probably due to a reduction in early steps of glucose metabolism as evidenced by the decrease in glucokinase and glucose-6-phosphate dehydrogenase activities leading to a fall in xylulose-5-phosphate concentration [107].

This type of regulation (i.e. through phosphorylation/dephosphorylation mechanisms) has also been described for other transcription factors. Indeed, fatty acids and/or their metabolites can affect either directly or indirectly the activity of various kinases (PKB, PKC, IKK) (reviewed in Ref. [18]) that in turn changes the degree of phosphorylation and activity of transcription factors such as NFκB [116] or nuclear receptors (reviewed in Ref. [26]), especially of PPARs (reviewed in Refs [27, 28]). For instance, it was shown that fatty acids activate protein kinase C (PKC) in a tissue-dependent manner (PKCθ in muscle and adipocyte [116, 117], PKCε in liver [118]) which, in turn, induces other kinases such as IκB kinase (IKK) or c-Jun N-terminal kinase (JNK) that contribute to the development of insulin resistance. Indeed, activation of IKK in the liver of high-fat diet mice has been clearly identified as a major determinant of insulin resistance through NFκB-regulated genes [119, 120].

Finally, regulation of gene expression by fatty acids can be exerted via fatty acid-binding proteins (FABP). At least nine different FABPs have been characterized in mammals (reviewed in Ref. [121]). Among this large family of proteins it was shown that the liver isotype (L-FABP) interacts physically with PPARα and therefore L-FABP is considered as a co-activator in PPARα-mediated gene expression [122]. Similarly, E-FABP (mainly expressed in adipose tissue and muscle) interacts with PPARβ and A-FABP (mainly expressed in adipose tissue) interacts with PPARγ [123], underlying the crucial role of these proteins as link between intracellular fatty acid level and the regulation of gene expression.

4.5
Conclusions

All the experimental data presented in this chapter emphasize the major role of dietary fat as a source of signal molecules for the regulation of an extensive network of genes involved in hepatic fatty acid metabolism. They also underline the great diversity of fatty acid-sensor proteins and the list of potential transcription factors involved in fatty acid-mediated gene expression is probably not closed. This review was focused on two main pathways by which fatty acids can control gene expression: (1) a direct binding on nuclear receptors (PPAR, LXR, HNF-4, RXR, FXR, ROR, etc.) and (2) an indirect action through changes in abundance or activity of transcription factors (SREBP, ChREBP, etc.). Knowledge of the mechanisms by which fatty acids control specific gene expression may provide insights into the development of new therapeutic strategies for a better management of whole-body lipid metabolism and the control of blood levels of triglycerides and cholesterol, important risk factors involved in several chronic diseases like insulin resistance, obesity and diabetes, metabolic syndrome, atherosclerosis and coronary heart diseases, inflammation, cancers, etc.

References

1 P.M. Kris-Etherton, W.S. Harris, L.J. Appel (**2002**) *Circulation* 106, 2747–2757.

2 D.E. Kelley, B.H. Goodpaster, L. Storlien (**2002**) *Annu Rev Nutr* 22, 325–346.

3 L. Kushi, E. Giovannucci (**2002**) *Am J Med* 113 (Suppl 9B), 63S–70S.

4 N.J. Faergeman, J. Knudsen (**1997**) *Biochem J* 323 (Pt 1), 1–12.

5 G. Iacomino, M.F. Tecce, C. Grimaldi, M. Tosto, G.L. Russo (**2001**) *Biochem. Biophys. Res. Commun.* 285, 1280–1289.

6 Y. Tabuchi, Y. Arai, T. Kondo, N. Takeguchi, S. Asano (**2002**) *Biochem. Biophys. Res. Commun.* 293, 1287–1294.

7 T. Nakagawa, B.D. Nelkin, S.B. Baylin, A. deBustros (**1988**) *Cancer Res.* 48, 2096–2100.

8 T.J. Smith, J.J. Piscatelli, V. Andersen, H.S. Wang, P. Lance (**1996**) *Hepathology* 23, 866–871.

9 C. Cherbuy, B. Darcy-Vrillon, M.T. Morel, J.P. Pégorier, P.H. Duée (**1995**) *Gastroenterology* 109, 1890–1899.

10 K. Nakano, T. Mizuno, Y. Sowa, T. Orita, T. Yoshino, Y. Okuyama, T. Fujita, N. Ohtani-Fujita, Y. Matsukawa, T. Tokino, H. Yamagishi, T. Oka, H. Nomura, T. Sakai (**1997**) *J. Biol. Chem.* 272, 22199–22206.

11 I.R. Sanderson, S. Naik (**2000**) *Annu. Rev. Nutr.* 20, 311–338.

12 I.R. Sanderson (**2004**) *J. Nutr.* 134, 2450S–2454S.

13 J.R. Davie (**2003**) *J. Nutr.* 133, 2485S–2493S.

14 S. Mei, A.D. Ho, U. Mahlknecht (**2004**) *Int. J. Oncol.* 25, 1509–1519.

15 T. Seo, W.S. Blaner, R.J.
Deckelbaum (2005) *Curr. Opin.
Lipidol.* 16, 11–18.

16 D.B. Jump, S.D. Clarke (1999) *Annu.
Rev. Nutr.* 19, 63–90.

17 D.B. Jump (2002) *Curr. Opin. Lipidol.*
13, 155–164.

18 D.B. Jump (2004) *Crit. Rev. Clin. Lab.
Sci.* 41, 41–78.

19 S.D. Clarke, R. Baillie, D.B. Jump,
M.T. Nakamura (1997) *Ann. NY Acad.
Sci.* 827, 178–187.

20 T. Hajri, N.A. Abumrad (2002) *Annu.
Rev. Nutr.* 22, 383–415.

21 T.M. Lewin, J.H. Kim, D.A. Granger,
J.E. Vance, R.A. Coleman (2001) *J.
Biol. Chem.* 276, 24674–24679.

22 J.P. Pégorier (1998) *Curr. Opin. Clin.
Nutr. Metab. Care* 1, 329–334.

23 J.F. Louet, F. Chatelain, J.F.
Decaux, E.D. Park, C. Kohl, T.
Pineau, J. Girard, J.P. Pégorier
(2001) *Biochem. J.* 354, 189–197.

24 B. Desvergne, W. Wahli (1999)
Endocr. Rev. 20, 649–688.

25 S.A. Khan, J.P. Vanden Heuvel
(2003) *J. Nutr. Biochem.* 14, 554–
567.

26 C. Rochette-Egly (2003) *Cell. Signal.*
15, 355–366.

27 C. Diradourian, J. Girard, J.P.
Pégorier (2005) *Biochimie* 87, 33–38.

28 L. Gelman, L. Michalik, B.
Desvergne, W. Wahli (2005) *Curr.
Opin. Cell. Biol.* 17, 1–7.

29 H. Escriva, F. Delaunay, V. Laudet
(2000) *Bioessays* 22, 717–727.

30 A. Pawar, D.B. Jump (2003) *J. Biol.
Chem.* 278, 35931–35939.

31 H.E. Xu, M.H. Lambert, V.G.
Montana, D.J. Parks, S.G.
Blanchard, P.J. Brown, D.D.
Sternbach, J.M. Lehmann, G.B.
Wisely, T.M. Willson, S.A. Kliewer,
M.V. Milburn (1999) *Mol. Cell.* 3,
397–403.

32 W. Feng, R.C. Ribeiro, R.L. Wagner,
H. Nguyen, J.W. Apriletti, R.J.
Fletterick, J.D. Baxter, P.J.
Kushner, B.L. West (1998) *Science*
280, 1747–1749.

33 R.T. Nolte, G.B. Wisely, S. Westin,
J.E. Cobb, M.H. Lambert, R.
Kurokawa, M.G. Rosenfeld, T.M.

Willson, C.K. Glass, M.V. Milburn
(1998) *Nature* 395, 137–143.

34 B.L. Kee, J. Arias, M.R. Montminy
(1996) *J. Biol. Chem.* 271, 2373–2375.

35 L. Xu, C. Glass, M.G. Rosenfeld
(1999) *Curr. Opin. Genet. Dev.* 9, 140–
147.

36 J. DiRenzo, M. Soderstrom, R.
Kurokawa, M.H. Ogliastro, M.
Ricote, S. Ingrey, A. Horlein, M.G.
Rosenfeld, C.K. Glass (1997) *Mol.
Cell. Biol.* 17, 2166–2176.

37 I.A. Jpenberg, E. Jeannin, W. Wahli,
B. Desvergne (1997) *J. Biol. Chem.*
272, 20108–20117.

38 A. Berger, D.M. Mutch, J.B.
German, M.A. Roberts (2002)
Genome Biol. 3, PREPRINT0004.

39 M. Takahashi, N. Tsuboyama-
Kasaoka, T. Nakatani, M. Ishii, S.
Tsutsumi, H. Aburatani, O. Ezaki
(2002) *Am. J. Physiol. Gastrointest.
Liver Physiol.* 282, G338–348.

40 L. Berthou, R. Saladin, P. Yaqoob,
D. Branellec, P. Caalder, J.C.
Fruchart, P. Denefle, J. Auwerx,
B. Staels (1995) *Eur. J. Biochem.* 232,
179–187.

41 E. Duplus, M. Glorian, C. Forest
(2000) *J. Biol. Chem.* 275, 30749–
30752.

42 C. Le May, M. Caüzac, C. Diradourian, D. Perdereau, J. Girard, A.F.
Burnol, J.P. Pégorier (2005) *J. Nutr.*
135, 2313–2319.

43 B. Ren, A.P. Thelen, J.M. Peters, F.J.
Gonzalez, D.B. Jump (1997) *J. Biol.
Chem.* 272, 26827–26832.

44 D.A. Pan, M.K. Mater, A.P. Thelen,
J.M. Peters, F.J. Gonzalez, D.B.
Jump (2000) *J. Lipid Res.* 41, 742–751.

45 H.P. Cho, M.T. Nakamura, S.D.
Clarke (1999) *J. Biol. Chem.* 274,
37335–37339.

46 H.P. Cho, M.T. Nakamura, S.D.
Clarke (1999) *J. Biol. Chem.* 274,
471–477.

47 D.J. Peet, B.A. Janowski, D.J.
Mangelsdorf (1998) *Curr. Opin.
Genet. Dev.* 8, 571–575.

48 J. Ou, H. Tu, B. Shan, A. Luk, R.A.
DeBose-Boyd, Y. Bashmakov, J.L.
Goldstein, M.S. Brown (2001) *Proc.
Natl Acad. Sci. USA* 98, 6027–6032.

49 A. Pawar, J. Xu, E. Jerks, D.J. Mangelsdorf, D.B. Jump (**2002**) *J. Biol. Chem.* 277, 39243–39250.

50 A. Pawar, D. Botolin, D.J. Mangelsdorf, D.B. Jump (**2003**) *J. Biol. Chem.* 278, 40736–40743.

51 A.C. Li, C.K. Glass (**2004**) *J. Lipid Res.* 45, 2161–2173.

52 T.T. Lu, J.J. Repa, D.J. Mangelsdorf (**2001**) *J. Biol. Chem.* 276, 37735–37738.

53 S.K. Cheema, L.B. Agellon (**2000**) *J. Biol. Chem.* 275, 12530–12536.

54 D.D. Patel, B.L. Knight, A.K. Soutar, G.F. Gibbons, D.P. Wade (**2000**) *Biochem. J.* 351 (Pt 3), 747–753.

55 M. Marrapodi, J.Y. Chiang (**2000**) *J. Lipid Res.* 41, 514–520.

56 G.P. Hayhurst, Y.H. Lee, G. Lambert, J.M. Ward, F.J. Gonzalez (**2001**) *Mol. Cell. Biol.* 21, 1393–1403.

57 F.M. Sladek, W. Zhong, E. Lai, J.E. Darnell, Jr (**1990**) *Genes Dev.* 4, 2353–2365.

58 L. Miquerol, S. Lopez, N. Cartier, M. Tulliez, M. Raymondjean, A. Kahn (**1994**) *J. Biol. Chem.* 269, 8944–8951.

59 N. Plengvidhya, A. Antonellis, L.T. Wogan, A. Poleev, M. Borgschulze, J.H. Warram, G.U. Ryffel, A.S. Krolewski, A. Doria (**1999**) *Diabetes* 48, 2099–2102.

60 R. Hertz, J. Magenheim, I. Berman, J. Bar-Tana (**1998**) *Nature* 392, 512–515.

61 A.A. Bogan, Q. Dallas-Yang, M.D. Ruse, Jr., Y. Maeda, G. Jiang, L. Nepomuceno, T.S. Scanlan, F.E. Cohen, F.M. Sladek (**2000**) *J. Mol. Biol.* 302, 831–851.

62 S. Dhe-Paganon, K. Duda, M. Iwamoto, Y.I. Chi, S.E. Shoelson (**2002**) *J. Biol. Chem.* 277, 37973–37976.

63 G.B. Wisely, A.B. Miller, R.G. Davis, A.D. Thornquest, Jr., R. Johnson, T. Spitzer, A. Sefler, B. Shearer, J.T. Moore, T.M. Willson, S.P. Williams (**2002**) *Structure (Camb.)* 10, 1225–1234.

64 C. Stehlin, J.M. Wurtz, A. Steinmetz, E. Greiner, R. Schule, D. Moras, J.P. Renaud (**2001**) *EMBO J.* 20, 5822–5831.

65 W. Bourguet, V. Andry, C. Iltis, B. Klaholz, N. Potier, A. Van Dorsselaer, P. Chambon, H. Gronemeyer, D. Moras (**2000**) *Protein Expr. Purif.* 19, 284–288.

66 R. Hertz, M. Seckbach, M.M. Zakin, J. Bar-Tana (**1996**) *J. Biol. Chem.* 271, 218–224.

67 F. Rajas, A. Gautier, I. Bady, S. Montano, G. Mithieux (**2002**) *J. Biol. Chem.* 277, 15736–15744.

68 R. Hertz, V. Sheena, B. Kalderon, I. Berman, J. Bar-Tana (**2001**) *Biochem. Pharmacol.* 61, 1057–1062.

69 J. Lengqvist, A. Mata De Urquiza, A.C. Bergman, T.M. Willson, J. Sjovall, T. Perlmann, W.J. Griffiths (**2004**) *Mol. Cell. Proteomics* 3, 692–703.

70 A.M. de Urquiza, S. Liu, M. Sjoberg, R.H. Zetterstrom, W. Griffiths, J. Sjovall, T. Perlmann (**2000**) *Science* 290, 2140–2144.

71 Y.Y. Fan, T.E. Spencer, N. Wang, M.P. Moyer, R.S. Chapkin (**2003**) *Carcinogenesis* 24, 1541–1548.

72 J.T. Goldstein, A. Dobrzyn, M. Clagett-Dame, J.W. Pike, H.F. DeLuca (**2003**) *Arch. Biochem. Biophys.* 420, 185–193.

73 A. Zhao, J. Yu, J.L. Lew, L. Huang, S.D. Wright, J. Cui (**2004**) *DNA Cell Biol.* 23, 519–526.

74 T. Hirose, R.J. Smith, A.M. Jetten (**1994**) *Biochem. Biophys. Res. Commun.* 205, 1976–1983.

75 F. Boukhtouche, J. Mariani, A. Tedgui (**2004**) *Arterioscler. Thromb. Vasc. Biol.* 24, 637–643.

76 E. Raspe, H. Duez, P. Gervois, C. Fievet, J.C. Fruchart, S. Besnard, J. Mariani, A. Tedgui, B. Staels (**2001**) *J. Biol. Chem.* 276, 2865–2871.

77 N. Vu-Dac, P. Gervois, T. Grotzinger, P. De Vos, K. Schoonjans, J.C. Fruchart, J. Auwerx, J. Mariani, A. Tedgui, B. Staels (**1997**) *J. Biol. Chem.* 272, 22401–22404.

78 P. Lau, S.J. Nixon, R.G. Parton, G.E. Muscat (**2004**) *J. Biol. Chem.* 279, 36828–36840.

79 T.F. Osborne (**2000**) *J. Biol. Chem.* 275, 32379–32382.

80 J.D. Horton, J.L. Goldstein, M.S. Brown (**2002**) *J. Clin. Invest.* 109, 1125–1131.

81 T. Yang, P.J. Espenshade, M.E. Wright, D. Yabe, Y. Gong, R. Aebersold, J.L. Goldstein, M.S. Brown (**2002**) *Cell* 110, 489–500.

82 D. Yabe, M.S. Brown, J.L. Goldstein (**2002**) *Proc. Natl Acad. Sci. USA* 99, 12753–12758.

83 D. Yabe, R. Komuro, G. Liang, J.L. Goldstein, M.S. Brown (**2003**) *Proc. Natl Acad. Sci. USA* 100, 3155–3160.

84 F. Foufelle, P. Ferre (**2002**) *Biochem. J.* 366, 377–391.

85 N. Yahagi, H. Shimano, A.H. Hasty, M. Amemiya-Kudo, H. Okazaki, Y. Tamura, Y. Iizuka, F. Shionoiri, K. Ohashi, J. Osuga, K. Harada, T. Gotoda, R. Nagai, S. Ishibashi, N. Yamada (**1999**) *J. Biol. Chem.* 274, 35840–35844.

86 M.K. Mater, A.P. Thelen, D. Pan, D.B. Jump (**1999**) *J. Biol. Chem.* 274, 32725–32732.

87 S.D. Clarke, M.K. Armstrong, D.B. Jump (**1990**) *J. Nutr.* 120, 225–231.

88 A. Katsurada, N. Iritani, H. Fukuda, Y. Matsumara, N. Nishimoto, T. Noguchi, T. Tanaka (**1990**) *Eur. J. Biochem.* 190, 427–433.

89 J. Lin, R. Yang, P.T. Tarr, P.H. Wu, C. Handschin, S. Li, W. Yang, L. Pei, M. Uldry, P. Tontonoz, C.B. Newgard, B.M. Spiegelman (**2005**) *Cell* 120, 261–273.

90 T.S. Worgall, R.A. Johnson, T. Seo, H. Gierens, R.J. Deckelbaum (**2002**) *J. Biol. Chem.* 277, 3878–3885.

91 J. Xu, M.T. Nakamura, H.P. Cho, S.D. Clarke (**1999**) *J. Biol. Chem.* 274, 23577–23583.

92 J. Xu, M. Teran-Garcia, J.H. Park, M.T. Nakamura, S.D. Clarke (**2001**) *J. Biol. Chem.* 276, 9800–9807.

93 J.R. Schultz, H. Tu, A. Luk, J.J. Repa, J.C. Medina, L. Li, S. Schwendner, S. Wang, M. Thoolen, D.J. Mangelsdorf, K.D. Lustig, B. Shan (**2000**) *Genes Dev.* 14, 2831–2838.

94 R.A. DeBose-Boyd, J. Ou, J.L. Goldstein, M.S. Brown (**2001**) *Proc. Natl Acad. Sci. USA* 98, 1477–1482.

95 X. Deng, L.M. Cagen, H.G. Wilcox, E.A. Park, R. Raghow, M.B. Elam (**2002**) *Biochem. Biophys. Res. Commun.* 290, 256–262.

96 S. Moriizumi, L. Gourdon, A.M. Lefrancois-Martinez, A. Kahn, M. Raymondjean (**1998**) *Gene Expr.* 7, 103–113.

97 A.K. Stoeckman, H.C. Towle (**2002**) *J. Biol. Chem.* 277, 27029–27035.

98 M.J. Diaz-Guerra, M.O. Bergot, A. Martinez, M.H. Cuif, A. Kahn, M. Raymondjean (**1993**) *Mol. Cell. Biol.* 13, 7725–7733.

99 M. Liimatta, H.C. Towle, S. Clarke, D.B. Jump (**1994**) *Mol. Endocrinol.* 8, 1147–1153.

100 K. Iizuka, R.K. Bruick, G. Liang, J.D. Horton, K. Uyeda (**2004**) *Proc. Natl Acad. Sci. USA* 101, 7281–7286.

101 R. Dentin, J.P. Pegorier, F. Benhamed, F. Foufelle, P. Ferre, V. Fauveau, M.A. Magnuson, J. Girard, C. Postic (**2004**) *J. Biol. Chem.* 279, 20314–20326.

102 H. Yamashita, M. Takenoshita, M. Sakurai, R.K. Bruick, W.J. Henzel, W. Shillinglaw, D. Arnot, K. Uyeda (**2001**) *Proc. Natl Acad. Sci. USA* 98, 9116–9121.

103 A.K. Stoeckman, L. Ma, H.C. Towle (**2004**) *J. Biol. Chem.* 279, 15662–15669.

104 S. Ishii, K. Iizuka, B.C. Miller, K. Uyeda (**2004**) *Proc. Natl Acad. Sci. USA* 101, 15597–15602.

105 T. Kawaguchi, K. Osatomi, H. Yamashita, T. Kabashima, K. Uyeda (**2002**) *J. Biol. Chem.* 277, 3829–3835.

106 Z. He, T. Jiang, Z. Wang, M. Levi, J. Li (**2004**) *Am. J. Physiol. Endocrinol. Metab.* 287, E424–430.

107 R. Dentin, F. Benhamed, J.P. Pégorier, F. Foufelle, B. Viollet, S. Vaulont, J. Girard, C. Postic (**2005**) *J. Clin. Invest.* 115, 2848–2854.

108 K. Uyeda, H. Yamashita, T. Kawaguchi (**2002**) *Biochem. Pharmacol.* 63, 2075–2080.

109 T. Kawaguchi, M. Takenoshita, T. Kabashima, K. Uyeda (**2001**) *Proc. Natl Acad. Sci. USA* 98, 13710–13715.

110 T. Kabashima, T. Kawaguchi, B.E. Wadzinski, K. Uyeda (**2003**) *Proc. Natl Acad. Sci. USA* 100, 5107–5112.

111 J.E. Tomlinson, R. Nakayama, D. Holten (**1988**) *J. Nutr.* 118, 408–415.

112 L.P. Stabile, D.L. Hodge, S.A. Klautky, L.M. Salati (**1996**) *Arch. Biochem. Biophys.* 332, 269–279.

113 L.P. Stabile, S.A. Klautky, S.M. Minor, L.M. Salati (**1998**) *J. Lipid Res.* 39, 1951–1963.

114 L.M. Salati, S.D. Clarke (**1986**) *Arch. Biochem. Biophys.* 246, 82–89.

115 G. Suchankova, M. Tekle, A.K. Saha, N.B. Ruderman, S.D. Clarke, T.W. Gettys (**2005**) *Biochem. Biophys. Res. Commun.* 326, 851–858.

116 Z. Gao, X. Zhang, A. Zuberi, D. Hwang, M.J. Quon, M. Lefevre, J. Ye (**2004**) *Mol. Endocrinol.* 18, 2024–2034.

117 Y. Le Marchand-Brustel, P. Gual, T. Gremeaux, T. Gonzalez, R. Barres, J.F. Tanti (**2003**) *Biochem. Soc. Trans.* 31, 1152–1156.

118 V.T. Samuel, Z.X. Liu, X. Qu, B.D. Elder, S. Bilz, D. Befroy, A.J. Romanelli, G.I. Shulman (**2004**) *J. Biol. Chem.* 279, 32345–32353.

119 M.C. Arkan, A.L. Hevener, F.R. Greten, S. Maeda, Z.W. Li, J.M. Long, A. Wynshaw-Boris, G. Poli, J. Olefsky, M. Karin (**2005**) *Nat. Med.* 11, 191–198.

120 D. Cai, M. Yuan, D.F. Frantz, P.A. Melendez, L. Hansen, J. Lee, S.E. Shoelson (**2005**) *Nat. Med.* 11, 183–190.

121 N.H. Haunerland, F. Spener (**2004**) *Prog. Lipid Res.* 43, 328–349.

122 C. Wolfrum, C.M. Borrmann, T. Borchers, F. Spener (**2001**) *Proc. Natl Acad. Sci. USA* 98, 2323–2328.

123 N.S. Tan, N.S. Shaw, N. Vinckenbosch, P. Liu, R. Yasmin, B. Desvergne, W. Wahli, N. Noy (**2002**) *Mol. Cell. Biol.* 22, 5114–5127.

5
Cellular Adaptation to Amino Acid Availability: Mechanisms Involved in the Regulation of Gene Expression

Alain Bruhat, Anne-Catherine Maurin, Céline Jousse,
Yoan Cherasse, and Pierre Fafournoux

5.1
Introduction

Regulation of metabolism is achieved by coordinated actions between cells and tissues and also by mechanisms operating at the cellular level. These mechanisms involve the conditional regulation of specific genes in the presence or absence of appropriate nutrients. In multicellular organisms, the control of gene expression involves complex interactions of hormonal, neuronal, and nutritional factors. Although not as widely appreciated, nutritional signals play an important role in controlling gene expression in mammals. It has been shown that major (carbohydrates, fatty acids, sterols) and minor (minerals, vitamins) dietary constituents participate in the regulation of gene expression [1–6]. However, the mechanisms involved in the amino acid control of gene expression have just begun to be understood in mammalian cells [7–9]. This review summarizes recent work on the effect of amino acid availability in the regulation of biological functions. On the basis of the physiological concepts of amino acid homeostasis, we will discuss specific examples of the role of amino acids in the regulation of physiological functions, particularly focusing on the mechanisms involved in the amino acid regulation of gene expression and protein turnover.

5.2
Regulation of Amino Acid Metabolism and Homeostasis in the Whole Animal

In contrast to other macronutrients (lipids or sugars) amino acids exhibit two important characteristics. First, in healthy adult humans, nine amino acids (valine, isoleucine, leucine, lysine, methionine, phenylalanine, threonine, histidine, and tryptophan) are indispensable (or essential). In addition, under a particular set of conditions certain dispensable (non-essential) amino acids may become indispensable. These amino acids are called "conditionally indispensable." For example, enough arginine is synthesized by the urea cycle to meet the needs of an adult but

Nutritional Genomics. Edited by Regina Brigelius-Flohé and Hans-Georg Joost
Copyright © 2006 WILEY-VCH Verlag GmbH & Co. KGaA, Weinheim
ISBN: 3-527-31294-3

not those of a growing child. Secondly, there are no large stores of amino acids. Consequently, when necessary, an organism has to hydrolyze muscle protein to produce free amino acids. This loss of protein will be at the expense of essential elements. Therefore, complex mechanisms that take into account these amino acid characteristics are needed to maintain the free amino acid pools.

5.2.1
Free Amino Acids Pools

The size of the pool of each amino acid is the result of a balance between input and removal. The metabolic outlets for amino acids are protein synthesis and amino acid degradation whereas the inputs are *de novo* synthesis (for non-essential amino acids), protein breakdown, and dietary supply. Changes in the rates of these systems lead to an adjustment in nitrogen balance. For example, a protein-containing meal given to an animal or a human subject has been reported to increase both the nitrogen balance and the level of amino acids in the plasma. In particular, the concentration of leucine and certain other amino acids approximately doubles in peripheral blood after a protein-rich meal [10] and reaches much higher concentrations within the portal vein [11]. It is now well established that the effect of a protein-rich meal on protein turnover is due to postprandial increases in the concentrations of circulating amino acids [12, 13].

Another example of an adjustment of the nitrogen balance is the adaptation to an amino acid-deficient diet. A dramatic diminution of the plasma concentrations of certain essential amino acids has been shown to occur following a dietary imbalance, a deficiency of any one of the essential amino acids or a deficient intake of protein [14–17]. Long-term feeding with such diets can lead to a negative nitrogen balance and clinical symptoms.

The examples cited above show that amino acid metabolism can be affected by various nutritional and/or pathological situations, with two major consequences: a large variation in blood amino acid concentrations and a negative nitrogen balance. In these situations, individuals have to adjust several of their physiological functions involved in the defense/adaptation to amino acid limitation by regulating numerous genes. In the next section the specific role of amino acids in the adaptation to amino acid-deficient diets will be considered.

5.3
Specific Examples of the Role of Amino Acids in the Adaptation to Protein Deficiency

5.3.1
Protein Undernutrition

Prolonged feeding on a low protein diet causes a fall in the plasma level of most essential amino acids. Protein undernutrition has its most devastating consequences during growth. For example, leucine and methionine concentrations can be reduced from about 100–150 μmol/l and 18–30 μmol/l to 20 μmol/l and 5 μmol/l, respectively, in plasma of children affected by kwashiorkor [14, 17]. It follows that individuals have to adjust several physiological functions in order to adapt to this amino acid deficiency. One of the main consequences of feeding a low protein diet is the dramatic inhibition of growth. Growth is controlled by a complex interaction of genetic, hormonal and nutritional factors. A large part of this control is due to growth hormone (GH) and insulin-like growth factors (IGFs). The biological activities of the IGFs are modulated by the IGF-binding proteins (IGFBPs) that specifically bind IGF-I and IGF-II [18, 19]. Strauss *et al.* [20] demonstrated that a dramatic overexpression of IGFBP-1 was responsible for growth inhibition in response to prolonged feeding on a low-protein diet. Known regulators of IGFBP-1 expression are GH, insulin, and glucose. However, the high IGFBP-1 levels found in response to a protein-deficient diet cannot be explained by these factors. It has been demonstrated that a fall in the amino acid concentration was directly responsible for IGFBP-1 induction [20, 21]. Therefore, amino acid limitation, as occurring during dietary protein deficiency, participates in the downregulation of growth through the induction of IGFBP-1.

5.3.2
Imbalanced Diet

In the event of a deficiency in one of the indispensable amino acids, the remaining amino acids are catabolized and body proteins are broken down to provide the limiting amino acid [22]. It follows that mammals (with the exception of the ruminants) need mechanisms that provide for selection of a balanced diet. The capacity to distinguish balance from imbalance among the amino acids in the diet and to select for the growth-limiting essential amino acid provides an adaptive advantage to animals.

After eating an amino acid-imbalanced diet, animals first recognize the amino acid deficiency and then develop a conditioned taste aversion. Recognition and anorexia resulting from an amino acid-imbalanced diet takes place very rapidly [23, 24]. The mechanisms that underlie the recognition of protein quality must act by the way of the amino acids resulting from intestinal digestion of proteins. It has been observed that a marked decrease in the blood concentration of the limiting amino acid can become apparent as early as few minutes after feeding an imbalanced diet. The anorectic response is correlated with a decreased concentration of the limiting amino acid in the plasma. Several lines of evidence suggest that the fall in the limiting amino acid concentration is detected in the brain. Gietzen [24, 25] reviews the evidences showing that a specific brain area, the anterior piriform cortex (APC), can sense the amino acid concentration. This recognition phase is associated with localized decreases in the concentration of the limiting amino acid and with important changes in protein synthesis rate and gene expression. Subsequent to recognition of the deficiency the second step, development of anorexia, involves another part of the brain.

These two examples demonstrate that a variation in blood amino acid concentration can activate, in target cells, several control processes that can specifically regulate the expression of target genes. Although the role of the amino acids considered to be regulators of genes expression is understood in only a few nutritional situations, recent progress has been made in understanding the mechanisms by which amino acid limitation controls the expression of several genes.

5.4
Amino Acid Control of Gene Expression

Although the molecular mechanisms involved in the control of gene expression by amino acid availability have just begun to be investigated in mammals, they have been extensively studied in yeast. After a summary of these processes, we will focus on the control of gene expression in mammalian cells.

5.4.1
Amino Acid Control of Gene Expression in Yeast

In yeast, several amino acid-sensing systems have been described (Fig. 5.1).

5.4.1.1   The Specific Control Process

It is well documented that numerous operons are regulated by the specific end products of the corresponding enzymes [26]. A small effector molecule can induce the transition of transcriptional activators from their inactive to their active form. For example, leucine biosynthesis is controlled by the transcriptional activator Leu3p in response to leucine availability. Leu3p is activated by the levels of the metabolic intermediate α-isopropylmalate, which serves as a sensor of leucine availability [27]. This type of regulation has also been described for the control of amino acid catabolism (proline for example) [28].

5.4.1.2   The General Control Process

1. *The GCN2 pathway*: In addition to specific control, yeast uses a general control process whereby a subset of genes is coordinately induced by starvation of the cell for one single amino acid. Free tRNAs accumulate and thus stimulate the activity of the protein kinase GCN2 (general control non-repressible-2) which phosphorylates the α-subunit of eIF-2, which in turn impairs the synthesis of the 43S preinitiation complex (Met-tRNA, GTP, eIF-2). Despite the strong inhibition of protein synthesis, the transcription factor GCN4 is translationally up-regulated. This control is due to the particular structure of the 5' untranslated region (UTR) of the *GCN4* mRNA [29, 30]. As a result, GCN4 induces more than 30 different genes involved in several different biosynthetic pathways.
2. *The TOR pathway*: The TOR (target of rapamycin) pathway is regulated by amino acid availability and is involved in the regulation of several cellular processes such as translation, transcription, and protein degradation [31] (see also

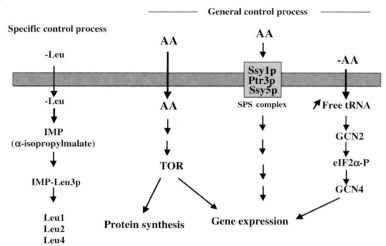

Fig. 5.1. Amino acid regulation of gene expression in yeast. Specific control: In response to leucine starvation, α-isopropylmalate, an intermediate molecule in the leucine biosynthesis pathway, accumulates in the cell and binds the transcription factor Leu3p. Leu3p then activates transcription of genes involved in the leucine biosynthesis. General control: (1) The GCN2 pathway: accumulation of free tRNA that occurs in response to starvation in any amino acid turns on a signaling pathway that leads to an increase of GCN4 translation. GCN4 is a transcription factor able to upregulate the expression of several genes involved in the amino acid synthesis. (2) The TOR pathway senses amino acid availability by mechanisms that remain to be identified and regulates transcription and translation of proteins involved in the control of cell growth. (3) The Ssy1p pathway: A protein complex located at the yeast membrane is able to detect the external amino acid concentration and activates an unknown pathway that regulates the expression of several genes.

chapter 10). The molecular mechanisms involved in the amino acid control of TOR activity remain to be identified.

3. *The SPS complex*: Recent advances in our understanding of nutrient sensing indicate that yeast cells possess an amino acid-sensing system localized at the plasma membrane that transduces information regarding the presence of extracellular amino acids. The primary amino acid sensor is a multimeric complex (SPS complex) that contains three proteins Ssy1p, Ptr3p, and Ssy5p. One of its components, Ssy1p, closely resembles amino acid permeases, a family of proteins that normally catalyze the transport of amino acids into the cell [32–34]. In response to a change in amino acid availability, a complex network of regulatory processes is activated by Ssy1p to modify the expression of target genes. The SPS complex is required for induction of a set of target genes (*BAP3, TAT2, CHA1*, etc.) by amino acids and is also required for the amino acid repression of another set of target genes (*DAL4, MET3, MMP1*, etc.) all encoding proteins involved in amino acid transport and metabolism [35].

5.4.2
Amino Acid Control of Gene Expression in Mammalian Cells

5.4.2.1 Genes Upregulated by Amino Acid Overload

Genes that are specifically upregulated in response to supraphysiological concentrations of amino acids have been described. For example, a high concentration of L-tryptophan enhances the expression of collagenase and of tissue inhibitors of metalloproteinase. In rat hepatocytes, Na^+-cotransported amino acids like glutamine, alanine, or proline stimulate acetyl-CoA carboxylase, glycogen synthetase, and arginino succinate synthetase activity. It was demonstrated that the swelling resulting from the addition of amino acids could be involved in the regulation of gene expression [36, 37], however, the molecular mechanisms involved in these processes are poorly understood.

5.4.2.2 Genes Upregulated by Amino Acid Starvation

In mammalian cells, specific mRNAs that are induced following amino acid deprivation have been reported [38]. Most of the molecular mechanisms involved in the amino acid regulation of gene expression have been obtained by studying the upregulation of C/EBP homologous protein (CHOP), asparagine synthetase (ASNS), and the cationic amino acid transporter (Cat1) genes.

5.5
Molecular Mechanisms Involved in the Mammalian Regulation of Gene Expression by Amino Acid Limitation

5.5.1
Post-transcriptional Regulation of Mammalian Genes Expression by Amino Acid Availability

It has been previously shown that amino acid depletion increased the stability of ASNS [39, 40] and CHOP mRNA [41]. Hatzoglou and collaborators studied the regulation of the cationic amino acid transporter (Cat1) expression by amino acid availability [42, 43]. They demonstrated that amino acid depletion initiates molecular events that lead to both an increase of the Cat1 mRNA stability and a specific activation of its translation. In a first paper, they showed that the transcription rate of the Cat1 gene remained unchanged during amino acid starvation whereas the mRNA level was dramatically increased [42]. They concluded that the Cat1 mRNA was stabilized by cis-acting RNA sequences within the 3′ UTR.

In a second paper, they have shown the presence of an internal ribosome entry site (IRES) located within the 5′ UTR of the Cat1 mRNA. This IRES is involved in the amino acid control of translation of the Cat1 transcript [43]. Under conditions of amino acid starvation, translation from this IRES is stimulated, whereas the cap-dependent protein synthesis is decreased. Another example of translation induced by amino acid starvation was reported for the branched-chain α-ketoacid dehydrogenase kinase, but the mechanism of translational control was not studied [44]. This mechanism of compensatory response allows translation of major proteins despite the inhibition of the "classic" translational apparatus.

5.5.2
Transcriptional Activation of Mammalian Genes by Amino Acid Starvation

It has been established that the increase in *CHOP* or *ASNS* mRNA following amino acid starvation is mainly due to increased transcription [41, 45]. By first identifying the genomic *cis*-elements and then the corresponding transcription factors responsible for regulation of these specific target genes, it is anticipated that one can progress backwards up the signal transduction pathway to understand the individual steps required.

5.5.2.1 Regulation of the Human *CHOP* Gene by Amino Acid Starvation
CHOP encodes a ubiquitous transcription factor that heterodimerizes avidly with the other members of the C/EBP [46] and jun/fos [47] families. The *CHOP* gene is tightly regulated by a wide variety of stresses in mammalian cells [48–50]. Leucine limitation in human cell lines leads to induction of *CHOP* mRNA and protein in a dose-dependent manner [41].

We have identified in the *CHOP* promoter a *cis*-positive element located between −313 and −295 that is essential for amino acid regulation of the gene transcription [51] (Fig. 5.2). This short sequence can regulate a basal promoter in response to starvation of several individual amino acids and then can be called amino acid response element (AARE). The sequence of the *CHOP* AARE region shows some homology with the specific binding sites of the C/EBP and ATF/CREB transcription factor families. Using gel-shift experiments and chromatin immunoprecipitation, we have shown that ATF2, ATF3, ATF4, and CCAAT/enhancer-binding protein beta (C/EBPβ) that belong to the ATF or C/EBP family have the ability to bind to the *CHOP* AARE. Among these factors ATF2 and ATF4 are involved in the amino acid control of *CHOP* expression: when knockout cell line for these two proteins were tested, amino acid regulation of *CHOP* expression was abolished [51, 52]. This work was enlarged to the regulation of other amino acid-regulated genes and confirms that ATF4 and ATF2 are key components of the amino acid control of gene expression [52].

5.5.2.2 Regulation of ASNS by Amino Acid Availability
ASNS is expressed in most mammalian cells as a housekeeping enzyme responsible for the biosynthesis of asparagine and glutamate from aspartate and glutamine [53]. The level of *ASNS* mRNA increases not only in response to asparagine starvation but also after leucine, isoleucine, or glutamine deprivation [40, 45, 54]. Kilberg's group has analyzed the regulation of the *ASNS* promoter by amino acid availability. They have characterized (Fig. 5.2) a nutrient-sensing regulatory unit (NSRU), which includes two *cis*-acting elements termed nutrient sensing response

(a)

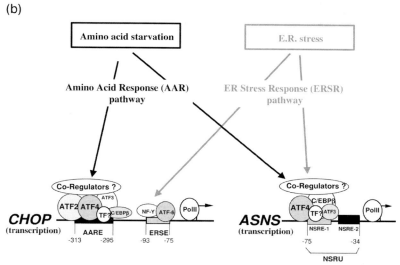

(b)

Fig. 5.2. *Cis*-acting elements required for induction of *CHOP* and *ASNS* genes following amino acid starvation or endoplasmic reticulum (ER) stress. (a) Sequence comparison of the *CHOP* amino acid response element (AARE) (−313 to −295) with the *ASNS* NSRE-1 (−57 to −75). Identical nucleotides are boxed in gray. The minimum *CHOP* AARE core sequence is underlined. (b) The *cis*-acting elements required for induction of the *CHOP* gene following amino acid starvation or the endoplasmic reticulum stress are located in sequences separated by several hundred base pairs. Transcriptional control of *CHOP* by ER stress involves the binding of ATF6 in the presence of NF-Y to the *cis*-acting

ER stress response element (ERSE) located between nucleotides −75 and −93. The amino acid response element used by the amino acid response pathway is contained within nucleotides −313 to −295. The *ASNS* NSRU includes NSRE-1 and NSRE-2 sequences (−75 to −34), which are required for activation of the gene following either amino acid limitation or ER stress. The transcription factors ATF2, ATF3, ATF4, and C/EBPβ could bind the *CHOP* AARE whereas C/EBPβ, ATF3 and ATF4 could bind NSRE-1. It is possible that other non-identified transcription factors (TF) or regulatory proteins (co-regulators) also bind the *CHOP* and *ASNS* transcriptional control elements used by the amino acid pathway.

elements (NSRE-1, NSRE-2) that are required to induce *ASNS* expression by either amino acid deprivation or the ER (endoplasmic reticulum) stress [55]. Although the NSRE-2-binding proteins are unknown, transient expression, gel-shift experiments and chromatin immunoprecipitation have demonstrated that activation of *ASNS* gene by either amino acid limitation or endoplasmic reticulum stress response (ERSR) involves ATF4 [56, 57], ATF3 [58], and CCAAT/C/EBPβ [59] action via the NSRE-1 site.

The comparison between *CHOP* and *ASNS* transcriptional control elements shows that *ASNS* NSRE-1 and *CHOP* AARE share nucleotide sequence and functional similarities (Fig. 5.2a). However, the *CHOP* AARE can function alone whereas *ASNS* NSRE-1 is functionally weak by itself and requires the presence of NSRE-2 [60]. The *ASNS* NSRE-2 has two properties: (1) it amplifies the NSRE-1 activity in response to amino acid starvation and (2) it confers a response to ER stress. For example, when cloned downstream of the *CHOP* AARE, it can confer ER stress responsiveness to the *CHOP* AARE.

5.5.3
Amino Acid Signaling Pathway

The eIF-2α kinase GCN2 is activated under conditions of nutrient deprivation. The role of GCN2 in response to amino acid starvation has been characterized extensively in yeast (see above). Another signaling event altered in response to nutrients involves the mammalian target of rapamycin (mTOR) protein kinase. Signaling downstream of mTOR is implicated in many aspects of cell growth including cell cycle control and ribosome biogenesis. mTOR activates both the ribosomal p70 S6 kinase (S6K1) and the mRNA cap-binding protein inhibitory protein 4E-BP1, and its pharmacological inhibition causes G1-phase cell cycle arrest [61, 62]. Recently, there have been reports of potential cross-talk between GCN2 and the TOR signaling pathway in yeast [63, 64]. Specifically, rapamycin releases TOR-directed phosphorylation of yeast GCN2, contributing to enhanced eIF-2α kinase activity. Although this finding has not yet been extended to mammals, it suggests that events downstream of mTOR, namely the phosphorylation of 4E-BP1 and S6K1, may be coordinated with eIF-2α phosphorylation via GCN2.

In mammalian cells, it appears that more than one amino acid signaling pathway independent of the ER stress pathway exists [65, 66]. However, the individual steps required for these pathways are not well understood.

5.5.3.1 ATF4 and the Amino Acid Signaling Pathways
Ron's group has revealed a signaling pathway for the regulation of gene expression in mammals that is homologous to the well-characterized yeast general control response to amino acid deprivation [67]. Its components include the mammalian homolog of the GCN2 kinase, the initiation factor eIF-2α and ATF4 (Fig. 5.3). Like the GCN4 transcript, the ATF4 mRNA contains an upstream open reading frame (uORF) in its 5′ UTR that allows translation when cap-dependent translation is inhibited. The authors showed that GCN2 activation, phosphorylation of eIF-2α, and translational activation of ATF4 are necessary for the induction of *CHOP* expression in response to leucine starvation. These data are in good agreement with the analysis of the *CHOP* and *ASNS* promoter, showing that ATF4 can bind to the promoter sequences involved in the response to amino acid starvation (Fig. 5.2).

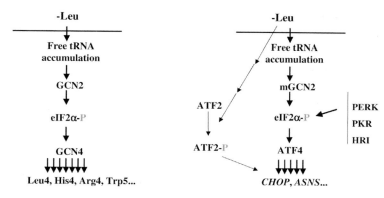

Yeast General control process

Amino Acid Regulation of gene expression in mammals

Fig. 5.3. Comparison between the general control process of gene expression by amino acid availability in yeast and the amino acid regulation of *CHOP* and *ASNS* expression in mammals. The mammalian pathway appears to be more complex than the yeast pathway;

for example, ATF2 needs to be phosphorylated to allow *CHOP* transcriptional induction in response to leucine starvation (–Leu). In addition, eIF-2α can be phosphorylated by four different kinases (GCN2, PERK, PKR, and HRI).

5.5.3.2 ATF2 and the Amino Acid Signaling Pathways

The transactivating capacity of ATF2 is activated via phosphorylation of the N-terminal residues Thr69, Thr71, and Ser90 [68, 69]. There are two lines of evidence suggesting that ATF2 phosphorylation belongs to the amino acid response pathway leading to the transcriptional activation of *CHOP* by amino acids: (i) leucine starvation induces ATF2 phosphorylation in human cell lines [52] and (ii) an ATF2-dominant negative mutant [70] in which the three residues cannot be phosphorylated inhibits the *CHOP* promoter activity enhanced by leucine starvation [51]. These data suggest that a specific amino acid-regulated pathway that leads to the transcriptional activation of *CHOP* may involve a phosphorylation of prebound ATF2 rather than an increase in ATF2 binding. However, the identity of the kinase(s) involved in ATF2 phosphorylation by amino acid starvation remain to be discovered (Fig. 5.3).

It appears that at least two different pathways that lead to ATF2 phosphorylation and to ATF4 expression are necessary to induce *CHOP* expression in response to one stimulus (amino acid starvation). In addition, ATF4 and ATF2 belong to the b-ZIP transcription factor family. These proteins have the ability to interact with several transcription factors to bind the target DNA sequence. In the case of amino acid regulation of *CHOP* expression, we have no evidence that ATF2 and ATF4 form a dimer that binds the AARE sequence, but they could be included in a larger regulatory protein complex. However, it has been shown that ATF2 is able to interact with the co-activator of transcription p300 which has histone acetyltransferase activity [71] and also with JDP2, a repressor that recruits a histone deacetylase complex [72], suggesting that this transcription factor could modulate transcription by interacting directly or indirectly with the chromatin structure.

5.6
Role of the GCN2/ATF4 Pathway in Nutrition: Amino Acid Deficiency Sensing by mGCN2 Triggers Food Aversion

Food intake results from a complex behavioral pattern in which innate factors play an important role, particularly in the case of omnivores. A remarkable example of an innate mechanism governing food choice is the fact that omnivorous animals will consume substantially less of an otherwise identical experimental meal lacking a single essential amino acid [24, 73]. Although it seems likely that the signaling pathway leading to this response comprises the sensing of amino acid variations, the basis for this innate aversive response is poorly understood. As described above, previous studies have implicated the anterior piriform cortex (APC) in controlling food intake according to amino acid levels [24, 74]. Furthermore, it has been reported recently that consumption of an amino acid-imbalanced diet rapidly elevates levels of phosphorylated eIF-2α in APC neurons [75].

The concentration of an essential amino acid in the blood decreases rapidly when the amino acid is missing in the diet. As a consequence, the protein kinase GCN2, which is ubiquitously expressed, could be activated in most tissues. Its only known substrate is the Ser51 of the α-subunit of eIF-2. Therefore GCN2 could be an important sensor of amino acid homeostasis inside cells and could activate downstream rectifying responses mediated by phosphorylated eIF-2α. The latter affects gene expression programs at the level of mRNA translation and transcription [67, 76].

Recent results [77] establishes that the aversive response of wild-type mice to a diet deficient in one essential amino acid is blunted in GCN2$^{-/-}$ mice whereas serum amino acid levels are decreased to similar levels by the imbalanced diet in both genotypes. These results indicate an altered response to amino acid deficiency in mice lacking GCN2 activity. Moreover, we confirmed the previously described increase in phosphorylated eIF-2α levels in the APC of wild-type animals after consumption of an imbalanced meal, and further showed that no such signal occurs in the GCN2$^{-/-}$. This observation indicates a role for GCN2 in mediating eIF-2α phosphorylation in the APC of mice fed an imbalanced diet. Using conditional GCN2-knockout mice, we further demonstrated that GCN2 ablation specifically in the brain also impairs the aversive response to an imbalanced diet. Thus, even if the consumption of an imbalanced meal also activates GCN2 and promotes eIF-2α in peripheral tissues, particularly in the liver, our observation implicates brain GCN2 signaling in initiating the aversive response. This example highlights the need to relate physiological observations to molecular events described *in vitro* to improve our knowledge of the physiological consequences of the nutritional status.

5.7
Conclusion

The idea that amino acids can regulate gene expression is now well established. Amino acids by themselves can play, in concert with hormones, an important role in the control of gene expression; however, the underlying processes have only begun to be discovered. Amino acid availability can modify the expression of target genes at the level of transcription, mRNA stability, and translation (for a recent review see Ref. [78]).

Defining the precise cascade of molecular events by which the cellular concentration of an individual amino acid regulates gene expression will be an important contribution to our understanding of metabolite control in mammalian cells. These studies will provide insight into the role of amino acids in the regulation of cellular functions such as cell division, protein synthesis or proteolysis.

References

1 TOWLE, H. C. (**1995**) *J Biol Chem* 270, 23235–23238.
2 PÉGORIER, J. P. (**1998**) *Curr Opin Clin Nutr Metab Care* 1, 329–334.
3 FOUFELLE, F., GIRARD, J., FERRE, P. (**1998**) *Curr Opin Clin Nutr Metab Care* 1, 323–328.
4 VAULONT, S., VASSEUR-COGNET, M., KAHN, A. (**2000**) *J Biol Chem* 275, 31555–31558.
5 DUPLUS, E., GLORIAN, M., FOREST, C. (**2000**) *J Biol Chem* 275, 30749–30752.
6 GRIMALDI, P. A. (**2001**) *Curr Opin Clin Nutr Metab Care* 4, 433–437.
7 KILBERG, M. S., HUTSON, R. G., LAINE, R. O. (**1994**) *FASEB J* 8, 13–19.
8 FAFOURNOUX, P., BRUHAT, A., JOUSSE, C. (**2000**) *Biochem J* 351, 1–12.
9 BRUHAT, A., FAFOURNOUX, P. (**2001**) *Curr Opin Clin Nutr Metab Care* 4, 439–443.
10 AOKI, T. T., BRENNAN, M. F., MULLER, W. A., SOELDNER, J. S., ALPERT, J. S., SALTZ, S. B., KAUFMANN, R. L., TAN, M. H., CAHILL, G. F., JR. (**1976**) *Am J Clin Nutr* 29, 340–350.
11 FAFOURNOUX, P., REMESY, C., DEMIGNE, C. (**1990**) *Am J Physiol* 259, E614–625.
12 YOSHIZAWA, F., ENDO, H., IDE, H., YAGASAKI, K., FUNABIKI, R. (**1995**) *Nutr Biochem* 6, 130–136.
13 SVANBERG, E., JEFFERSON, L. S., LUNDHOLM, K., KIMBALL, S. R. (**1997**) *Am J Physiol* 272, E841–847.
14 GRIMBLE, R. F., WHITEHEAD, R. G. (**1970**) *Lancet* 1, 918–920.
15 PENG, Y., HARPER, A. E. (**1970**) *J Nutr* 100, 429–437.
16 OZALP, I., YOUNG, V. R., NAGCHAUDHURI, J., TONTISIRIN, K., SCRIMSHAW, N. S. (**1972**) *J Nutr* 102, 1147–1158.
17 BAERTL, J. M., PLACKO, R. P., GRAHAM, G. G. (**1974**) *Am J Clin Nutr* 27, 733–742.
18 LEE, P. D., CONOVER, C. A., POWELL, D. R. (**1993**) *Proc Soc Exp Biol Med* 204, 4–29.
19 STRAUS, D. S. (**1994**) *FASEB J* 8, 6–12.
20 STRAUS, D. S., BURKE, E. J., MARTEN, N. W. (**1993**) *Endocrinology* 132, 1090–1100.
21 JOUSSE, C., BRUHAT, A., FERRARA, M., FAFOURNOUX, P. (**1998**) *Biochem J* 334, 147–153.
22 MUNRO, H. N. (**1976**) *Proc Nutr Soc* 35, 297–308.
23 ROGERS, Q. R., LEUNG, P. M. B. (**1977**) The control of food intake: when and how are amino acids involved? In: *The Chemical Senses and Nutrition* (KARE, M. R., MALLER, O., Eds), New York: Academic Press, p. 213.

24 GIETZEN, D. W. (**1993**) *J Nutr* 123, 610–625.

25 GIETZEN, D. W. (**2000**) Amino acid recognition in the central nervous system. In: *Neural and Metabolic Control of Macronutrient Intake* (BERTHOUD, H. R., SEELEY, R. J., Eds), New York: CRC Press, pp. 339–357.

26 STRUHL, K. (**1987**) *Cell* 49, 295–297.

27 SZE, J. Y., WOONTNER, M., JAEHNING, J. A., KOHLHAW, G. B. (**1992**) *Science* 258, 1143–1145.

28 MARCZAK, J. E., BRANDRISS, M. C. (**1991**) *Mol Cell Biol* 11, 2609–2619.

29 DEVER, T. E., FENG, L., WEK, R. C., CIGAN, A. M., DONAHUE, T. F., HINNEBUSCH, A. G. (**1992**) *Cell* 68, 585–596.

30 HINNEBUSCH, A. G. (**1993**) *Mol Microbiol* 10, 215–223.

31 ROHDE, J. R., CAMPBELL, S., ZURITA-MARTINEZ, S. A., CUTLER, N. S., ASHE, M., CARDENAS, M. E. (**2004**) *Mol Cell Biol* 24, 8332–8341.

32 FORSBERG, H., LJUNGDAHL, P. O. (**2001**) *Curr Genet* 40, 91–109.

33 FORSBERG, H., GILSTRING, C. F., ZARGARI, A., MARTINEZ, P., LJUNGDAHL, P. O. (**2001**) *Mol Microbiol* 42, 215–228.

34 IRAQUI, I., VISSERS, S., BERNARD, F., DE CRAENE, J. O., BOLES, E., URRESTARAZU, A., ANDRE, B. (**1999**) *Mol Cell Biol* 19, 989–1001.

35 KODAMA, Y., OMURA, F., TAKAHASHI, K., SHIRAHIGE, K., ASHIKARI, T. (**2002**) *Curr Genet* 41, 63–72.

36 WATFORD, M. (**1990**) *Trends Biochem Sci* 15, 329–330.

37 HAUSSINGER, D. (**1996**) *Biochem J* 313 (Pt 3), 697–710.

38 MARTEN, N. W., BURKE, E. J., HAYDEN, J. M., STRAUS, D. S. (**1994**) *FASEB J* 8, 538–544.

39 GUERRINI, L., GONG, S. S., MANGASARIAN, K., BASILICO, C. (**1993**) *Mol Cell Biol* 13, 3202–3212.

40 GONG, S. S., GUERRINI, L., BASILICO, C. (**1991**) *Mol Cell Biol* 11, 6059–6066.

41 BRUHAT, A., JOUSSE, C., WANG, X. Z., RON, D., FERRARA, M., FAFOURNOUX, P. (**1997**) *J Biol Chem* 272, 17588–17593.

42 AULAK, K. S., MISHRA, R., ZHOU, L., HYATT, S. L., DE JONGE, W., LAMERS, W., SNIDER, M., HATZOGLOU, M. (**1999**) *J Biol Chem* 274, 30424–30432.

43 FERNANDEZ, J., YAMAN, I. I., MISHRA, R., MERRICK, W. C., SNIDER, M. D., LAMERS, W. H., HATZOGLOU, M. (**2001**) *J Biol Chem* 376, 12285–12291.

44 DOERING, C. B., DANNER, D. J. (**2000**) *Am J Physiol Cell Physiol* 279, C1587–1594.

45 HUTSON, R. G., KILBERG, M. S. (**1994**) *Biochem J* 304, 745–750.

46 FAWCETT, T. W., EASTMAN, H. B., MARTINDALE, J. L., HOLBROOK, N. J. (**1996**) *J Biol Chem* 271, 14285–14289.

47 UBEDA, M., VALLEJO, M., HABENER, J. F. (**1999**) *Mol Cell Biol* 19, 7589–7599.

48 LUETHY, J. D., HOLBROOK, N. J. (**1992**) *Cancer Res* 52, 5–10.

49 SYLVESTER, S. L., AP RHYS, C. M., LUETHY-MARTINDALE, J. D., HOLBROOK, N. J. (**1994**) *J Biol Chem* 269, 20119–20125.

50 WANG, X. Z., LAWSON, B., BREWER, J. W., ZINSZNER, H., SANJAY, A., MI, L. J., BOORSTEIN, R., KREIBICH, G., HENDERSHOT, L. M., RON, D. (**1996**) *Mol Cell Biol* 16, 4273–4280.

51 BRUHAT, A., JOUSSE, C., CARRARO, V., REIMOLD, A. M., FERRARA, M., FAFOURNOUX, P. (**2000**) *Mol Cell Biol* 20, 7192–7204.

52 AVEROUS, J., BRUHAT, A., JOUSSE, C., CARRARO, V., THIEL, G., FAFOURNOUX, P. (**2004**) *J Biol Chem* 279, 5288–5297.

53 ANDRULIS, I. L., CHEN, J., RAY, P. N. (**1987**) *Mol Cell Biol* 7, 2435–2443.

54 HUTSON, R. G., KITOH, T., MORAGA AMADOR, D. A., COSIC, S., SCHUSTER, S. M., KILBERG, M. S. (**1997**) *Am J Physiol* 272, C1691–1699.

55 BARBOSA-TESSMANN, I. P., CHEN, C., ZHONG, C., SIU, F., SCHUSTER, S. M., NICK, H. S., KILBERG, M. S. (**2000**) *J Biol Chem* 275, 26976–26985.

56 SIU, F., BAIN, P. J., LEBLANC-CHAFFIN, R., CHEN, H., KILBERG, M. S. (**2002**) *J Biol Chem* 277, 24120–24127.

57 CHEN, H., PAN, Y. X., DUDENHAUSEN, E. E., KILBERG, M. S. (**2004**) *J Biol Chem* 279, 50829–50839.

58 PAN, Y., CHEN, H., SIU, F., KILBERG, M. S. (**2003**) *J Biol Chem* 278, 38402–38412.

59 SIU, F., CHEN, C., ZHONG, C., KILBERG, M. S. (**2001**) *J Biol Chem* 276, 48100–48107.

60 BRUHAT, A., AVEROUS, J., CARRARO, V., ZHONG, C., REIMOLD, A. M., KILBERG, M. S., FAFOURNOUX, P. (**2002**) *J Biol Chem* 25, 25.

61 PANWALKAR, A., VERSTOVSEK, S., GILES, F. J. (**2004**) *Cancer* 100, 657–666.

62 PROUD, C. G. (**2004**) *Biochem Biophys Res Commun* 313, 429–436.

63 KUBOTA, H., OBATA, T., OTA, K., SASAKI, T., ITO, T. (**2003**) *J Biol Chem* 278, 20457–20460.

64 CHERKASOVA, V. A., HINNEBUSCH, A. G. (**2003**) *Genes Dev* 17, 859–872.

65 BAIN, P. J., LeBlanc-CHAFFIN, R., CHEN, H., PALII, S. S., LEACH, K. M., KILBERG, M. S. (**2002**) *J Nutr* 132, 3023–3029.

66 JOUSSE, C., BRUHAT, A., FERRARA, M., FAFOURNOUX, P. (**2000**) *J Nutr* 130, 1555–1560.

67 HARDING, H. P., NOVOA, I. I., ZHANG, Y., ZENG, H., WEK, R., SCHAPIRA, M., RON, D. (**2000**) *Mol Cell* 6, 1099–1108.

68 GUPTA, S., CAMPBELL, D., DERIJARD, B., DAVIS, R. J. (**1995**) *Science* 267, 389–393.

69 LIVINGSTONE, C., PATEL, G., JONES, N. (**1995**) *EMBO J* 14, 1785–1797.

70 SANO, Y., HARADA, J., TASHIRO, S., GOTOH-MANDEVILLE, R., MAEKAWA, T., ISHII, S. (**1999**) *J Biol Chem* 274, 8949–8957.

71 KAWASAKI, H., SONG, J., ECKNER, R., UGAI, H., CHIU, R., TAIRA, K., SHI, Y., JONES, N., YOKOYAMA, K. K. (**1998**) *Genes Dev* 12, 233–245.

72 JIN, C., LI, H., MURATA, T., SUN, K., HORIKOSHI, M., CHIU, R., YOKOYAMA, K. K. (**2002**) *Mol Cell Biol* 22, 4815–4826.

73 HARPER, A. E., BENEVENGA, N. J., WOHLHUETER, R. M. (**1970**) *Physiol Rev* 50, 428–558.

74 LEUNG, P. M., ROGERS, Q. R. (**1971**) *Am J Physiol* 221, 929–935.

75 GIETZEN, D. W., ROSS, C. M., HAO, S., SHARP, J. W. (**2004**) *J Nutr* 134, 717–723.

76 HINNEBUSCH, A. G. (**1994**) *Semin Cell Biol* 5, 417–426.

77 MAURIN, A. C., JOUSSE, C., AVEROUS, J., PARRY, L., BRUHAT, A., CHERASSE, Y., ZENG, H., ZHANG, Y., HARDING, H. P., RON, D., FAFOURNOUX, P. (**2005**) *Cell Metabolism* 1, 273–277.

78 KILBERG, M. S., PAN, Y. X., CHEN, H., LEUNG-PINEDA, V. (**2005**) *Annu Rev Nutr* 25, 59–85.

6
Transcriptional Regulation of Hepatic Genes by Insulin and Glucose

Catherine Postic and Fabienne Foufelle

6.1
Introduction

Glucose is used continuously at a high rate in mammals by organs such as the brain (120 g/day in humans), red blood cells, and renal medulla. When a meal that contains carbohydrate is absorbed, it induces several metabolic events aimed at decreasing endogenous glucose production by the liver (glycogenolysis and gluconeogenesis) and increasing glucose uptake and storage in the form of glycogen in the liver and muscle. If glucose is delivered into the portal vein in large quantities and once the hepatic glycogen stores are repleted, glucose can be converted in the liver into lipids (lipogenesis), which are exported as very low-density lipoprotein (VLDL) and ultimately stored as triglycerides in adipose tissue. Conversely, if glucose availability in the diet is reduced, glucose-utilizing pathways are inhibited and glucose-producing pathways are activated.

The regulation of metabolic pathways involves the rapid modulation of the activity of specific proteins (enzymes, transporters) but also on a longer term basis changes in their quantity. This is mainly achieved by modulating their transcription rate. In the liver, the expression of several genes encoding enzymes involved in carbohydrate and lipid metabolism is induced by a high-carbohydrate diet, including glucokinase (GK) [1], pyruvate kinase (L-PK) [2] for glycolysis, ATP citrate lyase [3], stearoyl-CoA desaturase [4], acetyl-CoA carboxylase (ACC) [5], and fatty acid synthase (FAS) [6] for lipogenesis, glucose-6-phosphate dehydrogenase (G6PDH) [7] and 6-phosphogluconate dehydrogenase (6PGDH) for the pentose phosphate pathway. For most of the genes involved in glucose carbon utilization, the induction of their expression by a carbohydrate-rich diet is powerful (from 4- to 25-fold), rapid (in the 1–2 h range) and involves a transcriptional mechanism. Conversely, a high-carbohydrate diet inhibits the expression of gluconeogenic enzymes such as phosphoenolpyruvate carboxykinase (PEPCK) [8] and glucose-6-phosphatase (G6Pase) [9]. The inhibition is rapid but also easily reversible once the carbohydrate availability decreases.

Absorption of carbohydrate in the diet is concomitant with increases in the concentrations of substrates such as glucose but also with changes in the concentra-

Nutritional Genomics. Edited by Regina Brigelius-Flohé and Hans-Georg Joost
Copyright © 2006 WILEY-VCH Verlag GmbH & Co. KGaA, Weinheim
ISBN: 3-527-31294-3

tions of pancreatic hormones, insulin, and glucagon. Until recently, it was thought that insulin was the main regulator of glycolytic and lipogenic gene transcription. However, it has been shown, using primary cultures of hepatocytes, that glucose plays an important role in the regulation of gene transcription (see review in Refs [10, 11]). From these studies, different kinds of gene regulation have emerged: purely insulin-sensitive genes such as GK which can be induced by a high insulin concentration independently from the presence of glucose [12] and genes which require both insulin and a high glucose concentration in order to be induced, such as L-PK, FAS, ACC, and stearoyl-CoA desaturase [13–16]. PEPCK expression can be independently downregulated by insulin [17] or glucose [18, 19]. Finally, G6Pase expression is decreased by insulin but paradoxically increased by a high glucose concentration [9, 20]. In this review, we will focus on the mechanisms by which glucose and insulin can modulate the expression of genes encoding metabolic enzymes in the liver and we will specially develop the roles of transcription factors sterol regulatory element-binding protein (SREBP) and carbohydrate responsive element-binding protein (ChREBP) in these pathways.

6.2
Transcriptional Regulation by SREBP-1c

A pathway by which insulin can control gene expression has been discovered through the study of the transcription factor SREBP. SREBPs are basic helix-loop-helix leucine zipper (bHLH-LZ) transcription factors synthetized as inactive precursors bound to the membranes of the endoplasmic reticulum (ER) (see also chapters 4 and 12). Each SREBP precursor is organized into three domains: (a) an N-terminal domain that contains the transcription factor itself, (b) a central domain of 80 amino acids containing two transmembrane sequences separated by 31 amino acids that are in the lumen of the ER, and (c) a C-terminal regulatory domain. Upon activation, the ER-anchored SREBP precursor undergoes a sequential two-step cleavage process to release the N-terminal active domain, designated as the nuclear form of SREBP. Three isoforms of SREBP have been identified: SREBP-2 and SREBP-1a are mainly involved in the regulation of cholesterol biosynthesis and uptake and SREBP-1c is implicated in the regulation of fatty acid biosynthesis [21].

Recently, the SREBP-1c isoform has emerged as a major mediator of insulin action on glycolytic and lipogenic gene expression in the liver [22–24]. The first insights into the transcriptional regulation by SREBP-1c came from fasting/refeeding experiments in rodents, which showed that changes in the nutritional status regulate the expression of SREBP-1c itself in the liver. SREBP-1c expression is low during fasting but increases markedly when animals are fed with a high-carbohydrate diet [25]. In contrast, such manipulations induce only minor effects on the expression of the other SREBP isoforms. Subsequent experiments in isolated hepatocytes showed that the transcription of SREBP-1c is induced by insulin and inhibited by glucagon [23]. This induction of SREBP-1c transcription leads to a parallel increase in expression of both the ER membrane-bound precursor and the nuclear form of the transcription factor [26]. Experiments showing that SREBP-1c mRNA expression is decreased in the livers of streptozotocin (STZ) diabetic rats and nearly normalized by insulin treatment, confirmed the role of insulin on SREBP-1c transcription *in vivo* [27].

To become transcriptionally active, the SREBP precursor must undergo proteolytic cleavage in the Golgi apparatus to liberate its N-terminal domain, which constitutes the mature transcription factor. The group of Brown and Goldstein has demonstrated that the processing of SREBP-2 and SREBP-1a is controlled by the cellular sterol content. When cells are depleted of sterols, the SREBP-2 and SREBP-1a precursor proteins are cleaved twice by distinct proteases, releasing the mature transcriptionally active portion from its membrane tether. The processed SREBPs then translocate to the nucleus where they activate a number of target genes involved in cholesterol and fatty acid metabolism through binding to sterol regulatory elements in the promoters of target genes. Two proteins are essential to this cleavage process: SREBP cleavage-activating protein (SCAP) and insulin-induced gene (Insig). SCAP is a large integral membrane protein of the ER that interacts with newly synthesized SREBP precursor and escorts it to the Golgi apparatus [28]. However, SCAP can also interact with Insig, another ER protein that is deeply embedded in the membranes. Insig functions to retain the SCAP-SREBP complex within the ER [29].

In a recent study, we have demonstrated that the cleavage of the SREBP-1c isoform is under the control of insulin [30]. We show that insulin *per se* is able to rapidly accumulate SREBP-1c in the nucleus, independent of any effect on SREBP-1c transcription. The mechanisms by which insulin acts on SREBP-1c cleavage are not known. So far, by inducing the transcription of the SREBP-1c gene and by increasing the proteolytic cleavage of this specific isoform, insulin promotes the accumulation of SREBP-1c within the nucleus (Fig. 6.1).

In order to demonstrate the involvement of SREBP-1c in mediating the effects of insulin on hepatic gene expression, two different forms of SREBP-1c were overexpressed in hepatocytes using an adenoviral vector: (i) a dominant positive version of SREBP-1c, which corresponds to the mature nuclear form of SREBP-1c, directly imported into the nucleus; and (ii) a dominant negative form of SREBP-1c, which is the mature form of SREBP-1c containing a mutation in the basic domain that abolishes the binding of SREBP-1c to its binding sites, but still allows dimerization, leading to decreased availability of endogenous SREBP-1c. Overexpression of the dominant negative form of SREBP-1c in hepatocytes counteracts the stimulatory effect of insulin on GK expression. Conversely, overexpression of the active form of SREBP-1c bypasses the insulin requirement for GK expression [22]. Similar results were obtained for genes that require for their full expression both the presence of insulin and a high glucose concentration [23], namely those encoding FAS and ACC, demonstrating that SREBP-1c is the mediator of positive insulin action on glycolytic and lipogenic genes.

Kim and co-workers have identified two functional sterol regulatory elements (SRE) in the rat GK promoter [31]. The physiological *in vivo* interaction between

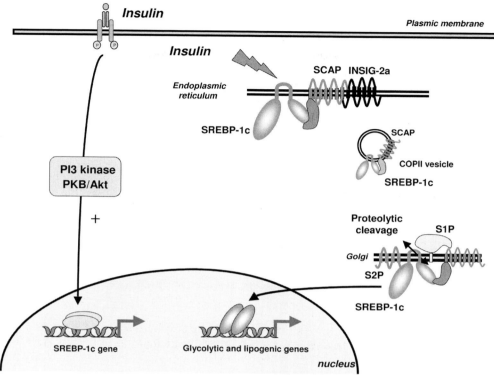

Fig. 6.1. Mechanisms involved in the activation of sterol regulatory element-binding protein 1c (SREBP-1c) by insulin in the liver. After binding on its receptor, insulin activates the transcription of the SREBP-1c gene through a phosphoinositide 3-kinase/protein kinase B pathway. This is followed by the synthesis of the precursor form of SREBP-1c, which is anchored in the endoplasmic reticulum membrane. Insulin also activates the proteolytic cleavage of this precursor form. The mechanism involved in the acute effect of insulin is not known but insulin could act on the interaction between SREBP cleavage-activating protein (SCAP) and insulin-induced gene 2 (insig-2) retention protein. The consequence is the relocalization of the SREBP/SCAP complex from the endoplasmic reticulum to the Golgi. Here, the precursor proteins are cleaved twice by distinct proteases (S1P, S2P), releasing the mature transcriptionally active portion from its membrane tether. The processed SREBPs then translocate to the nucleus where they activate a number of target genes involved in fatty acid metabolism through binding as homodimers to sterol regulatory elements in the promoters of target genes.

the SREBP-1c protein and SREs of the GK promoter was confirmed by chromatin immunoprecipitation (ChIP) assay using primary cultures of hepatocytes, demonstrating the direct involvement of SREBP-1c on GK gene expression. SREBP-1c is also able to induce lipogenic genes by its capacity to bind to SREs present in their promoters [32–34]. In addition, transgenic mice that overexpress SREBP-1c in the liver exhibit liver steatosis and increased mRNA of most lipogenic genes [35]. Consistent with these observations, SREBP-1c gene knockout mice have an impaired ability to fully induce lipogenic gene expression after high-carbohydrate feeding [36].

All together these results indicate that SREBP-1c plays a major role in the long-term control of glucose and lipid homeostasis by insulin, through the regulation of glycolytic and lipogenic gene expression. However, SREBP-1c activity alone does not appear to fully account for the stimulation of glycolytic and lipogenic gene expression in response to carbohydrate since SREBP-1c gene deletion in mice only results in a 50% reduction in fatty acid synthesis [36]. Indeed, the induction of glycolytic and lipogenic genes in response to a high-carbohydrate diet, although significantly diminished, is not completely suppressed in SREBP-1c-knockout mice [36]. In addition, we and others have provided evidence that SREBP-1c expression is not sufficient by itself to account for the glucose/insulin induction of glycolytic and lipogenic genes in primary cultured hepatocytes [33, 37, 38]. We have demonstrated, using hepatic GK-knockout mice (hGK-KO) [39], that over-expression of a constitutive active form of SREBP-1c in hGK-KO hepatocytes cultured in the presence of high glucose concentration (25 mmol/l) did not fully induce glycolytic and lipogenic genes compared with what was observed in control hepatocytes [38]. Therefore, glucose metabolism via GK and SREBP-1c exert a synergistic effect on the expression of glycolytic and lipogenic genes.

6.3
Transcriptional Regulation by Glucose: A Role for ChREBP

As mentioned above, glycolytic and lipogenic enzymes such as L-PK, ACC, and FAS require both glucose and insulin for their full expression. A number of arguments suggest that glucose must be metabolized in order to have its transcriptional effect [40]. For instance, in the liver, the glucose effect requires the presence of GK, the enzyme responsible for glucose phosphorylation to glucose-6-phosphate (G6P) [14, 41]. Since GK expression is strongly activated by insulin, the insulin dependency of these genes was in part explained by the necessity of GK induction by the hormone in order to allow glucose metabolism. However, some evidence also argues for a direct effect of insulin through the insulin-responsive transcription factor SREBP-1c (see above) (Fig. 6.2).

The questions that then arise concern the nature of the glucose signal, the transduction mechanism from the glucose metabolite to the transcriptional machinery, the glucose response element on the gene promoter, and the transcription factor involved. Although it had been established that a metabolite of glucose, and not glucose *per se*, was responsible for this glucose signal, both G6P [10, 14, 42] and xylulose 5-phosphate (X5P) [43, 44] have been proposed as critical metabolites (see also chapter 4). In previous works it has been proposed that G6P could be the metabolite acting as a signal on lipogenic enzymes gene transcription [14, 42]. This proposal was based on the fact that in adipocytes and in a β-cell line (INS1), the effect of glucose on gene transcription is mimicked by 2-deoxyglucose, a glucose analog, of which metabolism stops after its phosphorylation into 2-deoxyglucose-6-

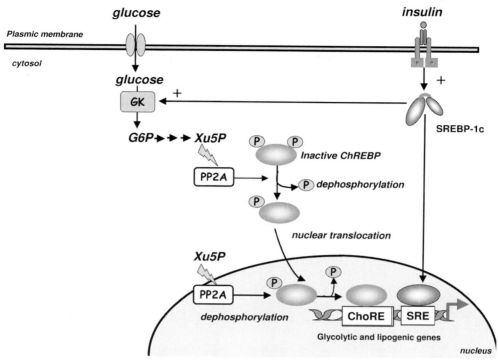

Fig. 6.2. Mechanisms involved in the activation of carbohydrate responsive element-binding protein (ChREBP) by glucose metabolism in liver. In liver, the glucose effect requires the presence of GK, the enzyme responsible for glucose phosphorylation into glucose 6-phosphate (G6P), an essential step for glucose metabolism as well as for the induction of ChREBP expression and function. Under basal conditions, ChREBP is localized in the cytosol and its nuclear translocation is rapidly induced under high glucose concentrations. The nuclear translocation of ChREBP is controlled by dephosphorylation of several serine/threonine residues. While the dephosphorylation of Ser196 allows ChREBP translocation in the nucleus, the dephosphorylation of Ser568 and Thr666 alleviates DNA-binding inhibition. Protein phosphatase 2A (PP2A), which was shown to be selectively activated by xylulose 5-phosphate (X5P), is believed to be responsible for both cytosolic and nuclear dephosphorylation of ChREBP. Then ChREBP binds its response element (ChoRE) to activate glycolytic and lipogenic gene expression. Together ChREBP and SREBP-1c provide a pair of transcription factors that function in synergy through distinct binding response elements to coordinately regulate glycolytic and lipogenic genes.

phosphate [42, 45]. Interestingly enough, in GLUT2-null mice, fasted animals have a paradoxical increase in L-PK expression and glycogen content and this is concomitant with concentrations of G6P which remain at a high level [46]. Another interesting situation is found when rats are injected with an inhibitor of the G6P translocator activity, a component of the glucose-6-phosphatase system. This induces a large increase in hepatic G6P, glycogen and triglyceride concentrations, and an activation of FAS and ACC gene expression [47]. On the other hand, X5P, an intermediate of the non-oxidative branch of the metabolic pathway, has also been proposed as the metabolite signal [43]. This was based mainly on the fact that xylitol, a precursor of X5P, is able to stimulate the transcription of a reporter gene driven by the L-PK promoter in hepatocytes and that X5P has been shown to specifically activate phosphatase 2A mediated-dephosphorylation. This phosphatase is involved in the dephosphorylation of transcription factors [48], including carbohydrate responsive element-binding protein (ChREBP), as we will discuss below. Definitive evidence for the involvement of one or the other metabolite will be given only when the full transduction machinery and the exact role of the signal metabolite in this context has been elucidated.

Regardless of the metabolite involved, it is now clear that increased metabolism through GK is required to initiate glucose signaling. However, the intracellular mechanism of the glucose-signaling pathway is not fully understood. In fact, the transcriptional effect of glucose has been extensively studied in several laboratories (see Ref. [49] for review). Glucose or carbohydrate response elements (ChoRE) that mediate the transcriptional response of glucose have been identified in the promoters of most lipogenic genes through promoter-mapping analysis [50–52]. This element is composed of two E-box (CACGTG) or E-box-like sequences separated by 5 bp. The presence of E-box motifs in these response elements suggests that a bHLH protein family member recognizes the ChoRE and mediates the response to glucose. Various transcription factors, including members of the upstream stimulatory factor (USF) family [49] or chicken ovalbumin upstream promoter/transcription factor II (COUP-TFII), an orphan receptor of the steroid/thyroid hormone receptor superfamily [53] (see chapter 2), have been previously proposed as potential candidates to mediate the transcriptional effects of glucose. Although data from gene knockout mice have revealed that endogenous USFs are important for the normal activation of several diet-regulated genes [54–56], these transcription factors cannot explain, by themselves, the transcriptional regulation of glucose responsive genes. Indeed, most genes whose promoters include USF-binding sites are not regulated by glucose and, more importantly, USF expression as well as its DNA-binding activity are not regulated by the glucose diet [57].

The recent identification of a glucose responsive bHLH-LZ transcription factor named ChREBP [58] (see also chapters 4 and 12) has recently shed light on the possible mechanism whereby glucose affects gene transcription. ChREBP is a large protein (864 amino acids and $M_r = 94\,600$) that contains several domains including a nuclear localization signal (NLS) near the N-terminus, polyproline domains, a bHLH-LZ, and a leucine-zipper-like (Zip-like) domain. ChREBP also contains several potential phosphorylation sites for cAMP-dependent protein kinase (PKA) and AMP-activated protein kinase (AMPK) [59]. ChREBP is regulated in a reciprocal manner by glucose and cAMP [59]. Under basal conditions ChREBP is localized in the cytosol, and its nuclear translocation is rapidly induced under high glucose concentrations. Nuclear translocation of ChREBP is controlled by dephosphorylation of several serine/threonine residues. Ser196 is the target of PKA phosphorylation, and its dephosphorylation allows ChREBP translocation in the nucleus. Two other residues, Ser568 and Thr666, are dephosphorylated in the nucleus, thus alleviating DNA-binding inhibition. Protein phosphatase 2A (PP2A), which was shown to be selectively activated by X5P, is believed to be responsible for both cytosolic and nuclear dephosphorylation of ChREBP [44] (Fig. 6.2). Overexpression of ChREBP in primary hepatocytes induces the activity of the ChoRE-containing L-PK promoter only under high glucose conditions [58]. Interestingly, ChREBP does not homodimerize or bind to the ChoRE as an homodimer [60–62]. In fact, using the yeast two-hybrid system, Towle and co-workers have identified a bHLH-LZ protein that interacts with the bHLH-LZ domain of ChREBP, named Mlx (Max-like protein X) [62]. Mlx is a member of the Myc/Max/Mad family of transcription factors that can serve as a common interaction partner of a transcription factor network [63]. The fact that ChREBP interacts with Mlx suggests that a network of transcription factors or cofactors may be required to fully regulate glucose responsive gene expression in liver.

The discovery of ChREBP and its potential role in glucose action prompted us to perform a series of experiments in liver of both control and hepatic GK-knockout mice (hGK-KO) [39]. This mouse model has allowed us to selectively dissociate the effects of insulin and of glucose metabolism via GK on glycolytic and lipogenic gene expression [39]. Through the use of hGK-KO mice, we have shown that the acute stimulation by high-carbohydrate refeeding of L-PK, FAS, and ACC requires GK expression. In fact, the loss of glucose effect observed in hGK-KO hepatocytes is correlated with a decrease in ChREBP gene expression, suggesting that glucose metabolism via GK is necessary for both expression and function of ChREBP in liver [38]. To address the direct role of ChREBP in mediating glucose signaling in liver, we have used the siRNA approach to silence ChREBP gene expression in control hepatocytes. Our studies have revealed, for the first time in a physiological context, that ChREBP mediates the glucose effect on both glycolytic and lipogenic gene expression and that this transcription factor is a key determinant of lipid synthesis in liver [38]. Our results were later confirmed by the global inactivation of ChREBP gene expression in mice (ChREBP$^{-/-}$ mice) [61, 64]. Hepatocytes from ChREBP$^{-/-}$ mice do not support a glucose response when transfected with a ChoRE-containing promoter, but this response can be reversed by the addition of a ChREBP expressing vector [64]. Finally, ChREBP was shown to directly bind to the promoter sequences of lipogenic genes using chromatin immunoprecipitation assays [64].

Altogether, these studies suggest that ChREBP is in fact the long-sought glucose response element-binding protein. The evidence that ChREBP functions with its obligatory partner Mlx was recently confirmed [65]. Through the use of an adenovirus expressing a dominant negative form of Mlx, Towle and co-workers have demonstrated that the inhibition of Mlx directly interferes with the endogenous ChREBP/Mlx complex and abrogates the glucose response of the ACC reporter gene in primary cultures of hepatocytes [65]. This blunted glucose response can, however, be partially restored when ChREBP is overexpressed. The fact that this rescue only occurs at high glucose concentrations of recombinant ChREBP adenovirus suggests that sufficient ChREBP needs to be provided in the cell in order to titrate out the dominant negative effect of Mlx. These results clearly show that Mlx is a functional partner of ChREBP and support an obligatory role for Mlx in the glucose responsive complex. However, whether Mlx is also directly regulated by glucose in liver or is rather a silent partner remains to be determined.

6.4
Conclusion

With the discovery of two key transcription factors, SREBP-1c and ChREBP, our understanding of the long-term regulation of glucose and lipid metabolism in liver has made considerable progress. Together ChREBP and SREBP-1c provide a pair of transcription factors that function in synergy through distinct binding response elements to regulate glycolytic and lipogenic genes. Such a system of regulation provides a means of using glucose for lipid storage only when appropriate conditions (high glucose and insulin concentrations) are met, allowing for a fine utilization of glucose and lipid synthesis. Although the role of ChREBP in regulating lipogenic gene expression has been now clearly established, its role in the physiopathology of obesity and/or insulin resistance remains to be elucidated. Experiments from our laboratories are currently in process to analyze the involvement of ChREBP in the physiopathology of obesity and type II diabetes.

Acknowledgments

We would like to thank, Renaud Dentin, Marc Foretz, Bronwyn Begarty, Isabelle Hainault, and Fadila Benhamed from INSERM Unit 671 and from the Department of Endocrinology (Institut Cochin, INSERM U567 CNRS UMR8104) who have performed the studies presented. Finally, we are indebted to Professor Pascal Ferré and Professor Jean Girard for their constant interest and friendly support.

References

1 P.B. Iynedjian, C. Ucla, B. Mach (**1987**) *J. Biol. Chem.* 262, 6032–6038.

2 S. Vaulont, A. Munnich, J.F. Decaux, A. Kahn (**1986**) *J. Biol. Chem.* 261, 7621–7625.

3 N.A. Elshourbagy, J.C. Near, P.J. Kmetz, G.M. Sathe, C. Southan, J.E. Strickler, M. Gross, J.F. Young, T.N. Wells, P.H. Groot (**1990**) *J. Biol. Chem.* 265, 1430–1435.

4 J.M. Ntambi (**1992**) *J. Biol. Chem.* 267, 10925–10930.

5 A. Katsurada, N. Iritani, H. Fukuda, Y. Matsumara, N. Nishimoto, T. Noguchi, T. Tanaka (**1990**) *Eur. J. Biochem.* 190, 435–441.

6 A. Katsurada, N. Iritani, H. Fukuda, Y. Matsumara, N. Nishimoto, T. Noguchi, T. Tanaka (**1990**) *Eur. J. Biochem.* 190, 427–435.

7 A. Katsurada, N. Iritani, H. Fukuda, Y. Matsumura, T. Noguchi, T. Tanaka (**1989**) *Biochim. Biophys. Acta* 1006, 104–110.

8 D.K. Granner, T.L. Andreone, K. Sasaki, E.G. Beale (**1983**) *Nature* 305, 549–551.

9 D. Argaud, T.L. Kirby, C.B.

NEWGARD, A.J. LANGE (**1997**) *J. Biol. Chem.* 272, 12854–12861.

10 J. GIRARD, P. FERRÉ, F. FOUFELLE (**1997**) *Annu. Rev. Nutr.* 17, 325–352.

11 H.C. TOWLE, E.N. KAYTOR, H.M. SHIH (**1997**) *Annu. Rev. Nutr.* 17, 405–433.

12 P.B. IYNEDJIAN, D. JOTTERAND, T. NOUSPIKEL, M. ASFARI, P.R. PILOT (**1989**) *J. Biol. Chem.* 264, 21824–21829.

13 J.F. DECAUX, B. ANTOINE, A. KAHN (**1989**) *J. Biol. Chem.* 264, 11584–11590.

14 C. PRIP-BUUS, D. PERDEREAU, F. FOUFELLE, J. MAURY, P. FERRÉ, J. GIRARD (**1995**) *Eur. J. Biochem.* 230, 309–315.

15 B.L. O'CALLAGHAN, S.H. KOO, Y. WU, H.C. FREAKE, H.C. TOWLE (**2001**) *J. Biol. Chem.* 276, 16033–16039.

16 K.M. WATERS, J.M. NTAMBI (**1994**) *J. Biol. Chem.* 269, 27773–27777.

17 K. SASAKI, T.P. CRIPE, S.R. KOCH, T.L. ANDREONE, D.D. PETERSEN, E.G. BEALE, D.K. GRANNER (**1984**) *J. Biol. Chem.* 259, 15242–15251.

18 D.K. SCOTT, R.M. O'DOHERTY, J.M. STAFFORD, C.B. NEWGARD, D.K. GRANNER (**1998**) *J. Biol. Chem.* 273, 24145–24151.

19 F. COURNARIE, D. AZZOUT-MARNICHE, M. FORETZ, C. GUICHARD, P. FERRÉ, F. FOUFELLE (**1999**) *FEBS Lett.* 460, 527–532.

20 E. VAN SCHAFTINGEN, I. GERIN (**2002**) *Biochem. J.* 362, 513–532.

21 M.S. BROWN, J.L. GOLDSTEIN (**1997**) *Cell* 89, 331–340.

22 M. FORETZ, C. GUICHARD, P. FERRÉ, F. FOUFELLE (**1999**) *Proc. Natl Acad. Sci. USA* 96, 12737–12742.

23 M. FORETZ, C. PACOT, I. DUGAIL, P. LEMARCHAND, C. GUICHARD, X. LE LIEPVRE, C. BERTHELIER–LUBRANO, B. SPIEGELMAN, J.B. KIM, P. FERRE, F. FOUFELLE (**1999**) *Mol. Cell. Biol.* 19, 3760–3768.

24 F. FOUFELLE, P. FERRÉ (**2002**) *Biochem. J.* 366, 377–391.

25 J.D. HORTON, Y. BASHMAKOV, I. SHIMOMURA, H. SHIMANO (**1998**) *Proc. Natl Acad. Sci. USA* 95, 5987–5992.

26 D. AZZOUT-MARNICHE, D. BÉCARD, C. GUICHARD, M. FORETZ, P. FERRÉ, F. FOUFELLE (**2000**) *Biochem. J.* 350, 389–393.

27 I. SHIMOMURA, Y. BASHMAKOV, S. IKEMOTO, J.D. HORTON, M.S. BROWN, J.L. GOLDSTEIN (**1999**) *Proc. Natl Acad. Sci. USA* 96, 13656–13661.

28 J. SAKAI, A. NOHTURFFT, D. CHENG, Y.K. HO, M.S. BROWN, J.L. GOLDSTEIN (**1997**) *J. Biol. Chem.* 272, 20213–20221.

29 T. YANG, P.J. ESPENSHADE, M.E. WRIGHT, D. YABE, Y. GONG, R. AEBERSOLD, J.L. GOLDSTEIN, M.S. BROWN (**2002**) *Cell* 110, 489–500.

30 B.D. HEGARTY, A. BOBARD, I. HAINAULT, P. FERRÉ, P. BOSSARD, F. FOUFELLE (**2005**) *Proc. Natl Acad. Sci. USA* 102, 791–796.

31 S.Y. KIM, H.I. KIM, T.H. KIM, S.S. IM, S.K. PARK, I.K. LEE, K.S. KIM, Y.H. AHN (**2004**) *J. Biol. Chem.* 279, 30823–30829.

32 M.M. MAGANA, T.F. OSBORNE (**1996**) *J. Biol. Chem.* 271, 32689–93264.

33 S.H. KOO, A.K. DUTCHER, H.C. TOWLE (**2001**) *J. Biol. Chem.* 276, 9437–9445.

34 M.K. BENNETT, J.M. LOPEZ, H.B. SANCHEZ, T.F. OSBORNE (**1995**) *J. Biol. Chem.* 270, 25578–25583.

35 H. SHIMANO, J.D. HORTON, I. SHIMOMURA, R.E. HAMMER, M.S. BROWN, J.L. GOLDSTEIN (**1997**) *J. Clin. Invest.* 99, 846–854.

36 G. LIANG, J. YANG, J.D. HORTON, R.E. HAMMER, J.L. GOLDSTEIN, M.S. BROWN (**2002**) *J. Biol. Chem.* 277, 9520–9528.

37 A.K. STOECKMAN, H.C. TOWLE (**2002**) *J. Biol. Chem.* 277, 27029–27035.

38 R. DENTIN, J. GIRARD, C. POSTIC (**2005**) *Biochimie* 87, 81–86.

39 C. POSTIC, M. SHIOTA, K.D. NISWENDER, T.L. JETTON, Y. CHEN, J.M. MOATES, K.D. SHELTON, J. LINDNER, A.D. CHERRINGTON, M.A. MAGNUSON (**1999**) *J. Biol. Chem.* 274, 305–315.

40 F. FOUFELLE, J. GIRARD, P. FERRÉ (**1996**) *Adv. Enzyme Regul.* 36, 199–226.

41 B. DOIRON, M.H. CUIF, A. KAHN, M.J. DIAZ-GUERRA (**1994**) *J. Biol. Chem.* 269, 10213–10216.

42 F. Foufelle, B. Gouhot, J.P. Pégorier, D. Perdereau, J. Girard, P. Ferré (**1992**) *J. Biol. Chem.* 267, 20543–20546.

43 B. Doiron, M.H. Cuif, R. Chen, A. Kahn (**1996**) *J. Biol. Chem.* 271, 5321–5324.

44 T. Kabashima, T. Kawaguchi, B.E. Wadzinski, K. Uyeda (**2003**) *Proc. Natl Acad. Sci. USA* 100, 5107–5112.

45 T. Brun, E. Roche, K.H. Kim, M. Prentki (**1993**) *J. Biol. Chem.* 268, 18905–18911.

46 R. Burcelin, M. del Carmen Munoz, M.T. Guillam, B. Thorens (**2000**) *J. Biol. Chem.* 275, 10930–10936.

47 R.H. Bandsma, C.H. Wiegman, A.W. Herling, H.J. Burger, A. ter Harmsel, A.J. Meijer, J.A. Romijn, D.J. Reijngoud, F. Kuipers (**2001**) *Diabetes* 50, 2591–2597.

48 M. Nishimura, K. Uyeda (**1995**) *J. Biol. Chem.* 270, 26341–26346.

49 S. Vaulont, M. Vasseur-Cognet, A. Kahn (**2000**) *J. Biol. Chem.* 275, 31555–31558.

50 K.S. Thompson, H.C. Towle (**1991**) *J. Biol. Chem.* 266, 8679–8682.

51 H.M. Shih, Z. Liu, H.C. Towle (**1995**) *J. Biol. Chem.* 270, 21991–21997.

52 C. Rufo, M. Teran-Garcia, M.T. Nakamura, S.H. Koo, H.C. Towle, S.D. Clarke (**2001**) *J. Biol. Chem.* 276, 21969–21975.

53 D.Q. Lou, M. Tannour, L. Selig, D. Thomas, A. Kahn, M. Vasseur-Cognet (**1999**) *J. Biol. Chem.* 274, 28385–28394.

54 V.S. Vallet, M. Casado, A.A. Henrion, D. Bucchini, M. Raymondjean, A. Kahn, S. Vaulont (**1998**) *J. Biol. Chem.* 273, 20175–20179.

55 V.S. Vallet, A.A. Henrion, D. Bucchini, M. Casado, M. Raymondjean, A. Kahn, S. Vaulont (**1997**) *J. Biol. Chem.* 272, 21944–21949.

56 M. Casado, V.S. Vallet, A. Kahn, S. Vaulont (**1999**) *J. Biol. Chem.* 274, 2009–2013.

57 E.N. Kaytor, H. Shih, H.C. Towle (**1997**) *J. Biol. Chem.* 272, 7525–7531.

58 H. Yamashita, M. Takenoshita, M. Sakurai, R.K. Bruick, W.J. Henzel, W. Shillinglaw, D. Arnot, K. Uyeda (**2001**) *Proc. Natl Acad. Sci. USA* 98, 9116–91121.

59 K. Uyeda, H. Yamashita, T. Kawaguchi (**2002**) *Biochem. Pharmacol.* 63, 2075–2080.

60 S. Cairo, G. Merla, F. Urbinati, A. Ballabio, A. Reymond (**2001**) *Hum. Mol. Genet.* 10, 617–627.

61 K. Iizuka, R.K. Bruick, G. Liang, J.D. Horton, K. Uyeda (**2004**) *Proc. Natl Acad. Sci. USA* 101, 7281–7286.

62 A.K. Stoeckman, L. Ma, H.C. Towle (**2004**) *J. Biol. Chem.* 279, 15662–15669.

63 G. Meroni, S. Cairo, G. Merla, S. Messali, R. Brent, A. Ballabio, A. Reymond (**2000**) *Oncogene* 19, 3266–3277.

64 S. Ishii, K. Iizuka, B.C. Miller, K. Uyeda (**2004**) *Proc. Natl Acad. Sci. USA* 101, 15597–15602.

65 L. Ma, N.G. Tsatsos, H.C. Towle (**2005**) *J. Biol. Chem.* 280, 12019–12027.

7

PPARs: Lipid Sensors that Regulate Cell Differentiation Processes

Frédéric Varnat, Liliane Michalik, Béatrice Desvergne, and Walter Wahli

7.1
Introduction

Differentiation is the process by which unspecialized cells become specialized for particular functions. To differentiate, cells have to exit the cell cycle and express a set of genes encoding specialized proteins. In this context the nuclear receptors peroxisome proliferator-activated receptors (PPARs) play major roles. The three closely related PPAR isotypes, PPARα (NR1C1), PPARβ/δ (NR1C2), and PPARγ (NR1C3), are coded by distinct genes and belong to the nuclear receptor super-family [1]. PPARs are inducible transcription factors activated by fatty acids as well as naturally occurring fatty acid-derived eicosanoids. They heterodimerize with the retinoid X receptor (RXR, NR2B) and bind to specific peroxisome proliferator-responsive elements (PPREs) located in the promoter of their target genes [2]. The pleiotropic functions that PPARs exert in metabolism have been extensively re-viewed [3, 4]. We will discuss herein the involvement of PPARγ and PPARβ in cell differentiation as it occurs in various organs such as adipose tissue, bones, skin, intestine, and placenta.

7.2
Peroxisome Proliferator-activated Receptor Gamma

7.2.1
PPARγ in Adipocyte Differentiation and Survival

PPARγ has been clearly linked to the adipocyte differentiation program [5]. At the protein level, there are two PPARγ isoforms, PPARγ1 and PPARγ2 (see also chap-ters 2, 4 and 12). Both are produced from the same gene by alternative promoter usage and mRNA splicing. PPARγ1 is mainly expressed in adipose tissues but is also detected at lower levels in several other organs, such as in the colon, spleen,

Nutritional Genomics. Edited by Regina Brigelius-Flohé and Hans-Georg Joost
Copyright © 2006 WILEY-VCH Verlag GmbH & Co. KGaA, Weinheim
ISBN: 3-527-31294-3

retina, hematopoietic cells, liver, and skeletal muscle, whereas PPARγ2 expression is restricted to the adipose tissue (reviewed in Ref. [2]).

The major role of PPARγ and its ligands in determining adipocyte differentiation was initially demonstrated using cell culture [5]. PPARγ is a late marker of adipocyte differentiation and its artificial expression is sufficient to force fibroblasts into the adipogenic program [5]. PPARγ is also able to induce the transdifferentiation of cultured myoblasts into adipocytes [6]. Whereas PPARγ-null mice are not viable, due to defects in placenta formation [7], the lack of PPAR$\gamma^{-/-}$ adipocytes in chimeric PPAR$\gamma^{+/+}$:PPAR$\gamma^{-/-}$ mice has demonstrated the importance, *in vivo*, of PPARγ for adipogenesis [8]. Indeed, a tetraploid-rescued PPARγ-null mutant mouse exhibited severe adipose tissue dystrophy at birth [7]. The important role of PPARγ in the mature adipose tissue is now confirmed by the characterization of mice carrying an adipose tissue-specific deletion of PPARγ. As the expression of the cAMP response element (CRE) enzyme, which is responsible for the gene deletion by a recombination event, is under the control of the aP2 (adipocyte fatty acid-binding protein) promoter in these mice, the ablation occurs after adipogenesis has started. This deletion results in a severe reduction of adipocyte number, both in white and brown adipose tissues, whereas small and likely nascent adipocytes remain relatively abundant. Furthermore, the selective spatio-temporal ablation of PPARγ in adipocytes of adult mice by using the tamoxifen-dependent Cre-ER(T2) recombination system revealed that mature PPARγ-null white and brown adipocytes die within a few days after deletion of the gene and are replaced by newly formed PPARγ-positive adipocytes [9]. These observations have demonstrated that PPARγ is essential for the *in vivo* survival of mature adipocytes, in addition to its already well-established requirement for their differentiation [10, 11].

While the causal link between PPARγ-mediated induction of the phosphatidylinositol 3-kinase (PI3K) signaling in human adipocytes [12] and adipogenesis remains to be established, different reports draw attention to the role of PPARγ in cell cycle control (see Fig. 7.1). Indeed, PPARγ downregulates *PP2A* gene expression. The phosphatase encoded by this gene is responsible for the activation of the phosphorylated transcription factor E2F, which is required for initiation of DNA synthesis in the S-phase. Thus, the PPARγ-mediated inhibition of E2F activation results in withdrawal from the cell cycle of the mesenchymal precursor cells [10]. In addition, PPARγ induces the expression of the two cyclin-dependent kinase (Cdk) inhibitors (CKI), p18 and p21, resulting in the inhibition of the Cdk activities that are necessary for cell cycle progression [11]. Finally, PPARγ also enhances adipogenic differentiation by directly controlling the expression of C/EBPα which is a major pro-adipogenic transcription factor. C/EBPα in turn directly activates PPARγ expression thus constituting a positive feedback loop (Fig. 7.1) [5]. Based on these observations, it is not surprising that several PPARγ anticarcinogenic effects have been reported (reviewed in Ref. [13]). Among those, the most striking are clinical trials in which a specific PPARγ ligand efficiently induced cell cycle arrest and differentiation of liposarcoma cells in several patients [14].

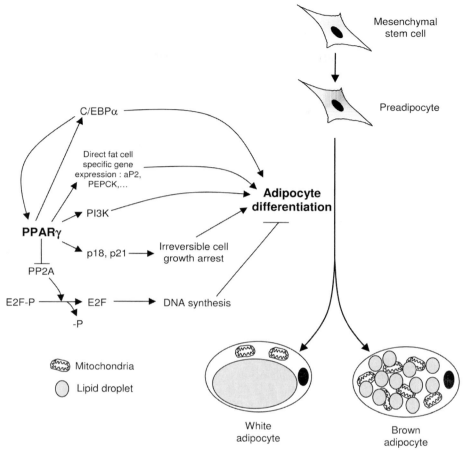

Fig. 7.1. PPARγ induces adipogenic differentiation via diverse and complementary mechanisms. It induces adipocyte differentiation by acting on the expression levels of the cell cycle inhibitors, the cyclin-dependent kinase (Cdk) inhibitors (CKIs) p18 and p21, leading to an irreversible cell growth arrest. It also inhibits the initiation of DNA synthesis necessary for the S-phase entry by maintaining the transcription factor E2F in its inactive phosphorylated form. In this case, PPARγ acts as a repressor of the phosphatase PP2A that is required for the dephosphorylation of E2F (-P: phosphate group). PPARγ also enhances pro-adipogenic pathways such as C/EBPα and phosphatidylinositol 3-kinase (PI3K) signaling. Finally, PPARγ can directly control the expression of fat cell-specific genes by binding to a peroxisome proliferator-responsive element (PPRE) in their promoter region (aP2, PEPCK, etc.).

7.2.2
PPARγ and Osteoblast Differentiation

Bone tissue undergoes continuous renewal in vertebrates. This process is due to the combined production of new osteoblasts by mesenchymal progenitors and to the degradation of extracellular bone matrix by osteoclasts, which derive from the monocyte/macrophage lineage. PPARγ is of special interest in bone homeostasis as its insufficiency in heterozygous PPARγ-deficient mice increases bone mass by enhancing the differentiation of mesenchymal progenitor cells into osteoblasts. Accordingly, PPARγ-deficient embryonic stem (ES) cells in culture spontaneously differentiate into osteoblasts, while they fail to differentiate into adipocytes [15]. *In vivo* observations using X-ray and three-dimensional computerized tomography (3D-CT) reveal that PPARγ heterozygous mice exhibit an unusual high bone mass due to the accumulation of osteoblasts. Consistently, cultured bone marrow cells from these PPARγ heterozygous mice are prone to osteoblastogenesis and have a low potency to differentiate into adipocytes compared with the wild-type cells. These PPARγ heterozygous cells have an increased expression of the transcription factors Runx2 (Cbfa1) and Osterix (Sp7), which are regulators of osteoblastic differentiation [15]. The converse experiment, i.e. analysis of the action of PPARγ activators on bone density in wild-type mice as well as in bone marrow cell-derived cultures, confirmed the PPARγ-dependent reduction of osteoblastogenesis via decreased expression of Runx2 and Osterix [16]. Thus, it appears that low PPARγ activity is an important factor for the differentiation of bone mesenchymal cells (Fig. 7.2).

These observations point to PPARγ as a regulator shifting cell differentiation towards adipogenesis rather than osteoblastogenesis. This raises the interesting question of the use of PPARγ agonists or antagonists in therapeutic approaches. Thiazolidinediones (TZDs) are PPARγ agonists that were developed for the treatment of insulin resistance. The mode of action of these compounds is not fully understood, but likely involves a redistribution of triglycerides from inappropriate muscle accumulation towards storage in adipose tissue. Amazingly, decreasing the activity of PPARγ, either through invalidation of one PPARγ allele or via a PPARγ antagonist, also precludes the development of a high-fat diet-induced insulin resistance. Thus, the use of selective PPARγ antagonists could prevent adipocyte differentiation, increase bone mass and would as well be potentially valuable tools against obesity and type II diabetes [17].

7.2.3
PPARγ and Colonocyte Differentiation

PPARγ expression is increased in differentiated colon epithelial cells [18] and many reports show that highly selective PPARγ agonists such as TZDs promote colonocyte differentiation [19]. Similar to what was observed and discussed above in the context of adipocyte differentiation, PPARγ is able to enhance colonocyte differentiation by controlling the expression of cell cycle genes. More precisely in

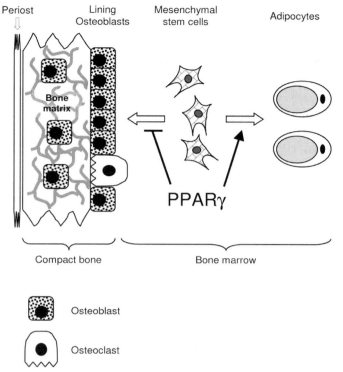

Fig. 7.2. Schematic representation of the role of PPARγ in bone tissue homeostasis. The compact bone is mainly composed of mature osteoblasts, which secrete an abundant extracellular calcified matrix. It is renewed continuously by the formation of new osteoblasts from mesenchymal stem cells, combined with a continuous degradation of the bone matrix by osteoclasts that derive from the monocyte/macrophage cell lineage. In the bone marrow, PPARγ favors adipogenic differentiation to the detriment of osteoblast differentiation. Thus, the disruption of PPARγ signaling, for instance by PPARγ antagonists, might offer a possibility to explore ways of increasing bone mass.

these cells, PPARγ is a positive regulator of the transcription factor TSC-22, which induces p21 that in turn reduces cell proliferation [20]. Similarly, it also induces the expression of an inhibitor of RegIA (a cell proliferator enhancer) [21] thereby inhibiting cell multiplication. Together with its ability to promote epithelial cell differentiation, these properties of PPARγ point to its possible use as target for drugs against colon cancer [19, 22, 23].

Interestingly, the K422Q mutation in the ligand-binding domain of PPARγ revealed an epithelial-specific mechanism of PPARγ-mediated differentiation [24]. Enterocyte cell lines bearing this mutation do not differentiate upon treatment with PPARγ ligands. Furthermore, this mutation alters neither the DNA-binding nor the ligand-dependent gene transactivation capacity of PPARγ. Introduction of wild-type PPARγ in the mutant cells restores their capacity to differentiate into enterocytes. Surprisingly, this K422Q PPARγ mutant is able to induce adipogenic differentiation, which strongly suggests that the lysine 422 residue is crucial for the interaction with cofactors that are required for epithelial but not for adipogenic differentiation [24].

7.3
Peroxisome Proliferator-activated Receptor Beta/Delta

7.3.1
PPARβ/δ, Cell Survival, and Cell Differentiation

Like PPARα and PPARγ, PPARβ/δ binds fatty acids and, therefore, is also most likely a sensor of dietary lipids and lipid derivatives (reviewed in Ref. [25]). The first evidence for a role of PPARβ/δ in differentiation was shown in preadipocytes where PPARβ/δ mediates long-chain fatty acid effects on the expression of adipose-related genes [26]. Together with two additional transcription factors, C/EBPβ and C/EBPδ, PPARβ/δ appears to be implicated in the induction of PPARγ expression [27, 28]. In turn, high expression of PPARγ and C/EBPα in adipocytes triggers and maintains the terminal differentiation program. The implication of PPARβ/δ in differentiation has also been proposed in other cell types. For example, its expression is strongly increased upon differentiation of human primary macrophages or monocyte/macrophage cell lines [29]. Activation of PPARβ/δ using a selective agonist also promotes oligodendrocyte differentiation in cell culture [30], consistent with a myelination defect in the corpus callosum of PPARβ/δ-null mice [31]. However, the roles of PPARβ/δ in cell differentiation have been best studied in keratinocytes and more recently in the gut and the placenta [32, 33].

7.3.2
PPARβ/δ and Skin Epithelium Differentiation

Skin epithelium and hair follicles are mainly composed of keratinocytes. The epidermis is a pluristratified epithelium where keratinocytes proliferate and finally differentiate in order to form a cornified layer at the skin surface. Hair follicles are associated with the sebaceous glands consisting of sebocytes that are responsible for the production of sebum. Keratinocytes and sebocytes both arise from multipotent epithelial stem cells located in the bulge region of the hair follicle (Fig. 7.3a). PPARα, β/δ, and γ ligands have been shown to enhance rodent and human keratinocyte differentiation [34–37]. These activators also induce rodent sebocyte differentiation [38, 39] but, so far, only PPARβ/δ activators have been reported to stimulate the differentiation of human sebocytes [40]. Interestingly, whereas the interfollicular epidermis expresses very low levels of PPARs, the production of PPARβ/δ and to a lesser extent that of PPARα, is reactivated upon wounding. In the context of this challenge, PPARβ/δ expression is triggered by inflammatory

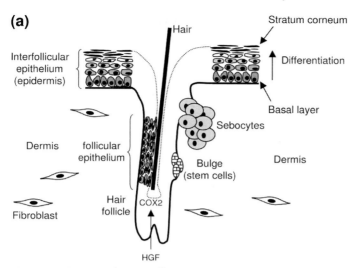

Fig. 7.3. PPARs are enhancers of keratinocyte and sebocyte differentiation. (a) The interfollicular squamous epithelium of the mammalian skin is composed of keratinocytes that differentiate as they migrate towards the stratum corneum. The hair follicle is also composed of several layers of keratinocytes, which give rise to the hair. Epithelial stem cells are located in the bulge region and are able to differentiate into keratinocytes and sebocytes.

cytokines, such as tumor necrosis factor alpha (TNFα), which activate the stress-associated protein kinase cascade, that in turn activates the transcription factor complex AP-1 and enhances its specific binding on the PPARβ/δ promoter [37, 41]. The first important function that PPARβ/δ exerts in this context is to promote cell survival, via transcriptional upregulation of *ILK* (integrin-linked kinase) and *PDK1* (phosphoinositide-dependent protein kinase-1) and indirect inhibition of *PTEN* (phosphatase and tensin homolog deleted on chromosome 10), leading to an increased PKB/Akt1 activity. PPARβ/δ is also necessary for the differentiation of primary keratinocytes in inflammatory situations, but the key target genes in this event are not yet identified (Fig. 7.3b) [37, 42].

Later on during healing, the activation of the transforming growth factor beta (TGFβ) signaling pathway leads to the inhibition of the AP-1-mediated increased expression of PPARβ/δ in keratinocytes. This decrease of PPARβ/δ expression to its basal level is due to the activation of Smad3 that interacts with Smad4. This complex enters the cell nucleus, binds to AP-1 and prevents its binding to the PPARβ promoter (Fig. 7.3b) [43]. Thus, this coordinated interplay between different cytokines time-regulates PPARβ/δ expression and activation to promote the healing program, involving cell survival, proliferation, and differentiation.

In contrast to the interfollicular epidermis, PPARβ/δ remains highly expressed in follicular keratinocytes throughout hair follicle development. Hair follicle morphogenesis depends on a fine-tuned balance between cell proliferation and apoptosis, which involves epithelial/mesenchymal interactions. Interestingly,

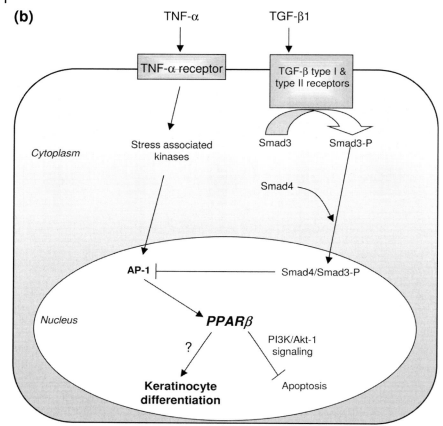

Fig. 7.3. (b) Early after an injury, tumor necrosis factor alpha (TNFα) is abundantly produced by inflammatory cells and keratinocytes, which activates a cascade of stress-associated protein kinases. The resulting activation of the transcription factor complex AP-1 leads to the activation of PPARβ/δ expression. Later on, transforming growth factor beta 1 (TGFβ1) produced by macrophages and myofibroblasts binds to its receptor (type I and II) at the keratinocyte surface and activates the transcription factor Smad3 by phosphorylation. The Smad3/Smad4 complex subsequently enters the nucleus and interacts with AP-1, preventing its binding to the PPARβ/δ promoter, which results in a reduction of PPARβ/δ expression.

PPARβ/δ-deficient mice exhibit significant retardation of postnatal hair follicle morphogenesis, particularly at the hair peg stage, mainly due to premature increased apoptosis of follicular keratinocytes [44]. The time-regulated activation of the PPARβ protein in follicular keratinocytes involves the upregulation of the cyclooxygenase-2 (COX-2) enzyme levels by a mesenchymal paracrine factor, the hepatocyte growth factor. The resulting enhanced activity of COX-2 in hair follicle keratinocytes then produces PPARβ ligands [18], and the subsequent PPARβ/δ-

PPARβ +/+ 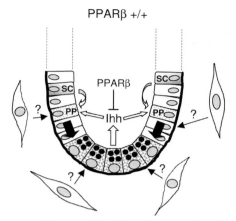 PPARβ -/-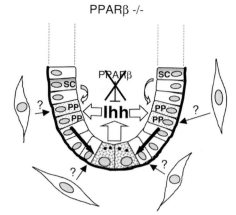

Fig. 7.4. PPARβ/δ regulates Paneth cell homeostasis in the small intestine. In the crypt of the small intestine, PPARβ/δ is a negative regulator of Indian hedgehog (Ihh) expression, which is mainly expressed by the mature Paneth cells. In PPARβ/δ-null animals, mature Paneth cells express higher amounts of Ihh protein than wild-type cells. This high level of Ihh in null animals indicates to Paneth cell precursors (PP) a false signal of plenty of mature Paneth cells, which leads to an inhibition of the differentiation of these precursors into mature cells (black arrows). This results in an accumulation of precursor cells and reduced amounts of mature cells in the crypts of the small intestine. Question marks indicate the presence of mesenchymal-derived signals that are likely participating to the overall process but are not yet clearly identified.

mediated activation of the PKB/Akt1 signaling pathway protects the hair peg keratinocytes against apoptosis, allowing for normal hair follicle differentiation (Fig. 7.3a).

7.3.3
PPARβ/δ and Differentiation of the Intestinal Epithelium

The intestinal epithelium is a fast self-renewing tissue whose maintenance and renewal is controlled by a small group of multipotent stem cells located in the intestinal crypts. These stem cells give rise to four differentiated cell types: enterocytes, goblet cells, enteroendocrine cells in the small intestine and the colon, and Paneth cells in the small intestine.

Paneth cells differentiate while migrating towards the crypt basis [45]. They are specialized in the defense against different pathogens and establish the antimicrobial barrier by sensing bacteria and secreting bactericidal peptides such as α-defensins, also called cryptdins in the mouse [46]. In addition, they have been implicated in intestinal angiogenesis [47], and are present in abnormally high amounts in intestinal metaplasia and in colon cancer [48]. Recent work in our laboratory showed that PPARβ/δ is involved in Paneth cell differentiation. PPARβ/δ acts on this process by negatively regulating the expression of the secreted protein Indian hedgehog (Ihh), which in the small intestine is mainly produced by mature

Paneth cells at the bottom of the crypts. The responding cells are few and located slightly higher, at the mid-crypt level. They express the Ihh receptor called Ptch-1 and most likely correspond to Paneth cell precursors as they also express lysosyme, while not bearing other characteristics of mature Paneth cells. Observations made in PPARβ/δ-null mice allowed the proposal of a regulatory negative feedback loop for the control of Paneth cell homeostasis. Secretion of high levels of Ihh by mature Paneth cells would limit the differentiation of their precursors since this signal would indicate sufficient amounts of differentiated cells (Fig. 7.4). By lowering *Ihh* gene expression, PPARβ/δ would favor precursor differentiation. Accordingly PPARβ/δ deletion results in fewer mature Paneth cells per crypt but a higher *Ihh* expression and a higher number of precursor cells in the small intestinal crypts (Fig. 7.4) [32].

7.3.4
PPARβ/δ and Trophoblastic Giant Cell Differentiation

The placenta is a crucial extra-embryonic organ where PPARβ/δ and γ play important roles. Indeed, null mice for PPARβ/δ or γ exhibit an embryonic lethality around E9.5 [7] due to an alteration of placenta development. Whereas this phenotype is fully penetrant in PPARγ homozygous conceptus, a few PPARβ/δ homozygous embryos survive this initial perturbation and are born alive, which has allowed the generation of mouse PPARβ/δ-null lines [31, 41, 49]. A close look at the placental alterations observed in the PPARγ- and PPARβ/δ-null embryos suggests a different role for each isotype during placenta development.

Placentas of PPARγ-null embryos exhibit specific alterations of the labyrinth, which normally is a structure with extensive villous branching (Fig. 7.5a) [50]. In PPARγ-null placentas, the invaginating chorionic villi are surrounded by an hypertrophied trophoblast tissue resulting in the absence of placental vasculature [7, 33]. The molecular mechanism underlying this alteration is so far unexplored. In comparison, histological studies revealed that the placenta of PPARβ/δ-null embryos exhibit an abnormally loose connection with the maternal decidua at E9.5 [49]. A systematic analysis showed that the placentas of PPARβ/δ-null embryos present alterations of the spongiotrophoblast and labyrinth layers. However, the most dramatically affected layer is the trophoblastic giant cell layer, which is a source of growth hormones (e.g. placental lactogen), angiogenic factors (e.g. VEGF and proliferin) and constitutes an interface between the maternal decidua and the placenta. The trophoblastic tissue also gives rise to the chorionic trophoblast, which subsequently differentiates into the labyrinth. Thus, the development of the giant cell layers is crucial for the overall placental development and is a major process altered in the PPARβ-null placenta.

The mechanism by which PPARβ acts on giant cell differentiation was further explored in the trophoblastic rat cell line Rcho-1. Interestingly, PPARβ/δ acts on giant cell differentiation via its positive action on the PI3K/Akt1 signaling pathway (Fig. 7.5b) [33], in a manner reminiscent of that seen in the context of keratinocyte survival during skin wound healing [42].

(a)

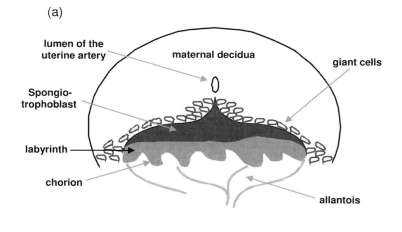

lumen of the
uterine artery

maternal decidua

giant cells

Spongio-
trophoblast

labyrinth

chorion

allantois

(b)

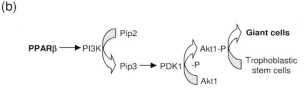

PPARβ ⟶ PI3K

Pip2

Pip3 ⟶ PDK1

Akt1-P

-P

Akt1

Giant cells

Trophoblastic
stem cells

Fig. 7.5. PPARβ/δ and PPARγ isotypes are required for the mouse placental development. (a) Scheme of the structure of the mouse placenta at E9.5. At this stage, the placenta is formed by three layers: the giant cell layer, which establishes the contact with the maternal decidua, the spongiotrophoblast, and the labyrinth where fetal and maternal blood vessels are in close contact. (b) PPARβ/δ acts on giant cell differentiation by activating the PI3K/Akt-1 signaling pathway.

7.4
Conclusion

In this chapter we have highlighted some important links between PPAR activity and cell differentiation. Since PPARs are controlled by lipid signals, a major un-answered question is how the generation of these signals is integrated into growth and differentiation programs that often involve interactions between tissues of dif-ferent embryonic origin. The effect of the hepatic growth factor (HGF) from the dermis on COX-2 expression in the epidermis and finally on the PPAR-dependent maturation of hair follicles, as reported herein, is an illustration of such intricate connections. Furthermore, the now well-recognized role of PPARs in energy ho-meostasis at the cellular and systemic levels suggests an association between the energy status of the cell and the organism, growth factor signaling, and PPAR activity. This association confers major functions to the lipid-activatable PPARs in the regulation of many vital pathways. As reviewed herein, effects of these recep-tors on cell survival or apoptosis, migration, proliferation, and differentiation have been documented for several cell types. Obviously, the underlying molecular mechanisms need now to be analyzed in depth since many facets of these regulations that are crucial for animal development and survival are very little explored so far.

Acknowledgments

The work performed in the authors' laboratory was supported by the Swiss National Science Foundation.

References

1 Nuclear Receptors Nomenclature Committee (**1999**) A unified nomenclature system for the nuclear receptor superfamily. *Cell* 97, 161–163.

2 Desvergne, B., W. Wahli (**1999**) Peroxisome proliferator-activated receptors: nuclear control of metabolism. *Endocr Rev* 20, 649–688.

3 Kersten, S., B. Desvergne, W. Wahli (**2000**) Roles of PPARs in health and disease. *Nature* 405, 421–424.

4 Desvergne, B., L. Michalik, W. Wahli (**2004**) Be fit or be sick: peroxisome proliferator-activated receptors are down the road. *Mol Endocrinol* 18, 1321–1332.

5 Rosen, E.D., C.J. Walkey, P. Puigserver, B.M. Spiegelman (**2000**) Transcriptional regulation of adipogenesis. *Genes Dev* 14, 1293–1307.

6 Hu, E., P. Tontonoz, B.M. Spiegelman (**1995**) Transdifferentiation of myoblasts by the adipogenic transcription factors PPAR gamma and C/EBP alpha. *Proc Natl Acad Sci USA* 92, 9856–9860.

7 Barak, Y., M.C. Nelson, E.S. Ong, Y.Z. Jones, P. Ruiz-Lozano, K.R. Chien, A. Koder, R.M. Evans (**1999**) PPAR gamma is required for placental, cardiac, and adipose tissue development. *Mol Cell* 4, 585–595.

8 Rosen, E.D., P. Sarraf, A.E. Troy, G. Bradwin, K. Moore, D.S. Milstone, B.M. Spiegelman, R.M. Mortensen (**1999**) PPAR gamma is required for the differentiation of adipose tissue in vivo and in vitro. *Mol Cell* 4, 611–617.

9 Imai, T., R. Takakuwa, S. Marchand, E. Dentz, J.M. Bornert, N. Messaddeq, O. Wendling, M. Mark, B. Desvergne, W. Wahli, P. Chambon, D. Metzger (**2004**) Peroxisome proliferator-activated receptor gamma is required in mature white and brown adipocytes for their survival in the mouse. *Proc Natl Acad Sci USA* 101, 4543–4547.

10 Altiok, S., M. Xu, B.M. Spiegelman (**1997**) PPARgamma induces cell cycle withdrawal: inhibition of E2F/DP DNA-binding activity via downregulation of PP2A. *Genes Dev* 11, 1987–1998.

11 Morrison, R.F., S.R. Farmer (**1999**) Role of PPARgamma in regulating a cascade expression of cyclin-dependent kinase inhibitors, p18(INK4c) and p21(Waf1/Cip1), during adipogenesis. *J Biol Chem* 274, 17088–17097.

12 Rieusset, J., C. Chambrier, K. Bouzakri, E. Dusserre, J. Auwerx, J.P. Riou, M. Laville, H. Vidal (**2001**) The expression of the p85alpha subunit of phosphatidylinositol 3-kinase is induced by activation of the peroxisome proliferator-activated receptor gamma in human adipocytes. *Diabetologia* 44, 544–554.

13 Michalik, L., B. Desvergne, W. Wahli (**2004**) Peroxisome-proliferator-activated receptors and cancers: complex stories. *Nat Rev Cancer* 4, 61–70.

14 Demetri, G.D., C.D. Fletcher, E. Mueller, P. Sarraf, R. Naujoks, N. Campbell, B.M. Spiegelman, S. Singer (**1999**) Induction of solid

tumor differentiation by the peroxisome proliferator-activated receptor-gamma ligand troglitazone in patients with liposarcoma. *Proc Natl Acad Sci USA* 96, 3951–3956.

15 Akune, T., S. Ohba, S. Kamekura, M. Yamaguchi, U.I. Chung, N. Kubota, Y. Terauchi, Y. Harada, Y. Azuma, K. Nakamura, T. Kadowaki, H. Kawaguchi (**2004**) PPARgamma insufficiency enhances osteogenesis through osteoblast formation from bone marrow progenitors. *J Clin Invest* 113, 846–855.

16 Ali, A.A., R.S. Weinstein, S.A. Stewart, A.M. Parfitt, S.C. Manolagas, R.L. Jilka (**2005**) Rosiglitazone causes bone loss in mice by suppressing osteoblast differentiation and bone formation. *Endocrinology* 146, 1226–1235.

17 Rieusset, J., F. Touri, L. Michalik, P. Escher, B. Desvergne, E. Niesor, W. Wahli (**2002**) A new selective peroxisome proliferator-activated receptor gamma antagonist with antiobesity and antidiabetic activity. *Mol Endocrinol* 16, 2628–2644.

18 Lefebvre, M., B. Paulweber, L. Fajas, J. Woods, C. McCrary, J.F. Colombel, J. Najib, J.C. Fruchart, C. Datz, H. Vidal, P. Desreumaux, J. Auwerx (**1999**) Peroxisome proliferator-activated receptor gamma is induced during differentiation of colon epithelium cells. *J Endocrinol* 162, 331–340.

19 Koeffler, H.P. (**2003**) Peroxisome proliferator-activated receptor gamma and cancers. *Clin Cancer Res* 9, 1–9.

20 Gupta, R.A., P. Sarraf, J.A. Brockman, S.B. Shappell, L.A. Raftery, T.M. Willson, R.N. DuBois (**2003**) Peroxisome proliferator-activated receptor gamma and transforming growth factor-beta pathways inhibit intestinal epithelial cell growth by regulating levels of TSC-22. *J Biol Chem* 278, 7431–7438.

21 Gupta, R.A., J.A. Brockman, P. Sarraf, T.M. Willson, R.N. DuBois (**2001**) Target genes of peroxisome proliferator-activated receptor gamma in colorectal cancer cells. *J Biol Chem* 276, 29681–29687.

22 Osawa, E., A. Nakajima, K. Wada, S. Ishimine, N. Fujisawa, T. Kawamori, N. Matsuhashi, T. Kadowaki, M. Ochiai, H. Sekihara, H. Nakagama (**2003**) Peroxisome proliferator-activated receptor gamma ligands suppress colon carcinogenesis induced by azoxymethane in mice. *Gastroenterology* 124, 361–367.

23 Sarraf, P., E. Mueller, D. Jones, F.J. King, D.J. DeAngelo, J.B. Partridge, S.A. Holden, L.B. Chen, S. Singer, C. Fletcher, B.M. Spiegelman (**1998**) Differentiation and reversal of malignant changes in colon cancer through PPARgamma. *Nat Med* 4, 1046–1052.

24 Gupta, R.A., P. Sarraf, E. Mueller, J.A. Brockman, J.J. Prusakiewicz, C. Eng, T.M. Willson, R.N. DuBois (**2003**) Peroxisome proliferator-activated receptor gamma-mediated differentiation: a mutation in colon cancer cells reveals divergent and cell type-specific mechanisms. *J Biol Chem* 278, 22669–22677.

25 Michalik, L., B. Desvergne, W. Wahli (**2003**) Peroxisome proliferator-activated receptors beta/delta: emerging roles for a previously neglected third family member. *Curr Opin Lipidol* 14, 129–135.

26 Amri, E.Z., F. Bonino, G. Ailhaud, N.A. Abumrad, P.A. Grimaldi (**1995**) Cloning of a protein that mediates transcriptional effects of fatty acids in preadipocytes. Homology to peroxisome proliferator-activated receptors. *J Biol Chem* 270, 2367–2371.

27 Clarke, S.L., C.E. Robinson, J.M. Gimble (**1997**) CAAT/enhancer binding proteins directly modulate transcription from the peroxisome proliferator-activated receptor gamma 2 promoter. *Biochem Biophys Res Commun* 240, 99–103.

28 Bastie, C., D. Holst, D. Gaillard, C. Jehl-Pietri, P.A. Grimaldi (**1999**) Expression of peroxisome proliferator-activated receptor PPARdelta promotes induction of PPARgamma and adipocyte differentiation in 3T3C2

fibroblasts. *J Biol Chem* 274, 21920–21925.

29 Vosper, H., G.A. Khoudoli, C.N. Palmer (**2003**) The peroxisome proliferator activated receptor delta is required for the differentiation of THP-1 monocytic cells by phorbol ester. *Nucl Recept* 1, 9.

30 Saluja, I., J.G. Granneman, R.P. Skoff (**2001**) PPAR delta agonists stimulate oligodendrocyte differentiation in tissue culture. *Glia* 33, 191–204.

31 Peters, J.M., S.S. Lee, W. Li, J.M. Ward, O. Gavrilova, C. Everett, M.L. Reitman, L.D. Hudson, F.J. Gonzalez (**2000**) Growth, adipose, brain, and skin alterations resulting from targeted disruption of the mouse peroxisome proliferator-activated receptor beta(delta). *Mol Cell Biol* 20, 5119–5128.

32 Varnat, F., B. Bordier-ten Heggeler, N. Boucard, I. Corthésy, W. Wahli, B. Desvergne (**2005**) PPAR-beta mediates the intestinal innate immunity by controlling the Paneth cell differentiation. Thesis work, presented on 17 December 2004.

33 Nadra, K. (**2005**) Role of PPAR-beta in placenta development. Thesis work, presented on 25 February 2005.

34 Komuves, L.G., K. Hanley, A.M. Lefebvre, M.Q. Man, D.C. Ng, D.D. Bikle, M.L. Williams, P.M. Elias, J. Auwerx, K.R. Feingold (**2000**) Stimulation of PPARalpha promotes epidermal keratinocyte differentiation in vivo. *J Invest Dermatol* 115, 353–360.

35 Mao-Qiang, M., A.J. Fowler, M. Schmuth, P. Lau, S. Chang, B.E. Brown, A.H. Moser, L. Michalik, B. Desvergne, W. Wahli, M. Li, D. Metzger, P.H. Chambon, P.M. Elias, K.R. Feingold (**2004**) Peroxisome-proliferator-activated receptor (PPAR)-gamma activation stimulates keratinocyte differentiation. *J Invest Dermatol* 123, 305–312.

36 Schmuth, M., C.M. Haqq, W.J. Cairns, J.C. Holder, S. Dorsam, S. Chang, P. Lau, A.J. Fowler, G.

Chuang, A.H. Moser, B.E. Brown, M. Mao-Qiang, Y. Uchida, K. Schoonjans, J. Auwerx, P. Chambon, T.M. Willson, P.M. Elias, K.R. Feingold (**2004**) Peroxisome proliferator-activated receptor (PPAR)-beta/delta stimulates differentiation and lipid accumulation in keratinocytes. *J Invest Dermatol* 122, 971–983.

37 Tan, N.S., L. Michalik, N. Noy, R. Yasmin, C. Pacot, M. Heim, B. Fluhmann, B. Desvergne, W. Wahli (**2001**) Critical roles of PPAR beta/delta in keratinocyte response to inflammation. *Genes Dev* 15, 3263–3277.

38 Rosenfield, R.L., A. Kentsis, D. Deplewski, N. Ciletti (**1999**) Rat preputial sebocyte differentiation involves peroxisome proliferator-activated receptors. *J Invest Dermatol* 112, 226–232.

39 Kim, M.J., D. Deplewski, N. Ciletti, S. Michel, U. Reichert, R.L. Rosenfield (**2001**) Limited cooperation between peroxisome proliferator-activated receptors and retinoid X receptor agonists in sebocyte growth and development. *Mol Genet Metab* 74, 362–369.

40 Chen, W., C.C. Yang, H.M. Sheu, H. Seltmann, C.C. Zouboulis (**2003**) Expression of peroxisome proliferator-activated receptor and CCAAT/enhancer binding protein transcription factors in cultured human sebocytes. *J Invest Dermatol* 121, 441–447.

41 Michalik, L., B. Desvergne, N.S. Tan, S. Basu-Modak, P. Escher, J. Rieusset, J.M. Peters, G. Kaya, F.J. Gonzalez, J. Zakany, D. Metzger, P. Chambon, D. Duboule, W. Wahli (**2001**) Impaired skin wound healing in peroxisome proliferator-activated receptor (PPAR)alpha and PPARbeta mutant mice. *J Cell Biol* 154, 799–814.

42 Di-Poi, N., N.S. Tan, L. Michalik, W. Wahli, B. Desvergne (**2002**) Antiapoptotic role of PPARbeta in keratinocytes via transcriptional control of the Akt1 signaling pathway. *Mol Cell* 10, 721–733.

43 Tan, N.S., L. Michalik, N. Di-Poi,

C.Y. NG, N. MERMOD, A.B. ROBERTS, B. DESVERGNE, W. WAHLI (**2004**) Essential role of Smad3 in the inhibition of inflammation-induced PPARbeta/delta expression. *EMBO J* 23, 4211–4221.

44 DI-POI, N., C.Y. NG, N.S. TAN, Z. YANG, B.A. HEMMINGS, B. DESVERGNE, L. MICHALIK, W. WAHLI (**2005**) Epithelium-mesenchyme interactions control the activity of peroxisome proliferator-activated receptor beta/delta during hair follicle development. *Mol Cell Biol* 25, 1696–1712.

45 KARAM, S.M. (**1999**) Lineage commitment and maturation of epithelial cells in the gut. *Front Biosci* 4, D286–298.

46 GANZ, T. (**2003**) Microbiology: Gut defence. *Nature* 422, 478–479.

47 STAPPENBECK, T.S., L.V. HOOPER, J.I. GORDON (**2002**) Developmental regulation of intestinal angiogenesis by indigenous microbes via Paneth cells. *Proc Natl Acad Sci USA* 99, 15451–15455.

48 PORTER, E.M., C.L. BEVINS, D. GHOSH, T. GANZ (**2002**) The multifaceted Paneth cell. *Cell Mol Life Sci* 59, 156–170.

49 BARAK, Y., D. LIAO, W. HE, E.S. ONG, M.C. NELSON, J.M. OLEFSKY, R. BOLAND, R.M. EVANS (**2002**) Effects of peroxisome proliferator-activated receptor delta on placentation, adiposity, and colorectal cancer. *Proc Natl Acad Sci USA* 99, 303–308.

50 ROSSANT, J., J.C. CROSS (**2001**) Placental development: lessons from mouse mutants. *Nat Rev Genet* 2, 538–548.

8
Advances in Selenoprotein Expression: Patterns and Individual Variations

Catherine Méplan, Vasileios Pagmantidis, and John E. Hesketh

8.1
Introduction: Selenium, Nutrigenomics, and Health

Nutrigenomics is the study of nutrient–gene interactions using the techniques of functional genomics. Nutrient–gene interactions reflect both how nutrients influence gene expression and conversely how genetic factors influence individual nutritional requirements. At the global level, nutrigenomics is concerned with characterizing patterns of gene expression in response to nutrients. At the individual level, nutrigenomics can help to determine the differences in requirements for population subgroups based on genetic variations, gender, and life-stage. Postgenomic approaches have already provided important insights into selenium metabolism and selenoprotein expression; for example, bioinformatics has helped to identify novel selenoprotein genes, knockout mice have given functional information, and transfected cells models have proven to be useful to understand several mechanisms of regulation. In this chapter we will review the mechanisms of selenoprotein expression and their regulation and then highlight how functional genomic techniques can contribute to the understanding of interactions between selenium and genes, both in terms of response to dietary factors and identification of single-nucleotide polymorphisms (SNPs) and their functions.

Selenium (Se) was first demonstrated to be an essential nutrient for animals in 1957 when it was found to be a component of "factor 3" that protects vitamin E-deficient rats against necrotic liver degeneration [1]. Subsequently, selenium was shown to be essential for human health when Keshan disease, an endemic cardiomyopathy, and Kashin-Beck disease, a deforming arthritis, were identified in areas of China where the soil is extremely low in selenium and when Keshan disease was shown to be due to a combination of severe selenium deficiency and viral infection (see Refs [2, 3]). Since then, selenium has been suggested to play a role in a number of physiological and pathological processes: in immune function, in viral suppression, and acquired immune deficiency syndrome (AIDS) [4, 5], in male fertility, in thyroid function, and as an anticancer agent [6]. Many of the physiological roles of selenium are attributed to its role, as the amino acid selenocysteine (Sec), in the selenium-containing selenoproteins. In the past decade, selenium and the

Nutritional Genomics. Edited by Regina Brigelius-Flohé and Hans-Georg Joost
Copyright © 2006 WILEY-VCH Verlag GmbH & Co. KGaA, Weinheim
ISBN: 3-527-31294-3

selenoproteins have received considerable attention for their role in health and disease [7, 8].

Human dietary selenium intakes vary geographically from high to low. Severe deficiency, such as that found in the Keshan area of China is rare, but suboptimal selenium intake, which is more common, has been linked to cancer susceptibility and selenium supplementation reported to reduce cancer morbidity [6]. Indeed, there is evidence that selenium intakes are diminishing in several European countries [7, 9]. For example in the UK, during the 1970s and 1980s, average intakes were around 60 µg/day. By the 1990s, selenium intakes had dropped by half to about 30 µg/day, well below the current daily Reference Nutrient Intake (RNI) for selenium in the UK – 1.0 µg/kg body weight, which is equivalent to 75 and 60 µg/day for men and women respectively. Thus, although selenium intakes are not low enough to cause overt deficiency they may not be sufficient for optimal health. Therefore, it is important to understand how selenium regulates selenoproteins and other genes, for example in respect to antioxidant protection and cell growth control.

8.2
Selenoproteins

In 1973 selenium was shown to be an essential component of mammalian glutathione peroxidase [10, 11] and subsequently the gene sequences for two selenoproteins, mammalian glutathione peroxidase and bacterial formate dehydrogenase, were reported [12, 13]. DNA and protein sequence alignments revealed that a selenocysteine (Sec) residue in the protein coincided with an in-frame TGA codon in the corresponding gene. Since these early observations, the presence of Sec has been demonstrated to be a feature shared by all selenoproteins and in all cases Sec incorporation occurs during translation at a UGA codon [14]. Selenoproteins have evolved to take advantage of the fact that the pK_a of Sec (5.2) is lower than physiological pH [15] so that this amino acid residue is ideally suited to participate in redox reactions. In several selenoproteins, Sec is present within a CXXU or UXXC motif (where C is a cysteine, X any amino acid and U Sec) at the active site of the selenoenzyme and it is essential for activity. The CXXU or UXXC motif is very similar to the classical CXXC redox motif found in redox-active proteins and this motif provides the redox active center of some selenoproteins. A catalytical triad of Sec, glutamine, and tryptophan is present in glutathione peroxidases [16]. Genes homologous to selenoprotein genes but in which Sec is replaced by cysteine occur, in particular in bacteria deficient in Sec-incorporation mechanisms (see Section 8.3.1). However, these homologs exhibit poor catalytic activity, showing that the Sec-containing motif is much more powerful than its cysteine counterpart [8].

Selenoprotein genes have been identified both in prokaryotes (archaea, bacteria) and in eukaryotes (plants, animals). Currently 25 mammalian selenoprotein genes have been identified [17], although not all of them have yet been characterized. A list of well-known selenoproteins and some of their features is presented in Table 8.1. The glutathione peroxidase (GPx) family is the most extensively studied so far

Table 8.1. Main mammalian selenoproteins, their function, tissue distribution, and subcellular localization.

Selenoproteins	Function	Tissue distribution	Subcellular localization	References — Knockout mice studies	References — Transgenic/over expression studies	Splice variants
Cytoplasmic GPx1	Antioxidant, selenium stores?	Ubiquitous	Cytosol, mitochondria	[3, 96, 122–125]	[126, 127]	2 variants (predicted)
Gastrointestinal GPx2	Antioxidant?	GI tract	Cytosol	[96, 128]		
Plasma GPx3	Plasma and extracellular antioxidant?	Plasma, kidney, thyroid, heart, placenta, lung, GI	Extracellular			
Phospholipid-hydroperoxide glutathione peroxidase GPx4	Prevents lipid peroxidation in cell membrane, part of the mitochondrial capsule in sperm	Various tissues, testes	Cell membrane, variant 2: nucleus, variant 1: mitochondria + cytosol	[97]	[98]	2 variants [78]
Sperm nucleus GPx6	Sperm maturation, prevents apoptosis	Testes, spermatozoa	Sperm nucleus			
TR1	Redox control, reduction Trx development	Ubiquitous, skin, astrocytes, neurons	Mainly cytosol, other compartments	[99]	[129]	6 variants [69–74]

TR2	Prevents apoptosis, hematopoiesis, heart development and function	Ubiquitous, heart	Mitochondria	[99, 100]	
TGR, Thioredoxin-glutathione reductase	Membrane-associated	Testes	Cytosol		
Selenoprotein P	Selenium transport, brain, antioxidant	Plasma, brain, liver, thyroid	Extracellular	[101–103]	
SPS2 (Selenophosphate synthetase 2)	Converts selenite into monoselenophosphate	Ubiquitous, liver, testis	Cytosol		2 variants (predicted)
IDI (deiodinase 1)	Converts thyroxine (T4) to bioactive T3	Thyroid, liver, kidney, pituitary	Plasma membrane		2 variants (predicted)
IDII (deiodinase 2)	Converts thyroxine (T4) to bioactive T3	Thyroid, heart, brain, brown adipose tissue	Endoplasmic reticulum	[130, 131]	3 variants [77]
IDIII (deiodinase 3)	Converts thyroxine (T4) to bioinactive reverse T3 (rT3)	Skin, CNS, placenta, fetal liver	Plasma membrane		3 variants [75, 76]
15kDa Selenoprotein (Sep15)	Unknown	Liver, prostate, testes, brain, kidney	Endoplasmic reticulum		2 variants (predicted)
Selenoprotein W	GSH-dependent antioxidant?	Ubiquitous, heart, skeletal muscle		[32]	

and five glutathione peroxidases have been identified: cytosolic glutathione peroxidase GPx1, gastrointestinal glutathione peroxidase GPx2, plasma glutathione peroxidase GPx3, phospholipid hydroperoxide glutathione peroxidase GPx4, and GPx6 (identified *in silico*) [17]. Glutathione peroxidases have antioxidant actions, protecting cells from oxidative stress, but also obviously have more specific functions [16].

The thioredoxin reductase (TR) family of selenoproteins, TR1, TR2, and TR3, comprises another important set of selenoenzymes. Mammalian TRs have many diverse cellular functions, including reduction of ribonucleotides to deoxyribonucleotides, which are essential for the synthesis of DNA, maintenance of redox state in the cell, the regulation of activity of transcription factors [18], and the regeneration of the redoxactive systems. TRs act in conjunction with their redox partner and substrate thioredoxin. Another important class of selenoproteins is the iodothyronine deiodinase family (IDs), IDI, IDII, and IDIII, involved in thyroid hormone metabolism [19].

Selenoprotein P (SelP) is an extracellular, monomeric protein containing 10 Sec residues (human, murine, and rat). SelP is ubiquitously expressed in mammalian tissues [20, 21]. In human plasma, SelP accounts for over 50% of the total selenium concentration. Because of the abundance of Sec residues and the fact that SelP is a major selenium component in the plasma, it has been suggested that SelP may be a useful marker of selenium status [22]. Although the function of SelP is not fully understood, it has been suggested to have a dual function [20], serving both as an extracellular antioxidant defense, especially against peroxynitrite (a reactive intermediate formed *in vivo* by reaction of nitrogen monoxide with the superoxide anion) [23, 24], and also as a selenium-transport protein, transporting hepatic selenium from the liver to the other organs [25–27]. The latter function has recently been confirmed in SelP-knockout mice (see Section 8.5.3).

Selenoproteins not yet fully characterized include the 15 kDa selenoprotein and selenoprotein W (SelW). The gene encoding the 15 kDa selenoprotein is localized on chromosome 1p31, a genetic locus commonly mutated or deleted in human cancers. This location has led to the suggestion that this protein may play a role in cancer etiology [28]. SelW was first isolated from rat muscle [29] and the cDNA sequence confirmed the presence of a Sec-UGA codon [30]. Expression of SelW responds to selenium supplementation and functional studies suggest that it has an antioxidant function. For example, overexpression of SelW in CHO and H1299 human lung cancer cells markedly reduced the sensitivity of both cell lines to H_2O_2 cytotoxicity [31].

8.3
Selenoprotein Synthesis

8.3.1
Role of SECIS and 3′ Untranslated Region (UTR)

The synthesis of selenoproteins involves the incorporation of Sec into the protein sequence during translation (see Fig. 8.1). Critically, this incorporation occurs at a

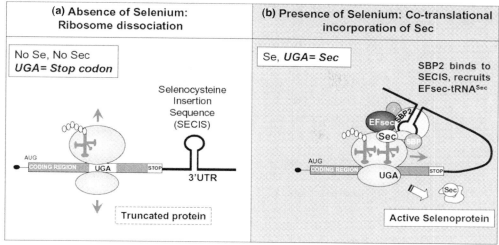

(a) Absence of Selenium: Ribosome dissociation

No Se, No Sec
UGA= Stop codon

Selenocysteine Insertion Sequence (SECIS)

AUG
CODING REGION UGA STOP
3'UTR

Truncated protein

(b) Presence of Selenium: Co-translational incorporation of Sec

Se, **UGA= Sec**

SBP2 binds to SECIS, recruits EFsec-tRNA^{Sec}

AUG
CODING REGION UGA STOP

Active Selenoprotein

Fig. 8.1. Co-translational selenocysteine incorporation in selenoproteins at a UGA codon, in the presence of a SECIS structure. The figure shows a model of Sec incorporation into selenoproteins during their translation. In both panels a scheme of a typical selenoprotein mRNA is shown, with from left to right the AUG (initiation codon), a gray rectangle (coding region of the mRNA, containing an in-frame Sec-coding UGA), a 3′ UTR (black line), containing a stem–loop structure (selenenocysteine-insertion sequence or SECIS structure). (a) In the absence of selenium, Sec is missing and when the ribosome reaches the Sec-coding UGA codon, it recognizes it as a stop codon and dissociates, ending prematurely the translation and resulting in the synthesis of an inactive-truncated protein. (b) In the presence of selenium several factors, including SBP2, bind to the SECIS structure, recruit Sec-tRNA bound to the elongation factor EFsec to the ribosome, and allow the incorporation of Sec within the protein sequence resulting in the production of an active selenoprotein. (?) stands for factors not yet identified. SBP (SECIS-binding protein 1) is also known to play a role in this mechanism.

UGA codon, which usually encodes a signal to terminate translation. This recoding of the UGA codon requires the presence in the selenoprotein mRNA of a Sec insertion sequence (SECIS) structure. In eukaryotes this SECIS is located in the 3′ UTR of the mRNA. Sec incorporation also requires protein factors such as the SBP2 (SECIS-binding protein 2) and the specific elongation factor EFSec, which recruits a specific tRNA charged with a Sec to the translation machinery. With the exception of the selenoprotein P mRNA, all selenoprotein mRNAs are characterized by the presence of one in-frame Sec-coding UGA and one single SECIS element located in their 3′ UTR. The selenoprotein P mRNA is remarkable in that it possesses 9–17 Sec-encoding UGA codons; for example the human SelP contains 10 Sec and two SECIS structures in its 3′ UTR [20].

In eukaryotes, the SECIS element is a sequence of 50–60 nucleotides, in the 3′ UTR of selenoprotein mRNAs. The length of the 3′ UTRs in selenoprotein mRNAs shows a wide variation and the position of the SECIS within the 3′ UTR is not constant (Fig. 8.2). Furthermore, as shown in Fig. 8.2, the distance between

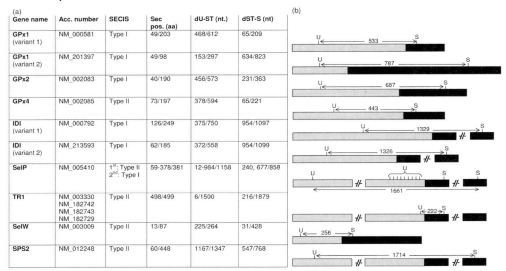

(a)

Gene name	Acc. number	SECIS	Sec pos. (aa)	dU-ST (nt.)	dST-S (nt)
GPx1 (variant 1)	NM_000581	Type I	49/203	468/612	65/209
GPx1 (variant 2)	NM_201397	Type I	49/98	153/297	634/823
GPx2	NM_002083	Type I	40/190	456/573	231/363
GPx4	NM_002085	Type II	73/197	378/594	65/221
IDI (variant 1)	NM_000792	Type I	126/249	375/750	954/1097
IDI (variant 2)	NM_213593	Type I	62/185	372/558	954/1099
SelP	NM_005410	1st: Type II 2nd: Type I	59-378/381	12-984/1158	240, 677/858
TR1	NM_003330 NM_182742 NM_182743 NM_182729	Type II	498/499	6/1500	216/1879
SelW	NM_003009	Type II	13/87	225/264	31/428
SPS2	NM_012248	Type II	60/448	1167/1347	547/768

Fig. 8.2. Variations in UGA/SECIS distances between the different selenoprotein mRNAs. (a) The position (pos) of the Sec amino acid, the distances between Sec and stop codon and between the stop codon and the SECIS element: expressed as ratios to the total number of amino acids, the total nucleotides in the coding region and the total nucleotides in the 3′ UTR respectively. (b) The gray areas indicate the coding region and the black areas represent the 3′ UTR of the corresponding selenoprotein mRNAs; the diagrams are in scale apart from where -//- indicates a break. The corresponding nucleotide distances between the UGA codon and the SECIS element are indicated by arrows. The sizes indicated in this figure were obtained from the NCBI database (www.ncbi.nih.gov). U, UGA codon; S, SECIS element; ST, STOP codon; d, distance; aa, amino acid; nt, nucleotide.

the UGA codon encoding Sec and the SECIS varies greatly from one selenoprotein mRNA to another [32], suggesting that this distance does not play a regulatory role in Sec incorporation. On the contrary, the prokaryotic SECIS sequences are longer than their eukaryotic counterparts and are located immediately downstream of the UGA codon [33]. Experimental displacement of the SECIS sequence compromises Sec incorporation [34], showing the importance in the case of bacterial mRNAs of the position of SECIS in relation to the UGA codon.

Chemical and enzymatic probing analysis of secondary structures of rat and human IDI and rat GPx1 SECIS elements revealed specific structural features of SECIS elements in eukaryotes and a general model for the SECIS structure was confirmed by sequence alignment from several selenoproteins [35, 36]. The SECIS element adopts a particular stem–loop structure with two helices (helix I and II) separated by an internal loop and a terminal loop after helix II. Only few short sequences are conserved between the different SECISs. However they all contain three specific, highly conserved, features: (1) a SECIS core, located at the base of the stem–loop and composed of two sequences, in 5′ a quartet AUAG sequence and in 3′ a GA, which form a stem containing non-Watson–Crick G-A/A-G tandem pairs; (2) a stretch of two to three adenosines in a loop or a bulge; and (3) a stem formed of sequences with great internal complementarity that separates the SECIS core at the base from the loop at the top of the stem [35, 36]. A second form (SECIS type II) has also been described and contains an additional bulge within helix II where the invariant adenosines are located [37].

The remarkable conservation of the SECIS structural motifs has made it possible to develop software that can predict the existence of SECIS structures, for example SECISearch2.0 software available at http://genome.unl.edu/SECISearch.html. In conjunction with other approaches, this software was used by several groups to predict new selenoprotein genes and allowed the identification of several new potential selenoproteins [17]. First, mammalian gene sequences were screened using the SECISearch2.0 software to determine candidates containing potential SECIS elements; second, the gene sequences were screened using the *geneid* software for the presence of in-frame UGA codons; finally, homologous genes with cysteine in the place of the suspected Sec were identified in SECIS-deficient species [17]. All characterized selenoprotein mRNAs respond to this model except SelP, which in mammals possesses two SECIS structures and several in-frame UGA codons, and TR2 which has a unique SECIS element.

8.3.2
Role of *trans*-acting Factors

8.3.2.1 Transfer RNA

A specific tRNA, tRNA$^{(ser)sec}$, is responsible for the incorporation of Sec into the nascent polypeptide and as a result, is required for biosynthesis of the whole family of selenoproteins. Unlike other amino acids which are synthesized prior to being loaded onto their corresponding tRNA, the synthesis of Sec occurs directly on its tRNA in both prokaryotes [38] and mammals [39]. The tRNA carrying a seryl residue serves as an acceptor for selenophosphate (synthesized from selenide and ATP in the presence of selenophosphate synthetase 2), the active form of selenium, and the serine is converted to Sec. Further nucleoside methylations are required for maturation of the tRNA (reviewed in Ref. [8]) and two isoforms of tRNA$^{(ser)sec}$, corresponding to methylation variants, have been described. The methylation responds to selenium status and interestingly, the level of selenium-induced methylated form of the tRNA$^{(ser)sec}$ correlates with the synthesis of certain selenoproteins such as GPx1, GPx3, SelR, SelT, and SelW [40, 41] while the isoform lacking the methyl group is required for the biosynthesis of other selenoproteins such as TR1 and TR3 [41–43]. The tRNA$^{(ser)sec}$ gene, *trsp*, is present in a single copy in the genomes of mammals and in most species where it has been identified, except for the zebrafish which contains two copies of the gene [44].

8.3.2.2 RNA-binding Proteins

The SECIS-specific binding protein 2 (SBP2) binds to SECIS elements in 3′ UTRs of selenoprotein mRNAs [45, 46] and it is proposed to form a complex with the 3′ UTR and the tRNA$^{(ser)sec}$-specific elongation factor EFsec (described below). SBP2 does not discriminate between the two types of SECIS and there is no clear evidence of SBP2 binding with a different affinity to the various selenoprotein mRNAs [47]. However, competition experiments suggest that SBP2 can bind preferentially to the first SECIS element of selenoprotein P mRNA and the GPx4 SECIS element [48]. SBP2 binds to the RNA backbone phosphate groups of the non-Watson–Crick base pair quartet at the core of the SECIS element and to the lower helix and internal loop [47]. The SECIS-binding domain of SBP2 shares homology with the L7A/L30 RNA-binding family [49] and is separated from the Sec-incorporation domain [47, 50]. SBP2 is considered to be the limiting *trans*-acting factor in the Sec incorporation machinery [48, 51].

Contrary to the prokaryotic Sec-incorporation machinery in which a unique protein, SelB, exerts the double role of binding to the SECIS element and of recruiting the tRNA$^{(ser)sec}$ to the ribosome, the mammalian homolog SelB/EFsec or EFsec is unable to bind to any SECIS structure. As illustrated schematically in Fig. 8.1, this specific elongation factor is specific for the tRNA$^{(ser)sec}$ and is thought to interact with the SBP2/SECIS complex in order to deliver the tRNA$^{(ser)sec}$ to the ribosome [52, 53]; EFsec binding to SBP2 has been shown to require the formation of a complex of the elongation factor with the tRNA$^{(ser)sec}$ [54] suggesting that EFsec binding to tRNA$^{(ser)sec}$ occurs prior to formation of the complex with SBP2 and the SECIS. Unlike SBP2, EFsec is not a limiting factor in the Sec-incorporation process [48].

8.3.2.3 **Selenophosphate Synthetases**

As described above, the biosynthesis of Sec involves reaction of seryl-tRNA with selenophosphate. The conversion of selenide into selenophosphate is catalyzed, in the presence of ATP, by a group of enzymes called selenophosphate synthetases [8]. Two selenophosphate synthetases, SPS1 and SPS2, were identified in mammals and *Drosophila*. The SPS2 protein is a homolog of the *selD* gene product of *Escherichia coli* [55, 56] in which the catalytic Cys17 is replaced by Sec. The presence of a Sec in the protein suggests that SPS2 could be involved in an autoregulation of its own synthesis. On the contrary, the SPS1 enzyme, in which the Sec is replaced in the human homolog by a threonine, has a poor catalytic activity. To better characterize the role of mammalian selenophosphate synthetases, Tamura *et al.* [57] recently performed an *in vivo* complementation assay using a *selD*-deficient *E. coli* strain transformed with vector expressing SPS1 or SPS2 cDNAs cloned from a lung adenocarcinoma cell line. The authors observed that SPS2 was able to complement efficiently the *selD* mutant in a selenite-containing medium. In contrast, only a weak complementation was observed with SPS1 except when cells were grown in the presence of Sec. The authors concluded that the role of SPS1 could be to recycle Sec, whereas that of SPS2 is to convert selenite into Sec [57].

8.4
Control of Selenoprotein Gene Expression

The regulation of the expression of selenoproteins occurs at transcriptional and post-transcriptional levels using both mechanisms that are common to the regulation of expression of other genes and mechanisms that are specific to selenoproteins due to the unique factors involved in Sec incorporation during translation.

8.4.1
Selenoprotein mRNAs: Transcriptional Regulation, Alternative Splicing, and mRNA Stability

Selenoprotein mRNA levels vary significantly in response to various stimuli or conditions, as illustrated by the examples below. Overall, selenoenzyme genes are upregulated by oxidative stress; this reflects the fact that their gene products display critical antioxidant properties. However, the gene regulation mechanisms differ from one gene to another. Characterization of the plasma glutathione peroxidase (GPx3) promoter region revealed the presence of a binding site for Sp1 transcription factor as well as a metal responsive element (MRE) and an antioxidant responsive element (ARE), known to be induced in response to oxidative/xenobiotic stress. A functional ARE has recently been identified in the promoter of GPx2 which responded to Nrf2 activators and to Nrf2 itself. Also endogenous GPx2 was upregulated by Nrf2 activators like sulforaphane or curcumin [58]. This adds GPx2 to the list of phase 2 enzymes which are part of the adaptive response (see below). Moreover, hypoxia was shown to be a strong transcriptional regulator of GPx3 with the identification of a binding site for the hypoxia-inducible factor 1 (HIF-1) [59], making GPx3 the only member of the GPx family so far to be regulated by hypoxia. In contrast, the cellular glutathione peroxidase (GPx1) was shown to be upregulated by increasing oxygen levels through two oxygen-responsive elements (ORE1 and ORE2) [60].

Regulation of the *TXNRD1* gene, encoding TR1 protein, appears to be particularly complex as the core promoter of the human gene contains both typical housekeeping gene features, including a lack of CCAAT and TATA boxes and the presence of binding sites for Oct-1 and Sp1/Sp3 transcription factors [61], as well as a cluster of AU-rich instability elements located in the 3′ UTR of the TR1 mRNA [62]. AU-rich elements (AUREs) are responsible for rapid mRNA turnover and are normally located in mRNAs of cytokines, proto-oncogenes, and transcription factors. In response to various stimuli, factors bind to AUREs and inhibit mRNA degradation, enabling a quick increase in gene expression. The combination of regulation by promoter elements and AU-rich instability motifs is specific for the TR1 isoform; as the 3′ UTR of the mitochondrial form, TR2, has no AURE. Further analysis of the TR1 promoter region showed that mutations at Oct-1 and Sp1/Sp3 motifs abolished *TXNRD1* gene expression by only 50%, suggesting that other elements control its expression (reviewed in Ref. [63]).

Recently, two putative antioxidant responsive elements (AREs) were identified in the *TXNRD1* promoter and they appear to account for *TXNRD1* upregulation in response to either sulforaphane, an anticarcinogenic isothiocyanate found in cruciferous vegetables [64–66], or to cadmium, a carcinogenic metal [67]. Cadmium-induced ARE-dependent transcription of TR1 is mediated by the transcription factor Nrf2, known to be involved in adaptation to oxidative/xenobiotic stress. Moreover, other studies have shown that Nrf2 was also upregulated by sulforaphane, suggesting that Nrf2 could be responsible for TR1 increased expression [68].

Another level of complexity in the *TXNRD1* promoter regulation was added when rare alternative transcripts, TR1α6 (found in testes) and TR1γ2 (skeletal muscle), were identified and suggested to be initiated from alternative promoters. These promoters may be activated under rare conditions or in specific tissues [69]. Moreover, an extensive alternative splicing pattern in the 5′ region of mammalian TR1 gene has been described. Indeed, the description of TR1 isoforms that differ in the origin of their N-terminal regions suggested that they resulted from alternative transcriptional start sites [63, 70–72]. More recently, several groups have demonstrated the existence of TR1 splice variants, revealing a complex regulation of gene transcription. Using a comparative genomic analysis approach, six human TR1 N-terminal splice variants that may have specific cellular functions have been identified [73]. TR1b, a novel human TR1 alternative splice variant, contains a 52-amino-acid N-terminal extension [74]. The N-terminal extension results from the formation of an in-frame start codon by recombination of exons 1 and 2 with exon 4 and the additional sequence reveals the presence of a consensus LLxxL motif which allows the binding of TR1b to both estrogen receptors (ER). Co-transfection of TR1b and ER genes results in a redistribution of TR1b from the cytoplasm to the nucleus where it co-localizes with ER and enhances *trans*-activation of a reporter gene. Alternative splicing of the *TXNRD1* gene has also been observed in the rat, where tissue-specific TR1 RNA processing generates kidney and liver-specific 5′ UTRs of TR1 that may be responsible for regulation of TR1 expression [71]. Understanding the differences between these variants, their tissue specificity, and their roles, will contribute to a better understanding of TR1 function.

Alternative splicing has also been observed in the human *DIO3* gene encoding IDIII, and at least three transcripts were identified resulting from a tissue-specific RNA processing [75]. Moreover, in both human and mouse, Hernandez and co-workers identified and partially characterized a gene, *DIO3OS*, that overlaps with the *DIO3* and *Dio3* promoter regions. Interestingly, this gene is transcribed in an antisense orientation, and is expressed in most tissues where it is, in its turn, subjected to alternative splicing [75]. In the rat, four IDIII transcripts have been identified and shown to be expressed specifically in different regions of the central nervous system [76]. Multiple transcript variants have also been described for IDII [77] and GPx4 [78] and alternatively spliced isoforms of GPx1, SPS2, IDI, Sep15, and GPx5 are predicted in the genome database (http://www.ncbi.nlm.nih.gov; see Table 8.1). However, the biological significance of these isoforms remains to be elucidated.

Overall, it is emerging that alternative splicing plays a determining role in the regulation of selenoprotein expression, location and function, and contributes to increase the overall variety of selenoproteins.

8.4.2
Response of Selenoprotein Synthesis to Selenium Supply: Prioritization

Dietary selenium intake influences the synthesis of selenoproteins and there is now strong evidence showing that there is a hierarchy for synthesis of different selenoproteins within a single tissue or cell type. In other words, when selenium

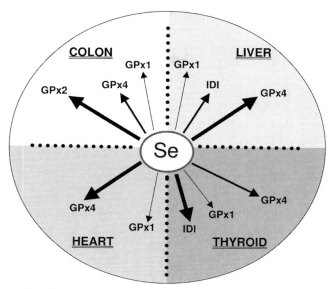

Fig. 8.3. The available selenium is differentially prioritized to the selenoproteins in different tissues under conditions of low selenium supply. The sensitivity of expression to selenium depletion is illustrated schematically by the thickness of the arrows. Note that some selenoproteins are more sensitive than others and this is tissue dependent.

supply is low the synthesis of some selenoproteins is dramatically reduced whilst that of others is only marginally affected [79–82]. For example, in the liver of severely selenium-deficient rats, GPx1 activity is almost totally lost whilst GPx4 activity is partially conserved; in less severely deficient rats, GPx4 activity is little affected whilst GPx1 activity is still dramatically lower [79]. Furthermore, the extent and pattern of the changes are different depending on the tissue, as illustrated schematically in Fig. 8.3. For example, there is evidence that the response of GPxs to selenium deficiency is tissue specific (reviewed in Ref. [17]). Severe selenium deficiency causes almost total loss of GPx1 activity in liver and heart and although IDI activity decreases by 95% in liver, it increases by 15% in the thyroid. GPx4 activity decreases by 75% in the liver, 60% in the heart and is unchanged in the thyroid [79]. The physiological effects of altered selenium intake are determined by the modification of the pattern of selenoprotein expression and a clearer picture of the overall functional effects of changes in selenium intake would be obtained if the effects on a much wider range, preferably all, of the selenoproteins could be defined.

The incorporation of Sec into the nascent polypeptide of selenoproteins is complex and the Sec-incorporation machinery involves a limiting factor, SBP2, a hierarchy of protein complexes formed between different *trans*-acting components, as well as generation of tRNA$^{(ser)sec}$. The complexity of this process provides a wide range of potential sites at which expression could be regulated. It has been shown that differences in 3′ UTRs among selenoprotein mRNAs influence Sec incorporation. Even when selenium supply is adequate, the efficiency of selenoprotein mRNA translation is affected by the 3′ UTR, as indicated by the observations that IDI and IDIII activities are altered by exchanging their native 3′ UTR for those of other selenoproteins [83, 84]. However, there is further influence of the 3′ UTR when selenium supply is low. The use of cells transfected with chimeric constructs containing the IDI-coding region linked to different 3′ UTRs has shown that the nature of the 3′ UTR affects the efficiency at which transcripts are translated under conditions of low selenium supply (see Section 8.5.1). Thus, for example, translation of transcripts containing the GPx4 3′ UTR is affected less by low selenium conditions than translation of IDI coding region linked to GPx1 3′ UTR [85]; reporter transcripts with the GPx2 3′ UTR are less sensitive to low selenium than with the GPx1 3′ UTR [82, 86]. The mechanism behind such effects is not known but they illustrate how the 3′ UTR of different selenoprotein mRNAs can influence the response to selenium levels.

8.5
Nutrigenomic or Functional Genomic Studies

8.5.1
Arrays: Looking at Patterns of Gene Expression

Array technology provides an opportunity to study expression of a range, or all, of the selenoprotein genes at the RNA level (Fig. 8.4). However, such an approach

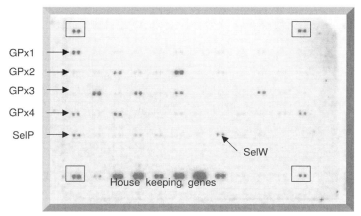

Fig. 8.4. A typical hybridization profile using a "selenoprotein array." RNA from Caco-2 cells was reverse transcribed into ^{32}P-labeled cDNA and hybridized with the array. Specific hybridization was detected by Phosphorimager. Every gene is spotted in duplicate. Orientation markers are marked with square boxes. Housekeeping genes are shown at the bottom of the array. Examples of selenoprotein genes detected are indicated.

does not take into account alterations at the level of translation of the mRNA into protein. Since Sec is incorporated during translation, studies of selenoprotein expression at the RNA level may not reflect variations in protein expression. However, a number of studies indicate that expression of individual selenoprotein mRNAs mirrors changes in protein expression [79, 80]. In the case of both GPx1 and SelW, selenium deficiency leads to a large decrease in mRNA levels associated with instability of the mRNA [85, 87, 88]; GPx1 mRNA becomes more unstable under conditions of low selenium supply or when the translation is inhibited by puromycin, suggesting that modulation of translation by selenium supply has subsequent effects on mRNA stability. The stability of GPx2 mRNA has been correlated with the resistance of GPx2 expression to selenium depletion [89]. In general, although one must bear in mind the caveats associated with selenium being incorporated during protein synthesis, mRNA expression studies can be informative about responses of selenoprotein expression to selenium supply. Potentially, gene array technology may prove useful in the study of patterns of selenoprotein expression.

Using a custom-made macroarray, a recent study assessed the effects of selenium depletion on expression of selenoproteins in the human gastrointestinal cell line Caco-2 [88]. This approach demonstrated that selenium depletion had most marked effects on expression of GPx1 and SelW and the effect on SelW was supported by northern hybridization analysis using RNA from colon samples from selenium-deficient rats. These observations regarding SelW expression illustrate the usefulness of an array approach to compare effects on a range of selenoproteins (e.g. GPx1, 2, 3, 4, SelP, SelW) and therefore identify novel effects on selenoprotein genes that are sensitive to nutritional regulation. Alteration of selenoprotein expression also has secondary effects on expression of other genes (for example, overexpression of GPx4 decreases expression of cell adhesion molecules in smooth muscle cells [90]) and it is likely that these secondary changes are part of the functional changes that occur due to altered selenium intake. Microarray techniques incorporating thousands of genes offer the opportunity to identify such secondary indirect changes. Recently a study was carried out using rat oligonucleotide arrays from the MRC-Wellcome Centre to analyze the global changes in gene expression in the colon from severely selenium-deficient rats. Rats were fed diets that were either severely selenium deficient or selenium sufficient [79, 88] and the RNA extracted from mucosal scrapings. RNA (75 μg) pooled from samples from either group was used to make cy3/cy5-labeled cDNAs for microarray analysis. After hybridization, arrays were scanned and analyzed using Genepix software. After removal of outliers, approximately 18 genes showed a greater than 1.5-fold change in expression level (see Table 8.2) [91]. Interestingly, these included genes the products of which are involved in lipid metabolism and vascular remodeling and angiogenesis. The functional significance of these changes remains to be explored but they point to novel aspects of the response of the colon to selenium depletion.

Comparison of northern blot, microarray, and macroarray data for a number of selenoproteins (Table 8.3) shows that the relative effects of selenium depletion on SelW, GPx1 and GPx2 were comparable using the three methods: SelW and GPx1

Table 8.2. Effect of selenium depletion on gene expression in rat colon using microarrays.

Accession number	Gene name	log. −Se/+Se	Function
D14437	Calponin	−2.142	Muscle contraction
X95849	Novel gene expressed in circadian manner	−2.042	
AB017820	Klotho	−2.019	"Anti-aging hormone"
X06801	Vascular alpha-actin	−1.814	
U95157	Ryanodine receptor type II	−1.783	Ca^{2+} sensing
AF007554	Mucin 1	−1.766	Lubrication and hydration of cell surfaces
AF039308	Glutaminyl cyclase	1.543	Formation of biologically active peptide hormones
AF078811	Tropic1808	1.999	
S73894	Endogenous vascular elastase	2.093	Vascular remodeling
NM_017307	Solute carrier gene encoding mitochondrial protein	2.099	
NM_012582	Haptoglobin	2.143	Hemoglobin (Hb)-binding plasma protein
NM_019321	Mast cell protease 4	2.360	Chymotrypsin-like serine protease
AB025023	Midkine	2.404	Angiogenesis, cell growth
X61925	Lipase	2.914	Hydrolysis of triglycerides and phospholipids
J00771	Pancreatic ribonuclease	4.714	Cytotoxicity to angiogenesis
X59014	Elastase IV	5.395	
NM_016998	Carboxypeptidase A1	6.128	Role in intestinal diseases
NM_013161	Pancreatic lipase	6.141	Dietary lipid digestion.

Significant change corresponds to genes with log ratio values ≥ 2.5 standard deviations (SD) from the mean. The results are the average values from two experiments. The SD values between the two experiments for each gene are ≤ 0.3. Selenoprotein genes were not included in the cut-off limits. Negative values indicate a decrease in expression between the control and the test populations, whereas positive values indicate an increase. Each successive integer step indicates a log 2 increase or decrease. Therefore a log −Se/+selenium of 2, represents a 2-fold change etc.

Table 8.3. Comparison of results from microarray, macroarray, and northern hybridization analysis of selenoprotein gene expression.

Gene name	Accession number	Microarray Rat colon log −Se/+Se	Macroarray Caco-2 cells −Se/+Se	Northern Rat colon −Se/+Se	Northern Caco-2 cells −Se/+Se
SelW	NM_013027	−0.690	0.27	Undetectable in -Se	N/A
GPx1	M21210	−0.460	0.46	0.17	0.39
tRNA[sec]-associated protein	AF181856	−0.020	1.20	N/A	N/A
SelP	NM_019192	0.125	0.90	N/A	N/A
TR2	AF072865	0.725	0.73	N/A	N/A
GPx2	NM_002083	N/A	1.63	1.22	1.26

Expression of several selenoprotein mRNAs under Se-adequate (+Se) or Se-deficient (−Se) conditions is shown for Caco-2 cells and rat colon. mRNA levels were measured by different techniques. The pattern of expression in Se-depletion is similar whether measured by northern hybridization, macroarray or microarray. Results taken from Refs [88, 91]. N/A stands for not applicable.

were highly sensitive, GPx4 less so, and GPx2 showed little change. Not only does this give credibility to the array data but interestingly it shows the potential of using arrays to define patterns of selenoprotein expression.

8.5.2
Transfected Cell Models to Study UTR Function

It is notable that the mRNAs of the different selenoproteins differ both in length of the 3′ UTR and the position of the Sec-encoding UGA within the coding region. As a result, the distance, in terms of nucleotides, between UGA and SECIS can vary quite considerably within certain limits (see Fig. 8.2), as originally shown by Berry et al. [83]. This ability of the Sec-incorporation mechanism to function over a range of UGA–SECIS distances means that it is possible to produce chimeric gene constructs in which the 3′ UTR from one selenoprotein is exchanged for another [83, 86, 92] without losing Sec incorporation. Studies with cells transfected with such constructs have shown that a heterologous 3′ UTR or SECIS is functional [83, 86] and that synthesis of the selenoprotein from such a construct is sensitive to selenium supply [92].

The deiodinase gene (IDI) has also proved effective in reporter constructs to study the ability of different 3′ UTR sequences to direct Sec incorporation at a UGA codon, and to study its modulation by selenium supply. As discussed in Section 8.4.2, in severe selenium deficiency expression of GPx1 is much more affected than GPx4 expression in rat liver [79]. To better understand the mechanisms underlying this differential response, the influence of GPx1 and GPx4 3′ UTRs on expression of the IDI reporter was investigated in transfected H4 hepatoma cells [92]. The expression of IDI from a chimeric construct with IDI-coding region linked to either GPx1 or GPx4 3′ UTR showed that both 3′ UTRs are functional in these transcripts; in addition the ability of the GPx4 3′ UTR to direct Sec incorporation into the IDI reporter is less affected by selenium depletion than that of GPx1 3′ UTR [92]. This observation reflects the responses of the levels of these two GPxs to selenium depletion in rat liver *in vivo* [79]. The mechanism behind this effect is not known but we have proposed that it reflects the better ability of the GPx4 3′ UTR to form complexes with appropriate *trans*-acting factors involved in Sec incorporation, and in this way compete better for available selenium in the form of Sec [93].

It would be expected that expression of an IDI transgene would cause the cell to utilize some of the available selenium for IDI synthesis, thus potentially depleting selenium normally available for synthesis of the endogenous selenoproteins, such as those that protect cells from oxidative stress. Indeed, H4 hepatoma cells transfected with IDI are more sensitive to the pro-oxidant *tert*-butyl hydroperoxide under both selenium-depleted ($ID_{50} = 24$ µmol/l) and selenium-supplemented culture conditions ($ID_{50} = 37$ µmol/l) compared with untransfected control cells ($ID_{50} = 35$ and 51 µmol/l under selenium-depleted and -supplemented conditions, respectively) (Duperrier, Bermano, and Hesketh, unpublished observations). These data are consistent with the transgenic expression of IDI putting the cells under greater oxidative stress by competing with other selenoproteins for selenium availability.

In addition, 3′ UTRs from selenoprotein genes have been shown to be capable of promoting read-through of a UGA codon between two coding regions in synthetic reporter constructs. A model system was developed in which the expression of the luciferase gene, cloned downstream of the β-galactosidase gene and separated from it by an in-frame UGA, was driven by a downstream SECIS element in a selenoprotein 3′ UTR [82, 89]. In this model, in the presence of selenium the SECIS in the 3′ UTR will promote read-through of the UGA but there is no read-through in the absence of selenium. Experiments with this reporter system have shown that the sensitivity of read-through to selenium is different for GPx1 and GPx4 3′ UTRs [82]. A similar expression system was used to assess the ability of 3′ UTR sequences from the 15 kDa selenoprotein gene to drive read-through at the UGA codon [94] and study the functionality of SNPs in the 3′ UTR.

Overall, a variety of studies show that reporter genes have provided effective models to study the influence of the 3′ UTR on selenoprotein synthesis and its modulation by selenium supply.

8.5.3
Knockout Mice Models of Selenoproteins

Transgenic technologies have also been able to produce a number of knockout mice in which specific selenoproteins, GPx1, GPx1 and 2, or SelP are not expressed. The use of such models has brought new insights to the understanding of selenoproteins *in vivo*, opening new areas for future investigation. Beck and colleagues have used the GPx1-knockout mouse to investigate the role of GPx1 in viral infection. These knockout mice infected with a benign coxsackie virus show severe myocarditis, whereas the wild-type mice show only a mild inflammation [3]; the effect of GPx1 knockout parallels that of dietary selenium deficiency. In addition, mice in which both GPx1 and GPx2 (gastrointestinal) were knocked out have been shown to be more susceptible to colitis [95] and to bacteria-induced inflammation and cancer [96]. Thus, studies with both GPx1 and GPx1/2 knockout mice indicate a role for the GPx selenoproteins in modulating inflammatory responses.

It has not been possible to produce GPx4-knockout mice as they die *in utero* between embryonic stages E7.5 and E8.5 [97]. The lethal phenotype is rescued when a human GPx4 transgene is overexpressed in the knockout mice [98], showing that expression of GPx4 is essential to mouse development. In addition, GPX4$^{+/-}$ mice show increased sensitivity to oxidative stress with reduced survival to gamma-irradiation [97] whereas overexpression of human GPx4 in transgenic mouse embryonic fibroblasts protects the cells against oxidative stress-induced apoptosis and reduces chemically induced liver damage and lipid peroxidation in transgenic animals [98].

Thioredoxin reductase-knockout mice also die during embryogenesis around stage E10.5 for TR1$^{-/-}$ [99] and E13.5 for TR2$^{-/-}$ [100]. TR2-null embryos die of a severe anemia, reduced hematopoiesis, and altered cardiac development. Heart-specific deletion of the *TR2* gene is fatal, revealing that TR2 is crucial for development and proper function of the heart [100]. In contrast, deletion of the *TR1* gene has shown that TR1 function is essential for embryogenesis in most tissues but dispensable for cardiac development [99]. Moreover, overexpression of a dominant-negative mutant of TR1 increases myocardial oxidative stress and ROS-induced cardiac hypertrophy compared with the non-transgenic mice, whereas overexpression of a wild-type TR1 protects the mice against the development of induced-cardiac hypertrophy. These data suggest that the dominant-negative TR1 might have repressed TR2 function in the heart [97].

For many years it has been suggested that SelP has a function in selenium transport between tissues. The production of SelP-knockout mice has provided the first strong evidence that this is the case. The SelP-knockout mice exhibit low selenium concentrations in the brain, plasma, kidney, and testes whereas the liver selenium content is elevated suggesting that SelP transports hepatic selenium to other organs. The mice lose weight, develop neurological disorders, and the males are infertile [101, 102]. In addition, feeding a selenium-enriched diet to these mice can partially rescue expression of selenoproteins and neurological defects of the knockout [103, 104]. However liver-specific deletion of the *Trsp* gene, which encodes tRNA[(ser)sec], dramatically decreases plasma SelP levels but does not affect brain selenium levels or brain function [43]. Recently, Schweizer *et al.*, using liver-specific *trsp*[−/−] mice, showed that hepatic SelP was required for selenium supply to the kidney but not to the brain, as local expression of SelP could maintain its selenium levels and selenoprotein synthesis [105]. Taken together these data suggest that SelP has a role in local selenium supply in the brain and a more general function in transport of hepatic selenium to other organs.

Further insights in the understanding of the tRNA[(ser)sec] functions have come from the development of knockout and transgenic models for the *Trsp* gene. A complete deletion of the *trsp* gene is embryonic lethal suggesting that selenoproteins are essential for normal development [106, 107]. Conditional deletion of the gene affects selenoprotein expression in a tissue-specific manner [107]. Overexpression of wild-type transgene has little effect on selenoprotein expression, but overexpression of a transgene carrying a mutation for a methylation site required for tRNA[(ser)sec] maturation alters the expression of certain selenoproteins in a tissue-specific manner [41]. Recently, Carlson *et al.* rescued selenoprotein expression in *trsp*-null mice by overexpression of either a wild-type or methylation mutant transgene. Whereas the wild-type transgene was able to rescue overall selenoprotein synthesis, only TR1 and TR2 expression could be completely rescued by the methylation mutant. A partial rescue of GPx2, GPx4, SelP, and Sep15 synthesis was achieved but only a poor synthesis of GPx1, GPx3, SelR, SelT, and SelW was observed. These data confirm that the different methylation isoforms of tRNA[(ser)sec] are selectively required for the synthesis of the different selenoproteins [43].

8.6
Nutrigenetics of Selenium Metabolism

Many genes in the human genome exhibit polymorphic allelic variants that are present at stable frequencies within the population. The majority of these are variations at single positions (SNPs). Such SNPs can occur in the gene region corresponding to the protein-coding region, in which case the allelic variation may be silent (no amino acid change) or the amino acid may be altered, with or without functional consequences. SNPs may also occur in regulatory gene regions, for example in the promoter, 5′ UTR, or 3′ UTR and in these cases they may potentially influence regulation of expression of the gene. In selenoprotein genes, since the 3′ UTR and SECIS are critical for selenium incorporation, SNPs in the gene region corresponding to the 3′ UTR are of particular interest since they may potentially affect synthesis and activity, and ultimately influence individual requirements for selenium. Any functional effect of a single nucleotide change is likely to be subtle, and therefore phenotypic physiological consequences may be masked at high selenium intakes and only observable at suboptimal intake. SNPs are likely to be important in the etiology of multifactorial disease and the combination of a SNP in a selenoprotein gene in conjunction with low selenium intake has the potential to influence the mechanisms of protecting cells from oxidative stress, inflammatory processes, thyroid metabolism, fertility, etc.

A well-characterized SNP has been found in the GPx1 gene at codon 198 within the protein-coding region [108]. Originally detected in a loss of heterozygosity study and identified by bioinformatics [109], it is a T to C variation that has also been described in Scandinavian, Afro-Caribbean, Caucasian, and Japanese populations [108, 110, 111]. The common, and thus presumably wild-type, homozygous form is a C that leads to incorporation of a Pro at codon 198. The alternative homozygous form (T) only occurs at 7–11% in healthy Caucasian populations [108, 112] but at higher frequency (~15%) in healthy Afro-Caribbeans [110, 113], resulting in change of codon 198 to that for Leu. The Leu allele has been suggested to have lower activity in transfected cell lysates and increased association with lung, breast, and bladder cancer [110, 113, 114], although the association with breast cancer has not been confirmed in a recent larger study [115]. In addition, the mean intima-media thickness of the carotid arteries and prevalence of cardiovascular disease was higher in Leu/Pro heterozygotes than in the Pro/Pro homozygotes in a population of Japanese type 2 diabetic patients [111]. Interestingly a haplotype study reported the association of the Leu allele with bladder cancer to be influenced by a SNP within the manganese superoxide dismutase gene which codes for another antioxidant defense protein [114].

A range of SNPs have been detected in the coding region of GPx4 gene but their functionality is still open to debate and there is no association of any of the allelic variations with male infertility [116], despite GPx4 being a key structural component of the sperm. In addition, bioinformatics predicts two SNPs in the region of the GPx4 gene corresponding to the 3′ UTR – at positions 718 and 738, both close to the predicted SECIS structure. Sequencing of this region in a cohort of Caucasian subjects confirmed a T/C variation at position 718 but did not confirm the second predicted SNP at 738 [93]. Both T and C variants at position 718 are common in this polymorphism with the frequency distribution being approximately 22–24% for TT, 26–33% for CC, and 42–52% for CT heterozygotes in healthy Caucasian subjects [93, 112]. Similar distributions have been found in Chinese (18% TT, 32% CC, 50% CT) and South Asian (18% TT, 36% CC, 46% CT) populations [117]; these distributions are in Hardy–Weinberg equilibrium. In the original study of this SNP it was observed that individuals who were TT at position 718 had different levels of 5′ lipoxygenase metabolites compared with those who were homozygous CC [93], providing indirect evidence that the SNP was functional. Recently, association studies in patients with ulcerative colitis and colon cancer have shown that the frequency of the TT genotype is much lower in these patients compared to healthy controls or patients with adenomatous polyps [112, 118], again suggesting that this SNP is functional and also providing the first evidence that it may be a factor in disease.

SNPs have been identified in GPx2 but their function remains unknown [119]. Incorporation of selenium into selenoproteins requires any dietary selenium to be transported to the tissues and then to be used to synthesize tRNA$^{(ser)sec}$. Variation in the efficiency of these processes would be expected to lead to interindividual variation in selenium metabolism, and ultimately dietary selenium requirements. We have therefore embarked upon a study to define SNPs in the genes encoding SelP and SPS2: SelP since it has recently been strongly implicated in selenium transport throughout the body [120] and SPS2 because it is involved in selenoprotein synthesis through synthesis of Sec. A number of SNPs have been detected, in most of which one allelic variant occurs at very low frequency, but two SNPs, one in SelP and one in SPS2 have been found in which two variants are found at >15% frequency either in Caucasian or South Asian subjects. Functional studies of these SNPs are in progress. Recently, a SNP has been described in the SelP promoter region [119].

Two polymorphic variants, a C/T substitution at position 811 and a G/A at position 1125, have been identified in the 3′ UTR of the Sep15 mRNA [94]. The SNP at position 1125 is located in the SECIS structure and therefore may influence the efficiency of the Sec incorporation into the protein during translation. Studies examining the chemopreventive effect of selenium in mesothelioma cells found that individuals with the A1125 variant were less responsive to the protective effect of dietary selenium supplementation [121]. Furthermore, although the function of this protein is not known, studies with breast cancer patients show an association of alleles at the two positions and transfection studies show that together these two variations affect 3′ UTR function in response to selenium depletion, with the CG variants enhancing read-through [94].

Thus, to summarize, a number of SNPs have been identified in the glutathione peroxidase and other selenoprotein genes, some within the coding region and others within the 3′ UTRs. There is some evidence from a variety of studies that some of these may be functional and that they may be associated with reduced protection from oxidative stress and susceptibility to a range of disease. However, the evidence for the importance of these SNPs is still limited and more detailed studies are required to define the functionality and relation to diseases, and in the latter case the interaction with nutritional factors such as dietary selenium.

8.7
Conclusions and Future Perspectives

Nutrigenomics is contributing to the study of selenium physiology and selenoproteins in several ways. It can assist in studying how expression changes in response to nutrition. The expression of selenoprotein genes changes in response to a variety of factors, but particularly selenium supply. Gene arrays are allowing us to move from studies of single genes to studies of a range of selenoproteins so as to get a picture of the pattern of gene expression response of selenium incorporation. Future studies should expand these approaches to include arrays that can take the complexity of alternative transcripts into account and also develop proteomic methods. Specific arrays can be targeted towards the "selenoproteome" whereas microarrays with a wide range of genes can define secondary effects and so give a broader picture of how selenium affects cell and tissue physiology. In addition, a variety of functional genomic techniques (transfected cells, knockout mice, overexpression studies) are beginning to provide key information on selenoprotein function and on the mechanisms by which selenium supply regulates expression.

Studies with different reporter systems are showing that the 3′ UTR is important in determining the response of selenoprotein synthesis to selenium supply but this is only part of the story and other factors, possibly tRNA methylation, also play a role. It thus appears that the regulation is complex – a complexity increased by alternative splicing. Further studies are needed to define the precise mechanisms of regulation of the different selenoproteins and their isoforms in different tissues.

The overall picture emerging is of a family of genes regulated by a variety of mechanisms in a complex manner. Subtle differences in gene sequence play an important role in determining the response of expression to selenium. All eukayotic selenoprotein mRNAs contain a SECIS in the 3′ UTR but the 3′ UTRs vary considerably in length, position of SECIS, etc. These differences are critical to determining the response of expression to selenium. Furthermore, there is evidence that single base changes (SNPs) within 3′ UTR sequences can influence regulation of selenoprotein expression.

By expanding the study of these and other SNPs, nutrigenomics can contribute to describing how individual genetic differences influence response to dietary selenium. We still know relatively little about the genetic variation, in terms of SNPs, in selenoprotein genes in the human population. However, we now appreciate that SNPs in the selenoprotein genes exist, both in the coding regions and the untranslated regions and promoters. Some of these appear to have functional effects. Thus, genetic variation in these genes may mean that individuals differ in their requirements for selenium for optimal health but this requires further definition of SNPs in genes such as SelP, SPS2, SBP2, and tRNA$^{(ser)sec}$, the products of which are involved in selenium metabolism, the coding regions of selenoprotein genes themselves and in the gene regions corresponding to 3′ UTRs. There are tantalizing suggestions from some studies that some of the SNPs already identified in selenoprotein genes are possibly associated with disease but larger and wider studies are needed to determine whether such SNPs could be reliable markers of predisposition to disease. Recent studies have implicated SNPs in selenoprotein S as an influence on the inflammatory response (Curran J. E. et al. Nature Genetics epub Oct 9th 2005) and identified mutations in the SBP2 gene that result in familial thyroid disorders (Dumitrescu A. M. et al. Nature Genetics epub Oct 16th 2005).

Acknowledgments

The authors' work is supported by the Food Standard Agency, the World Cancer Research Fund, and the Biotechnology and Biological Science Reseach Council.

References

1 SCHWARZ, K., FOLTZ, C. M. (1958) *J Biol Chem* 233, 245–51.
2 RAYMAN, M. P. (2002) *Proc Nutr Soc* 61, 203–15.
3 BECK, M. A., ESWORTHY, R. S., HO, Y. S., CHU, F. F. (1998) *FASEB J* 12, 1143–9.
4 LEVANDER, O. A. (2000) *J Nutr* 130, 485S–488S.

5 GLADYSHEV, V. N., STADTMAN, T. C., HATFIELD, D. L., JEANG, K. T. (**1999**) *Proc Natl Acad Sci USA* 96, 835–9.

6 CLARK, L. C., COMBS, G. F., JR., TURNBULL, B. W., SLATE, E. H., CHALKER, D. K., CHOW, J., DAVIS, L. S., GLOVER, R. A., GRAHAM, G. F., GROSS, E. G., KRONGRAD, A., LESHER, J. L., JR., PARK, H. K., SANDERS, B. B., JR., SMITH, C. L., TAYLOR, J. R. (**1996**) *JAMA* 276, 1957–63.

7 RAYMAN, M. P. (**2000**) *Lancet* 356, 233–41.

8 HATFIELD, D. L., GLADYSHEV, V. N. (**2002**) *Mol Cell Biol* 22, 3565–76.

9 COMBS, G. F., JR. (**2001**) *Nutr Cancer* 40, 6–11.

10 FLOHE, L., GUNZLER, W. A., SCHOCK, H. H. (**1973**) *FEBS Lett* 32, 132–4.

11 ROTRUCK, J. T., POPE, A. L., GANTHER, H. E., SWANSON, A. B., HAFEMAN, D. G., HOEKSTRA, W. G. (**1973**) *Science* 179, 588–90.

12 CHAMBERS, I., FRAMPTON, J., GOLD-FARB, P., AFFARA, N., MCBAIN, W., HARRISON, P. R. (**1986**) *EMBO J* 5, 1221–7.

13 ZINONI, F., BIRKMANN, A., STADT-MAN, T. C., BOCK, A. (**1986**) *Proc Natl Acad Sci USA* 83, 4650–4.

14 BERRY, M. J., LARSEN, P. R. (**1993**) *Am J Clin Nutr* 57, 249S–255S.

15 GROMER, S., JOHANSSON, L., BAUER, H., ARSCOTT, L. D., RAUCH, S., BALLOU, D. P., WILLIAMS, C. H., JR., SCHIRMER, R. H., ARNER, E. S. (**2003**) *Proc Natl Acad Sci USA* 100, 12618–23.

16 BRIGELIUS-FLOHE, R. (**1999**) Free Radical Biology and Medicine 27, 951–965.

17 KRYUKOV, G. V., CASTELLANO, S., NOVOSELOV, S. V., LOBANOV, A. V., ZEHTAB, O., GUIGO, R., GLADYSHEV, V. N. (**2003**) *Science* 300, 1439–43.

18 HIROTA, K., MURATA, M., SACHI, Y., NAKAMURA, H., TAKEUCHI, J., MORI, K., YODOI, J. (**1999**) *J Biol Chem* 274, 27891–7.

19 BECKETT, G. J., ARTHUR, J. R. (**2005**) *J Endocrinol* 184, 455–65.

20 BURK, R. F., HILL, K. E., MOTLEY, A. K. (**2003**) *J Nutr* 133, 1517S–20S.

21 EVENSON, J. K., WHEELER, A. D.,

22 PERSSON-MOSCHOS, M., HUANG, W., SRIKUMAR, T. S., AKESSON, B., LINDEBERG, S. (**1995**) *Analyst* 120, 833–6.

23 BURK, R. F., HILL, K. E. (**1994**) *J Nutr* 124, 1891–7.

24 MOSTERT, V. (**2000**) *Arch Biochem Biophys* 376, 433–8.

25 MOTCHNIK, P. A., TAPPEL, A. L. (**1990**) *J Inorg Biochem* 40, 265–9.

26 SAITO, Y., TAKAHASHI, K. (**2002**) *Eur J Biochem* 269, 5746–51.

27 SAITO, Y., SATO, N., HIRASHIMA, M., TAKEBE, G., NAGASAWA, S., TAKAHASHI, K. (**2004**) *Biochem J* 381, 841–6.

28 KUMARASWAMY, E., MALYKH, A., KOROTKOV, K. V., KOZYAVKIN, S., HU, Y., KWON, S. Y., MOUSTAFA, M. E., CARLSON, B. A., BERRY, M. J., LEE, B. J., HATFIELD, D. L., DIAMOND, A. M., GLADYSHEV, V. N. (**2000**) *J Biol Chem* 275, 35540–7.

29 VENDELAND, S. C., BEILSTEIN, M. A., CHEN, C. L., JENSEN, O. N., BAROFSKY, E., WHANGER, P. D. (**1993**) *J Biol Chem* 268, 17103–7.

30 VENDELAND, S. C., BEILSTEIN, M. A., YEH, J. Y., REAM, W., WHANGER, P. D. (**1995**) *Proc Natl Acad Sci USA* 92, 8749–53.

31 JEONG, D., KIM, T. S., CHUNG, Y. W., LEE, B. J., KIM, I. Y. (**2002**) *FEBS Lett* 517, 225–8.

32 LOW, S. C., BERRY, M. J. (**1996**) *Trends Biochem Sci* 21, 203–8.

33 ZINONI, F., HEIDER, J., BOCK, A. (**1990**) *Proc Natl Acad Sci USA* 87, 4660–4.

34 HEIDER, J., BOCK, A. (**1992**) *J Bacteriol* 174, 659–63.

35 WALCZAK, R., WESTHOF, E., CARBON, P., KROL, A. (**1996**) *RNA* 2, 367–79.

36 WALCZAK, R., HUBERT, N., CARBON, P., KROL, A. (**1997**) *Biomed Environ Sci* 10, 177–81.

37 GRUNDNER-CULEMANN, E., MARTIN, G. W., 3rd, HARNEY, J. W., BERRY, M. J. (**1999**) *RNA* 5, 625–35.

38 LEINFELDER, W., STADTMAN, T. C., BOCK, A. (**1989**) *J Biol Chem* 264, 9720–3.

39 LEE, B. J., WORLAND, P. J., DAVIS, J. N., STADTMAN, T. C., HATFIELD, D. L. (**1989**) *J Biol Chem* 264, 9724–7.

40 CHITTUM, H. S., BAEK, H. J., DIAMOND, A. M., FERNANDEZ-SALGUERO, P., GONZALEZ, F., OHAMA, T., HATFIELD, D. L., KUEHN, M., LEE, B. J. (**1997**) *Biochemistry* 36, 8634–9.

41 MOUSTAFA, M. E., CARLSON, B. A., EL-SAADANI, M. A., KRYUKOV, G. V., SUN, Q. A., HARNEY, J. W., HILL, K. E., COMBS, G. F., FEIGENBAUM, L., MANSUR, D. B., BURK, R. F., BERRY, M. J., DIAMOND, A. M., LEE, B. J., GLADYSHEV, V. N., HATFIELD, D. L. (**2001**) *Mol Cell Biol* 21, 3840–52.

42 JAMESON, R. R., DIAMOND, A. M. (**2004**) *RNA* 10, 1142–52.

43 CARLSON, B. A., NOVOSELOV, S. V., KUMARASWAMY, E., LEE, B. J., ANVER, M. R., GLADYSHEV, V. N., HATFIELD, D. L. (**2004**) *J Biol Chem* 279, 8011–7.

44 XU, X. M., ZHOU, X., CARLSON, B. A., KIM, L. K., HUH, T. L., LEE, B. J., HATFIELD, D. L. (**1999**) *FEBS Lett* 454, 16–20.

45 COPELAND, P. R., FLETCHER, J. E., CARLSON, B. A., HATFIELD, D. L., DRISCOLL, D. M. (**2000**) *EMBO J* 19, 306–14.

46 LESCURE, A., ALLMANG, C., YAMADA, K., CARBON, P., KROL, A. (**2002**) *Gene* 291, 279–85.

47 FLETCHER, J. E., COPELAND, P. R., DRISCOLL, D. M., KROL, A. (**2001**) *RNA* 7, 1442–53.

48 LOW, S. C., GRUNDNER-CULEMANN, E., HARNEY, J. W., BERRY, M. J. (**2000**) *EMBO J* 19, 6882–90.

49 ALLMANG, C., CARBON, P., KROL, A. (**2002**) *RNA* 8, 1308–18.

50 COPELAND, P. R., STEPANIK, V. A., DRISCOLL, D. M. (**2001**) *Mol Cell Biol* 21, 1491–8.

51 MEHTA, A., REBSCH, C. M., KINZY, S. A., FLETCHER, J. E., COPELAND, P. R. (**2004**) *J Biol Chem* 279, 37852–9.

52 TUJEBAJEVA, R. M., COPELAND, P. R., XU, X. M., CARLSON, B. A., HARNEY, J. W., DRISCOLL, D. M., HATFIELD, D. L., BERRY, M. J. (**2000**) *EMBO Rep* 1, 158–63.

53 FAGEGALTIER, D., HUBERT, N., YAMADA, K., MIZUTANI, T., CARBON,

P., KROL, A. (**2000**) *EMBO J* 19, 4796–805.

54 ZAVACKI, A. M., MANSELL, J. B., CHUNG, M., KLIMOVITSKY, B., HARNEY, J. W., BERRY, M. J. (**2003**) *Mol Cell* 11, 773–81.

55 GUIMARAES, M. J., PETERSON, D., VICARI, A., COCKS, B. G., COPELAND, N. G., GILBERT, D. J., JENKINS, N. A., FERRICK, D. A., KASTELEIN, R. A., BAZAN, J. F., ZLOTNIK, A. (**1996**) *Proc Natl Acad Sci USA* 93, 15086–91.

56 WILTING, R., SCHORLING, S., PERSSON, B. C., BOCK, A. (**1997**) *J Mol Biol* 266, 637–41.

57 TAMURA, T., YAMAMOTO, S., TAKAHATA, M., SAKAGUCHI, H., TANAKA, H., STADTMAN, T. C., INAGAKI, K. (**2004**) *Proc Natl Acad Sci USA* 101, 16162–7.

58 BANNING, A., DEUBEL, S., KLUTH, D., ZHOU, Z., BRIGELIUS-FLOHÉ, R. (**2005**) *Mol Cell Biol* 25, 4914–4923.

59 BIERL, C., VOETSCH, B., JIN, R. C., HANDY, D. E., LOSCALZO, J. (**2004**) *J Biol Chem* 279, 26839–45.

60 MERANTE, F., ALTAMENTOVA, S. M., MICKLE, D. A., WEISEL, R. D., THATCHER, B. J., MARTIN, B. M., MARSHALL, J. G., TUMIATI, L. C., COWAN, D. B., LI, R. K. (**2002**) *Mol Cell Biochem* 229, 73–83.

61 RUNDLOF, A. K., CARLSTEN, M., ARNER, E. S. (**2001**) *J Biol Chem* 276, 30542–51.

62 GASDASKA, J. R., HARNEY, J. W., GASDASKA, P. Y., POWIS, G., BERRY, M. J. (**1999**) *J Biol Chem* 274, 25379–85.

63 RUNDLOF, A. K., ARNER, E. S. (**2004**) *Antioxid Redox Signal* 6, 41–52.

64 HINTZE, K. J., WALD, K. A., ZENG, H., JEFFERY, E. H., FINLEY, J. W. (**2003**) *J Nutr* 133, 2721–7.

65 HINTZE, K. J., KECK, A. S., FINLEY, J. W., JEFFERY, E. H. (**2003**) *J Nutr Biochem* 14, 173–179.

66 ZHANG, J., SVEHLIKOVA, V., BAO, Y., HOWIE, A. F., BECKETT, G. J., WILLIAMSON, G. (**2003**) *Carcinogenesis* 24, 497–503.

67 SAKURAI, A., NISHIMOTO, M., HIMENO, S., IMURA, N., TSUJIMOTO,

M., KUNIMOTO, M., HARA, S. (**2004**) *J Cell Physiol* 203, 529–37.

68 ZHANG, D. D., LO, S. C., CROSS, J. V., TEMPLETON, D. J., HANNINK, M. (**2004**) *Mol Cell Biol* 24, 10941–53.

69 RUNDLOF, A. K., JANARD, M., MIRANDA-VIZUETE, A., ARNER, E. S. (**2004**) *Free Radic Biol Med* 36, 641–56.

70 OSBORNE, S. A., TONISSEN, K. F. (**2001**) *BMC Genomics* 2, 10.

71 RUNDLOF, A. K., CARLSTEN, M., GIACOBINI, M. M., ARNER, E. S. (**2000**) *Biochem J* 347, 661–8.

72 SUN, Q. A., ZAPPACOSTA, F., FACTOR, V. M., WIRTH, P. J., HATFIELD, D. L., GLADYSHEV, V. N. (**2001**) *J Biol Chem* 276, 3106–14.

73 SU, D., GLADYSHEV, V. N. (**2004**) *Biochemistry* 43, 12177–88.

74 DAMDIMOPOULOS, A. E., MIRANDA-VIZUETE, A., TREUTER, E., GUSTAFSSON, J. A., SPYROU, G. (**2004**) *J Biol Chem* 279, 38721–9.

75 HERNANDEZ, A., MARTINEZ, M. E., CROTEAU, W., ST GERMAIN, D. L. (**2004**) *Genomics* 83, 413–24.

76 TU, H. M., LEGRADI, G., BARTHA, T., SALVATORE, D., LECHAN, R. M., LARSEN, P. R. (**1999**) *Endocrinology* 140, 784–90.

77 OHBA, K., YOSHIOKA, T., MURAKI, T. (**2001**) *Mol Cell Endocrinol* 172, 169–75.

78 MAIORINO, M., SCAPIN, M., URSINI, F., BIASOLO, M., BOSELLO, V., FLOHE, L. (**2003**) *J Biol Chem* 278, 34286–90.

79 BERMANO, G., NICOL, F., DYER, J. A., SUNDE, R. A., BECKETT, G. J., ARTHUR, J. R., HESKETH, J. E. (**1995**) *Biochem J* 311, 425–30.

80 LEI, X. G., EVENSON, J. K., THOMPSON, K. M., SUNDE, R. A. (**1995**) *J Nutr* 125, 1438–46.

81 VILLETTE, S., BERMANO, G., ARTHUR, J. R., HESKETH, J. E. (**1998**) *FEBS Lett* 438, 81–4.

82 WINGLER, K., BOCHER, M., FLOHE, L., KOLLMUS, H., BRIGELIUS-FLOHE, R. (**1999**) *Eur J Biochem* 259, 149–57.

83 BERRY, M. J., BANU, L., CHEN, Y. Y., MANDEL, S. J., KIEFFER, J. D., HARNEY, J. W., LARSEN, P. R. (**1991**) *Nature* 353, 273–6.

84 SALVATORE, D., LOW, S. C., BERRY, M.,

MAIA, A. L., HARNEY, J. W., CROTEAU, W., ST GERMAIN, D. L., LARSEN, P. R. (**1995**) *J Clin Invest* 96, 2421–30.

85 BERMANO, G., ARTHUR, J. R., HESKETH, J. E. (**1996**) *FEBS Lett* 387, 157–60.

86 MULLER, C., WINGLER, K., BRIGELIUS-FLOHE, R. (**2003**) *Biol Chem* 384, 11–18.

87 GU, Q. P., REAM, W., WHANGER, P. D. (**2002**) *Biometals* 15, 411–20.

88 PAGMANTIDIS, V., BERMANO, G., VILLETTE, S., BROOM, I., ARTHUR, J., HESKETH, J. (**2005**) *FEBS Lett* 579, 792–6.

89 WINGLER, K., MULLER, C., SCHMEHL, K., FLORIAN, S., BRIGELIUS-FLOHE, R. (**2000**) *Gastroenterology* 119, 420–30.

90 BANNING, A., SCHNURR, K., BOL, G. F., KUPPER, D., MULLER-SCHMEHL, K., VIITA, H., YLA-HERTTUALA, S., BRIGELIUS-FLOHE, R. (**2004**) *Free Radic Biol Med* 36, 135–44.

91 PAGMANTIDIS, V., VILLETTE, S., HESKETH, J. E. (**2004**) Proceedings of the Annual Spring Meeting, German Society for Biochemistry and Molecular Biology. 10.1240/sav_gbm_2004_m_000474.

92 BERMANO, G., ARTHUR, J. R., HESKETH, J. E. (**1996**) *Biochem J* 320, 891–5.

93 VILLETTE, S., KYLE, J. A., BROWN, K. M., PICKARD, K., MILNE, J. S., NICOL, F., ARTHUR, J. R., HESKETH, J. E. (**2002**) *Blood Cells Mol Dis* 29, 174–8.

94 HU, Y. J., KOROTKOV, K. V., MEHTA, R., HATFIELD, D. L., ROTIMI, C. N., LUKE, A., PREWITT, T. E., COOPER, R. S., STOCK, W., VOKES, E. E., DOLAN, M. E., GLADYSHEV, V. N., DIAMOND, A. M. (**2001**) *Cancer Res* 61, 2307–10.

95 ESWORTHY, R. S., MANN, J. R., SAM, M., CHU, F. F. (**2000**) *Am J Physiol Gastrointest Liver Physiol* 279, G426–36.

96 CHU, F. F., ESWORTHY, R. S., CHU, P. G., LONGMATE, J. A., HUYCKE, M. M., WILCZYNSKI, S., DOROSHOW, J. H. (**2004**) *Cancer Res* 64, 962–8.

97 YANT, L. J., RAN, Q., RAO, L., VAN REMMEN, H., SHIBATANI, T., BELTER, J. G., MOTTA, L., RICHARDSON, A., PROLLA, T. A. (**2003**) *Free Radic Biol Med* 34, 496–502.

98 RAN, Q., LIANG, H., GU, M., QI, W., WALTER, C. A., ROBERTS, L. J., II, HERMAN, B., RICHARDSON, A., VAN REMMEN, H. (**2004**) *J Biol Chem* 279, 55137–46.

99 JAKUPOGLU, C., PRZEMECK, G. K., SCHNEIDER, M., MORENO, S. G., MAYR, N., HATZOPOULOS, A. K., DE ANGELIS, M. H., WURST, W., BORNKAMM, G. W., BRIELMEIER, M., CONRAD, M. (**2005**) *Mol Cell Biol* 25, 1980–8.

100 CONRAD, M., JAKUPOGLU, C., MORENO, S. G., LIPPL, S., BANJAC, A., SCHNEIDER, M., BECK, H., HATZOPOULOS, A. K., JUST, U., SINOWATZ, F., SCHMAHL, W., CHIEN, K. R., WURST, W., BORNKAMM, G. W., BRIELMEIER, M. (**2004**) *Mol Cell Biol* 24, 9414–23.

101 HILL, K. E., ZHOU, J., MCMAHAN, W. J., MOTLEY, A. K., ATKINS, J. F., GESTELAND, R. F., BURK, R. F. (**2003**) *J Biol Chem* 278, 13640–6.

102 SCHOMBURG, L., SCHWEIZER, U., HOLTMANN, B., FLOHE, L., SENDTNER, M., KOHRLE, J. (**2003**) *Biochem J* 370, 397–402.

103 HILL, K. E., ZHOU, J., MCMAHAN, W. J., MOTLEY, A. K., BURK, R. F. (**2004**) *J Nutr* 134, 157–61.

104 SCHWEIZER, U., SCHOMBURG, L., SAVASKAN, N. E. (**2004**) *J Nutr* 134, 707–10.

105 SCHWEIZER, U., STRECKFUSS, F., PELT, P., CARLSON, B. A., HATFIELD, D. L., KOHRLE, J., SCHOMBURG, L. (**2005**) *Biochem J* 386, 221–6.

106 BOSL, M. R., TAKAKU, K., OSHIMA, M., NISHIMURA, S., TAKETO, M. M. (**1997**) *Proc Natl Acad Sci USA* 94, 5531–4.

107 KUMARASWAMY, E., CARLSON, B. A., MORGAN, F., MIYOSHI, K., ROBINSON, G. W., SU, D., WANG, S., SOUTHON, E., TESSAROLLO, L., LEE, B. J., GLADYSHEV, V. N., HENNIGHAUSEN, L., HATFIELD, D. L. (**2003**) *Mol Cell Biol* 23, 1477–88.

108 FORSBERG, L., DE FAIRE, U., MARKLUND, S. L., ANDERSSON, P. M., STEGMAYR, B., MORGENSTERN, R. (**2000**) *Blood Cells Mol Dis* 26, 423–6.

109 MOSCOW, J. A., HE, R., GUDAS, J. M., COWAN, K. H. (**1994**) *Gene* 144, 229–36.

110 HU, Y. J., DIAMOND, A. M. (**2003**) *Cancer Res* 63, 3347–51.

111 HAMANISHI, T., FURUTA, H., KATO, H., DOI, A., TAMAI, M., SHIMOMURA, H., SAKAGASHIRA, S., NISHI, M., SASAKI, H., SANKE, T., NANJO, K. (**2004**) *Diabetes* 53, 2455–60.

112 QATATSHEH, A., SEAL, C. J., JOWETT, S. L., WELFARE, M. R., HESKETH, J. E. (**2005**) *Proc Nutr Soc* 64, 20A.

113 RATNASINGHE, D., TANGREA, J. A., ANDERSEN, M. R., BARRETT, M. J., VIRTAMO, J., TAYLOR, P. R., ALBANES, D. (**2000**) *Cancer Res* 60, 6381–3.

114 ICHIMURA, Y., HABUCHI, T., TSUCHIYA, N., WANG, L., OYAMA, C., SATO, K., NISHIYAMA, H., OGAWA, O., KATO, T. (**2004**) *J Urol* 172, 728–32.

115 COX, D. G., HANKINSON, S. E., KRAFT, P., HUNTER, D. J. (**2004**) *Cancer Epidemiol Biomarkers Prev* 13, 1821–2.

116 MAIORINO, M., BOSELLO, V., URSINI, F., FORESTA, C., GAROLLA, A., SCAPIN, M., SZTAJER, H., FLOHE, L. (**2003**) *Biol Reprod* 68, 1134–41.

117 MÉPLAN, C., CROSLEY, L. K., MATHERS, J. C., ARTHUR, J. R., HESKETH, J. E. (**2005**) Proceedings of 12th Symposium on Trace Elements in Man And Animals. http://tema12.com, p 112.

118 BERMANO, G., PAGMANTIDIS, V., HOLLOWAY, N., KADRI, S., MOWAT, A., BROOM, J., HESKETH, J. E. (**2004**) *J Nutr* 134, 3526S.

119 AL-TAIE, O. H., UCEYLER, N., EUBNER, U., JAKOB, F., MORK, H., SCHEURLEN, M., BRIGELIUS-FLOHE, R., SCHOTTKER, K., ABEL, J., THALHEIMER, A., KATZENBERGER, T., ILLERT, B., MELCHER, R., KOHRLE, J. (**2004**) *Nutr Cancer* 48, 6–14.

120 BURK, R. F., HILL, K. E. (**2005**) *Annu Rev Nutr* 25, 215–34.

121 APOSTOLOU, S., KLEIN, J. O., MITSUUCHI, Y., SHETLER, J. N., POULIKAKOS, P. I., JHANWAR, S. C., KRUGER, W. D., TESTA, J. R. (**2004**) *Oncogene* 23, 5032–40.

122 CHENG, W. H., HO, Y. S., ROSS, D. A., VALENTINE, B. A., COMBS, G. F., LEI, X. G. (**1997**) *J Nutr* 127, 1445–50.

123 Esworthy, R. S., Ho, Y. S., Chu, F. F. (**1997**) *Arch Biochem Biophys* 340, 59–63.

124 Klivenyi, P., Andreassen, O. A., Ferrante, R. J., Dedeoglu, A., Mueller, G., Lancelot, E., Bogdanov, M., Andersen, J. K., Jiang, D., Beal, M. F. (**2000**) *J Neurosci* 20, 1–7.

125 Lei, X. G. (**2001**) *Biofactors* 14, 93–9.

126 Cheng, W. H., Ho, Y. S., Ross, D. A., Han, Y., Combs, G. F., Jr., Lei, X. G. (**1997**) *J Nutr* 127, 675–80.

127 Furling, D., Ghribi, O., Lahsaini, A., Mirault, M. E., Massicotte, G. (**2000**) *Proc Natl Acad Sci USA* 97, 4351–6.

128 Esworthy, R. S., Aranda, R., Martin, M. G., Doroshow, J. H., Binder, S. W., Chu, F. F. (**2001**) *Am J Physiol Gastrointest Liver Physiol* 281, G848–55.

129 Yamamoto, M., Yang, G., Hong, C., Liu, J., Holle, E., Yu, X., Wagner, T., Vatner, S. F., Sadoshima, J. (**2003**) *J Clin Invest* 112, 1395–406.

130 Schneider, M. J., Fiering, S. N., Pallud, S. E., Parlow, A. F., St Germain, D. L., Galton, V. A. (**2001**) *Mol Endocrinol* 15, 2137–48.

131 Ng, L., Goodyear, R. J., Woods, C. A., Schneider, M. J., Diamond, E., Richardson, G. P., Kelley, M. W., Germain, D. L., Galton, V. A., Forrest, D. (**2004**) *Proc Natl Acad Sci USA* 101, 3474–9.

9
PPARs in Atherosclerosis

Hélène Duez, Jean-Charles Fruchart, and Bart Staels

9.1
Introduction

Atherosclerosis is a complex chronic inflammatory disease of the vascular wall associated with metabolic abnormalities (i.e. lipid and glucose disorders). Peroxisome
proliferator-activated receptors (PPARs) are "nutrient" nuclear receptors that regulate expression of their target genes upon activation by their natural (fatty acids
and derivatives) or synthetic (fibrates or thiazolidinediones) ligands. PPARα induces fatty acid oxidation and plays a crucial role in lipid and lipoprotein metabolism
while PPARγ promotes adipogenesis and fatty acid storage in adipose tissue. Both
PPARα and PPARγ play essential roles in glucose homeostasis and insulin sensitization. The role of PPARδ is more elusive but recent data highlight its role in fatty
acid metabolism and energy expenditure. PPARs are also involved in the control of
vascular wall inflammation and macrophage lipid metabolism and thus exert
beneficial effects on the different stages of atherosclerotic plaque development. Together these data identify PPARs as pharmacologically relevant targets for the
development of new drugs in the treatment of both metabolic disorders (dyslipidemia, type 2 diabetes) and vascular dysfunction predisposing to atherosclerosis and
coronary heart disease.

9.2
Atherosclerosis

Atherosclerosis is a complex chronic inflammatory disease of the vascular wall
initiated by the accumulation of modified low-density lipoproteins (LDLs) (e.g. oxidized low-density lipoprotein (oxLDL)) in the subendothelial space of the vessels.
This results in the activation of endothelial cells (ECs), which in turn secrete
chemoattractant (such as the monocyte chemoattractant protein 1 (MCP-1)) and
adhesion molecules (such as selectins or the vascular cell adhesion molecule 1
(VCAM-1)), leading to the recruitment of circulating monocytes and T lymphocytes

at the endothelial surface and their penetration into the subendothelial space. Trapped monocytes subsequently differentiate into macrophages that take up cholesterol from modified trapped lipoproteins via scavenger receptors SR-A or CD36, thus potentially forming foam cells. However, excess cholesterol can also be removed from macrophages via efflux through the ATP-binding cassette ABCA1 or ABCG1 transmembrane transporters and transported back to the liver through the "reverse cholesterol transport" pathway for removal via bile and feces. Lipid-loaded activated macrophages secrete proinflammatory molecules and promote the activation of ECs, leading to the recruitment of additional monocytes. These fatty streaks may evolve into more complex lesions with the proliferation of activated smooth muscle cells (SMCs) and their migration from the media to the intima and the formation of a neo-intima. Activation of the different cell types within the vascular wall further leads to the release of proinflammatory cytokines, resulting in chronic inflammation. Death of lipid-rich cells and the accumulation of debris and extracellular matrix proteins secreted by SMCs lead to the formation of a lipid-rich necrotic core surrounded by a fibrous cap. Chronic inflammation within such advanced plaque, combined with the secretion of metalloproteinases (MMPs) and expression of procoagulant factors (such as tissue factor (TF)) next results in plaque vulnerability and rupture, and acute occlusion by thrombosis leading to myocardial infarction and stroke [1].

Several epidemiologic studies have demonstrated the importance of lipid disorders (i.e. hypertriglyceridemia, increased plasma levels of atherogenic lipoproteins, such as LDL and small dense LDL, or decreased anti-atherogenic high-density lipoprotein (HDL)) as risk factors in the development of atherosclerosis [2–5]. In addition, dyslipidemia is commonly associated with a constellation of other worsening metabolic abnormalities such as elevated blood glucose and insulin resistance, hypertension, a proinflammatory state, and obesity, clustered in the so-called metabolic syndrome, which individually and collectively contribute to increase the risk of coronary atherosclerosis [6–8].

Recent work indicates that peroxisome proliferator-activated receptors (PPARs) modulate lipid and glucose metabolism and also influence systemic and vascular inflammation. Indeed, a large body of evidence has highlighted the beneficial use of PPARα and PPARγ ligands to reduce cardiovascular disease risk and the development of atherosclerosis by modulating the onset of these metabolic and inflammatory abnormalities. The hypolipidemic fibrate drugs are synthetic PPARα ligands which have successfully been used in the treatment of dyslipidemia and have been shown to reduce the risk of cardiovascular events [9]. Activation of PPARγ by its insulin-sensitizing thiazolidinedione (TZDs or glitazones) ligands leads to insulin sensitization, glucose lowering, and triglyceride clearance. Here, we will review the biological consequences of PPAR activation and their potential as pharmacological targets for the treatment of cardiovascular disease.

9.3
Peroxisome Proliferator-activated Receptors: Expression, Ligands, and Transcriptional Activity

PPARs are ligand-activated transcription factors that belong to the nuclear hormone receptor superfamily (see chapter 2). The PPAR family consists of three different receptors: PPARα, PPARβ/δ (hereafter referred to as PPARδ), and PPARγ. They exhibit different expression patterns, PPARα being mainly expressed in tissues exhibiting high rates of β-oxidation such as liver, heart, kidney, muscle, and in steroidogenic tissues, while PPARγ is predominantly found in adipose tissues where it plays a critical role in adipogenesis (Table 9.1). PPARδ has a more ubiquitous expression pattern. In addition, both PPARs are expressed in several immunological cells and vascular wall cells such as monocytes/macrophages and macrophage-derived foam cells, lymphocytes, ECs, and SMCs [10] (Table 9.1).

PPARs exert diverse biological functions in response to activation by a large variety of ligands. They primarily act as lipid sensors. As such, dietary and endogeneous fatty acids (FAs) and their derivatives, such as eicosanoids, are natural PPAR ligands. In addition, fibrates, which are potent hypolipidemic drugs, and the insulin-sensitizing thiazolidinediones (TZDs or glitazones) are synthetic ligands for PPARα and PPARγ, respectively [11]. After activation by their ligands, they heterodimerize with the nuclear receptor RXR (retinoid X receptor) and bind to PPAR response elements (PPREs) in the promoter region of their target genes, allowing their transcriptional regulation. These response elements consist of a direct repeat of the AGGTCA core motif spaced by one (DR-1) or two (DR-2) nucleotides (Fig. 9.1). PPARs are also able to repress expression of their target genes in both DNA-independent and -dependent manners by interfering with other signaling pathways or recruiting co-repressors to the promoter (Fig. 9.1).

9.4
Physiological Functions of Peroxisome Proliferator-activated Receptors

9.4.1
Metabolic Effects of PPAR Activation

9.4.1.1 Metabolic effects of PPARα activation

PPARα upregulates the expression of numerous genes involved in the mitochondrial and peroxisomal uptake and β-oxidation of FA, thereby stimulating their utilization in the liver [12] (Fig. 9.2). Furthermore, PPARα has also been shown to increase triglyceride (TG)-rich lipoprotein catabolism by inducing lipoprotein lipase (LPL) activity, and to decrease plasma apoC-III levels. As a result, administration of the fibrate PPARα ligands leads to lowered TG plasma levels and reduced very low-density lipoprotein (VLDL) production [13]. PPARα activation was also demonstrated to induce expression of apolipoprotein (apo)A-V, which is known as a potent

Table 9.1. Peroxisome proliferator-activated receptors: major sites of expression, naturally occurring and synthetic ligands and major biological functions

	PPARα (NR1C1)	PPARβ/δ (NR1C2)	PPARγ (NR1C3)
Expression	Liver, muscle, kidney, heart; Different cell types of the vascular wall (macrophages, ECs, SMCs, etc.)	Ubiquitous (skin, adipose tissue, small intestine, etc.); Macrophages, ECs	Brown and white adipose tissue; Different cell types of the vascular wall (macrophages, ECs, SMCs)
Ligands			
Natural ligands	FAs and derivatives (eicosanoides, LTB4); 8-HETE, 9-HODE, 13-HODE	FAs	FAs and derivatives ((15dPGJ2), 9-HODE, etc.)
Synthetic ligands	Fibrates		Thiazolidinediones (TZDs)
Biological functions	β-Oxidation; Lipid and lipoprotein metabolism; Glucose homeostasis; Inflammatory response; Macrophage lipid metabolism	FA oxidation (muscle, adipose tissue); Lipid metabolism; Energy homeostasis; Inflammatory response	Control of adipogenesis; Glucose homeostasis; Inflammatory response; Macrophage lipid metabolism

FAs, Fatty acids; HETE, hydroxyeicosatetraenoic acid; HODE, hydroxyoctadecadienoic acid; 15dPGJ2, 15-deoxyΔ12,14 prostaglandin J2. Only clinically used synthetic ligands have been reported.

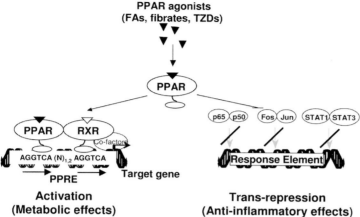

Fig. 9.1. Peroxisome proliferator-activated receptors: transcriptional activity. Upon activation by their ligands, PPARs heterodimerize with another nuclear receptor, RXR, and then bind to specific PPAR response elements (PPREs) in the promoter of their target genes. In addition, PPARs can interact with other transcription factors in a DNA-binding independent manner and exhibit anti-inflammatory properties by repressing gene expression.

TG determinant [14]. Fibrates also decrease the small dense LDL lipoprotein fraction and stimulate the appearance of larger and less atherogenic particles. In humans, PPARα activation also increases the expression the two major protein constituents of HDL, apoA-I and apoA-II, leading to improved HDL-cholesterol plasma levels in humans [13].

Growing evidence points to a role of PPARα in glucose metabolism and insulin signaling, although conflicting data have been reported. On the one hand, PPARα activation has been shown to enhance insulin sensitivity in high-fat diet-induced insulin resistance [15] and in lipodystrophic mice [16]. This was thought to be a secondary effect of FA oxidation, alleviating the inhibitory effect of FFAs on peripheral glucose utilization. On the other hand, PPARα-deficient mice are resistant to diet-induced insulin resistance [17, 18] while cardiac-specific PPARα overexpression leads to altered FA and glucose utilization as seen in diabetic conditions [19]. In a recent study, Finck *et al.* have found that muscle-specific PPARα overexpression leads to decreased basal and insulin-stimulated muscle glucose uptake and reduced insulin-stimulated whole-body glucose utilization, suggesting that muscle PPARα overexpressing mice become insulin resistant despite a lower weight gain on high-fat diet and smaller fat depots [20]. This was correlated with a reduced expression of the muscle glucose transporter Glut4 and genes encoding glucose-handling proteins, whereas expression of genes involved in FA utilization were markedly increased. However, insulin still suppressed endogenous glucose production, indicating a peripheral dysfunction. By contrast, PPARα-deficient mice were consistently shown to be resistant to high-fat diet-induced insulin resistance even

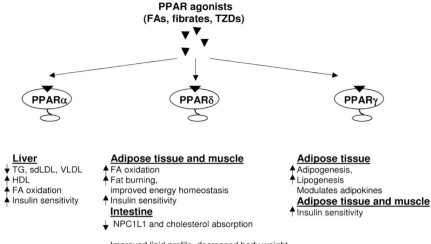

Fig. 9.2. Metabolic actions of peroxisome proliferator-activated receptors. PPARα regulates the expression of genes involved in fatty acid (FA) oxidation, VLDL synthesis and HDL metabolism, resulting in decreased triglycerides (TG), small dense LDL (sdLDL) and increased HDL plasma levels. PPARδ stimulates FA oxidation primarily in muscle, but also in adipose tissue, and regulates energy utilization. In the intestine it downregulates NPC1L1 expression and cholesterol absorption. PPARδ may improve the lipid profile. PPARγ modulates insulin signaling, cytokine production and free fatty acid (FFA) metabolism in adipose tissue. In addition, PPARγ may play a direct role in insulin sensitivity in muscle and liver.

though they developed an obese phenotype [20]. In contrast, when assessing high fat diet-induced insulin resistance by the same hyperinsulinemic euglycemic clamp method, Haluzik *et al.* failed to demonstrate any difference between PPARα-deficient and wild-type mice [21]. Thus, the role of PPARα in glucose metabolism is still controversial and further studies are required to clarify its *in vivo* function and the potential consequences of its activation.

9.4.1.2 Metabolic effects of PPARδ activation

PPARδ is thought to regulate whole-body lipid metabolism. Administration of a synthetic PPARδ ligand to obese Rhesus monkeys resulted in increased HDL-cholesterol, and lowered LDL-cholesterol and TG plasma levels [22] (Fig. 9.2). Consistently, treatment with a PPARδ ligand improved the lipid profile in db/db mice [23]. Interestingly, a recent report established that a synthetic PPARδ activator increases plasma HDL-cholesterol levels in mice due to reduced intestinal expression of Niemann–Pick C1-like-1 (NPC1L1) transporter and cholesterol absorption [24]. PPARδ may also regulate FA oxidation and energy expenditure. Adipose tissue-specific overexpression of a PPARδ/VP16 fusion protein that is constitutively active resulted in increased FA oxidation and improved lipid plasma levels [25]. In the same constitutive report, PPARδ activation promoted fat burning and conferred protection against the development of high-fat diet-induced obesity [25]. In addition, PPARδ also affects muscle FA metabolism and energy homeostasis [25–27] and PPARδ overexpression in skeletal muscle led to enhanced expression of genes involved in FA utilization, resistance to obesity, and improved insulin sensitivity after a high-fat diet challenge. Finally, PPARδ activation increases cardiac FA utilization [28] ensuring a normal cardiac function [29]. Together these data strongly indicate the involvement of PPARδ in lipid metabolism and support a protective role for this receptor against the development of obesity.

9.4.1.3 Metabolic effects of PPARγ activation

PPARγ plays a prominent role in adipogenesis by controlling the adipocyte differentiation program [30–33] (Fig. 9.2). It participates in the control of adipocyte lipid metabolism by regulating genes involved in adipose FA uptake and storage such as aP2 (fatty acid-binding protein), LPL, PEPCK (phosphoenolpyruvate carboxykinase), and CD36 (see Ref. [33] and references therein). In addition, PPARγ is a major regulator of glucose homeostasis and insulin sensitivity. Indeed, TZD treatment ameliorates insulin sensitivity in patients whereas mutations in the PPARγ gene led to hyperlipidemia and insulin resistance [34–36]. This may be secondary to FA storage in adipose tissue, thereby improving muscle insulin sensitivity and glucose disposal, and reducing pancreatic lipotoxicity induced by elevated free fatty acid (FFA) plasma levels. PPARγ is also a crucial modulator of endocrine function of adipose tissue, regulating leptin (decrease), resistin (decrease), and adiponectin (increase) synthesis/release by adipocytes [37, 38]. It also decreases cytokine (tumor necrosis factor alpha (TNFα), plasminogen activator inhibitor 1 (PAI-1), and interleukin 6 (IL-6)) production (see Ref. [39] and references inside). It is believed that together with a redistribution of fat cells towards smaller insulin-sensitive adipocytes, these actions participate in the insulin-sensitizing and glucose-lowering effect of PPARγ activation.

TZDs also improve insulin sensitivity in muscle and liver, although this benefit was lost *in vivo* in lipodystrophic mice, suggesting that adipose tissue is a major site of PPARγ insulin-sensitizing action [40]. Liver-specific PPARγ deficiency in lipoatrophic A-ZIP/F-1 (AZIP) mice resulted in hyperlipidemia, elevated glucose and insulin plasma levels, and TZDs were inefficient in this model [41]. However, in mice with adipose tissue, liver-specific PPARγ deficiency results in higher fasting and postprandial triglyceride levels and increased serum glucose and insulin concentrations, but TZD response was normal. These observations corroborate the notion that TZD action most likely occurs via adipose tissue PPARγ activation [41]. However, TZDs were inefficient when liver PPARγ was disrupted in AZIP mice, also leading to worsened hyperlipidemia and muscle insulin resistance. In line with this, adipose tissue-specific PPARγ deficiency in mice led to hyperlipidemia, steatosis, and insulin resistance at the hepatic level but whole-body insulin sensitivity was maintained [42]. Muscle-specific loss of PPARγ on the other hand resulted in impaired whole-body insulin sensitivity even though it is still debatable whether or not TZD treatment improved insulin sensitivity when these mice were challenged with a high fat diet [43, 44]. This indicates that different tissues participate in the insulin-sensitizing action of PPARγ.

9.4.2
Vascular Effects of PPAR Activation

PPARs are expressed in the different cell types of the vascular wall (i.e. ECs, vascular smooth muscle cells, monocytes/macrophages, and T lymphocytes [10]) and regulate some aspects of their physiology, thus modulating atherosclerotic lesion formation (Fig. 9.3).

9.4.2.1   Vascular Actions of PPARα
In ECs, PPARα interferes with other signaling pathways (i.e. NFκB or AP-1 signaling) and decreases cytokine-induced expression of adhesion molecules such as VCAM-1, thereby impeding leukocyte recruitment and adhesion to the endothe-

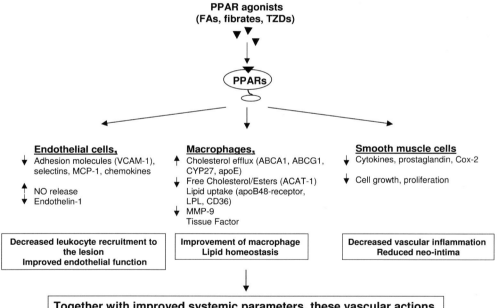

Fig. 9.3. Vascular effects of peroxisome proliferator-activated receptor activation. PPARα and PPARγ regulate macrophage lipid metabolism, promoting cholesterol efflux thus limiting foam cell formation. PPARs reduce the recruitment of leukocytes to the lesion by diminishing expression of chemoattractant and adhesion molecules. Moreover, all three PPARs decrease vascular inflammation and may affect plaque stability by modulating the expression of metalloproteinase and coagulation factors.

lium [45]. In contrast, PPARα does not appear to modify cytokine-induced expression of ICAM-1 and E-selectin [45], monocyte chemoattractant protein 1 (MCP-1) [46] or interferon gamma (IFNγ)-induced expression of the CXC-chemokines IP-10, Mig, and I-Tac, although conflicting results have been reported [47]. Beside its role in decreasing leukocyte recruitment to the lesion, PPARα activation has also been shown to substantially reduce vascular inflammation. In human aortic SMCs, PPARα activation by fibrates inhibits IL-1β-induced production of IL-6 and prostaglandins, and cyclooxygenase-2 (COX-2) expression [48]. Furthermore, aortas from PPARα-deficient mice displayed exacerbated response to inflammatory stimuli [49]. In addition, PPARα ligands decreased TNFα expression in macrophages. PPARα activators reduce TF and MMP expression in monocytes and macrophages, thus potentially affecting the stability and thrombogenicity of the plaque [50, 51]. Together this indicates a vascular anti-inflammatory action of PPARα. Moreover, PPARα agonists induce endothelial nitric oxide synthase (eNOS) expression and release of the dilatation mediator nitric oxide (NO), and decrease thrombin- and oxLDL-induced expression of endothelin-1 (ET-1) in endothelial cells, suggesting that PPARα activators may influence vascular tone and ameliorate endothelial dysfunction [10].

The role of PPARα in monocyte/macrophage lipid homeostasis has been extensively studied. Treatment with PPARα ligands was shown to increase the expression of the HDL receptor CLA-1/SRB-I and ABCA1, a membrane transporter controlling apoA-I-mediated cholesterol efflux [52, 53]. In addition, PPARα regulates macrophage intracellular cholesterol metabolism by decreasing the intracellular cholesteryl ester:free cholesterol ratio via reduced ACAT-1 activity [54]. Moreover, PPARα activators reduce the expression of the apoB48-remnant receptor in differentiated macrophages, thereby reducing the uptake of glycated LDL and TG-rich remnant lipoproteins [55, 56]. PPARα activation has been shown to decrease secretion and activity of LPL [56] although conflicting data have also been reported [57].

PPARα ligands have also been shown to enhance the production of reactive oxygen species (ROS) via the induction of NADPH oxidase expression and activity in macrophages [58]. This results in the formation of oxLDL metabolites and may lead to the generation of endogenous PPARα ligands. Counterbalancing a potential pathologic effect raised by the formation of oxidized LDL, PPARα activation by these metabolites promotes potentially compensatory anti-inflammatory actions. In the same line, it has previously been shown that lipolysis of TG-rich lipoproteins generates PPARα ligands [59]. Together these data suggest a beneficial effect of PPARα activation by increasing the free cholesterol pool available for cholesterol efflux and promoting its removal from macrophages, and thus regulating the overall lipid metabolism of these cells to avoid accumulation, potentially leading to the regression of fatty streaks.

9.4.2.2 Vascular Actions of PPARδ

In ECs, PPARδ activation was shown to reduce cytokine-induced VCAM-1 expression and MCP-1 release [60], which should result in a beneficial effect on leukocyte chemotaxis and recruitment to the lesions. In addition, PPARδ is likely to play a role in macrophage lipid homeostasis, since it increases ABCA1 gene expression and apoA-I-mediated cholesterol efflux [22]. By contrast, PPARδ may also promote triglyceride uptake from VLDL, and cholesterol loading and storage in macrophages via the induction of scavenger receptors CD36 and SRA, as well as adipose differentiation-related protein (ADRP) and fatty acid-binding protein (FABP) expression [61, 62]. However, Lee *et al.* reported that PPARδ activation did not affect macrophage cholesterol homeostasis [63]. The same conclusion was drawn from studies using PPARδ activator-treated peritoneal macrophage transferred in hypercholesterolemic mice [64]. In addition, recent reports suggest a role for PPARδ in the control inflammation. Indeed, PPARδ activators decrease lipopolysaccharide (LPS)-induced iNOS and COX-2 expression in macrophages [65]. Although deletion of the PPARδ gene in macrophages results in a lower inflammatory response due to decreased expression of IL-1β, MMP-9 and MCP-1, PPARδ activation was also shown to decrease the production of inflammatory molecules [63]. Therefore PPARδ seems to influence both inflammatory markers and macrophage lipid metabolism. However, recent studies in LDL receptor-deficient mice did not reveal any atheroprotective activity upon treatment with a selective potent PPARδ agonist [64].

9.4.2.3 Vascular Functions of PPARγ

Like the other PPAR isotypes, PPARγ also regulates macrophage lipid homeostasis, inducing both potentially anti- and pro-atherogenic responses. On the one hand, PPARγ activators promote cholesterol removal from macrophages via the induction of CLA-1/SRB-I, ABCA1, ABCG1, CYP27, and apoE expression [66–69] and reduce the accumulation of intracellular cholesterol esters via decreased scavenger receptor A (SRA) I/II activity [70]. PPARγ activators diminish glycated LDL uptake [56] and, like PPARα, PPARγ decreases macrophage apoB48 receptor expression and TG accumulation [55]. In addition, both PPARα and γ also control the expression of adiponectin receptors in macrophages, resulting in an enhanced adiponectin-induced decrease of cholesteryl ester content [71]. On the other hand, however, PPARγ may also exert deleterious effects by increasing the expression of the oxLDL scavenger receptor CD36 [72], which may promote foam cell formation. Moreover, components of these modified lipoproteins are PPARγ ligands leading to the formation of a deleterious vicious circle. However, PPARγ activation does not promote lipid accumulation and transformation of macrophages into foam cells *in vitro* [53], and *in vivo* studies suggest an overall beneficial effect of PPARγ activators on fatty streak formation and atherosclerosis development (see below). Finally, PPARγ activators have been shown to normalize insulin signaling and decrease the expression of scavenger receptors, such as CD36, in macrophages isolated from insulin-resistant mice [73].

PPARγ inhibits IFNγ-induced expression of IP-10, MIG, and I-TAC, but not MCP-1, thereby interfering with immune cell recruitment [46]. It also reduces vascular inflammation by reducing macrophage cytokine (TNFα, IL-1β and IL-6) production and expression of iNOS, SRA, and MMP-9 [74]. In ECs, PPARγ inhibits expression of ET-1 [75] and increases NO release, thus having a vasorelaxant effect.

This was further demonstrated by *in vivo* results showing that TZD treatment decreases hypertension in hypertensive rats [76]. PPARγ activators inhibit VSMC growth, proliferation, and migration [77] and decrease the production of matrix-degrading enzymes [78]. PPARγ activation in VSMCs leads also to a diminished expression of the angiotensin-II type 1 receptor [79]. Finally, PPARγ has been proposed to play a major role in lymphocytes and to shift the inflammatory response away from Th1 (for review see Ref. [10]).

9.5
PPARs and Atherosclerosis: Insights from Animal Models

9.5.1
PPARα

The above-mentioned reports have highlighted the potential beneficial role of PPARα in regulating lipid metabolism, limiting or preventing macrophage cholesterol engorgement and interfering with inflammatory reactions, prompting investigators to assess the effect of PPARα activation on the development of atherosclerosis in mice (Table 9.2). Surprisingly, conflicting data have been reported so far. PPARα-deficiency in apoE$^{-/-}$ mice resulted in enhanced insulin sensitivity and reduced atherosclerotic lesion development [18]. In the same line, ciprofibrate treatment was reported to aggravate atherosclerosis development in apoE-deficient mice [80]. On the other hand, we observed that fenofibrate administration resulted in reduced atherogenesis in the descending aortas in Western diet-fed apoE$^{-/-}$ mice although lesion size at the aortic arch was not reduced [81]. Since plasma lipids did not display major alterations, a direct vascular effect of PPARα likely contributed to its actions on atherosclerosis development *in vivo*. Interestingly, fenofibrate treatment led to a more pronounced effect, with a markedly decreased atherosclerotic lesion size in the aortic sinus in apoE$^{-/-}$ mice overexpressing the human apoA-I gene, suggesting that lipid improvement further enhances the beneficial effect of PPARα activation [81]. In a recent study, Li *et al.* reported a strong reduction of atherosclerotic lesion size and vascular inflammatory markers in high cholesterol-fed LDL-R$^{-/-}$ mice after treatment with the synthetic PPARα agonist GW7647, and the effect was consistently more pronounced in the descending aorta compared to the aortic arch [64]. The quantitative difference between these two studies is likely to be due to the difference in the mouse models with an absence of lipid effects in the apoE$^{-/-}$ models while PPARα activation led to a reduction of insulin levels and an improved lipid profile in LDL-R-deficient mice. Differences in PPARα ligands may also have influenced the extent of lesion reduction.

In an attempt to further assess the effect of PPARα activation on *in vivo* foam cell formation, Li *et al.* used peritoneal macrophage and bone marrow transplantation experiments in LDL-R-deficient mice and found that PPARα activation inhibited lipid accumulation in a PPARα- and LXR-dependent manner, but independent of ABCA1. However, whether this effect on peritoneal macrophages may reflect the

Table 9.2. *In vivo* studies examining the effect of peroxisome proliferator-activated receptor activation on the development of atherosclerosis

Target	Mouse model	Treatment	Metabolic effects	Inflammatory markers	Lesion	Reference
PPARα	PPARα$^{-/-}$ × apoE$^{-/-}$	No	Increased TG, TC, improved glucose levels, decreased blood pressure	Decreased MCP-1	Decreased lesion size	18
	apoE$^{-/-}$	Fenofibrate	No improvement	Decreased MCP-1	Decreased cholesterol content in descending aortas	81
	apoE$^{-/-}$ × hapoA-I Tg	Fenofibrate	Increased apoA-I, slightly increased HDL-cholesterol levels		Decreased lesion area in aortic arch	81
	apoE$^{-/-}$	Ciprofibrate	Increased TC, decreased TG		Increased lesion area	80
	LDL-R$^{-/-}$	GW7647	Improved lipid profile, improved insulin sensitivity	Reduced aortic inflammation	Decreased lesion size, reduced foam cell formation	64
PPARδ	LDL-R$^{-/-}$	GW0742	Improved lipid profile	Reduced aortic inflammation	No effect	64
PPARγ	LDL-R$^{-/-}$ mice	Rosiglitazone, GW7845	Improved lipid and glucose metabolism but gender-specific differences	Decreased MCP-1, TNFα, VCAM	Reduced lesions in male	64, 82
	ApoE$^{-/-}$ mice	Rosiglitazone	Increased TC and TG, unchanged glucose and insulin		Reduced lesions	84
	LDLR$^{-/-}$ mice	Troglitazone HF-fed mice	Decreased insulin, slightly improved systemic parameters	Decreased MCP-1-mediated leukocyte recruitment	Reduced lesions	83
	ApoE$^{-/-}$ mice	Troglitazone	Increased HDL, reduced insulin		Reduced lesions	85

hapoA-I Tg, human apoA-I transgenic mice; apoE$^{-/-}$, apoE-deficient mice; LDL-R$^{-/-}$, LDL receptor-deficient mice; HF, high fat.

in vivo situation in the vascular wall needs to be clarified. Together these data indicate a beneficial effect of PPARα activation on atherosclerosis development even though its action on glucose homeostasis and diet-induced insulin resistance is still to be clearly established.

9.5.2
PPARδ

A few data exist with regard to the influence of PPARδ activation on atherosclerosis *in vivo*. Although PPARδ activation has clearly been shown to improve lipid profile [22, 23], transplantation of bone marrow from PPARδ-deficient mice in apoE-deficient recipient mice resulted in less atherosclerosis, suggesting a negative impact of macrophage PPARδ [63]. Li *et al.* have recently reported the absence of benefit from PPARδ activation on the size of atherosclerotic lesions in high cholesterol-fed LDL-R-deficient mice although the lipid profile was slightly improved and the expression of vascular inflammatory markers was significantly reduced [64] (Table 9.2). The lack of any obvious beneficial effect on macrophage lipid metabolism or whole-body glucose homeostasis may have impeded the positive impact of the PPARδ ligand on plasma lipids.

9.5.3
PPARγ

PPARγ ligands have been tested *in vivo* with regard to their effects on atherosclerotic lesion development. In contrast to PPARα, PPARγ activation yields more consistent results with an overall protective effect of PPARγ activation, though CD36 expression is generally increased [82–85]. As expected, PPARγ activation led to attenuated expression of vascular inflammatory markers and an improved macrophage cholesterol homeostasis, which might beneficially influence atherosclerosis development [82] (Table 9.2). This was further confirmed by wild-type or PPARγ-deficient bone marrow transplantation to LDL-R-deficient recipient mice, showing the anti-atherogenic role of PPARγ in cells of the immune system, such as macrophages [67]. Li *et al.* recently reported that PPARγ activation increased macrophage ABCA1 gene expression and cholesterol efflux *in vitro*, but expression of ABCA1 remains unchanged in the aortas of treated LDL-R-deficient mice [64]. This study further suggests that macrophage lipid metabolism improvement *in vivo* may occur in an ABCA1/LXR-independent pathway and may rather rely on ABCG1 induction [64]. These results demonstrate that PPARγ activation positively affects vascular inflammation and macrophage lipid homeostasis, resulting in a reduction of atherosclerotic lesion size *in vivo*.

Taken together, these studies indicate that PPARα and γ activators are efficient at reducing atherosclerosis and this benefit is likely to derive from both improvement of whole-body lipid and glucose homeostasis, vascular anti-inflammatory action, and inhibition of macrophage foam cell formation after PPAR activation. The potential anti-atherosclerotic effect of PPARδ activation, however, is still debatable but the improvement of lipid profile and glucose utilization, as well as a protection against the development of obesity, are promising observations.

9.6
PPAR Ligands in Clinical Trials

9.6.1
PPARα

The relevance of PPARα as a pharmacological target has so far been highlighted by the use of fibrates in the treatment of dyslipidemia. Several clinical trials (i.e. the BECAIT [86], DAIS [87], and LOCAT [88] studies) have shown a reduction of the progression of coronary atherosclerosis after fibrate treatment. A decreased incidence of coronary artery disease (CAD) was also demonstrated in the Helsinky Heart Study [89] and VA-HIT trial [90]. This benefit may be due to an improved lipid profile (decreased TG, increased HDL-cholesterol) but also to anti-inflammatory actions of PPARα. Indeed, fenofibrate treatment was shown to reduce plasma levels of C-reactive protein (CRP) and IL-6 in dyslipidemic patients with CAD [48] and a reduction of CRP was also demonstrated in diabetic patients after gemfibrozil administration [91]. Moreover, the DAIS study has shown a stronger benefit of fibrate treatment in patients with insulin resistance, which shows a lesser lipid response to treatment [92].

9.6.2
PPARγ

Although no data are available on the effect of TZDs on clinical endpoints, several clinical studies have demonstrated the benefit of PPARγ activation on endothelial function in a manner that may be independent of any glucose-lowering effect [93, 94]. In addition, PPARγ activators inhibit inflammatory markers in humans. As such, treatment of type 2 diabetic patients with TZDs significantly reduced serum levels of CRP and MMP-9 [95] as well as SAA, TNFα [96] and sCD40L levels [97]. Clinical data have already shown that TZDs inhibit carotid intimal thickening [98]. TZDs also reduce neo-intimal proliferation after coronary stent implantation [99].

9.7
Therapeutic Perspective: Selective PPAR Modulators (SPPARMs) and PPARα/γ Co-agonists

The PPAR/RXR heterodimer interacts with co-activators that act as bridges with the basal transcription machinery and play crucial roles in the transmission of regulatory signals. Different ligands for the same nuclear receptor can induce different receptor conformations in a way that is unique to each ligand, leading to dif-

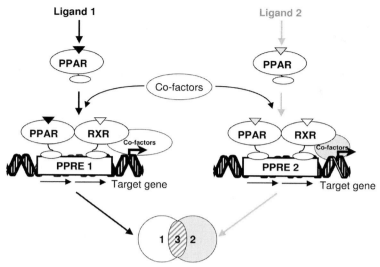

Common/Differential biological response

Fig. 9.4. The SPPARM concept. Different ligands for the same nuclear receptor can induce different receptor conformations in a way that is unique to each ligand, leading to different co-activator recruitment depending on the cell type or promoter context. This results in differential regulation of different/overlapping genes, leading to differential biological responses.

ferential co-activator recruitment. This concept has been called the selective PPAR modulator (SPPARM) concept by analogy with selective estrogen receptor modulators (SERMs) and may explain differences in biological activity with different ligands able to activate, or repress, specific genes depending on the cell type (Fig. 9.4). For instance, compared with other glitazones, troglitazone behaves as a partial or full agonist for PPARγ depending on the cellular environment and promoter context, leading to selective PPARγ modulation and differential downstream effects on gene expression [100]. This may be of great pharmacologic interest since ligands may be developed having the favorable PPARγ effects on glucose metabolism without stimulating adipocyte differentiation.

We have recently shown that this concept pertains for PPARα also and that it may contribute to the differential effects of fenofibrate (increase) and gemfibrozil (no effect) on human apoA-I gene expression [101]. Indeed, whereas fenofibrate behaves as a full agonist, gemfibrozil acts as a partial agonist due to a differential recruitment of co-activators to the human apoA-I promoter, leading to an increase of apoA-I gene expression after fenofibrate, but not gemfibrozil, treatment.

Since PPARα and PPARγ exert both overlapping and different beneficial actions on both the lipid profile, insulin sensitivity, and vascular inflammation and lipid metabolism, combined PPARα/γ activation might lead to complementary and/or synergistic favorable metabolic and anti-inflammatory effects as well as attenuation of side effects such as body weight gain as seen with PPARγ activators. Studies conducted in rodents have shown that several of these dual agonists improve insulin sensitivity, as well as fatty acid, glucose, and lipoprotein metabolism [102–106]. In humans, clinical data confirmed the beneficial effects of PPARα/γ co-agonists on insulin sensitivity, HDL and TG levels, and inflammatory markers even if undesirable side effects are still to overcome.

9.8
Conclusions

Taken together, the studies reviewed here demonstrate an overall benefit of PPAR activation on the lipid profile, glucose metabolism and insulin sensitivity, obesity, and energy homeostasis, as well as a direct vascular anti-inflammatory action. Both systemic and vascular beneficial effects of PPAR activation contribute to a decreased risk of cardiovascular disease. Thus PPARs have emerged over the last years as interesting pharmacological targets to correct abnormalities related to atherosclerosis and, more generally the metabolic syndrome, and efforts are currently been made to develop new efficient (dual) agonists.

Acknowledgments

This work was supported by grants from INSERM, the Région Nord Pas de Calais (Génopôle no. 01360124) and the Leducq Foundation.

References

1 Lusis, A.J. (**2000**) *Nature* 407, 233–241.

2 Assmann, G., P. Cullen, F. Jossa, B. Lewis, M. Mancini (**1999**) *Arterioscler Thromb Vasc Biol* 19, 1819–1824.

3 Gordon, D.J., J.L. Probstfield, R.J. Garrison, J.D. Neaton, W.P. Castelli, J.D. Knoke, D.R. Jacobs, Jr., S. Bangdiwala, H.A. Tyroler (**1989**) *Circulation* 79, 8–15.

4 Genest, J.J., J.R. McNamara, D.N. Salem, E.J. Schaefer (**1991**) *Am J Cardiol* 67, 1185–1189.

5 Ginsberg, H.N. (**2001**) *Am J Cardiol* 87, 1174–1180.

6 Ridker, P.M. (**2003**) *Am J Cardiol* 92, 17K–22K.

7 Lakka, H.M., D.E. Laaksonen, T.A. Lakka, L.K. Niskanen, E. Kumpusalo, J. Tuomilehto, J.T. Salonen (**2002**) *JAMA* 288, 2709–2716.

8 Grundy, S.M. (**2004**) *J Clin Endocrinol Metab* 89, 2595–2600.

9 Robins, S.J. (**2001**) *J Cardiovasc Risk* 8, 195–201.

10 Marx, N., H. Duez, J.C. Fruchart, B. Staels (**2004**) *Circ Res* 94, 1168–1178.

11 Willson, T.M., W. Wahli (**1997**) Peroxisome proliferator-activated receptor agonists. *Curr Opin Chem Biol* 1, 235–241.

12 Desvergne, B., W. Wahli (**1999**) *Endocr Rev* 20, 649–688.

13 Staels, B., J. Dallongeville, J. Auwerx, K. Schoonjans, E. Leitersdorf, J.C. Fruchart (**1998**) *Circulation* 98, 2088–2093.

14 Vu-Dac, N., P. Gervois, H. Jakel, M.

Nowak, E. Bauge, H. Dehondt, B. Staels, L.A. Pennacchio, E.M. Rubin, J. Fruchart-Najib, J.C. Fruchart (**2003**) *J Biol Chem* 278, 17982–17985.

15 Guerre-Millo, M., P. Gervois, E. Raspe, L. Madsen, P. Poulain, B. Derudas, J.M. Herbert, D.A. Winegar, T.M. Willson, J.C. Fruchart, R.K. Berge, B. Staels (**2000**) *J Biol Chem* 275, 16638–16642.

16 Chou, C.J., M. Haluzik, C. Gregory, K.R. Dietz, C. Vinson, O. Gavrilova, M.L. Reitman (**2002**) *J Biol Chem* 277, 24484–24489.

17 Guerre-Millo, M., C. Rouault, P. Poulain, J. Andre, V. Poitout, J.M. Peters, F.J. Gonzalez, J.C. Fruchart, G. Reach, B. Staels (**2001**) *Diabetes* 50, 2809–2814.

18 Tordjman, K., C. Bernal-Mizrachi, L. Zemany, S. Weng, C. Feng, F. Zhang, T.C. Leone, T. Coleman, D.P. Kelly, C.F. Semenkovich (**2001**) *J Clin Invest* 107, 1025–1034.

19 Finck, B.N., J.J. Lehman, T.C. Leone, M.J. Welch, M.J. Bennett, A. Kovacs, X. Han, R.W. Gross, R. Kozak, G.D. Lopaschuk, D.P. Kelly (**2002**) *J Clin Invest* 109, 121–130.

20 Finck, B.N., C. Bernal-Mizrachi, D. Han, T. Coleman, N. Sambandam, L.L. LaRiviere, J.O. Holloszy, C. Semenkovich, D. Kelly (**2005**) *Cell Metab* 1, 133–144.

21 Haluzik, M., O. Gavrilova, D. LeRoith (**2004**) *Endocrinology* 145, 1662–1667.

22 Oliver, W.R., Jr., J.L. Shenk, M.R. Snaith, C.S. Russell, K.D. Plunket, N.L. Bodkin, M.C. Lewis, D.A. Winegar, M.L. Sznaidman, M.H. Lambert, H.E. Xu, D.D. Sternbach, S.A. Kliewer, B.C. Hansen, T.M. Willson (**2001**) *Proc Natl Acad Sci USA* 98, 5306–5311.

23 Leibowitz, M.D., C. Fievet, N. Hennuyer, J. Peinado-Onsurbe, H. Duez, J. Bergera, C.A. Cullinan, C.P. Sparrow, J. Baffic, G.D. Berger, C. Santini, R.W. Marquis, R.L. Tolman, R.G. Smith, D.E. Moller, J. Auwerx (**2000**) *FEBS Lett* 473, 333–336.

24 van der Veen, J.N., J.K. Kruit, R. Havinga, J.F. Baller, G. Chimini, S. Lestavel, B. Staels, P.H. Groot, A.K. Groen, F. Kuipers (**2005**) *J Lipid Res* 46, 526–534.

25 Wang, Y.X., C.H. Lee, S. Tiep, R.T. Yu, J. Ham, H. Kang, R.M. Evans (**2003**) *Cell* 113, 159–170.

26 Tanaka, T., J. Yamamoto, S. Iwasaki, H. Asaba, H. Hamura, Y. Ikeda, M. Watanabe, K. Magoori, R.X. Ioka, K. Tachibana, Y. Watanabe, Y. Uchiyama, K. Sumi, H. Iguchi, S. Ito, T. Doi, T. Hamakubo, M. Naito, J. Auwerx, M. Yanagisawa, T. Kodama, J. Sakai (**2003**) *Proc Natl Acad Sci USA* 100, 15924–15929.

27 Dressel, U., T.L. Allen, J.B. Pippal, P.R. Rohde, P. Lau, G.E. Muscat (**2003**) *Mol Endocrinol* 17, 2477–2493.

28 Gilde, A.J., K.A. Van Der Lee, P.H. Willemsen, G. Chinetti, F.R. van der Leij, G.J. Van Der Vusse, B. Staels, M. Van Bilsen (**2003**) *Circ Res* 92, 518–524.

29 Cheng, L., G. Ding, Q. Qin, Y. Huang, W. Lewis, N. He, R.M. Evans, M.D. Schneider, F.A. Brako, Y. Xiao, Y.E. Chen, Q. Yang (**2004**) *Nat Med* 10, 1245–1250.

30 Barak, Y., M.C. Nelson, E.S. Ong, Y.Z. Jones, P. Ruiz-Lozano, K.R. Chien, A. Koder, R.M. Evans (**1999**) *Mol Cell* 4, 585–595.

31 Rosen, E.D., P. Sarraf, A.E. Troy, G. Bradwin, K. Moore, D.S. Milstone, B.M. Spiegelman, R.M. Mortensen (**1999**) *Mol Cell* 4, 611–617.

32 Kubota, N., Y. Terauchi, H. Miki, H. Tamemoto, T. Yamauchi, K. Komeda, S. Satoh, R. Nakano, C. Ishii, T. Sugiyama, K. Eto, Y. Tsubamoto, A. Okuno, K. Murakami, H. Sekihara, G. Hasegawa, M. Naito, Y. Toyoshima, S. Tanaka, K. Shiota, T. Kitamura, T. Fujita, O. Ezaki, S. Aizawa, T. Kadowaki (**1999**) *Mol Cell* 4, 597–609.

33 Lazar, M.A. (**2005**) PPAR gamma, 10 years later. *Biochimie* 87, 9–13.

34 Fonseca, V.A., T.R. Valiquett, S.M. Huang, M.N. Ghazzi, R.W. Whitcomb (**1998**) *J Clin Endocrinol Metab* 83, 3169–3176.

35 BARROSO, I., M. GURNELL, V.E.
CROWLEY, M. AGOSTINI, J.W.
SCHWABE, M.A. SOOS, G.L. MASLEN,
T.D. WILLIAMS, H. LEWIS, A.J.
SCHAFER, V.K. CHATTERJEE,
S. O'RAHILLY (**1999**) *Nature* 402,
880–883.

36 MEIRHAEGHE, A., L. FAJAS, N.
HELBECQUE, D. COTTEL, J. AUWERX,
S.S. DEEB, P. AMOUYEL (**2000**) *Int J
Obes Relat Metab Disord* 24, 195–199.

37 STEPPAN, C.M., S.T. BAILEY, S. BHAT,
E.J. BROWN, R.R. BANERJEE, C.M.
WRIGHT, H.R. PATEL, R.S. AHIMA,
M.A. LAZAR (**2001**) *Nature* 409, 307–
312.

38 MAEDA, N., M. TAKAHASHI, T.
FUNAHASHI, S. KIHARA, H.
NISHIZAWA, K. KISHIDA, H.
NAGARETANI, M. MATSUDA, R.
KOMURO, N. OUCHI, H. KURIYAMA, K.
HOTTA, T. NAKAMURA, I. SHIMOMURA,
Y. MATSUZAWA (**2001**) *Diabetes* 50,
2094–2099.

39 RANGWALA, S.M., M.A. LAZAR (**2004**)
Trends Pharmacol Sci 25, 331–336.

40 CHAO, L., B. MARCUS-SAMUELS, M.M.
MASON, J. MOITRA, C. VINSON, E.
ARIOGLU, O. GAVRILOVA, M.L.
REITMAN (**2000**) *J Clin Invest* 106,
1221–1228.

41 GAVRILOVA, O., M. HALUZIK, K.
MATSUSUE, J.J. CUTSON, L. JOHNSON,
K.R. DIETZ, C.J. NICOL, C. VINSON,
F.J. GONZALEZ, M.L. REITMAN (**2003**)
J Biol Chem 278, 34268–34276.

42 HE, W., Y. BARAK, A. HEVENER, P.
OLSON, D. LIAO, J. LE, M. NELSON,
E. ONG, J.M. OLEFSKY, R.M. EVANS
(**2003**) *Proc Natl Acad Sci USA* 100,
15712–15717.

43 NORRIS, A.W., L. CHEN, S.J. FISHER,
I. SZANTO, M. RISTOW, A.C. JOZSI,
M.F. HIRSHMAN, E.D. ROSEN, L.J.
GOODYEAR, F.J. GONZALEZ, B.M.
SPIEGELMAN, C.R. KAHN (**2003**) *J Clin
Invest* 112, 608–618.

44 HEVENER, A.L., W. HE, Y. BARAK,
J. LE, G. BANDYOPADHYAY, P. OLSON,
J. WILKES, R.M. EVANS, J. OLEFSKY
(**2003**) *Nat Med* 9, 1491–1497.

45 MARX, N., G.K. SUKHOVA, T. COLLINS,
P. LIBBY, J. PLUTZKY (**1999**) *Circulation*
99, 3125–3131.

46 MARX, N., F. MACH, A. SAUTY,
J.H. LEUNG, M.N. SARAFI, R.M.
RANSOHOFF, P. LIBBY, J. PLUTZKY,
A.D. LUSTER (**2000**) *J Immunol* 164,
6503–6508.

47 LEE, H., W. SHI, P. TONTONOZ, S.
WANG, G. SUBBANAGOUNDER, C.C.
HEDRICK, S. HAMA, C. BORROMEO,
R.M. EVANS, J.A. BERLINER, L. NAGY
(**2000**) *Circ Res* 87, 516–521.

48 STAELS, B., W. KOENIG, A. HABIB, R.
MERVAL, M. LEBRET, I.P. TORRA, P.
DELERIVE, A. FADEL, G. CHINETTI,
J.C. FRUCHART, J. NAJIB, J. MACLOUF,
A. TEDGUI (**1998**) *Nature* 393, 790–
793.

49 DELERIVE, P., K. DE BOSSCHER,
S. BESNARD, W. VANDEN BERGHE,
J.M. PETERS, F.J. GONZALEZ, J.C.
FRUCHART, A. TEDGUI, G. HAEGEMAN,
B. STAELS (**1999**) *J Biol Chem* 274,
32048–32054.

50 MARX, N., N. MACKMAN, U.
SCHÖNBECK, N. YILMAZ, V. HOMBACH,
P. LIBBY, J. PLUTZKY (**2001**) *Circulation*
103, 213–219.

51 NEVE, B., D. CORSEAUX, G. CHINETTI,
C. ZAWADZKI, J.C. FRUCHART, P.
DURIEZ, B. STAELS, B. JUDE (**2001**)
Circulation 103, 207–212.

52 CHINETTI, G., S. GRIGLIO, M.
ANTONUCCI, I.P. TORRA, P. DELERIVE,
Z. MAJD, J.C. FRUCHART, J. CHAPMAN,
J. NAJIB, B. STAELS (**1998**) *J Biol Chem*
273, 25573–25580.

53 CHINETTI, G., S. LESTAVEL, V. BOCHER,
A.T. REMALEY, B. NEVE, I.P. TORRA, E.
TEISSIER, A. MINNICH, M. JAYE, N.
DUVERGER, H.B. BREWER, J.C.
FRUCHART, V. CLAVEY, B. STAELS
(**2001**) *Nat Med* 7, 53–58.

54 CHINETTI, G., S. LESTAVEL, J.C.
FRUCHART, V. CLAVEY, B. STAELS
(**2003**) *Circ Res* 92, 212–217.

55 HARAGUCHI, G., Y. KOBAYASHI, M.L.
BROWN, A. TANAKA, M. ISOBE, S.H.
GIANTURCO, W.A. BRADLEY (**2003**) *J
Lipid Res* 44, 1224–1231.

56 GBAGUIDI, F.G., G. CHINETTI, D.
MILOSAVLJEVIC, E. TEISSIER, J.
CHAPMAN, G. OLIVECRONA, J.C.
FRUCHART, S. GRIGLIO, J. FRUCHART-
NAJIB, B. STAELS (**2002**) *FEBS Lett* 512,
85–90.

57 Li, L., M.C. Beauchamp, G. Renier (**2002**) *Atherosclerosis* 165, 101–110.

58 Teissier, E., A. Nohara, G. Chinetti, R. Paumelle, B. Cariou, J.C. Fruchart, R.P. Brandes, A. Shah, B. Staels (**2004**) *Circ Res* 95, 1174–1182.

59 Ziouzenkova, O., S. Perrey, L. Asatryan, J. Hwang, K.L. MacNaul, D.E. Moller, D.J. Rader, A. Sevanian, R. Zechner, G. Hoefler, J. Plutzky (**2003**) *Proc Natl Acad Sci USA* 100, 2730–2735.

60 Rival, Y., N. Beneteau, T. Taillandier, M. Pezet, E. Dupont-Passelaigue, J.F. Patoiseau, D. Junquero, F.C. Colpaert, A. Delhon (**2002**) *Eur J Pharmacol* 435, 143–151.

61 Vosper, H., L. Patel, T.L. Graham, G.A. Khoudoli, A. Hill, C.H. Macphee, I. Pinto, S.A. Smith, K.E. Suckling, C.R. Wolf, C.N. Palmer (**2001**) *J Biol Chem* 276, 44258–44265.

62 Chawla, A., C.H. Lee, Y. Barak, W. He, J. Rosenfeld, D. Liao, J. Han, H. Kang, R.M. Evans (**2003**) *Proc Natl Acad Sci USA* 100, 1268–1273.

63 Lee, C.H., A. Chawla, N. Urbiztondo, D. Liao, W.A. Boisvert, R.M. Evans, L.K. Curtiss (**2003**) *Science* 302, 453–457.

64 Li, A.C., C.J. Binder, A. Gutierrez, K.K. Brown, C.R. Plotkin, J.W. Pattison, A.F. Valledor, R.A. Davis, T.M. Willson, J.L. Witztum, W. Palinski, C.K. Glass (**2004**) *J Clin Invest* 114, 1564–1576.

65 Welch, J.S., M. Ricote, T.E. Akiyama, F.J. Gonzalez, C.K. Glass (**2003**) *Proc Natl Acad Sci USA* 100, 6712–6717.

66 Chinetti, G., F.G. Gbaguidi, S. Griglio, Z. Mallat, M. Antonucci, P. Poulain, J. Chapman, J.C. Fruchart, A. Tedgui, J. Najib-Fruchart, B. Staels (**2000**) *Circulation* 101, 2411–2417.

67 Chawla, A., W.A. Boisvert, C.H. Lee, B.A. Laffitte, Y. Barak, S.B. Joseph, D. Liao, L. Nagy, P.A. Edwards, L.K. Curtiss, R.M. Evans, P. Tontonoz (**2001**) *Mol Cell* 7, 161–171.

68 Akiyama, T.E., S. Sakai, G. Lambert, C.J. Nicol, K. Matsusue, S. Pimprale, Y.H. Lee, M. Ricote, C.K. Glass, H.B. Brewer, Jr., F.J. Gonzalez (**2002**) *Mol Cell Biol* 22, 2607–2619.

69 Szanto, A., S. Benko, I. Szatmari, B.L. Balint, I. Furtos, R. Ruhl, S. Molnar, L. Csiba, R. Garuti, S. Calandra, H. Larsson, U. Diczfalusy, L. Nagy (**2004**) *Mol Cell Biol* 24, 8154–8166.

70 Argmann, C.A., C.G. Sawyez, C.J. McNeil, R.A. Hegele, M.W. Huff (**2003**) *Arterioscler Thromb Vasc Biol* 23, 475–482.

71 Chinetti, G., C. Zawadski, J.C. Fruchart, B. Staels (**2004**) *Biochem Biophys Res Commun* 314, 151–158.

72 Tontonoz, P., L. Nagy, J.G. Alvarez, V.A. Thomazy, R.M. Evans (**1998**) *Cell* 93, 241–252.

73 Liang, C.P., S. Han, H. Okamoto, R. Carnemolla, I. Tabas, D. Accili, A.R. Tall (**2004**) *J Clin Invest* 113, 764–773.

74 Chinetti, G., J.C. Fruchart, B. Staels (**2001**) *Z Kardiol* 90 (Suppl 3), 125–132.

75 Delerive, P., F. Martin-Nizard, G. Chinetti, F. Trottein, J.C. Fruchart, J. Najib, P. Duriez, B. Staels (**1999**) *Circ Res* 85, 394–402.

76 Iglarz, M., R.M. Touyz, F. Amiri, M.F. Lavoie, Q.N. Diep, E.L. Schiffrin (**2003**) *Arterioscler Thromb Vasc Biol* 23, 45–51.

77 Goetze, S., X.P. Xi, H. Kawano, T. Gotlibowski, E. Fleck, W.A. Hsueh, R.E. Law (**1999**) *J Cardiovasc Pharmacol* 33, 798–806.

78 Marx, N., U. Schonbeck, M.A. Lazar, P. Libby, J. Plutzky (**1998**) *Circ Res* 83, 1097–1103.

79 Takeda, K., T. Ichiki, T. Tokunou, Y. Funakoshi, N. Iino, K. Hirano, H. Kanaide, A. Takeshita (**2000**) *Circulation* 102, 1834–1839.

80 Fu, T., P. Kashireddy, J. Borensztajn (**2003**) *Biochem J* 373, 941–947.

81 Duez, H., Y.S. Chao, M. Hernandez, G. Torpier, P. Poulain, S. Mundt, Z. Mallat, E. Teissier, C.A. Burton, A. Tedgui, J.C. Fruchart, C. Fievet,

S.D. WRIGHT, B. STAELS (**2002**) *J Biol Chem* 277, 48051–48057.

82 LI, A.C., K.K. BROWN, M.J. SILVESTRE, T.M. WILLSON, W. PALINSKI, C.K. GLASS (**2000**) *J Clin Invest* 106, 523–531.

83 COLLINS, A.R., W.P. MEEHAN, U. KINTSCHER, S. JACKSON, S. WAKINO, G. NOH, W. PALINSKI, W.A. HSUEH, R.E. LAW (**2001**) *Arterioscler Thromb Vasc Biol* 21, 365–371.

84 CLAUDEL, T., M.D. LEIBOWITZ, C. FIEVET, A. TAILLEUX, B. WAGNER, J.J. REPA, G. TORPIER, J.M. LOBACCARO, J.R. PATERNITI, D.J. MANGELSDORF, R.A. HEYMAN, J. AUWERX (**2001**) *Proc Natl Acad Sci USA* 98, 2610–2615.

85 CHEN, Z., S. ISHIBASHI, S. PERREY, J. OSUGA, T. GOTODA, T. KITAMINE, Y. TAMURA, H. OKAZAKI, N. YAHAGI, Y. IIZUKA, F. SHIONOIRI, K. OHASHI, K. HARADA, H. SHIMANO, R. NAGAI, N. YAMADA (**2001**) *Arterioscler Thromb Vasc Biol* 21, 372–377.

86 ERICSSON, C.G., J. NILSSON, L. GRIP, B. SVANE, A. HAMSTEN (**1997**) *Am J Cardiol* 80, 1125–1129.

87 INVESTIGATORS, D.A.I.S. (**2001**) *Lancet* 357, 905–910.

88 FRICK, M.H., M. SYVANNE, M.S. NIEMINEN, H. KAUMA, S. MAJAHALME, V. VIRTANEN, Y.A. KESANIEMI, A. PASTERNACK, M.R. TASKINEN (**1997**) *Circulation* 96, 2137–2143.

89 FRICK, M.H., O. ELO, K. HAAPA, O.P. HEINONEN, P. HEINSALMI, P. HELO, J.K. HUTTUNEN, P. KAITANIEMI, P. KOSKINEN, V. MANNINEN (**1987**) *N Engl J Med* 317, 1237–1245.

90 RUBINS, H.B., S.J. ROBINS, D. COLLINS, C.L. FYE, J.W. ANDERSON, M.B. ELAM, F.H. FAAS, E. LINARES, E.J. SCHAEFER, G. SCHECTMAN, T.J. WILT, J. WITTES (**1999**) *N Engl J Med* 341, 410–418.

91 DESPRES, J.P., I. LEMIEUX, A. PASCOT, N. ALMERAS, M. DUMONT, A. NADEAU, J. BERGERON, D. PRUD'HOMME (**2003**) *Arterioscler Thromb Vasc Biol* 23, 702–703.

92 ROBINS, S.J., H.B. RUBINS, F.H. FAAS, E.J. SCHAEFER, M.B. ELAM, J.W. ANDERSON, D. COLLINS (**2003**) *Diabetes Care* 26, 1513–1517.

93 CABALLERO, A.E., R. SAOUAF, S.C. LIM, O. HAMDY, K. ABOU-ELENIN, C. O'CONNOR, F.W. LOGERFO, E.S. HORTON, A. VEVES (**2003**) *Metabolism* 52, 173–180.

94 PISTROSCH, F., J. PASSAUER, S. FISCHER, K. FUECKER, M. HANEFELD, P. GROSS (**2004**) In type 2 diabetes, rosiglitazone therapy for insulin resistance ameliorates endothelial dysfunction independent of glucose control. *Diabetes Care* 27, 484–490.

95 HAFFNER, S.M., A.S. GREENBERG, W.M. WESTON, H. CHEN, K. WILLIAMS, M.I. FREED (**2002**) *Circulation* 106, 679–684.

96 MARX, N., J. FROEHLICH, L. SIAM, J. ITTNER, G. WIERSE, A. SCHMIDT, H. SCHARNAGL, V. HOMBACH, W. KOENIG (**2003**) *Arterioscler Thromb Vasc Biol* 23, 283–288.

97 MARX, N., A. IMHOF, J. FROEHLICH, L. SIAM, J. ITTNER, G. WIERSE, A. SCHMIDT, W. MAERZ, V. HOMBACH, W. KOENIG (**2003**) *Circulation* 107, 1954–1957.

98 MINAMIKAWA, J., S. TANAKA, M. YAMAUCHI, D. INOUE, H. KOSHIYAMA (**1998**) *J Clin Endocrinol Metab* 83, 1818–1820.

99 TAKAGI, T., T. AKASAKA, A. YAMAMURO, Y. HONDA, T. HOZUMI, S. MORIOKA, K. YOSHIDA (**2000**) *J Am Coll Cardiol* 36, 1529–1535.

100 CAMP, H.S., O. LI, S.C. WISE, Y.H. HONG, C.L. FRANKOWSKI, X. SHEN, R. VANBOGELEN, T. LEFF (**2000**) *Diabetes* 49, 539–547.

101 DUEZ, H., B. LEFEBVRE, P. POULAIN, I.P. TORRA, F. PERCEVAULT, G. LUC, J.M. PETERS, F.J. GONZALEZ, R. GINESTE, S. HELLEBOID, V. DZAVIK, J.C. FRUCHART, C. FIEVET, P. LEFEBVRE, B. STAELS (**2005**) *Arterioscler Thromb Vasc Biol* 25, 585–591.

102 ETGEN, G.J., B.A. OLDHAM, W.T. JOHNSON, C.L. BRODERICK, C.R. MONTROSE, J.T. BROZINICK, E.A. MISENER, J.S. BEAN, W.R. BENSCH, D.A. BROOKS, A.J. SHUKER, C.J. RITO, J.R. MCCARTHY, R.J. ARDECKY, J.S. TYHONAS, S.L. DANA, J.M. BILAKOVICS, J.R. PATERNITI, JR., K.M. OGILVIE, S.

Liu, R.F. Kauffman (**2002**) *Diabetes* 51, 1083–1087.

103 Ljung, B., K. Bamberg, B. Dahllof, A. Kjellstedt, N.D. Oakes, J. Ostling, L. Svensson, G. Camejo (**2002**) *J Lipid Res* 43, 1855–1863.

104 Ye, J.M., M.A. Iglesias, D.G. Watson, B. Ellis, L. Wood, P.B. Jensen, R.V. Sorensen, P.J. Larsen, G.J. Cooney, K. Wassermann, E.W. Kraegen

(**2003**) *Am J Physiol Endocrinol Metab* 284, E531–E540.

105 Chakrabarti, R., P. Misra, R.K. Vikramadithyan, M. Premkumar, J. Hiriyan, S.R. Datla, R.K. Damarla, J. Suresh, R. Rajagopalan (**2004**) *Eur J Pharmacol* 491, 195–206.

106 Hegarty, B.D., S.M. Furler, N.D. Oakes, E.W. Kraegen, G.J. Cooney (**2004**) *Endocrinology* 145, 3158–3164.

10
Protein Synthesis and Cancer

Andreas G. Bader and Peter K. Vogt

10.1
Introduction

Differential regulation of transcription is widely considered the primary cause of oncogenic cellular transformation and of cancer. This view is reinforced by micro-array technology characterizing gene expression profiles of cancer cells. However, cellular phenotypes are governed by the activity of proteins, and translation of mRNAs into polypeptides introduces another layer of controls that is subject to change in oncogenesis. Translation can be regulated globally or selectively; the first mechanism involves shared components of the protein-synthesizing machinery, the second depends on regulatory sequences encoded by a particular mRNA. mRNAs associate with mRNA-binding proteins to form messenger ribonucleo-protein particles (mRNPs). When actively translated, these mRNPs interact with multiple ribosomes, also referred to as polysomes. Monosomes, however, contain poorly translated mRNAs that remain as mRNPs and hardly associate with ribosomes. The polysome/monosome ratio is measurable and indicates translational activity. It is this ratio that changes during oncogenesis, resulting in an increase of growth-promoting proteins and a decline of growth-inhibitory proteins [1].

10.2
Translation Initiation

Translation is separated into three consecutive steps: initiation, elongation, and termination (reviewed in Refs [2, 3]). During initiation, mRNA together with eukaryotic translational initiation factors (eIFs) and ribosomal subunits assemble a translational complex that scans the mRNA until recognition of the start codon. During elongation, the coding sequence of the mRNA is translated into a polypeptide. Termination involves the recognition of the stop codon and the disassembly of the mRNA/ribosome complex and its components. The 5' end of virtually all mRNAs is protected by a "cap" with the chemical structure m^7GpppN (where m^7G represents 7-methylguanylate, p represents a phosphate group and N any base). Most

Nutritional Genomics. Edited by Regina Brigelius-Flohé and Hans-Georg Joost
Copyright © 2006 WILEY-VCH Verlag GmbH & Co. KGaA, Weinheim
ISBN: 3-527-31294-3

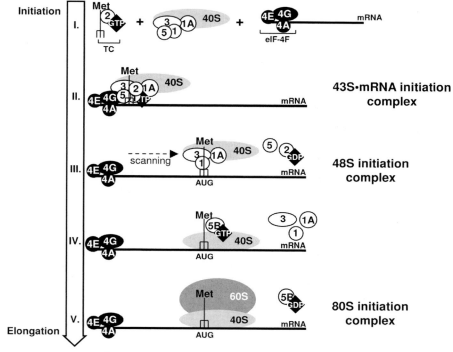

Fig. 10.1. Initiation of translation. See text for explanations. The initiator transfer RNA is symbolized by a fork. Met, methionine; TC, ternary complex; AUG, start codon.

eukaryotic mRNAs are translated in a cap-dependent fashion; a minority is translated by the use of internal ribosome entry sites (IRES).

The control of translation occurs usually at the initiation step of translation. In the current model of translation initiation, the ternary complex comprising eIF-2, GTP, and the initiator transfer RNA (Met-tRNA$_i$) associates with the 40S ribosomal subunit. This association is facilitated by eIFs 1, 1A, 3, and 5 (Fig. 10.1). Together, they constitute the 43S preinitiation complex. This complex is recruited to the cap of the mRNA by eIF-4F. eIF-4F is a trimeric complex consisting of eIF-4E, the cap-binding protein eIF-4A, an RNA-dependent ATPase with helicase activity, and eIF-4G, a scaffolding protein that makes direct contacts with eIF-4E, eIF-4A, and the multisubunit factor eIF-3. eIF-4G also interacts with the poly(A)-tail-binding protein (PABP), which leads to a circularization of the message.

The interaction between eIF-4E and PABP stimulates translation and may be a regulatory mechanism to ensure that only full-length mRNAs are translated efficiently. eIF-4A is a non-processive helicase whose effectiveness in unwinding RNA duplexes is negatively correlated with the stability of the duplex [4]. eIF-4A is required to break secondary structures at the 5′ terminus of the mRNA to enable the loading of the 43S preinitiation complex [5]. The RNA-binding protein eIF-4B stimulates the helicase activity of eIF-4A. With the support of eIFs 1, 1A, and F, the 43S complex scans the message in 5′ to 3′ direction until it encounters the initiation codon, usually the first AUG downstream of the cap. It is unclear whether the 43S preinitiation complex leaves eIF-4F behind at the cap as illustrated in Fig. 10.1, or whether the 43S complex stays associated with eIF-4F during the scanning process. Once the 43S complex has recognized the start codon by making base pair contacts between the AUG and the initiator transfer RNA of the ternary complex, eIF-5 triggers the hydrolysis of eIF-2·GTP to eIF-2·GDP. The 43S unit forms a stable complex with the mRNA that is referred to as the 48S initiation complex. eIF-2·GDP and eIF-5 disassociate from the complex, and so do eIFs 1, 1A, and 3. During this process, eIF-5B joins the complex in the GTP-bound form and recruits the 60S ribosomal subunit. Binding of the 60S unit induces the hydrolysis of eIF-5B·GTP. eIF-5B·GDP has much lower binding affinity and leaves the complex. The final 80S initiation complex is ready to enter the elongation phase of translation.

mRNAs contain 5′ and 3′ untranslated regions (UTRs) harboring regulatory sequences that govern efficiency of translation [3, 6–8]. Cap-dependent translation requires the cap structure at the 5′ terminus and the poly(A) tail located at 3′ terminus of the mRNA (Fig. 10.2a). The cap structure is the attachment site for the protein eIF-4E. The poly(A) tail has a regulatory function by recruiting PABP. Both the cap and the poly(A) tail domains regulate mRNA stability by protecting the mRNA from degradation. Internal mRNA sequences contribute to a more subtle regulation of translation.

Long 5′ UTRs and secondary structures impair loading of the 43S preinitiation complex and the scanning process along the mRNA. Secondary start codons and internal ribosome entry sites allow variable use of the mRNA. Translational control elements (TCEs) facilitate the interaction with regulatory RNA-binding proteins that either stimulate or repress translation. These structural mRNA characteristics – and various combinations thereof – determine the individual translational activity and constitute a roster of mRNAs with different preferences for translation.

Global control of translation takes place at the immediate early stage of translation initiation [2, 3]. One checkpoint is represented by eIF-2, a component of the ternary complex (Fig. 10.2b). eIF-2 alternates between an active GTP-bound state and an inactive GDP-bound form that needs to be recycled to enter a new round of translation. The corresponding guanine nucleotide exchange factor (GEF) is eIF-2B that catalyzes the reaction from eIF-2·GDP to eIF-2·GTP. eIF-2 consists of the three subunits α, β, and γ. Translation is negatively regulated by phosphorylation of the serine residue at position 51 in the α-subunit. Phosphorylation of Ser51 decreases the dissociation of eIF-2 from eIF-2B and stabilizes the eIF-2·GDP-eIF-2B complex. Since cellular eIF-2B levels are limited, eIF-2B is sequestered by eIF-2·GDP, and global translation is inhibited. A number of kinases are able to phosphorylate eIF-2 at Ser51; most of them are activated during cellular stress conditions. These include PKR (protein kinase activated by double-stranded RNA) that is activated upon viral infections, PERK (PKR-like endoplasmic reticu-

(a)

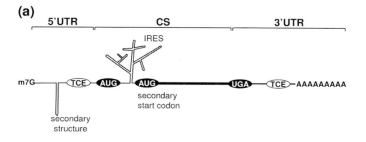

(b) **(c)**

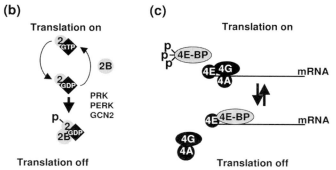

Fig. 10.2. Individual and global control of translation. (a) mRNA structures that regulate efficiency of translation: 5'-terminally located cap-structure (m7G) and 3' poly(A)-tail, long 5' untranslated region (5' UTR), secondary mRNA structures within the 5' UTR, translation control elements (TCE), internal ribosome entry site (IRES) and secondary start codon; CS, coding sequence; AUG, start codon; UGA, stop codon. (b) Control of translation by phosphorylation of eIF-2. During initiation of translation, eIF-2 is turned over from an active GTP-bound form to an inactive GDP-bound form. Under normal conditions, eIF-2 is recycled by eIF-2B to enter a new round of translation. Under stress situations, eIF-2 is phosphorylated at Ser51 (indicated by p) by PRK, PERK and GCN2 which leads to a stable eIF-2·GDP·eIF-2B complex formation and inhibition of translation. (c) Regulation of translation by eIF-4E-binding proteins. Hypophosphorylated 4E-BPs interact with eIF-4E and compete with eIF-4G for 4E-binding. eIF-4E bound to 4E-BP is unable to assemble a functional initiation complex and no longer initiates translation. p indicates multiple phosphorylation on 4E-BP.

lum kinase) that is stimulated during endoplasmic reticulum stress, and GCN2 (general control non-repressible-2) that is activated in response to amino acid deprivation.

A second mechanism regulating global translation involves the cap-binding protein eIF-4E (Fig. 10.2c) [2, 9]. eIF-4E is among the least abundant initiation factors and is therefore the rate-limiting factor of cap-dependent translation. Its affinity for the cap is very weak unless it is stabilized in a protein complex. Such stabilization is achieved through the interaction with the scaffold protein eIF-4G. However, eIF-4E-binding proteins (4E-BPs), a class of regulatory proteins that inhibit translation, compete with eIF-4G for the cap-binding protein eIF-4E. Since 4E-BPs and eIF-4G share the same eIF-4E interaction surface, 4E-BP·4E and 4G·4E complexes are mutually exclusive. In complex with 4E-BPs, eIF-4E remains "frozen" on the cap structure, unable to assemble a functional initiation complex. The availability of free eIF-4E determines the activity of cap-dependent translation, and therefore, the interaction between eIF-4E and 4E-BPs is tightly regulated. 4E-BPs can be phosphorylated on multiple sites in the course of intracellular signaling, resulting in reduced affinity for eIF-4E and subsequent dissociation from eIF-4E. In contrast, hypophosphorylated 4E-BPs have high affinity for eIF-4E and sequester the cap-binding protein.

10.3
Control of Translation by the Phosphoinositide 3-kinase Signaling Pathway

The phosphoinositide 3-kinase (PI3K) pathway is a major signaling cascade that governs the fate of translational activity in the cell (Fig. 10.3). It is regulated by signals that define the state of energy levels, amino acid availability, oxygen levels, and growth-stimulatory signals, all of which play a decisive role in determining how much protein and which proteins are going to be made (reviewed in Refs [10, 11]). Mammalian PI3Ks constitute a protein superfamily divided into three classes with a total of eight members [12]. We will concentrate here on class I_A, specifically the catalytic subunit p110α, and its regulatory subunit p85 because most studies on the effects of PI3K with respect to protein synthesis have been carried out with these isoforms. For simplicity, these isoforms will be referred to as PI3K. The catalytic subunit p110α has lipid kinase activity and catalyzes the phosphorylation of phosphatidylinositol 4,5-bisphosphate (PIP2) at position D3, generating phosphatidylinositol 3,4,5-trisphosphate (PIP3) [12]. The regulatory subunit p85 stimulates PI3K activity by transmitting signals from receptor tyrosine kinases (RTKs) to p110α. Activated RTKs bind to p85 that consequently recruits p110α to the plasma membrane and brings it into close proximity with its substrate. Negative regulation of PI3K signaling is mediated by the lipid phosphatase PTEN (phosphatase and tensin homolog deleted on chromosome 10) that antagonizes the PI3K-catalyzed reaction [13].

PIP3 is an important cellular second messenger. It binds to the pleckstrin homology domains of phosphoinositide-dependent kinase 1 (PDK1) and Akt, also referred to as protein kinase B (PKB), and anchors these proteins at the plasma membrane. This leads to a full activation of PDK1 and a partial activation of Akt. For complete kinase activity, Akt has to be phosphorylated by PDK1 and another kinase, termed PDK2 whose identity has been controversial. Located downstream of Akt is the Ser/Thr kinase TOR (target of rapamycin) which becomes activated by Akt either directly or indirectly via the tuberous sclerosis complex (TSC) and the Ras-like protein Rheb (ras homolog enriched in brain). The TSC protein complex is a heterodimer consisting of TSC-1 and TSC-2, also known as hamartin and tuberin, respectively. Together, they function as a GTPase-activating protein (GAP)

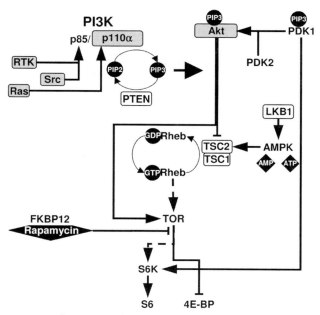

Fig. 10.3. The PI3K signaling pathway. Proto-oncoproteins are shadowed in gray, tumor suppressors are indicated by white boxes. Solid and dashed lines denote direct or indirect interactions, respectively. Arrowheads symbolize stimulatory regulation, bars represent an inhibitory effect of the signal. See text for detailed explanation of the pathway.

towards the small GTPase Rheb and ensure a rapid turn-over of the active GTP-bound form of Rheb. Phosphorylation of TSC-2 by Akt leads to an inhibition of GAP function and therefore provides continuous Rheb activity.

Since no GEF specific for Rheb has yet been identified, available data favor a model in which Rheb predominantly exists in the active GTP-bound state and has to be actively disabled by TSC to abrogate the signal [14, 15]. TSC is also regulated by Ser/Thr kinases LKB1 and AMPK (AMP-activated protein kinase) [16]. AMPK functions as an intracellular sensor of energy levels and becomes activated upon high levels of AMP. AMPK activity can also be induced by LKB1-mediated phosphorylation. As a consequence, AMPK directly phosphorylates TSC-2, which results in an activation of TSC GAP activity. Thus, PI3K, PDK1, Akt, and Rheb are positive regulators; PTEN, LKB1, AMPK, and TSC are negative regulators of PI3K signaling.

The Ser/Thr kinase TOR represents the interface that connects upstream signaling with downstream events regulating translation (reviewed in Refs [11, 17]). Among its downstream targets are p70 S6 kinase (S6K), 4E-BP1, eIF-4G, and translational elongation factor 2 kinase (eEF-2K). TOR also regulates translation indirectly by activating transcriptional initiation factor 1A (TIF-1A), a cofactor of RNA polymerase I, necessary for the generation of ribosomal RNAs. TOR has been placed downstream of Rheb. Recent evidence suggests that Rheb directly associates with TOR, but only Rheb in the GTP-bound state is able to activate TOR [18]. Despite the enormous size of TOR (about 2440 amino acids), only few interacting proteins have been isolated. TOR forms a complex with the novel protein raptor (regulatory-associated protein of TOR) and LST8 (lethal with sec-thirteen, also referred to as GβL), a protein with unknown function, originally identified genetically as a mutation that is synthetically lethal together with mutation sec13 that causes a sorting defect in the secretory pathway. TOR has autophosphorylation activity, yet, the phosphorylation of other target proteins may be regulated indirectly by protein phosphatase 2A (PP2A) – at least for a subset of TOR targets [19]. The recent discovery of a TOR signaling (TOS) motif in S6K and 4E-BP1, however, has refined our understanding of TOR-mediated phosphorylation [20]. The TOS motif in S6K and 4E-BP1 is a binding site for raptor and is required for TOR-dependent phosphorylation. These observations suggest a mechanism by which raptor bridges the association of TOR substrates with TOR, allowing the phosphorylation of these substrates. LST8 stabilizes the TOR–raptor interaction.

Downstream effectors of TOR signaling are S6K and 4E-BP1. Phosphorylation of 4E-BP1 frees eIF-4E from 4E-BP1 and enables cap-dependent translation. Phosphorylation of S6K stimulates kinase activity. S6K can also be phosphorylated directly by PDK1. Activated S6K phosphorylates the ribosomal protein S6 that has been implicated in the recruitment of 5′ TOP mRNAs to the ribosome. 5′ TOP mRNAs contain a 5′-terminal oligopyrimidine tract adjacent to the cap structure and encode components of the translational machinery, including ribosomal proteins, elongation factors, and PABP [21]. However, whether S6K is crucial for the regulation of these mRNAs remains controversial as $S6K1^{-/-}$ cells that show diminished levels of phosphorylated S6 are not impaired in the translation of 5′ TOP mRNAs [22]. Yet, the translation of these mRNAs is responsive to PI3K and may therefore be controlled by an alternative branch of PI3K signaling.

Recent studies on S6K revealed a role in a negative feedback loop to PI3K. S6K targets insulin receptor substrate 1 (IRS-1) that links signaling from the insulin receptor (InR) to the p85 subunit of PI3K. S6K-mediated phosphorylation of IRS-1 transcriptionally and functionally represses IRS-1 and thereby abrogates insulin signaling [23]. Another S6K target is eIF-4B, an accessory protein of the helicase eIF-4A [24]. The functional consequence of this phosphorylation remains to be clarified.

10.4
TOR and Nutrients

Survival of all cells depends on a steady supply of intracellular metabolites. Fast proliferating cells, including cancer cells, have a high demand for energy and precursor molecules to accommodate anabolic processes. Cells have at least two ways to control intracellular metabolic supply. One involves the uptake of extracellular nutrients, such as glucose, amino acids, and fatty acids; another one is the degradation of intracellular macromolecules to provide universal precursor molecules and ATP. Prolonged nutrient deprivation results in failure to maintain basic cellular functions and viability. This metabolic stress is characterized by an initial scavenging of intracellular biomolecules, also referred to as autophagy, which ultimately leads to cell death [25]. Therefore, growth factor signaling pathways and nutrient-sensing pathways are unavoidably connected to permit cell growth and proliferation under conditions when nutrients are abundant.

The PI3K signaling pathway and TOR in particular are critical for the regulation of nutrient uptake. TOR, a key player in controlling protein synthesis, functions also as a nutrient sensor and responds to both mitogenic signals and nutrients. While the exact connection from nutrients to TOR is poorly defined, current data demonstrate that nutrients stimulate TOR activity, and this activation might at least in part involve TSC and Rheb through the regulation of AMPK (Fig. 10.3). For instance, glucose, alanine, and other amino acids are catabolized to pyruvate which is fuel for the tricarboxylic acid cycle and oxidative decarboxylation to generate ATP [26]. High ATP levels in turn inactivate AMPK and consequently lead to stimulation of TOR activity, nutrient uptake, and metabolic processes. Since most biomolecules, such as sugars, fatty acids, nucleotides, and amino acids can be converted to ATP, high energy levels are an intracellular readout for nutrient abundance and cellular welfare.

Nutrient uptake can also be regulated by growth factor signaling pathways, including the PI3K pathway which has a profound impact on TOR activity. Inactivation of TOR or rapamycin treatment mimics nutrient deprivation in yeast, *Drosophila*, and mammalian cells [27–29]. Likewise, insulin treatment or activation of TOR induces the transcription of metabolic enzymes and regulates the abundance of amino acid permeases [30]. Akt blocks AMPK and directly activates the cardiac 6-phosphofructose-2-kinase, a kinase that regulates a rate-limiting step during mammalian glycolysis [31, 32]. Akt also stimulates the expression of the glucose transporters GLUT1 and GLUT4 and induces the translocation of GLUT4 to the plasma membrane [33–35]. Even in the absence of growth factors – under conditions when nutrient uptake is diminished and surface transporters are degraded – constitutively active Akt maintains transporters for glucose, low-density lipoprotein (LDL), iron, and amino acids – a phenotype which is completely inhibited by rapamycin, demonstrating a key role for TOR in the regulation of nutrient uptake [36]. Thus, the activities of Akt and TOR are critical for the determination of catabolic versus anabolic processes.

Cancer cells often show an increase in Akt activity (see below) and therefore predispose cells toward nutrient uptake. In fact, aberrant PI3K signaling uncouples growth factor signaling from nutrient-sensing pathways and promotes anabolism even when nutrients are low. Since Akt blocks autophagy as well as the degradation of proteins and of fatty acids, for reasons that remain largely unexplored, these cells are solely dependent on glucose as a source for energy supply [25]. It is this "glucose addiction" that may be useful in developing anticancer strategies.

10.5
Oncogenicity of Phosphoinositide 3-kinase Signaling

Many molecular participants of the PI3K signaling pathway induce aberrant cell growth once they have become disconnected from normal controls (Fig. 10.3). While constitutively active forms of PI3K and Akt have oncogenic activity, PTEN, TSC, and LKB1 function as tumor suppressors by antagonizing PI3K and Akt. P3k, a homolog of p110α, was identified as the oncogenic component of the avian sarcoma virus ASV16 that readily causes hemangiosarcomas in chickens [37]. Similarly, Akt is the oncogenic effector of AKT8, a murine retrovirus originally isolated from a spontaneous thymoma [38–40]. When overexpressed from an avian helper virus, retroviral Akt induces the formation of aggressive hemangiosarcomas in chickens indistinguishable from those caused by P3k [41]. Both oncoproteins are the result of gain-of-function mutations that lead to a continuous activation of the PI3K signaling pathway. P3k and Akt oncogenicity requires an intact kinase domain and a constitutive membrane address, such as a myristylation or farnesylation signal, or in the case of P3k a fusion with the retroviral Gag protein [41, 42]. A gain-of-function mutation in the pleckstrin homology domain of Akt is also effective [41].

Expression of a dominant-negative form of Akt in chicken embryo fibroblasts (CEF) interferes with oncogenic transformation by P3k, demonstrating that Akt function is critical for P3k-mediated oncogenesis [41]. Since there are no reports showing transforming activities of Rheb or TOR, other Akt targets are likely to play an essential role in transformation by P3k and Akt. Proteins that are directly phosphorylated and inactivated by Akt include negative regulators of survival and anti-apoptotic pathways (reviewed in Refs [43, 44]). Examples are caspase 9 and BAD, both of which are activated during apoptosis; GSK3β (glycogen synthase kinase 3β) that targets cyclin D1 for proteasomal degradation; MDM2 (murine double minute 2), a nuclear–cytoplasmic shuttling protein disabling p53 function; and the forkhead transcription factors FOXO1, FOXO3a, and FOXO4.

Signaling molecules that are located upstream of PI3K also have oncogenic potential (Fig. 10.3). Examples are receptor tyrosine kinases (RTKs), a class of integral membrane proteins that respond to extracellular signals. Binding of ligand induces autophosphorylation of an intracellular RTK domain that interacts with the regulatory subunit p85. RTKs that stimulate PI3K include estrogen receptor (ER), epidermal growth factor receptors (EGFRs), and insulin receptor (InR) that interacts with p85 indirectly through insulin receptor substrate 1 and 2 (IRS1, IRS2) [45–47]. A homolog of EGFR functions as a retroviral oncoprotein [48]. The Ras and Src proto-oncoproteins are also able to activate PI3K [12, 45]. While Src binds to p85, Ras stimulates PI3K activity by directly binding to the catalytic subunit p110α.

The strong transforming potential of Src and Ras may be explained by the fact that they are able to fire into two major signaling pathways: one mediated by mitogen-activated protein kinases (MAPKs) and another by PI3K. The PI3K pathway is of particular importance as a dominant-negative form of PI3K blocks transformation by Ras [49]. The PI3K inhibitor LY294002 weakens Src-dependent transformation and – when applied in conjunction with an inhibitor of the MAPK pathway – results in complete inhibition of transformation [50].

The role of translation in P3k-induced transformation became evident by studies with the immunosuppressant rapamycin. Rapamycin is a macrolide antibiotic derived from *Streptomyces hygroscopicus* [51, 52]. In complex with the cellular protein FKBP12 (FK506-binding protein), rapamycin binds to TOR and blocks TOR-dependent signaling (Fig. 10.3). While autophosphorylation activity of TOR remains unaffected, the rapamycin/FKBP12 complex interferes with the TOR–raptor interaction, inhibiting the phosphorylation of TOR substrates [53]. In chicken embryo fibroblasts, minimal amounts of rapamycin (1 ng/µl) are sufficient to completely inhibit oncogenic transformation by P3k or Akt [54]. This resistance to transformation is correlated with a constitutive downregulation of S6K and 4E-BP1. Rapamycin also induces reversion of P3k and Akt-transformed CEF to normal cell morphology. The effect of rapamycin on oncogenic transformation is also evident in PTEN-null mice or mice that are heterozygous for an inactivating mutation of PTEN [55, 56]. These PTEN-deficient mice develop tumors of the uterus and the adrenal medulla. Treatment with the rapamycin analog CCI-779 reduces neoplastic cell transformation and tumor size without affecting aberrant signaling upstream of TOR.

TOR does not directly function at the level of protein synthesis, yet there are emerging data that closely correlate oncogenesis with deregulated translation. Overexpression of the cap-binding protein eIF-4E transforms NIH-3T3 cells, causes tumors in nude mice and cooperates with c-Myc in the development of lymphomas in mice [57, 58]. The scaffold protein eIF-4G is similarly transforming [59]. A constitutively active form of 4E-BP1 that cannot be phosphorylated interferes with c-Myc-induced cell transformation [60]. Further evidence for an essential role of translation during oncogenic cell transformation is provided by the translational regulators Pdcd4 (programmed cell death 4) and the Y box-binding protein 1 (YB-1). Pdcd4 is a tumor suppressor that interferes with tetradecanoyl phorbol acetate-induced transformation in susceptible JB6 cells by inhibiting the helicase activity of eIF-4A [61]. YB-1 is a universal mRNA-binding protein that is transcriptionally repressed in P3k and Akt transformed CEF and, when overexpressed, specifically blocks cellular transformation by P3k or Akt [62]. The anti-oncogenic activity of YB-1 can be explained by inhibition of translation [63]. YB-1 inhibits translation at the initiation stage of translation [64]. However, the molecular mechanism of that inhibition is still unknown.

10.6
Aberrant Phosphoinositide 3-kinase Signaling in Human Cancer

The PI3K pathway is commonly deregulated in human cancer (reviewed in Refs [13, 16, 65, 66]). The catalytic subunit p110α of PI3K is overexpressed in ovarian and cervical carcinoma, and the corresponding gene, *PIK3CA*, is frequently mutated in solid tumors [67, 68]. The mutations are not random but occur on specific sites within the *PIK3CA* gene. Mutations at these sites result in a gain-of-function and render the protein oncogenic [69]. Akt is amplified and overexpressed in a large variety of cancers, including gastric, breast, ovarian, pancreatic, and prostate cancer. High eIF-4E levels are common for most carcinomas. eIF-4E is overexpressed in various lymphomas, cancers of the head and the neck, and in breast, gastrointestinal, and thyroid carcinomas. Increased eIF-4G levels are found in squamous cell carcinoma of the lung. eIF-2α is overexpressed in non-Hodgkin's lymphoma, thyroid carcinoma, and bronchioalveolar carcinoma of the lung.

More evidence for deregulated PI3K signaling in human cancer is provided by the tumor suppressor PTEN. PTEN is one of the most frequent targets of mutation in human cancers. It is inactivated in glioblastoma, prostate cancer, melanoma, endometrial carcinomas, and breast cancer. Loss of PTEN function has also been observed in other abnormalities, including Cowden's disease, Bannayan–Riley–Ruvalcaba syndrome, Proteus syndrome and Lhermitte–Duclos disease. These tumors are of benign nature and are broadly classified as hamartomas. Histologically similar tumors are also induced upon inactivating mutations of either TSC gene or LKB1 that causes the Peutz–Jeghers syndrome.

There is no evidence demonstrating that TOR may be directly involved in tumor development, and yet, inhibition of TOR function by rapamycin reduces aberrant cell growth of various human cancer cell lines. Rapamycin showed promising results in the treatment of transplantable human tumors [70, 71]. However, its poor aqueous solubility and chemical instability limited its use as a powerful anticancer drug. Consequently, rapamycin-derivatives with more favorable pharmacological kinetics were developed, such as CCI-779 (temsirolimus), Rad001 (everolimus), and AP23573 (reviewed in Refs [65, 72, 73]). CCI-779 is designed for oral and intravenous administration; Rad001 is for oral use only. Both drugs have shown promising antitumor activity in phase I trials and progressed to phase III evaluation. AP23573 has recently entered phase I trials. Updated information about these drugs in clinical trials is available online at the US National Cancer Institute clinical trial website http://www.cancer.gov.

Our current knowledge on PI3K signaling is fundamental for the design of new anticancer strategies. The development of small molecule inhibitors against targets that show a gain of function in the PI3K pathway will be an achievable goal. Candidate targets for small molecule drugs include p110α, Akt, eIF-4E, eIF-4G, 4E-BP1, and eIF-2α. Future drugs will be effective at a low molar range and will have a narrow-band target specificity. Consequently, these drugs will efficiently treat cancers with minimal adverse events.

Acknowledgments

The authors thank Sohye Kang and Li Zhao for critical comments on the manuscript. This is TSRI manuscript number 17187-MEM. This work was supported by grants from the National Cancer Institute, National Institutes of Health.

References

1 RAJASEKHAR, V.K., E.C. HOLLAND (**2004**) Postgenomic global analysis of translational control induced by oncogenic signaling. *Oncogene* 23, 3248–3264.

2 KAPP, L.D., J.R. LORSCH (**2004**) The molecular mechanics of eukaryotic translation. *Annu Rev Biochem* 73, 657–704.

3 GEBAUER, F., M.W. HENTZE (**2004**) Molecular mechanisms of translational control. *Nat Rev Mol Cell Biol* 5, 827–835.

4 ROGERS, G.W., JR., N.J. RICHTER, W.C. MERRICK (**1999**) Biochemical and kinetic characterization of the RNA helicase activity of eukaryotic initiation factor 4A. *J Biol Chem* 274, 12236–12244.

5 SVITKIN, Y.V., A. PAUSE, A. HAGHIGHAT, S. PYRONNET, G. WITHERELL, G.J. BELSHAM, N. SONENBERG (**2001**) The requirement for eukaryotic initiation factor 4A (eIF4A) in translation is in direct proportion to the degree of mRNA 5′ secondary structure. *RNA* 7, 382–394.

6 KUERSTEN, S., E.B. GOODWIN (**2003**) The power of the 3′ UTR: translational control and development. *Nat Rev Genet* 4, 626–637.

7 ZIMMER, S.G., A. DEBENEDETTI, J.R. GRAFF (**2000**) Translational control of malignancy: the mRNA cap-binding protein, eIF-4E, as a central regulator of tumor formation, growth, invasion and metastasis. *Anticancer Res* 20, 1343–1351.

8 GRAFF, J.R., S.G. ZIMMER (**2003**) Translational control and metastatic progression: enhanced activity of the mRNA cap-binding protein eIF-4E selectively enhances translation of metastasis-related mRNAs. *Clin Exp Metastasis* 20, 265–273.

9 GINGRAS, A.-C., B. RAUGHT, N. SONENBERG (**1999**) eIF4 initiation factors: effectors of mRNA recruitment to ribosomes and regulators of translation. *Annu Rev Biochem* 68, 913–963.

10 KIMBALL, S.R., L.S. JEFFERSON (**2005**) Role of amino acids in the translational control of protein synthesis in mammals. *Semin Cell Dev Biol* 16, 21–27.

11 HAY, N., N. SONENBERG (**2004**) Upstream and downstream of mTOR. *Genes Dev* 18, 1926–1945.

12 FRUMAN, D.A., R.E. MEYERS, L.C. CANTLEY (**1998**) Phosphoinositide kinases. *Annu Rev Biochem* 67, 481–507.

13 SIMPSON, L., R. PARSONS (**2001**) PTEN: life as a tumor suppressor. *Exp Cell Res* 264, 29–41.

14 IM, E., F.C. VON LINTIG, J. CHEN, S. ZHUANG, W. QUI, S. CHOWDHURY, P.F. WORLEY, G.R. BOSS, R.B. PILZ (**2002**) Rheb is in a high activation state and inhibits B-Raf kinase in mammalian cells. *Oncogene* 21, 6356–6365.

15 LI, Y., M.N. CORRADETTI, K. INOKI, K.L. GUAN (**2004**) TSC2: filling the GAP in the mTOR signaling pathway. *Trends Biochem Sci* 29, 32–38.

16 INOKI, K., M.N. CORRADETTI, K.L. GUAN (**2005**) Dysregulation of the TSC-mTOR pathway in human disease. *Nat Genet* 37, 19–24.

17 THOMAS, G., SABATINI, D.M., HALL, N.M. (editors) (**2004**) *TOR – Target of Rapamycin. Current Topics in Microbiology and Immunology* 279.

18 LONG, X., Y. LIN, S. ORTIZ-VEGA, K. YONEZAWA, J. AVRUCH (**2005**) Rheb binds and regulates the mTOR kinase. *Curr Biol* 15, 702–713.

19 PETERSON, R.T., B.N. DESAI, J.S. HARDWICK, S.L. SCHREIBER (**1999**) Protein phosphatase 2A interacts with the 70-kDa S6 kinase and is activated by inhibition of FKBP12-rapamycinassociated protein. *Proc Natl Acad Sci USA* 96, 4438–4442.

20 SCHALM, S.S., J. BLENIS (**2002**) Identification of a conserved motif required for mTOR signaling. *Curr Biol* 12, 632–639.

21 MEYUHAS, O. (**2000**) Synthesis of the translational apparatus is regulated at

the translational level. *Eur J Biochem*
267, 6321–6330.

22 STOLOVICH, M., H. TANG, E.
HORNSTEIN, G. LEVY, R. COHEN, S.S.
BAE, M.J. BIRNBAUM, O. MEYUHAS
(**2002**) Transduction of growth or
mitogenic signals into translational
activation of TOP mRNAs is fully
reliant on the phosphatidylinositol
3-kinase-mediated pathway but
requires neither S6K1 nor rpS6
phosphorylation. *Mol Cell Biol* 22,
8101–8113.

23 HARRINGTON, L.S., G.M. FINDLAY,
R.F. LAMB (**2005**) Restraining PI3K:
mTOR signalling goes back to the
membrane. *Trends Biochem Sci* 30, 35–
42.

24 RAUGHT, B., F. PEIRETTI, A.C.
GINGRAS, M. LIVINGSTONE, D.
SHAHBAZIAN, G.L. MAYEUR, R.D.
POLAKIEWICZ, N. SONENBERG, J.W.
HERSHEY (**2004**) Phosphorylation of
eucaryotic translation initiation factor
4B Ser422 is modulated by S6 kinases.
EMBO J 23, 1761–1769.

25 LUM, J.J., R.J. DEBERARDINIS, C.B.
THOMPSON (**2005**) Autophagy in
metazoans: cell survival in the land of
plenty. *Nat Rev Mol Cell Biol* 6, 439–
448.

26 TOKUNAGA, C., K. YOSHINO, K.
YONEZAWA (**2004**) mTOR integrates
amino acid- and energy-sensing
pathways. *Biochem Biophys Res
Commun* 313, 443–446.

27 ZHANG, H., J.P. STALLOCK, J.C. NG,
C. REINHARD, T.P. NEUFELD (**2000**)
Regulation of cellular growth by the
Drosophila target of rapamycin dTOR.
Genes Dev 14, 2712–2724.

28 BARBET, N.C., U. SCHNEIDER, S.B.
HELLIWELL, I. STANSFIELD, M.F. TUITE,
M.N. HALL (**1996**) TOR controls
translation initiation and early G1
progression in yeast. *Mol Biol Cell* 7,
25–42.

29 HARA, K., K. YONEZAWA, Q.P. WENG,
M.T. KOZLOWSKI, C. BELHAM, J.
AVRUCH (**1998**) Amino acid sufficiency
and mTOR regulate p70 S6 kinase
and eIF-4E BP1 through a common
effector mechanism. *J Biol Chem* 273,
14484–14494.

30 BECK, T., A. SCHMIDT, M.N. HALL
(**1999**) Starvation induces vacuolar
targeting and degradation of the
tryptophan permease in yeast. *J Cell
Biol* 146, 1227–1238.

31 DEPREZ, J., D. VERTOMMEN, D.R.
ALESSI, L. HUE, M.H. RIDER (**1997**)
Phosphorylation and activation of
heart 6-phosphofructo-2-kinase by
protein kinase B and other protein
kinases of the insulin signaling
cascades. *J Biol Chem* 272, 17269–
17275.

32 HAHN-WINDGASSEN, A., V. NOGUEIRA,
C.C. CHEN, J.E. SKEEN, N.
SONENBERG, N. HAY (**2005**) Akt
activates mTOR by regulating cellular
ATP and AMPK activity. *J Biol Chem*
280, 32081–32089.

33 BARTHEL, A., S.T. OKINO, J. LIAO, K.
NAKATANI, J. LI, J.P. WHITLOCK, JR.,
R.A. ROTH (**1999**) Regulation of
GLUT1 gene transcription by the
serine/threonine kinase Akt1. *J Biol
Chem* 274, 20281–20286.

34 KOHN, A.D., S.A. SUMMERS, M.J.
BIRNBAUM, R.A. ROTH (**1996**)
Expression of a constitutively active
Akt Ser/Thr kinase in 3T3-L1
adipocytes stimulates glucose uptake
and glucose transporter 4
translocation. *J Biol Chem* 271, 31372–
31378.

35 HAJDUCH, E., D.R. ALESSI, B.A.
HEMMINGS, H.S. HUNDAL (**1998**)
Constitutive activation of protein
kinase B alpha by membrane targeting
promotes glucose and system A amino
acid transport, protein synthesis, and
inactivation of glycogen synthase
kinase 3 in L6 muscle cells. *Diabetes*
47, 1006–1013.

36 EDINGER, A.L., C.B. THOMPSON (**2002**)
Akt maintains cell size and survival by
increasing mTOR-dependent nutrient
uptake. *Mol Biol Cell* 13, 2276–2288.

37 CHANG, H.W., M. AOKI, D. FRUMAN,
K.R. AUGER, A. BELLACOSA, P.N.
TSICHLIS, L.C. CANTLEY, T.M.
ROBERTS, P.K. VOGT (**1997**)
Transformation of chicken cells by the
gene encoding the catalytic subunit of
PI 3-kinase. *Science* 276, 1848–1850.

38 BELLACOSA, A., J.R. TESTA, S.P. STAAL,

P.N. Tsichlis (**1991**) A retroviral oncogene, akt, encoding a serine-threonine kinase containing an SH2-like region. *Science* 254, 274–277.

39 Staal, S.P., J.W. Hartley (**1988**) Thymic lymphoma induction by the AKT8 murine retrovirus. *J Exp Med* 167, 1259–1264.

40 Staal, S.P., J.W. Hartley, W.P. Rowe (**1977**) Isolation of transforming murine leukemia viruses from mice with a high incidence of spontaneous lymphoma. *Proc Natl Acad Sci USA* 74, 3065–3067.

41 Aoki, M., O. Batista, A. Bellacosa, P. Tsichlis, P.K. Vogt (**1998**) The akt kinase: molecular determinants of oncogenicity. *Proc Natl Acad Sci USA* 95, 14950–14955.

42 Aoki, M., C. Schetter, M. Himly, O. Batista, H.W. Chang, P.K. Vogt (**2000**) The catalytic subunit of phosphoinositide 3-kinase: require-ments for oncogenicity. *J Biol Chem* 275, 6267–6275.

43 Brazil, D.P., J. Park, B.A. Hemmings (**2002**) PKB binding proteins. Getting in on the Akt. *Cell* 111, 293–303.

44 Datta, S.R., A. Brunet, M.E. Greenberg (**1999**) Cellular survival: a play in three Akts. *Genes Dev* 13, 2905–2927.

45 Navolanic, P.M., L.S. Steelman, J.A. McCubrey (**2003**) EGFR family signaling and its association with breast cancer development and resistance to chemotherapy (Review). *Int J Oncol* 22, 237–252.

46 Simoncini, T., A. Hafezi-Moghadam, D.P. Brazil, K. Ley, W.W. Chin, J.K. Liao (**2000**) Interaction of oestrogen receptor with the regulatory subunit of phosphatidylinositol-3-OH kinase. *Nature* 407, 538–541.

47 Yenush, L., M.F. White (**1997**) The IRS-signalling system during insulin and cytokine action. *Bioessays* 19, 491–500.

48 Di Fiore, P.P., J.H. Pierce, M.H. Kraus, O. Segatto, C.R. King, S.A. Aaronson (**1987**) erbB-2 is a potent oncogene when overexpressed in NIH/3T3 cells. *Science* 237, 178–182.

49 Rodriguez-Viciana, P., P.H. Warne, A. Khwaja, B.M. Marte, D. Pappin, P. Das, M.D. Waterfield, A. Ridley, J. Downward (**1997**) Role of phosphoinositide 3-OH kinase in cell transformation and control of the actin cytoskeleton by Ras. *Cell* 89, 457–467.

50 Penuel, E., G.S. Martin (**1999**) Transformation by v-Src: Ras-MAPK and PI3K-mTOR mediate parallel pathways. *Mol Biol Cell* 10, 1693–1703.

51 Vezina, C., A. Kudelski, S.N. Sehgal (**1975**) Rapamycin (AY-22,989), a new antifungal antibiotic. I. Taxonomy of the producing streptomycete and isolation of the active principle. *J Antibiot (Tokyo)* 28, 721–726.

52 Hamilton, G.S., J.P. Steiner (**1998**) Immunophilins: beyond immunosuppression. *J Med Chem* 41, 5119–5143.

53 Oshiro, N., K. Yoshino, S. Hidayat, C. Tokunaga, K. Hara, S. Eguchi, J. Avruch, K. Yonezawa (**2004**) Dissociation of raptor from mTOR is a mechanism of rapamycin-induced inhibition of mTOR function. *Genes Cells* 9, 359–366.

54 Aoki, M., E. Blazek, P.K. Vogt (**2001**) A role of the kinase mTOR in cellular transformation induced by the oncoproteins P3k and Akt. *Proc Natl Acad Sci USA* 98, 136–141.

55 Neshat, M.S., I.K. Mellinghoff, C. Tran, B. Stiles, G. Thomas, R. Petersen, P. Frost, J.J. Gibbons, H. Wu, C.L. Sawyers (**2001**) Enhanced sensitivity of PTEN-deficient tumors to inhibition of FRAP/mTOR. *Proc Natl Acad Sci USA* 98, 10314–10319.

56 Podsypanina, K., R.T. Lee, C. Politis, I. Hennessy, A. Crane, J. Puc, M. Neshat, H. Wang, L. Yang, J. Gibbons, P. Frost, V. Dreisbach, J. Blenis, Z. Gaciong, P. Fisher, C. Sawyers, L. Hedrick-Ellenson, R. Parsons (**2001**) An inhibitor of mTOR reduces neoplasia and normalizes p70/S6 kinase activity in Pten+/− mice. *Proc Natl Acad Sci USA* 98, 10320–10325.

57 Lazaris-Karatzas, A., K.S. Montine, N. Sonenberg (**1990**) Malignant

transformation by a eukaryotic initiation factor subunit that binds to mRNA 5′ cap. *Nature* 345, 544–547.

58 RUGGERO, D., L. MONTANARO, L. MA, W. XU, P. LONDEI, C. CORDON-CARDO, P.P. PANDOLFI (**2004**) The translation factor eIF-4E promotes tumor formation and cooperates with c-Myc in lymphomagenesis. *Nat Med* 10, 484–486.

59 FUKUCHI-SHIMOGORI, T., I. ISHII, K. KASHIWAGI, H. MASHIBA, H. EKIMOTO, K. IGARASHI (**1997**) Malignant transformation by overproduction of translation initiation factor eIF4G. *Cancer Res* 57, 5041–5044.

60 LYNCH, M., C. FITZGERALD, K.A. JOHNSTON, S. WANG, E.V. SCHMIDT (**2004**) Activated eIF4E-binding protein slows G1 progression and blocks transformation by c-myc without inhibiting cell growth. *J Biol Chem* 279, 3327–3339.

61 YANG, H.S., A.P. JANSEN, A.A. KOMAR, X. ZHENG, W.C. MERRICK, S. COSTES, S.J. LOCKETT, N. SONENBERG, N.H. COLBURN (**2003**) The transformation suppressor Pdcd4 is a novel eukaryotic translation initiation factor 4A binding protein that inhibits translation. *Mol Cell Biol* 23, 26–37.

62 BADER, A.G., K.A. FELTS, N. JIANG, H.W. CHANG, P.K. VOGT (**2003**) Y box-binding protein 1 induces resistance to oncogenic transformation by the phosphatidylinositol 3-kinase pathway. *Proc Natl Acad Sci USA* 100, 12384–12389.

63 BADER, A.G., P.K. VOGT (**2005**) Inhibition of protein synthesis by Y box-binding protein 1 blocks oncogenic cell transformation. *Mol Cell Biol* 25, 2095–2106.

64 NEKRASOV, M.P., M.P. IVSHINA, K.G. CHERNOV, E.A. KOVRIGINA, V.M. EVDOKIMOVA, A.A. THOMAS, J.W. HERSHEY, L.P. OVCHINNIKOV (**2003**) The mRNA-binding protein YB-1 (p50) prevents association of the eukaryotic initiation factor eIF4G with mRNA and inhibits protein synthesis at the initiation stage. *J Biol Chem* 278, 13936–13943.

65 BJORNSTI, M.A., P.J. HOUGHTON (**2004**) The TOR pathway: a target for cancer therapy. *Nat Rev Cancer* 4, 335–348.

66 VIVANCO, I., C.L. SAWYERS (**2002**) The phosphatidylinositol 3-kinase AKT pathway in human cancer. *Nat Rev Cancer* 2, 489–501.

67 SAMUELS, Y., Z. WANG, A. BARDELLI, N. SILLIMAN, J. PTAK, S. SZABO, H. YAN, A. GAZDAR, S.M. POWELL, G.J. RIGGINS, J.K. WILLSON, S. MARKOWITZ, K.W. KINZLER, B. VOGELSTEIN, V.E. VELCULESCU (**2004**) High frequency of mutations of the PIK3CA gene in human cancers. *Science* 304, 554.

68 KANG, S., A.G. BADER, L. ZHAO, P.K. VOGT (**2005**) Mutated PI 3-kinases: cancer targets on a silver platter. *Cell Cycle* 4, 578–581.

69 KANG, S., A.G. BADER, P.K. VOGT (**2005**) Phosphatidylinositol 3-kinase mutations identified in human cancer are oncogenic. *Proc Natl Acad Sci USA* 102, 802–807.

70 ENG, C.P., S.N. SEHGAL, C. VEZINA (**1984**) Activity of rapamycin (AY-22,989) against transplanted tumors. *J Antibiot (Tokyo)* 37, 1231–1237.

71 HOUCHENS, D.P., A.A. OVEJERA, S.M. RIBLET, D.E. SLAGEL (**1983**) Human brain tumor xenografts in nude mice as a chemotherapy model. *Eur J Cancer Clin Oncol* 19, 799–805.

72 CARRAWAY, H., M. HIDALGO (**2004**) New targets for therapy in breast cancer: mammalian target of rapamycin (mTOR) antagonists. *Breast Cancer Res* 6, 219–224.

73 PANWALKAR, A., S. VERSTOVSEK, F.J. GILES (**2004**) Mammalian target of rapamycin inhibition as therapy for hematologic malignancies. *Cancer* 100, 657–666.

11

Mutations in the PPARγ Gene Relevant for Diabetes and the Metabolic Syndrome

Markku Laakso

11.1
Introduction

Within the next 10–15 years health care systems worldwide will face an epidemic of obesity, type 2 diabetes, and metabolic syndrome, a clustering of atherogenic risk factors. The reasons for this epidemic are high energy intake, sedentary lifestyle, lack of exercise, and poor diet. The development of insulin resistance is a critical step in this evolution. An epidemic of cardiovascular disease, particularly coronary heart disease, will follow in due course. Although lifestyle changes and environment are major triggers for this development, genetic factors play an important role. Type 2 diabetes and metabolic syndrome are likely to develop only in individuals genetically prone to these diseases. Major genes for type 2 diabetes and the metabolic syndrome, however, have not yet been identified.

The ability to maintain metabolic homeostasis in varying nutritional and environmental states is essential for adaptation and survival. To accomplish this, control mechanisms are needed, for example the transcriptional control through nuclear receptors. The peroxisome proliferator-activated receptors (PPARs) comprise a subfamily of the nuclear hormone receptor superfamily, the largest family of transcription factors [1] (see chapter 2). These ligand-activated transcription factors control energy, glucose, and lipid homeostasis, governing adipogenesis and body fat mass formation (see chapters 4, 12 and 17).

The PPAR family consists of three distinct members PPARα, PPARβ/δ, and PPARγ. All of these are activated by naturally occurring fatty acids or fatty acid derivatives. The first PPAR identified was mouse PPARα, characterized as the nuclear receptor activated by peroxisome proliferators [2]. PPARα is highly expressed in the liver, skeletal muscle, and the heart, PPARβ/δ is ubiquitously expressed having the highest expression in skin and skeletal muscle, whereas PPARγ is predominantly expressed in adipose tissue [3]. Upon binding the ligands, PPARγ undergoes a conformational change and heterodimerizes with the retinoid X receptor (RXR). This leads to the recruitment of cofactors, and regulation of the transcription of target genes through the binding to specific response elements (PPAR re-

Nutritional Genomics. Edited by Regina Brigelius-Flohé and Hans-Georg Joost
Copyright © 2006 WILEY-VCH Verlag GmbH & Co. KGaA, Weinheim
ISBN: 3-527-31294-3

sponse elements, PPREs), which consist of a direct repeat of the nuclear hexameric DNA core recognition motif spaced by one nucleotide [4].

The activity of PPARγ is regulated by the binding of endogenous or synthetic ligands. Endogenous ligands include small lipophilic ligands, mainly fatty acids, derived from nutrition or metabolism. Activation of PPARγ leads to adipocyte differentiation and fatty acid storage. Synthetic ligands include, for example, thiozolidinediones, drugs used in the treatment of type 2 diabetes.

11.2
Peroxisome Proliferator-activated Receptor Gamma: Gene Structure and Function

The human PPARγ gene is located on chromosome 3p25–p24 [5], and is composed of nine exons spanning more than 100 kb of genomic DNA [6]. There are four human PPARγ isoforms, PPARγ1–γ4, generated by alternative splicing and alternative promoter use [7]. PPARγ1 and γ3–γ4 mRNAs give rise to an identical protein product, whereas PPARγ2 mRNA is translated from exon B, yielding a protein with 28 additional amino acids in its N-terminus. Exons 1–6 of the PPARγ gene are shared by all four mRNA isoforms. PPARγ1 mRNA contains the untranslated exons A1 and A2, PPARγ3 contains the untranslated exon A2, and PPARγ4 mRNA initiates at exon 1.

The PPARγ gene consists of several functional domains (Fig. 11.1) [4, 8]: the N-terminal A/B domain harboring a ligand-independent transcriptional activation function (AF-1), the C region, comprising two zinc fingers and containing the DNA-binding domain (DBD); the D hinge region important for cofactor docking, and the C-terminal region (E) containing the ligand-binding domain (LBD) and the ligand-dependent activation domain AF-2, which is involved in the generation of the co-activator binding pocket [4, 8].

Co-activators play an important role in determining the genes targeted by PPARγ. Perhaps the most important co-activator of PPARγ is PPARγ co-activator 1α (PGC-1α) which is involved in the regulation of energy metabolism [9].

PPARγ regulates adipocyte differentiation and energy storage. It increases the expression of genes that promote fatty acid storage and represses genes inducing

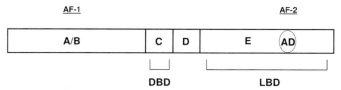

Fig. 11.1. Structural and functional organization of the PPARγ gene (modified from Ref. [4]). The AF-1 is a constitutively activated ligand-independent transactivation function. The C domain contains the DNA-binding domain (DBD). The D domain constitutes a cofactor-docking region. The C-terminal portion (E) encompasses the ligand-binding domain (LBD), a dimerization interface and the ligand-dependent activation domain AF-2. AD, activation domain.

lipolysis in adipocytes [10]. White adipose tissue is needed for proper glucose homeostasis as shown by lipodystrophy, which is associated with severe insulin resistance [11]. PPARγ agonists increase fat mass and improve glycemic control, indicating that white adipose tissue is a central tissue for glucose homeostasis. Furthermore, PPARγ agonists reduce free fatty acid levels, and induce adipogenesis and decrease glucose and triglyceride levels [10]. PPARγ also modulates the expression and action of adipokines (adipocyte-derived signaling molecules), tumor necrosis factor α, leptin, resistin, and adiponectin.

PPARγ may also play an important role in the pathogenesis of atherosclerosis, and it can directly modulate macrophage function and foam cell formation [12] (see also chapter 9). PPARγ is highly expressed in foam cells of early atherosclerotic lesions. On the other hand, PPARγ ligands reduce inflammatory cytokine production by macrophages, and inhibit atherosclerosis in mouse models [8].

11.3
Rare Mutations in the PPARγ Gene

Genetic studies have revealed several dominant-negative mutations in the PPARγ gene [13]. Figure 11.2 [8] shows mutations found in the ligand-independent activation domain and in the ligand binding domain.

11.3.1
Dominant-negative Mutations

Pro467Leu and Val290Met are the best characterized dominant-negative mutations of PPARγ (Fig. 11.2). The Pro467Leu mutation is located in helix 12 of the AF-2

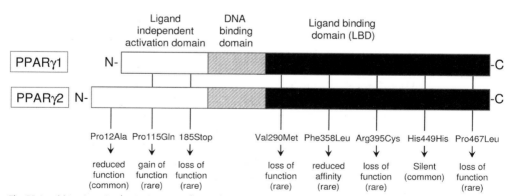

Fig. 11.2. Mutations in the PPARγ gene located in the ligand-independent activation domain and in the ligand-binding domain (modified from Ref. [8]). All the dominant-negative mutations are found in the ligand-binding domain.

motif of the LBD of PPARγ that is critical for mediating ligand-independent transcription and co-activator recruitment [14]. Mutation carriers had insulin resistance, a loss of limb and buttock subcutaneous adipose tissue and hepatic steatosis, and preserved abdominal subcutaneous and visceral adipose tissue. Plasma adiponectin levels were severely reduced, and these individuals had hypertension.

The Val290Met mutation is more proximally located within the LBD on helix 3. The net effect of this mutation, similar to Pro467Leu, is to disrupt the orientation of helix 12, which is important for the interaction of PPARγ with ligands and co-activators. *In vitro* studies have demonstrated that both Val290Met and Pro467Leu affect PPARγ signaling because both mutants exhibit a markedly impaired response to PPARγ agonist-mediated transcriptional transactivation.

The Arg395Cys substitution, also located in the LBD, results in impaired ligand binding and/or heterodimerization with RXRα. Data on the effect of this mutation on PPARγ transcriptional activity *in vitro* is missing.

11.3.2
Gain-of-function Mutations

A rare human PPARγ mutation, Pro115Gln, mapping to the N-terminal ligand-independent activation domain, prevents serine 112 phosphorylation and results in accelerated adipocyte differentiation and cellular triglyceride accumulation *in vitro* [15]. Mutation carriers were markedly obese, but had low levels of insulin as an indication of high insulin sensitivity. However, this has not been a consistent finding. A Ser122Ala mutation also inhibiting PPARγ phosphorylation in mice did not result in obesity, but improved insulin sensitivity with diet-induced obesity [16].

11.3.3
Other Mutations

A silent polymorphism in exon 6 of the PPARγ2 gene, C161T, is quite common in Western populations [8]. It is in strong linkage disequilibrium with the Pro12Ala variant (see below). The T allele has been associated with obesity, particularly in combination with the Ala allele of the Pro12Ala polymorphism of PPARγ2 [17]. Several variants in the promoter region of PPARγ2 have been described, but their association with insulin sensitivity remains to be determined.

11.4
Effect of the Pro12Ala Polymorphism on Clinical Features of the Metabolic Syndrome

The CCA to GCA substitution in exon B at codon 12 of the PPARγ2 gene leads to the Pro12Ala polymorphism [18]. This mutation is the most prevalent human PPARγ mutation. The frequency of this polymorphism varies between populations and ethnic groups, being relatively high (15%) in Caucasian populations, but low in African and Asian populations (1–3%) [8]. PPARγ2 has an extension of 28 amino acids at its N-terminus that renders its ligand-independent activation domain 5- to 10-fold more effective than that of PPARγ1. Whereas the effect of the Ala substitution on the structure of PPARγ2 remains unknown, the Ala allele shows decreased binding affinity to the cognate promoter element and reduced ability to transactivate responsive promoters [18]. These observations are in agreement with studies showing that heterozygous PPARγ mice exhibit increased insulin sensitivity compared with their wild-type littermates [13]. Furthermore, transcription of PPARγ target genes is less efficient in cells overexpressing the Ala allele of the PPARγ2 gene compared with the wild type, and in the presence of the Ala allele the ability to mediate thiazolidinedione-induced adipogenesis is reduced.

The Pro12Ala polymorphism of PPARγ influences body weight, insulin sensitivity, and glucose metabolism as well as lipid metabolism and the risk of atherosclerosis. In the following, these aspects are discussed separately.

11.4.1
Body Weight Regulation

PPARγ2 regulates adipocyte differentiation, and therefore it is expected that the Pro12Ala polymorphism has an impact on body weight regulation. A recent meta-analysis including data from 30 studies having 19 136 subjects showed that the Ala allele was associated with higher body mass index (BMI) in obese subjects (mean BMI $>$ 27 kg/m^2) [19]. In contrast, no association was found in normal weight subjects (BMI $<$ 27 kg/m^2). When the data analysis was done separately among all three genotypes, subjects with the Ala12Ala genotype had significantly higher BMI than subjects with the Pro12Ala or Pro12Pro genotypes. This indicates that in overweight subjects the Ala allele is associated with higher BMI under a recessive model.

There are only a few data from prospective population-based studies on the association of the Pro12Ala polymorphism and weight changes. In our prospective cohort of 119 non-diabetic subjects followed up to 10 years, subjects with the Pro12Ala or Ala12Ala genotypes gained significantly more weight than subjects with the Pro12Pro genotype (change in weight 5.6 versus 1.8%, respectively, $P = 0.013$) [20]. The mechanisms for these weight changes remain unknown but are in agreement with the hypothesis that high insulin sensitivity predisposes to weight gain. Taken together, these observations suggest that the effects of the Pro12Ala substitution on fat mass and BMI are subtle, and subject to modification by other genetic and environmental factors.

Because the Ala allele seems to be associated with weight gain in adulthood, we investigated the effects of the Pro12Ala polymorphism on weight at birth, 7 years, 20 years, and 41 years [21]. Our study indicated that the Ala allele was associated with high weight at birth and weight gain and high waist circumference in adulthood. In a separate Finnish cohort we also demonstrated that small body size at birth and insulin resistance was seen only in individuals with the Pro12Pro genotype [22].

11.4.2
Insulin Sensitivity and Type 2 Diabetes

The first study demonstrating that the 12Ala allele of PPARγ2 is associated with insulin sensitivity was published in 1998 [18]. Although the mechanisms behind this association remain speculative, lower transactivation capacity of the Ala variant may lead to less efficient stimulation of PPARγ target gene expression and predispose to lower levels of adipose tissue mass accumulation, which in turn may be responsible for improved insulin sensitivity. However, the association of the Ala allele and insulin sensitivity disappeared when adjusted for BMI, suggesting that the primary effect of this allele could be on body fat mass. Since 1998 several other studies, but not all, have confirmed our original findings. The Pro12Ala polymorphism does not modulate the response to thiazolidinedione treatment [23]. The relationship between PPARγ activity and insulin sensitivity seems to be a complex U-shaped curve. A modest reduction in receptor function and adipogenesis (Pro12Ala or heterozygous null mice) improves insulin sensitivity, whereas a modest increase in receptor activity (human Pro115Gln) may predispose to insulin resistance through promoting obesity [13].

Previously published studies on the relationship of insulin sensitivity with Pro12Ala polymorphism have been based on fasting insulin, insulin sensitivity index, or the measurement of whole-body insulin sensitivity using the euglycemic clamp technique. Because PPARγ2 expression is predominantly in the adipose tissue, the measurement of tissue-specific insulin sensitivity is necessary for the understanding of the mechanisms related to the effects of the Pro12Ala polymorphism. Indeed, using positron emission tomography we have been able to quantify insulin sensitivity in skeletal muscle and adipose tissue. Our study demonstrated that the Ala allele was associated with high insulin sensitivity in skeletal muscle [24]. This study implies that although PPARγ2 expression is <5% in skeletal muscle compared with adipose tissue, the association of high insulin sensitivity with the Ala allele is explained by elevated glucose uptake in skeletal muscle.

Several studies, including our original report [18], have indicated that the Pro12-Pro genotype is a genetic risk factor for the development of type 2 diabetes. Altshuler *et al.* [25] evaluated 16 published genetic associations to type 2 diabetes and related subphenotypes using a family-based design to control for population stratification, and replication samples to increase power. Only one association (i.e. that of the Pro12Ala polymorphism with type 2 diabetes) was found by analyzing more than 3000 individuals. There was a modest (1.25-fold) but significant ($P = 0.002$) increase in diabetes risk associated with the more common Pro allele (frequency of about 85%). Because the risk allele occurs at such high frequency, its modest effect translates into a large population attributable risk, and could influence as much as 25% of type 2 diabetes in the general population. It is generally accepted that the Pro12Pro genotype is the most important risk genotype for type 2 diabetes so far published.

11.4.3
Lipids and Lipoproteins

The effect of the Pro12Ala polymorphism on lipid and lipoprotein levels varied among the studies published. In our study, including almost 1000 non-diabetic individuals, the Ala allele was associated with low triglyceride and high high-density lipoprotein (HDL) levels [18], but several other studies have not reported these findings. In some studies elevated total and/or low-density lipoprotein (LDL)-cholesterol levels were found in carriers of the Ala allele. It remains to be proven that the Pro12Ala polymorphism has a major impact on lipid and lipoprotein levels.

With respect to uncommon mutations in the LBD of the PPARγ gene, hypertriglyceridemia and low HDL-cholesterol has been reported [13].

11.4.4
Elevated Blood Pressure

Elevated blood pressure is often found in subjects with type 2 diabetes and metabolic syndrome. Its pathophysiology is complex but may be in part related to insulin resistance. Early-onset hypertension has been found in individuals with the Pro467Leu, Val290Met, and Arg395Cys mutations of PPARγ2. In contrast, in carriers of the Phe358Leu mutation hypertension has not been always detected. There are no reliable data indicating whether or not the Pro12Ala substitution is associated with elevated blood pressure.

11.4.5
Atherosclerosis

Subjects with the Ala allele reportedly have significantly lower intima media thickness compared with carriers of the Pro12Pro genotype [26]. Our study provided evidence that PPARγ2 is expressed in atherosclerotic lesions and macrophages, indicating that the Pro12Ala polymorphism could potentially regulate atherosclerosis [26]. With respect to cardiovascular disease events, the Ala allele has been associated with lower incidence of myocardial infarction [27], but this findings has not been confirmed in all studies. Thiazolidinediones, ameliorating insulin resistance, hyperlipidemia, and hypertension have been demonstrated to reduce intima media thickness in subjects with type 2 diabetes [28]. Furthermore, these drugs have other anti-atherogenic effects, such as reduction of the release of inflammatory cytokines (tumor necrosis factor α, interleukin 6) from macrophages, and increase of the level of adiponectin. However, no definitive conclusions can be made on the association of the Pro12Ala polymorphism with adipocyte-derived hormones on the basis of published studies.

11.5
Interaction of the Pro12Ala Polymorphism with Other Genes

11.5.1
Insulin Receptor Substrate 1

In German normoglycemic subjects a gene–gene interaction was studied between the Pro12Ala polymorphism and the Gly972Arg polymorphism of insulin receptor substrate 1 (IRS-1) [29]. In that study insulin sensitivity was not different between the Pro12Pro and 12Ala allele carriers. However, insulin sensitivity was significantly higher in carriers of X12Ala (PPARγ2) + X972Arg (IRS-1) than in carriers of Pro12Pro + X972Arg. These results show that the Arg972 (IRS-1) background produced a marked difference in insulin sensitivity between X12Ala and Pro12Pro that was not present in the whole population or against the Gly972 background. Therefore, the Ala allele becomes particularly advantageous against the background of an additional, possibly disadvantageous genetic polymorphism.

We have recently demonstrated that serum adiponectin concentrations were significantly higher among non-diabetic subjects who simultaneously were carriers of the Ala12Ala and the Gly972Gly genotype, compared with other genotype combinations [30].

11.5.2
Peroxisome Proliferated-activated Receptor Co-activator 1 (PGC-1)

PGC-1α co-activates a series of nuclear receptors including PPARγ, and controls transcription of genes involved in adaptive thermogenesis, adipogenesis and oxidative metabolism [31]. We screened the Gly482Ser variant of the PGC-1α gene, which has been reported to be associated with type 2 in previous studies. Altogether 770 subjects with impaired glucose tolerance participating in the STOP-NIDDM trial [32], aiming to investigate the effect of acarbose on the prevention of diabetes, were included in data analyses. The 482Ser allele was associated with a 1.6-fold higher risk for type 2 diabetes compared with the Gly482Gly genotype in the placebo group. Subjects having both the Pro12Pro genotype of PPARγ2 and the 482Ser allele of PGC-1α had even higher risk of developing type 2 diabetes, although the interaction between these genes was not statistically significant [32].

11.6
Interaction of the Pro12Ala Polymorphism with Lifestyle Factors

11.6.1
Dietary Factors

Luen *et al.* [33] studied 592 non-diabetic participants who were genotyped for the Pro12Ala polymorphism. No difference in fasting insulin concentration or BMI between the Ala allele carriers and Pro homozygotes was found, but a strong inter-action was evident between the ratio of dietary polysaturated fat to saturated fat (P:S ratio) for both BMI ($P = 0.0038$) and fasting insulin ($P = 0.0097$). When the dietary P:S ratio was low, the BMI in Ala carriers was greater than that in Pro homozygotes, but when the dietary P:S ratio was high, the opposite was observed. Therefore, a strong gene–diet interaction was found between the Pro12Ala polymorphism and dietary patterns of fatty acid intake. However, subsequent publications have failed to confirm an interaction between the P:S ratio and the Pro12Ala polymorphism in modulating BMI.

We studied the effect of the Pro12Ala polymorphism on serum lipid and lipoprotein responses to *n*-3 fatty acid supplementation (fish oil). After the three-month study period, carriers of the Ala allele presented a greater decrease in serum triacylglycerol concentration in response to *n*-3 fatty acid supplementation than did subjects with the Pro12Pro genotype when the total dietary fat intake or the intake of saturated fatty acids was low [34]. Therefore, the Pro12Ala polymorphism may modify the inter-individual variability in serum triacylglycerol response to *n*-3 fatty acid supplementation.

11.6.2
Weight Loss and Physical Exercise

The association of the Pro12Ala polymorphism of the PPARγ2 gene with the incidence of type 2 diabetes was investigated in 522 subjects with impaired glucose tolerance in the Finnish Diabetes Prevention Study [35]. Subjects were randomized to either an intensive diet and exercise group or a control group. The risk for type 2 diabetes increased in subjects who gained weight or belonged to the control group. In the intervention group, subjects with the Ala12Ala genotype lost more weight during the follow-up than did subjects with other genotypes. None of subjects with the Ala12Ala genotype developed type 2 diabetes in this group. Therefore, this study shows that in carriers of the Ala12Ala genotype, beneficial changes in diet, increases in physical activity, and weight loss are particularly effective means to prevent type 2 diabetes. The evidence that the Ala allele can modulate weight changes is supported also by another study showing that after weight loss, weight regain was greater in women with the Ala allele compared with women with the Pro12Pro genotype [36].

A recent study has shown that sedentary men with the Pro12Ala variant have lower insulin action, but were particularly responsive to the improvement in insulin action by endurance training [37].

11.7
Concluding Remarks

Both genetic and pharmacological evidence supports the relationship of PPARγ with adipogenesis and insulin sensitivity. PPARγ activity in humans corresponds

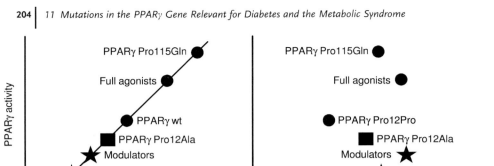

Fig. 11.3. The association of PPARγ activity with adipogenesis and insulin sensitivity (modified from Ref. [38]).

directly to adipose tissue mass and not necessarily to insulin sensitivity as shown in Fig. 11.3 [38]. Insulin sensitivity can be achieved by the inhibition of PPARγ, either by loss-of-function mutations or PPARγ antagonists, or by activation of PPARγ, either by gain-of-function mutations or full agonists.

From the genetic point of view the Pro12Ala polymorphism of the PPARγ2 gene is still the most promising candidate gene for type 2 diabetes. Although previous studies indicate that the presence of the Ala allele protects against environmental influences, such as high-fat diet and lack of exercise, and the Pro12Pro genotype is an important risk genotype for type 2 diabetes, further studies are needed, focusing particularly on gene–gene and gene–lifestyle/diet interactions. Although very uncommon, mutations in the PPARγ gene have been valuable for the understanding of the activation and inhibition of the PPAR system.

Genetic association studies published so far have often been based on a small sample size, subgroup analyses, and heterogeneous study populations. Furthermore, publication bias favoring the reporting of positive findings, and true differences between populations and ethnic group may have caused contradictory findings on the association of the Pro12Ala polymorphism and several features of the metabolic syndrome. Therefore, in order to clarify the role of the Pro12Ala polymorphism in the etiology and pathophysiology of type 2 diabetes and the metabolic syndrome further studies should include large population-based samples, careful phenotyping of study subjects and prospective study design.

References

1 MANGELSDORF, D.J., C. THUMMEL, M. BEATO, P. HERRLICH, G. SCHUTZ, K. UMESONO, B. BLUMBERG, P. KASTNER, E. MARK, P. CHAMBON, R.M. EVANS (1995) The nuclear receptor super-family: the second decade. *Cell* 83, 835–839.

2 ISSEMAN, I., S. GREEN (1990) Activation of a member of the steroid hormone receptor superfamily by peroxisome proliferators. *Nature* 347, 645–650.

3 CHINETTI, G., J.-C. FRUCHART, B. STAELS (2003) Peroxisome proliferators-activated receptors and inflammation: from basic science to clinical applications. *Int. J. Obes.* 27, S41–S45.

4 DEBRIL, M.-B., J.-P. RENAUD, L. FAJAS, J. AUWERX (2001) The pleiotropic functions of peroxisome proliferators-activated receptor γ. *J. Mol. Med.* 79, 30–47.

5 BEAMER, B.A., C. NEGRI, C.J. YEN, O. GAVRILOVA, J.M. RUMBERGER, M.J. DURCAN, D.P. YARNALL, A.I. HAWKINS, C.A. GRIFFIN, D.K. BURNS, J. ROTH, M. REITMAN, A.R. SHULDINER (1997) Chromosomal localization and partial genomic structure of the human peroxisome proliferator activated receptor-gamma (hPPAR gamma) gene. *Biochem. Biophys. Res. Commun.* 23, 756–759.

6 FAJAS, Ll., D. AUBOEUF, E. RASPE, K. SCHOONJANS, A.M. LEFEBVRE, R. SALADIN, J. NAJIB, M. LAVILLE, J.-C. FRUCHART, S. DEEB, A. VIDAL-PUIG, J. FLIER, M.R. BRIGGS, B. STAELS, H. VIDAL, J. AUWERX (1997) Organization, promoter analysis and expression of the human PPARγ gene. *J. Biol. Chem.* 272, 18779–18789.

7 ZHU, Y., C. QI, J.R. KORENBERG, X.-N. CHEN, D. NOYA, M.S. RAO, J.K. REDDY (1995) Structural organization of mouse peroxisome proliferators activated receptor γ (mPPARγ) gene: alternative promoter use and different slicing yield two mPPARγ isoforms. *Proc. Natl Acad. Sci. USA* 92, 7921–7925.

8 KNOUFF, C., J. AUWERX (2004) Peroxisome proliferators-activated receptor-γ calls for activation in moderation: lessons from genetics and pharmacology. *Endocr. Rev.* 25, 899–918.

9 PUIGSERVER, P., B.M. SPIEGELMAN (2003) Peroxisome proliferator-activated receptor-γ coactivator 1α (PGC-1α): transcriptional coactivator and metabolic regulator. *Endocr. Rev.* 24, 78–90.

10 AUWERX, J. (1999) PPARγ, the ultimate thrifty gene. *Diabetologia* 42, 1033–1049.

11 MOITRA, J., M.M. MASON, M. OLIVE, D. KRYLOV, O. GAVRILOVA, B. MARCUS-SAMUELS, L. FEIGENBAUM, E. LEE, T. AOYAMA, M. ECKHAUS, M.L. REITMAN, C. VINSON (1998) Life without fat: a transgenic mouse. *Gen. Dev.* 12, 3168–3181.

12 LEE, C.H., R.M. EVANS (2002) Peroxisome proliferators-activated receptor-γ in macrophage lipid homeostasis. *Trends Endocrinol. Metab.* 13, 331–335.

13 GURNELL, M., D.B. SAVAGE, V.K.K. CHATTERJEE, S. O'RAHILLY (2003) The metabolic syndrome: peroxisome proliferators-activated receptor γ and its therapeutic modulation. *J. Clin. Endocrinol. Metab.* 88, 2412–2421.

14 BARROSO, I., M. GURNELL, V.E. CROWLEY, M. AGOSTINI, J.W. SCHWABE, M.A. SOOS, G.L. MASLEN, T.D. WILLIAMS, H. LEWIS, A.J. SHAFER, V.K. CHATTERJEE, S. O'RAHILLY (1999) Dominant negative mutations in human PPARγ associated with severe insulin resistance, diabetes mellitus and hypertension. *Nature* 402, 880–883.

15 RISTOW, M., D. MULLER-WIELAND, A. PFEIFFER, W. KRONE, C.R. KAHN (1998) Obesity associated with a mutation in a genetic regulator of adipocyte differentiation. *N. Engl. J. Med.* 339, 953–959.

16 RANGWALA, S.M., J.S. SHAPIRO, A.S. RICH, J.K. KIM, G.I. SHULMAN, K.H.

KAESTNER, M.A. LAZAR (**2003**) Genetic modulation of PPARγ phosphorylation regulates insulin sensitivity. *Dev. Cell.* 5, 657–663.

17 VALVE, R., K. SIVENIUS, R. MIETTINEN, J. PIHLAJAMÄKI, A. RISSANEN, S.S. DEEB, J. AUWERX, M. UUSITUPA, M. LAAKSO (**1999**) Two polymorphisms in the peroxisome proliferator-activated receptor-γ gene are associated with severe overweight among obese women. *J. Clin. Endocrinol. Metab.* 84, 3708–3712.

18 DEEB, S.S., L. FAJAS, M. NEMOTO, J. PIHLAJAMÄKI, L. MYKKÄNEN, J. KUUSISTO, M. LAAKSO, W. FUJIMOTO, J. AUWERX (**1998**) A Pro12Ala substitution in PPARγ2 associated with decreased receptor activity, lower body mass index and improved insulin sensitivity. *Nat. Genet.* 20, 284–287.

19 MASUD, S., S. YE (**2003**) Effect of peroxisome proliferators activated receptor-γ gene Pro12Ala variant on body mass index: a meta-analysis. *J. Med. Genet.* 40, 773–780.

20 LINDI, V., K. SIVENIUS, L. NISKANEN, M. LAAKSO, M. UUSITUPA (**2001**) Effect of the Pro12Ala polymorphim of the PPAR-gamma2 gene on long-term weight change in Finnish non-diabetic subject. *Diabetologia* 44, 925–926.

21 PIHLAJAMÄKI, J., M. VANHALA, P. VANHALA, M. LAAKSO (**2004**) The Pro12Ala polymorphism of the *PPARγ2* regulates weight from birth to adulthood. *Obes. Res.* 12, 187–190.

22 ERIKSSON, J.G., V. LINDI, M. UUSITUPA, T.J. FORSEN, M. LAAKSO, C. OSMOND, D.J.P. BARKER (**2002**) The effect of the Pro12Ala polymorphism of the peroxisome proliferators-activated receptor-γ2 gene on insulin sensitivity and insulin metabolism interact with size at birth. *Diabetes* 51, 2321–2324.

23 SNITKER, S., R.M. WATANABE, I. ANI, A.H. XIANG, A. MARROQUIN, C. OCHOA, J. GOICO, A.R. SHULDINER, T.A. BUCHANAN (**2004**) Changes in insulin sensitivity in response to troglitazone do not differ between subjects with and without the common, functional Pro12Ala

peroxisome proliferators-activated receptor-γ2 gene variant. *Diabete Care* 27, 1365–1368.

24 VÄNTTINEN, M., P. NUUTILA, J. PIHLAJAMAKI, K. HALLSTEN, K.A. VIRTANEN, R. LAUTAMAKI, P. PELTONIEMI, J. KEMPPAINEN, T. TAKALA, A.P. VILJANEN, J. KNUUTI, M. LAAKSO (**2005**) The effect of the Ala12 allele of the peroxisome proliferator-activated receptor-gamma2 gene on skeletal muscle glucose uptake depends on obesity: a positron emission tomography study. *J. Clin. Endocrinol. Metab.* 90, 4249–4254.

25 ALTSHULER, D., J.N. HIRSCHHORN, M. KLANNEMARK, C.M. LINDGREN, M.-C. VOHL, J. NEMESH, C.R. LANE, S.F. SCHAFFNER, S. BOLK, C. BREWER, T. TUOMI, D. GAUDET, T.J. HUDSON, M. DALY, L. GROOP, E.S. LANDER (**2000**) The common PPARγ Pro12Ala polymorphism is associated with decreased risk of type 2 diabetes. *Nat. Genet.* 26, 76–80.

26 TEMELKOVA-KURKTSCHIEV, T., M. HANEFELD, G. CHINETTI, C. ZAWADZKI, S. HAULON, A. KUBASZEK, C. KOEHLER, W. LEONHARDT, B. STAELS, M. LAAKSO (**2004**) Ala12Ala genotype of the peroxisome proliferated-activated receptor γ2 protects against atherosclerosis. *J. Clin. Endocrinol. Metab.* 89, 4238–4242.

27 RIDKER, P.M., N.R. COOK, S. CHENG, H.A. ERLICH, K. LINDPAINTNER, J. PLUTZKY, R.Y. ZEE (**2003**) Alanine for proline substitution in the peroxisome proliferator-activated receptor gamma-2 (PPARG2) gene and the risk of incident myocardial infarction. *Arterioscler. Thromb. Vasc. Biol.* 23, 859–863.

28 KOSHIYAMA, H., D. SHIMONO, N. KAWAMURA, J. MINAMIKAWA, Y. NAKAMURA (**2001**) Inhibitory effect of pioglitazone on carotid wall thickness in type 2 diabetes. *J. Clin. Endocrinol. Metab.* 86, 3452–3456.

29 STUMVOLL, M., N. STEFAN, A. FRITSCHE, A. MADAUS, O. TSCHRITTER, M. KOCH, F. MACHICAO, H. HÄRING (**2001**) Interaction effect between

common polymorphisms in PPARγ2 (Pro12Ala) and insulin receptor substrate 1 (Gly972Arg) on insulin sensitivity. *J. Mol. Med.* 80, 33–38.

30 MOUSAVINASAB, F., T. TÄHTINEN, J. JOKELAINEN, P. KOSKELA, M. VANHALA, J. OIKARINEN, S. KEINÄNEN-KIUKAANNIEMI, M. LAAKSO (**2005**) Common polymorphisms in the PPARγ2 and IRS-1 genes and their interaction influence serum adiponectin concentration in young Finnish men. *Mol. Genet. Metab.* 84, 344–348.

31 PUIGSERVER, P., Z. WU, C.W. PARK, R. GRAVES, M. WRIGHT, B.M. SPIEGELMAN (**1998**) A cold-induced coactivator of nuclear receptors linked to adaptive thermogenesis. *Cell* 92, 829–839.

32 ANDRULIONYTE, L., J. ZACHAROVA, J.L. CHIASSON, M. LAAKSO, for the STOP-NIDDM Study Group (**2004**) Common polymorphisms of the PPAR-γ2 (Pro12Ala) and PGC-1α (Gly482Ser) genes are associated with the conversion from impaired glucose tolerance to type 2 diabetes in the STOP-NIDDM trial. *Diabetologia* 47, 2176–2184.

33 LUAN, J., P.O. BROWNE, A.-H. HARDING, D.J. HALSALL, S. O'RAHILLY, V.K.K. CHATTERJEE, N.J. WAREHAM (**2001**) Evidence for gene-nutrient interaction at the PPARγ locus. *Diabetes* 50, 686–689.

34 LINDI, V., U. SCHWAB, A. LOUHERANTA, M. LAAKSO, B. VESSBY, K. HERMANSON, L. STORLIEN, G. RICCARDI, A.A. RIVELLESE, M.I.J. UUSITUPA (**2003**) Impact of the

Pro12Ala polymorphism of the PPAR-gamma2 gene on serum triacylglycerol response to n-3 fatty acid supplementation. *Mol. Genet. Metab.* 79, 52–60.

35 LINDI, V.I., M.I.J. UUSITUPA, J. LINDSTRÖM, A. LOUHERANTA, J.G. ERIKSSON, T.T. VALLE, H. HÄMÄLÄINEN, P. ILANNE-PARIKKA, S. KEINÄNEN-KIUKAANNIEMI, M. LAAKSO, J. TUOMILEHTO, for the Finnish Diabetes Prevention Study Group (**2002**) Association of the Pro12Ala polymorphism in the PPAR-γ2 gene with 3-year incidence of type 2 diabetes and body weight change in the Finnish Diabetes Prevention Study. *Diabetes* 51, 2582–2586.

36 NICKLAS, B.J., E.F.C. VAN ROSSUM, D.M. BERMAN, A.S. RYAN, K.E. DENNIS, A.R. SHULDINER (**2001**) Genetic variation in the peroxisome proliferators-activated receptor-γ2 gene (Pro12Ala) affects metabolic responses to weight loss and subsequent weight regain. *Diabetes* 50, 2172–2176.

37 WEISS, E.P., O. KULAPUTANA, I.A. GHIU, J. BRANDAUER, C.R. WOHN, D.A. PHARES, A.R. SHULDINER, J.M. HAGBERG. (**2005**) Endurance training-induced changes in the insulin response to oral glucose are associated with the peroxisome proliferators-activated receptor-gamma2 Pro12Ala genotype in men but not in women. *Metabolism* 54, 97–102.

38 COCK, T.A., S.M. HOUTEN, J. AUWERX (**2004**) Peroxisome proliferators-activated receptor-γ: too much of a good thing causes harm. *EMBO Rep.* 5, 142–147.

12
Regulation of Lipogenic Genes in Obesity

Stéphane Mandard and Sander Kersten

Lipogenesis describes the process of fatty acid and triglyceride synthesis. It mainly occurs in liver and fat tissue and is under the coordinated control of hormonal, nutritional, and transcription factors. Several transcription factors have been identified as critical regulators that mediate the effect of hormones and nutrients on gene transcription. These include sterol regulatory element-binding protein 1c (SREBP-1c), CCAAT/enhancer-binding protein alpha (C/EBPα), and the nuclear hormone receptors liver X receptor (LXR), peroxisome proliferator-activated receptors gamma and alpha (PPARγ and PPARα), and estrogen-related receptor alpha (ERRα). The role of these transcription factors in these processes is reviewed and discussed in this chapter. Although lipogenesis may appear to be an attractive target for pharmacological treatment of obesity, recent insights into the metabolic consequences of non-adipose triglyceride storage has shifted attention to alternative targets.

12.1
Introduction

The growing prevalence of obesity worldwide has become an immediate public health concern. Nowadays, obesity occurs at a progressively younger age, which calls for urgent action to prevent a future epidemic of type 2 diabetes. The complexity of obesity as a metabolic disorder, often being associated with insulin resistance, dyslipidemia, and hypertension, and the poor response of this disease to treatment belie the notion that it is caused by a simple imbalance between energy consumption and energy expenditure. Accordingly, strategies aimed at reducing obesity should consider not only the complexity of regulation of energy storage and utilization, but also take into account the powerful evolutionary mechanisms that resist long-term weight loss.

Lipogenesis describes the process of fatty acid and triglyceride synthesis and is obviously of great relevance to obesity. An increase in lipogenesis will inevitably lead to fat accumulation if it is not associated with elevated fat utilization. Thus, it

Nutritional Genomics. Edited by Regina Brigelius-Flohé and Hans-Georg Joost
Copyright © 2006 WILEY-VCH Verlag GmbH & Co. KGaA, Weinheim
ISBN: 3-527-31294-3

appears that in principle the process of lipogenesis is an attractive nutritional and pharmacological target for obesity. This paper will review the most recent information on the molecular regulation of lipogenesis, with emphasis on the role of nuclear hormone and nutrient receptors as well as other transcription factors.

12.2
Transcriptional Regulation of Lipogenesis in Adipose Tissue

The effects of various nutrients and hormones on the expression of lipogenic genes in adipose tissue are mediated by a small number of transcription factors, including SREBP-1c. SREBP-1c (also known as adipocyte determination and differentiation-1) belongs to the basic helix-loop-helix leucine zipper (bHLH-LZ) family of transcription factors. It was initially discovered as a key player of adipocyte differentiation [1]. In addition, SREBP-1c plays a critical role in lipogenesis [2]. SREBP-1c is first synthesized as a precursor form (110 kDa) anchored in the endoplasmic reticulum, which then undergoes a proteolytic cleavage to generate the mature active form of SREBP-1c (50 kDa). In the nucleus, the mature form of SREBP-1c is able to bind to specific sequences (sterol-regulatory elements (SREs) and E-boxes) located in the promoter gene of SREBP target genes (Fig. 12.1).

SREBP-1c probably works in tandem with the adipogenic transcription factor PPARγ, which is a direct target of SREBP-1c in adipocytes and contains a sterol response element in its promoter [3]. It has been proposed that besides upregulating of PPARγ expression, SREBP-1c is able to induce PPARγ activity by increasing the production of an endogenous ligand, leading to the stimulation of adipogenesis and lipogenesis [4]. Next to PPARγ, important lipogenic genes such as acetyl-CoA carboxylase-1 and -2 (ACC), fatty acid synthase (FAS), stearoyl-CoA desaturase-1 (SCD-1), glycerol-3-phosphate acyltransferase (GPAT), and low-density lipoprotein receptor (LDL-R) have been identified as direct targets of SREBP-1c in mature 3T3-L1 adipocytes [5]. An interesting SREBP-1c target in adipocytes is insulin-induced gene 1 (INSIG-1), which is one of two recently discovered polytopic membrane proteins of the endoplasmic reticulum [6–8]. In the presence of sterols, INSIGs bind to the SREBP cleavage-activating protein (SCAP), a critical escort protein required for the cleavage and activation of the SREBP family of membrane-bound transcription factors. It appears that INSIG tethers the SCAP/SREBP complex to the endoplasmic reticulum. Recent *in vitro* studies have shown that stable overexpression of INSIG-1 in 3T3-L1 preadipocytes results in defective adipocyte differentiation, which is likely due to impaired fatty acid and triglyceride synthesis [9]. Although the relevance of the SREBP-1c-mediated upregulation of INSIG-1 in adipose tissue remains to be demonstrated, it can be speculated that it provides a feedback mechanism by which nuclear SREBP-1c can modulate its maturation and, by limiting the processing of SREBPs precursors, maintain a check on lipogenesis.

Apart from SREBP, several other nuclear factors play a critical role in the transcriptional control of lipogenesis in adipose tissue. Numerous loss- and gain-of-

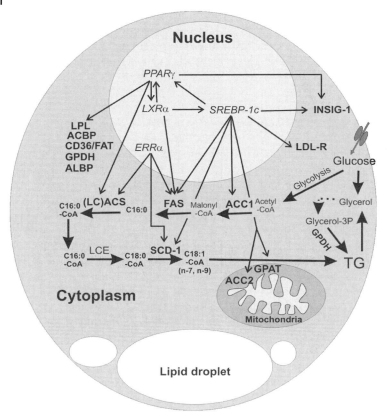

Fig. 12.1. Regulation of lipogenic gene expression in adipocytes. Schematic representation of regulation of genes involved in *de novo* fatty acid biosynthesis in adipocytes by nuclear hormone and nutrient receptors. Nuclear hormone and nutrient receptors are shown in italics. Genes under transcriptional control of nuclear hormone and nutrient receptors are shown in bold. ACBP, acyl-CoA-binding protein; ACC, acetyl-CoA carboxylase; ALBP, adipocyte lipid-binding protein; ERRα, estrogen related receptor α; FAS, fatty acid synthase; CD36/FAT, fatty acid translocase; glycerol 3-P, glycerol-3-phosphate; GPDH, glycerol 3-phosphate dehydrogenase; GPAT, glycerol-3-phosphate acyltransferase; INSIG-1, insulin-induced gene 1; (LC)ACS, (long-chain) acyl-CoA synthetase; LCE, long-chain fatty acid elongase; LDL-R, low-density lipoprotein receptor; LXRα, liver-X-receptor α; PPARγ, peroxisome proliferator-activated receptor γ; SCD-1, stearoyl-CoA desaturase-1; SREBP-1c, sterol responsive element-binding protein-1c; TG, triglycerides.

function experiments have shown that the nuclear hormone receptor PPARγ is essential for in adipocyte differentiation (reviewed in Ref. [10]), and also stimulates lipogenesis. Several lipo- and/or adipogenic genes have already been identified as direct PPARγ targets, including lipoprotein lipase, the scavenger receptor CD36/ fatty acid translocase (FAT), adipocyte lipid-binding protein, and cytosolic glycerol-3-phosphate dehydrogenase. Recently, *INSIG1* was reported as a novel PPARγ target gene as well (Fig. 12.1) [6]. Considering that INSIG-1 appears to inhibit lipo- and adipogenesis, it is hard to reconcile this observation with the pro-adipogenic role of PPARγ [9]. Likely, fine-tuning of adipogenesis is achieved by the balanced expression of several opposing factors that include PPARγ, SREBP-1c, and INSIG.

A recently identified direct target gene of PPARγ in adipocytes is acyl-CoA-binding protein (ACBP) [11]. It was proposed that ACBP would repress PPARγ-mediated *trans*-activation induced by exogenous fatty acids, thereby inhibiting 3T3-L1 adipogenesis. However, additional research is necessary to establish the potential effects of ACBP on PPARγ *trans*-activation.

Supporting a role of PPARγ in lipid storage in mature fat cells, it has been observed that both heterozygous PPARγ mutant mice and adipose-specific PPARγ-deficient mice exhibit smaller fat stores on a high fat diet [12–14]. In adipose-specific PPARγ-deficient mice, mRNA expression of genes involved in both lipo-genesis and adipogenesis were strongly downregulated, illustrating a requirement for PPARγ [14, 15]. Fibrosis, macrophage infiltration, and hypertrophy, implying loss of more than 80% of adipocytes, were evident in mutant fat. Thus, PPARγ is not only essential for the early steps of the adipogenesis program but also for post-differentiation and survival of mature white adipocytes *in vivo* [14, 16].

While it is clear that SREBP-1c and PPARγ are extremely important regulators of adipo- and lipogenesis, recent studies have drawn attention to other nuclear hormone receptors. One of these receptors is the estrogen-related receptor alpha (ERRα). Deletion of ERRα in mice reduces body weight and peripheral fat storage, while food consumption and energy expenditure are unaffected [17]. At the gene level, expression of several enzymes involved in lipogenesis was downregulated in white adipose tissue. In line with this, *de novo* lipogenesis was shown to be re-duced in the mutant animals, suggesting a stimulatory role of ERRα in lipogenesis. These data suggest that ERRα functions as a metabolic regulator with an important effect on fat synthesis.

12.3
Transcriptional Regulation of Hepatic Lipogenesis

Consumption of large amounts of carbohydrates stimulates the conversion of glu-cose to fatty acids in liver by upregulating glycolytic and lipogenic enzymes. The effects of carbohydrate feeding are mediated by insulin and glucose, which activate distinct signalling pathways. Upregulation of lipogenic enzymes by insulin par-tially occurs via the upstream stimulatory factor (USF). USFs are ubiquitous bHLH-LZ transcription factors that are able to form homo- and/or heterodimers. They modulate expression of genes such as FAS and ACC by direct binding to pro-moter gene sequences called E-boxes.

Studies with mice lacking *USF1* and/or *USF2* have provided compelling evi-dence for their role in the stimulatory effect of insulin and glucose on lipogenesis. In $USF1^{-/-}$ or $USF2^{-/-}$ mice, FAS expression in liver was strongly impaired after a fasting/refeeding cycle, demonstrating that USFs are critical factors for the transcriptional activation of FAS by diet [18]. At the molecular level, USFs exert their stimulatory effect on FAS transcription via an E-box motif located in the promoter of the FAS gene. This motif is likely shared with SREBP-1, which also is a potent activator of FAS gene expression. Studies with mice lacking or overexpressing SREBP-1c have indicated that this transcription factor is responsible for the coordinate induction of numerous lipogenic genes in liver, including ACC1 and 2, FAS, SCD-1, GPAT, ACL, malic enzyme, and long fatty acid elongase (LCE) (Fig. 12.2). Accordingly, overexpression SREBP-1c is associated with a dramatic build-up of

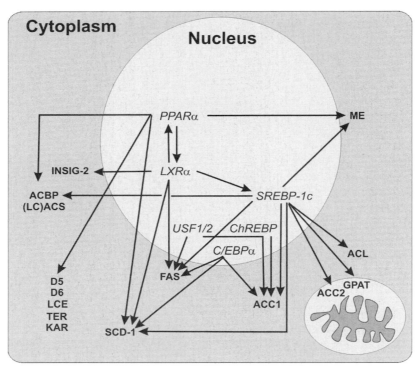

Fig. 12.2. Regulation of lipogenic gene expression in hepatocytes. Genes involved in *de novo* fatty acid biosynthesis in hepatocytes and regulated at the transcription level by nuclear hormone and nutrient receptors are listed. Nuclear hormone and nutrient receptors are shown in italics. Genes under transcriptional control of nuclear hormone and nutrient receptors are shown in bold. ACBP, acyl-CoA-binding protein; ACC, acetyl-CoA carboxylase; ACL, ATP citrate lyase; C/EBPα, CCAAT/enhancer-binding protein alpha; ChREBP, carbohydrate response element-binding protein; D5, delta-5 desaturase; D6, delta-6 desaturase; FAS, fatty acid synthase; GPAT, glycerol-3-phosphate acyltransferase; INSIG-2, insulin-induced gene 2; KAR, microsomal 3-keto acyl-CoA reductase; (LC)ACS, (long-chain) acyl-CoA synthetase; LCE, long-chain fatty acid elongase; LXRα: liver X receptor α; ME, malic enzyme; PPARα, peroxisome proliferator-activated receptor α; SCD-1, stearoyl-CoA desaturase-1; SREBP-1c, sterol responsive element-binding protein-1c. TER, *trans*-2,3-enoyl-CoA reductase; USF, upstream stimulatory factor.

hepatic triglycerides [19]. At the present time, SREs have been identified in the promoters of FAS [20], ACC [21], ACL [22], GPAT [23], Spot 14 (24), SCD-1 [25], and ACBP [26]. The last gene encodes a small intracellular protein that is able to bind long-chain fatty acyl-CoAs. Although the role of ACBP *in vivo* is not very clear, it can be speculated that ACBP participates in *de novo* fatty acid biosynthesis. Indeed, high expression levels of ACBP have been observed in both hepatocytes and adipocytes and in liver ACBP expression is stimulated by insulin and repressed by fasting.

The expression of many lipogenic enzymes is enhanced by insulin, yet optimal transcription of most lipogenic genes requires high carbohydrate levels as well. In concordance with this observation, glucose has been identified as a potent activator of lipogenesis not only by acting as substrate, but also as an important regulatory molecule. Recent data indicate that increased glucose metabolism activates an intracellular signaling pathway, probably involving xylulose 5-phosphate, that transcriptionally regulates genes encoding lipogenic enzymes via the carbohydrate response element-binding protein (ChREBP), a bHLH-LZ transcription factor. ChREBP was identified and purified by taking advantage of its binding to the carbohydrate response element (ChRE) within the promoter of the L-type pyruvate kinase gene. ChREBP is stimulated by high concentrations of glucose that promote both translocation of ChREBP from the cytosol to the nucleus coupled with binding of ChREBP to a ChRE [27]. Importantly, ChREBP becomes active in response to high glucose concentrations in liver independently of the insulin level. Mice with a targeted disruption of the ChREBP gene were recently generated, showing reduced expression of several glycolytic genes [28]. In addition, the mRNA levels of ACC1, ACC2, LCE, malic enzyme, SCD-1, and FAS were significantly lower in the ChREBP$^{-/-}$ mice compared with wild-type mice. Follow-up studies indicated that FAS and ACC are direct targets of ChREBP with a ChRE present in their promoters [29, 30]. These results suggest that ChREBP, apart from its predictable role in glucose utilization, is also of importance for fatty acid biosynthesis.

Knockout mice have also turned out to be an invaluable tool to demonstrate a critical role for C/EBPα in hepatic lipogenesis. Deletion of C/EBPα in combination with leptin deficiency caused diminished lipogenic gene expression and was associated with a significant decrease in hepatic triglyceride content [31]. Development of the fatty liver was exacerbated by high-fat feeding but much less so in liver-specific C/EBPα-null mice. Thus, in addition to its functional role in adipogenesis, C/EBPα stimulates hepatic lipogenesis (reviewed in Ref. [32]).

The liver X receptor (LXR) is a nuclear hormone receptor that is highly expressed in liver and that is activated by oxysterols. It was first shown to play an important role in the feed-forward control of bile acid synthesis from cholesterol by upregulating cholesterol 7-alpha hydroxylase expression. More recent studies indicate that LXRs are also potent stimulators of lipogenesis in liver in mice. Functional LXR response elements were identified in the promoters of the PPARγ, SREBP-1c, and FAS genes [33, 34]. Accordingly, activation of LXRs by synthetic agonists leads to a marked induction of lipogenic genes that is translated into an increase of both plasma triglyceride and phospholipid levels [34]. Studies with LXRs mutant null mice further established LXRs as key factors that mediate the effect of insulin on lipogenesis [35]. In cultured primary hepatocytes insulin increases the half-life of LXRα mRNA, which in turn increases LXRα protein.

While both insulin and LXRα stimulate SREBP-1c protein levels, they do so by different pathways. According to Hegarty *et al.* LXRα activation stimulates the production of the precursor SREBP-1c protein, which, however, is poorly transformed into the mature nuclear form [36]. In contrast, insulin was shown to efficiently and rapidly stimulate the cleavage and maturation of the precursor SREBP-1c form. The molecular explanation for this differential effect may lie with INSIG-2, which is upregulated by LXRα. According to this scenario, INSIG-2 retains the mature form of SREBP-1c in the endoplasmic reticulum, which thus escapes the maturation process.

The nuclear receptor PPARα plays a pivotal role in the adaptive response to fasting by upregulating many genes involved in hepatic fatty acid oxidation, ketogenesis, and gluconeogenesis. In this context, it is remarkable that PPARα upregulates several genes involved in fatty acid elongation and desaturation, including SCD-1, SCD-2, delta-5 desaturase, delta-6 desaturase, LCE, *trans*-2,3-enoyl-CoA reductase (TER) and microsomal 3-keto acyl-CoA reductase (KAR) (reviewed in Refs [37, 38] and unpublished data). The enzymes SCD-1 and SCD-2 catalyze the conversion of stearic acid into oleic acid and accordingly are essential for the synthesis of mono- and polyunsaturated fatty acids. An important question that arises is why a single factor such as PPARα would stimulate both fatty acid oxidation and fatty acid elongation/desaturation. It has been proposed that upregulation of desaturase expression by PPARα may generate unsaturated fatty acids, which are agonists for PPARα. Alternatively, it is possible that the high turnover of plasmatic membrane phospholipids requires a constant level of (unsaturated) free fatty acids to prevent cell death. The story is even more complex, since delta-5, -6, and -9 desaturase expression is also under the transcriptional control of SREBP-1c, which is generally considered to have a function completely opposite to that of PPARα [39].

PPARα mediates the stimulatory effect of polyunsaturated fatty acids (PUFAs) on hepatic fatty acid oxidation. However, PUFAs also inhibit hepatic lipogenesis by a process that is PPARα independent. Studies by a variety of groups have shown that PUFAs suppress lipogenesis by inhibiting expression of SREBP-1c [40–43]. At the molecular level, it has been reported that PUFAs are able to act as competitive LXRα antagonists, which would result in the downregulation of SREBP-1c and other lipogenic genes [44, 45]. However, others have argued that downregulation of SREBP-1c by PUFAs is independent of LXRα [46]. Further studies are required to determine the precise molecular mechanisms responsible for this discrepancy.

12.4
Hepatic Lipogenesis in Steatosis

A major metabolic consequence of obesity is insulin resistance, which is usually accompanied by storage of triglycerides in the liver. It is believed that the excess release of free fatty acids from adipose tissue lipolysis accounts for the triglyceride accumulation in liver. While PPARγ is barely expressed in liver under basal conditions, its expression is markedly increased in animal models of insulin resistance and fatty liver, suggesting a role for PPARγ in hepatic steatosis [47]. Indeed, specific deletion of hepatic PPARγ improves fatty liver in two mouse models of hepatic steatosis (*ob/ob* and AZIP), yet worsens hyperglycemia and insulin resistance [48, 49]. In addition, forced expression of PPARγ1 in liver of PPARα-deficient mice using adenovirus leads to the strong induction of both lipogenic and adipogenic genes (including adipocyte fatty acid-binding protein A (FABP)/aP2, adipsin (see below), LPL, CD36, FAS, SCD-1), indicating transformation of hepatocytes towards adipocyte-like cells [50]. Similarly, forced expression of PPARγ2 in hepatocytes was shown to lead to both hepatic lipid accumulation and induction of lipogenic and adipogenic gene expression [51]. Thus, overexpression of PPARγ appears to be both necessary and sufficient for hepatic steatosis. Thus, in certain pathological conditions, overexpression of PPARγ in liver may contribute to steatosis.

12.5
Nuclear Hormone Receptors and the Control of Adipocytokine Gene Expression

In the past few years it has become clear that fat tissue is not merely a storage depot for excess fat but also is an active endocrine tissue, which secretes a number of biologically active proteins with putative roles in metabolic syndrome. Several proteins secreted by adipocytes that affect satiety and/or energy homeostasis have been identified. These so-called adipocytokines include resistin, adiponectin, acylation-stimulating protein, fasting-induced adipose factor (FIAF), visfatin, leptin, adipsin, plasminogen activator inhibitor-1, renin angiotensin system, metallothioneins, and the inflammatory cytokines interleukin 6, tumor necrosis factor alpha (TNFα), and transforming growth factor beta. These molecules have been linked to a wide range of clinical abnormalities such as obesity, atherosclerosis, insulin resistance, and type 2 diabetes mellitus (reviewed in Ref. [52]). For example, adipose TNFα mRNA levels are increased in several animal models of genetically induced obesity and the same is true for the plasma resistin level. Both proteins seem to play a role in obesity-induced insulin resistance, at least in mice. While resistin and TNFα diminish insulin sensitivity, a large body of evidence in mice and human indicates that adiponectin improves insulin sensitivity. Indeed, plasma levels of adiponectin were found to be decreased in obese patients and were positively correlated with insulin resistance [53, 54].

An adipocytokine that has major effects on triglyceride synthesis in adipocytes is acylation-stimulating protein (ASP) [55]. The uptake and re-esterification of non-esterified fatty acids by adipose tissue is enhanced by ASP, probably by stimulating the activity of diacylglycerol acyltransferase (DGAT), a key enzyme in triglyceride synthesis. Moreover, ASP stimulates glucose uptake and transport in human adipocytes, thereby providing the substrates for further triglyceride synthesis.

An interesting newly (re)discovered adipocytokine is visfatin [56]. Injection of recombinant visfatin was shown to decrease blood glucose in insulin-resistant or -deficient mice. Furthermore, like insulin, visfatin stimulated glucose uptake by cultured adipocyte and muscle cells and decreased glucose output by hepatocytes. In adipocytes, visfatin was found to stimulate triglyceride synthesis and accumulation to a similar extent as insulin. The precise molecular mechanism behind these observations is unclear but is likely linked to the insulin pathway, since visfatin was shown to bind the insulin receptor.

Thiazolidinediones (TZDs), which are insulin-sensitizing drugs used in the treatment of type 2 diabetes, influence the expression of many of these adipocytokines both positively and negatively. Leptin, TNFα, and resistin are negative targets of TZDs [57]. Resistin might mediate the effect of TZD on insulin sensitivity, at least in mice. Adiponectin, in contrast, is a direct positive target gene of PPARγ [58], which provides a molecular explanation for the induction of adiponectin gene expression by TZDs. FIAF/ANGPTL4 is also a direct positive target of PPARγ, expression of which is upregulated by TZDs in both mouse and human adipocytes [59]. Visfatin might be regulated by PPARγ ligands as well, although this remains to be demonstrated. The possible stimulatory role of PPARγ on visfatin gene expression would be of interest in the context of type 2 diabetes, since recombinant visfatin protein was shown to promote glucose uptake by adipocytes and to suppress glucose release by hepatocytes [56]. In summary, PPARγ not only has a role in adipose lipogenesis and adipogenesis but also controls different endocrine pathways that are critical for whole-body energy homeostasis.

12.6
Targeting Lipogenesis for Obesity?

In principle, the process of lipogenesis appears as an attractive target for the treatment of obesity. However, two major issues may impose restrictions on the applicability of lipogenesis inhibitors in obesity management. First, it has long been believed that conversion of carbohydrates and amino acids to fatty acids is of minor significance in human individuals consuming a regular diet that is high in fat, which is in contrast to the situation in mice. According to this notion, inhibiting an enzyme such as FAS in liver and adipose tissue is unlikely to have major consequences in humans. However, recent studies suggests that *de novo* lipogenesis is more relevant than previously estimated and is especially important in patients with non-alcoholic fatty liver disease [60]. In normal subjects, *de novo* lipogenesis becomes much more important postprandially than in the fasted state [61]. These novel data suggest that lipogenesis may be an attractive target for obesity management after all. The second issue that questions the usefulness of targeting lipogenesis is that inhibition of triglyceride synthesis in fat tissue, without concomitantly activating energy expenditure, will merely cause a redistribution of fat storage from adipose tissue to other tissues such as liver and pancreas, a process which is now considered highly undesirable. Indeed, it seems that the effectiveness of thiazolidinediones toward improving insulin resistance is positively correlated with their effect on subcutaneous fat gain, suggesting that at least for diabetes management strategies should promote preferential storage of fat in subcutaneous stores.

12.7
Conclusions and Perspectives

In conclusion, lipogenesis is tightly controlled by nutritional, hormonal, and transcription factors. In the last few years, major advances have been made to identify the complex regulatory networks involved in the regulation of lipogenesis. One of the most pressing challenges that researchers will have to deal with is to extend their findings from rodents to humans. Finally, although lipogenesis may appear as an attractive target for the pharmacological treatment and dietary prevention of obesity, it has become increasingly clear that inhibiting triglyceride storage in subcutaneous fat tissue is more likely to do harm than any good.

Acknowledgments

The work of the authors is supported by the Netherlands Organization for Scientific Research (NWO), the Dutch Diabetes Foundation, the Wageningen Centre for Food Sciences (WCFS), and the Royal Netherlands Academy of Arts and Sciences. The authors would like to thank Rinke Stienstra for critical reading of the manuscript.

References

1 P. Tontonoz, J.B. Kim, R.A. Graves, B.M. Spiegelman (**1993**) *Mol. Cell. Biol.* 13, 4753–4759.

2 J.D. Horton, I. Shimomura (**1999**) *Curr. Opin. Lipidol.* 10, 143–150.

3 L. Fajas, K. Schoonjans, L. Gelman, J.B. Kim, J. Najib, G. Martin, J.C. Fruchart, M. Briggs, B.M. Spiegelman, J. Auwerx (**1999**) *Mol. Cell. Biol.* 19, 5495–5503.

4 J.B. Kim, H.M. Wright, M. Wright, B.M. Spiegelman (**1998**) *Proc. Natl Acad. Sci. USA* 9, 4333–4337.

5 S. Le Lay, I. Lefrere, C. Trautwein, I. Dugail, S. Krief (**2002**) *J. Biol. Chem.* 277, 35625–35634.

6 H.R. Kast-Woelbern, S.L. Dana, R.M. Cesario, L. Sun, L.Y. de Grandpre, M.E. Brooks, D.L. Osburn, A. Reifel-Miller, K.

Klausing, M.D. Leibowitz (**2004**) *J. Biol. Chem.* 279, 23908–23915.

7 T. Yang, P.J. Espenshade, M.E. Wright, D. Yabe, Y. Gong, R. Aebersold, J.L. Goldstein, M.S. Brown (**2002**) *Cell.* 110, 489–500.

8 D. Yabe, M.S. Brown, J.L. Goldstein (**2002**) *Proc. Natl Acad. Sci. USA* 99, 12753–12758.

9 J. Li, K. Takaishi, W. Cook, S.K. McCorkle, R.H. Unger (**2003**) *Proc. Natl Acad. Sci. USA* 100, 9476–9481.

10 E.D. Rosen, B.M. Spiegelman (**2001**) *J. Biol. Chem.* 276, 37731–37734.

11 T. Helledie, Grontved, S.S. Jensen, P. Kiilerich, L. Rietveld, T. Albrektsen, M.S. Boysen, J. Nohr, L.K. Larsen, J. Fleckner, H.G. Stunnenberg, K. Kristiansen, S.

MANDRUP (**2002**) *J. Biol. Chem.* 277, 26821–26830.

12 N. KUBOTA, Y. TERAUCHI, H. MIKI, H. TAMEMOTO, T. YAMAUCHI, K. KOMEDA, S. SATOH, R. NAKANO, C. ISHII, T. SUGIYAMA, K. ETO, Y. TSUBAMOTO, A. OKUNO, K. MURAKAMI, H. SEKIHARA, G. HASEGAWA, M. NAITO, Y. TOYOSHIMA, S. TANAKA, K. SHIOTA, T. KITAMURA, T. FUJITA, O. EZAKI, S. AIZAWA, T. KADOWAKI et al. (**1999**) *Mol. Cell.* 4, 597–609.

13 P.D. MILES, Y. BARAK, W. HE, R.M. EVANS, J.M. OLEFSKY (**2000**) *J. Clin. Invest.* 105, 287–292.

14 W. HE, Y. BARAK, A. HEVENER, P. OLSON, D. LIAO, J. LE, M. NELSON, E. ONG, J.M. OLEFSKY, R.M. EVANS (**2003**) *Proc. Natl Acad. Sci. USA* 100, 15712–15717.

15 J. ZHANG, M. FU, T. CUI, C. XIONG, K. XU, W. ZHONG, Y. XIAO, D. FLOYD, J. LIANG, E. LI, Q. SONG, Y.E. CHEN (**2004**) *Proc. Natl Acad. Sci. USA* 101, 10703–10708.

16 T. IMAI, R. TAKAKUWA, S. MARCHAND, E. DENTZ, J.M. BORNERT, N. MESSADDEQ, O. WENDLING, M. MARK, B. DESVERGNE, W. WAHLI, P. CHAMBON, D. METZGER (**2004**) *Proc. Natl Acad. Sci. USA* 101, 4543–4547.

17 J. LUO, R. SLADEK, J. CARRIER, J.A. BADER, D. RICHARD, V. GIGUERE (**2003**) *Mol. Cell. Biol.* 23, 7947–7956.

18 M. CASADO, V.S. VALLET, A. KAHN, S. VAULONT (**1999**) *J. Biol. Chem.* 274, 2009–2013.

19 I. SHIMOMURA, Y. BASHMAKOV, J.D. HORTON (**1999**) *J. Biol. Chem.* 274, 30028–30032.

20 M.K. BENNETT, J.M. LOPEZ, H.B. SANCHEZ, T.F. OSBORNE (**1995**) *J. Biol. Chem.* 270, 25578–25583.

21 M.M. MAGANA, S.S. LIN, K.A. DOOLEY, T.F. OSBORNE (**1997**) *J. Lipid Res.* 38, 1630–1638.

22 R. SATO, A. OKAMOTO, J. INOUE, W. MIYAMOTO, Y. SAKAI, N. EMOTO, H. SHIMANO, M. MAEDA (**2000**) *J. Biol. Chem.* 275, 12497–12502.

23 J. ERICSSON, S.M. JACKSON, J.B. KIM, B.M. SPIEGELMAN, P.A. EDWARDS (**1997**) *J. Biol. Chem.* 272, 7298–7305.

24 S.H. KOO, A.K. DUTCHER, H.C. TOWLE (**2001**) *J. Biol. Chem.* 276, 9437–9445.

25 D.E. TABOR, J.B. KIM, B.M. SPIEGELMAN, P.A. EDWARDS (**1999**) *J. Biol. Chem.* 274, 20603–20610.

26 M.B. SANDBERG, M. BLOKSGAARD, D. DURAN-SANDOVAL, C. DUVAL, B. STAELS, S. MANDRUP (**2005**) *J. Biol. Chem.* 280, 5258–5266.

27 H. YAMASHITA, M. TAKENOSHITA, M. SAKURAI, R.K. BRUICK, W.J. HENZEL, W. SHILLINGLAW, D. ARNOT, K. UYEDA (**2001**) *Proc. Natl Acad. Sci. USA* 98, 9116–9121.

28 K. IIZUKA, R.K. BRUICK, G. LIANG, J.D. HORTON, K. UYEDA (**2004**) *Proc. Natl Acad. Sci. USA* 101, 7281–7286.

29 S. ISHII, K. IIZUKA, B.C. MILLER, K. UYEDA (**2004**) *Proc. Natl Acad. Sci. USA* 101, 15597–15602.

30 H.M. SHIH, H.C. TOWLE (**1992**) *J. Biol. Chem.* 267, 13222–13228.

31 K. MATSUSUE, O. GAVRILOVA, G. LAMBERT, H.B. BREWER, JR., J.M. WARD, Y. INOUE, D. LEROITH, F.J. GONZALEZ (**2004**) *Mol. Endocrinol.* 18, 2751–2764.

32 E.D. ROSEN, C.J. WALKEY, P. PUIGSERVER, B.M. SPIEGELMAN (**2000**) *Genes Dev.* 14, 1293–1307.

33 J.J. REPA, G. LIANG, J. OU, Y. BASHMAKOV, J.M. LOBACCARO, I. SHIMOMURA, B. SHAN, M.S. BROWN, J.L. GOLDSTEIN, D.J. MANGELSDORF (**2000**) *Genes Dev.* 14, 2819–2830.

34 J.R. SCHULTZ, H. TU, A. LUK, J.J. REPA, J.C. MEDINA, L. LI, S. SCHWENDNER, S. WANG, M. THOOLEN, D.J. MANGELSDORF, K.D. LUSTIG, B. SHAN (**2000**) *Genes Dev.* 14, 2831–2838.

35 K.A. TOBIN, S.M. ULVEN, G.U. SCHUSTER, H.H. STEINEGER, S.M. ANDRESEN, J.A. GUSTAFSSON, H.I. NEBB (**2002**) *J. Biol. Chem.* 277, 10691–10697.

36 B.D. HEGARTY, A. BOBARD, I. HAINAULT, P. FERRE, P. BOSSARD, F. FOUFELLE (**2005**) *Proc. Natl Acad. Sci. USA* 102, 791–796.

37 S. MANDARD, M. MULLER, S. KERSTEN (**2004**) *Cell. Mol. Life Sci.* 61, 393–416.

38 S. KERSTEN, J. SEYDOUX, J.M. PETERS, F.J. GONZALEZ, B. DESVERGNE, W.

WAHLI (**1999**) *J. Clin. Invest.* 103, 1489–1498.

39 T. MATSUZAKA, H. SHIMANO, N. YAHAGI, M. AMEMIYA-KUDO, T. YOSHIKAWA, A.H. HASTY, Y. TAMURA, J. OSUGA, H. OKAZAKI, Y. IIZUKA, A. TAKAHASHI, H. SONE, T. GOTODA, S. ISHIBASHI, N. YAMADA (**2002**) *J. Lipid Res.* 43, 107–114.

40 M. SEKIYA, N. YAHAGI, T. MATSUZAKA, Y. NAJIMA, M. NAKAKUKI, R. NAGAI, S. ISHIBASHI, J. OSUGA, N. YAMADA, H. SHIMANO (**2003**) *Hepatology* 38, 1529–1539.

41 J. XU, M.T. NAKAMURA, H.P. CHO, S.D. CLARKE (**1999**) *J. Biol. Chem.* 274, 23577–23583.

42 N. YAHAGI, H. SHIMANO, A.H. HASTY, M. AMEMIYA-KUDO, H. OKAZAKI, Y. TAMURA, Y. IIZUKA, F. SHIONOIRI, K. OHASHI, J. OSUGA, K. HARADA, T. GOTODA, R. NAGAI, S. ISHIBASHI, N. YAMADA (**1999**) *J. Biol. Chem.* 274, 35840–35844.

43 S.D. CLARKE, D.B. JUMP (**1994**) *Annu. Rev. Nutr.* 14, 83–98.

44 J. OU, H. TU, B. SHAN, A. LUK, R.A. DEBOSE-BOYD, Y. BASHMAKOV, J.L. GOLDSTEIN, M.S. BROWN (**2001**) *Proc. Natl Acad. Sci. USA* 98, 6027–6032.

45 T. YOSHIKAWA, H. SHIMANO, N. YAHAGI, T. IDE, M. AMEMIYA-KUDO, T. MATSUZAKA, M. NAKAKUKI, S. TOMITA, H. OKAZAKI, Y. TAMURA, Y. IIZUKA, K. OHASHI, A. TAKAHASHI, H. SONE, J. OSUGA JI, T. GOTODA, S. ISHIBASHI, N. YAMADA (**2002**) *J. Biol. Chem.* 277, 1705–1711.

46 A. PAWAR, D. BOTOLIN, D.J. MANGELSDORF, D.B. JUMP (**2003**) *J. Biol. Chem.* 278, 40736–40743.

47 L. CHAO, B. MARCUS-SAMUELS, M.M. MASON, J. MOITRA, C. VINSON, E. ARIOGLU, O. GAVRILOVA, and M.L. REITMAN (**2000**) *J. Clin. Invest.* 106, 1221–1228.

48 K. MATSUSUE, M. HALUZIK, G. LAMBERT, S.H. YIM, O. GAVRILOVA, J.M. WARD, B. BREWER, JR., M.L. REITMAN, F.J. GONZALEZ (**2003**) *J. Clin. Invest.* 111, 737–747.

49 O. GAVRILOVA, M. HALUZIK, K. MATSUSUE, J.J. CUTSON, L. JOHNSON, K.R. DIETZ, C.J. NICOL, C. VINSON, F.J. GONZALEZ, M.L. REITMAN (**2003**) *J. Biol. Chem.* 278, 34268–34276.

50 S. YU, K. MATSUSUE, P. KASHIREDDY, W.Q. CAO, V. YELDANDI, A.V. YELDANDI, M.S. RAO, F.J. GONZALEZ, J.K. REDDY (**2003**) *J. Biol. Chem.* 278, 498–505.

51 S.E. SCHADINGER, N.L. BUCHER, B.M. SCHREIBER, S.R. FARMER (**2005**) *Am. J. Physiol. Endocrinol. Metab.* 288, E1195–1205.

52 M. FARAJ, H.L. LU, K. CIANFLONE (**2004**) *Biochem. Cell. Biol.* 82, 170–190.

53 A.S. LIHN, S.B. PEDERSEN, B. RICHELSEN (**2005**) *Obes. Rev.* 6, 13–21.

54 M.E. TRUJILLO, P.E. SCHERER (**2005**) *J. Intern. Med.* 257, 167–175.

55 K. CIANFLONE, Z. XIA, L.Y. CHEN (**2003**) *Biochim. Biophys. Acta* 1609, 127–143.

56 A. FUKUHARA, M. MATSUDA, M. NISHIZAWA, K. SEGAWA, M. TANAKA, K. KISHIMOTO, Y. MATSUKI, M. MURAKAMI, T. ICHISAKA, H. MURAKAMI, E. WATANABE, T. TAKAGI, M. AKIYOSHI, T. OHTSUBO, S. KIHARA, S. YAMASHITA, M. MAKISHIMA, T. FUNAHASHI, S. YAMANAKA, R. HIRAMATSU, Y. MATSUZAWA, I. SHIMOMURA (**2005**) *Science* 307, 426–430.

57 C.M. STEPPAN, S.T. BAILEY, S. BHAT, E.J. BROWN, R.R. BANERJEE, C.M. WRIGHT, H.R. PATEL, R.S. AHIMA, M.A. LAZAR (**2001**) *Nature* 409, 307–312.

58 M. IWAKI, M. MATSUDA, N. MAEDA, T. FUNAHASHI, Y. MATSUZAWA, M. MAKISHIMA, I. SHIMOMURA (**2003**) *Diabetes* 52, 1655–1663.

59 S. MANDARD, F. ZANDBERGEN, N.S. TAN, P. ESCHER, D. PATSOURIS, W. KOENIG, R. KLEEMANN, A. BAKKER, F. VEENMAN, W. WAHLI, M. MULLER, S. KERSTEN (**2004**) *J. Biol. Chem.* 279, 4411–34420.

60 K.L. DONNELLY, C.I. SMITH, S.J. SCHWARZENBERG, J. JESSURUN, M.D. BOLDT, E.J. PARKS (**2005**) *J. Clin. Invest.* 115, 1343–1351.

61 M.T. TIMLIN, E.J. PARKS (**2005**) *Am. J. Clin. Nutr.* 81, 35–42.

Part III
Nutrigenetics (Nutrient–Genotype Interactions)

13

The Genetics of Type 2 Diabetes

Catherine Stevenson, Inês Barroso, and Nicholas Wareham

13.1
Evidence for Genetic Factors

This chapter describes the evidence supporting the notion that type 2 diabetes is caused, at least in part, by genetic factors and reviews the ever-increasing volume of literature that has been produced from studies aimed at identifying what those genetic factors are. The pace of development in this field is such that only real-time reviews can capture the current state-of-the-art. However, we review published findings as at the beginning of 2005 and present some pointers as to how this field might develop in the future and where further advances are likely.

Overall there is considerable consensus that lifestyle factors such as diet and physical inactivity are critical to the development of type 2 diabetes, but that genetic factors clearly play a role in underlying susceptibility to the disease. Evidence for this comes from several sources: differences in disease prevalence between ethnic groups, segregation of the disease within families, and twin studies. This is supported by the identification of genes and chromosome regions contributing to diabetes risk.

13.1.1
Differences between Ethnic Groups

The prevalence of type 2 diabetes varies widely around the world, from the very low levels of about 1% seen in populations such as tribes of Mapuche Indian or the mainland Chinese population, to extremely high levels of over 50% in the Nauru and the Pima Indians in Arizona (Fig. 13.1). Part of the variability observed is clearly due to differences in environmental factors, but studies in admixed populations, where interbreeding has occurred between distinct ethnic groups, have provided additional evidence that ethnic-specific prevalence has a genetic basis. The prevalence of type 2 diabetes in the Pima Indians is inversely related to the extent of interbreeding with the European Americans [2]. Similarly, in studies of Pacific Islanders the prevalence of type 2 diabetes is higher in full-blooded Nauruans than in those with admixture [3].

Nutritional Genomics. Edited by Regina Brigelius-Flohé and Hans-Georg Joost
Copyright © 2006 WILEY-VCH Verlag GmbH & Co. KGaA, Weinheim
ISBN: 3-527-31294-3

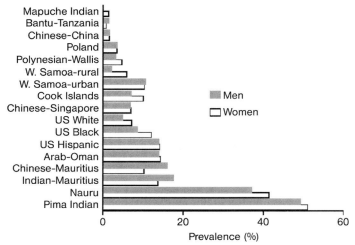

Fig. 13.1. Prevalence of diabetes in different countries and various ethnic groups. Adapted from Ref. [1].

The prevalence of type 2 diabetes varies between ethnic groups living in similar environments, which may suggest an unequal distribution of susceptibility genes. In the United States, the prevalence of diabetes varies considerably, being lowest in individuals of European ancestry (5%), 10% in African-Americans, 16% in Cubans, 24% in Mexican-Americans, 26% in Puerto Ricans, and over 50% in some Native American tribes such as the Pima Indians of Arizona [4]. The "thrifty genotype" hypothesis [5] provides one possible explanation for the very high prevalence of obesity and type 2 diabetes observed in the American Pima Indians, Australian Aborigines, and Pacific Islanders. The basis for the susceptibility to diabetes and obesity could be the result of an evolutionary advantageous thrifty genotype that promoted fat deposition and storage of calories in times of plenty, allowing survival in times of fast. With Westernization, these populations gained access to excess food, in particular, foods rich in fat and processed carbohydrates. This, accompanied by a move away from traditional ways of life and the acquisition of sedentary lifestyles, may have turned the previously advantageous metabolic profile into a disadvantage [6].

13.1.2
Segregation in Families

Family studies compare the degree of aggregation of the disease in relatives of probands with expected rates from the general population. In type 2 diabetes, it has been estimated that individuals who have a first-degree relative with the disease have a 3.5-fold greater risk of developing diabetes than those without [7].

This familial aggregation has been conclusively demonstrated in populations from North America [8–11], South America [12], Europe [13, 14], Asia [15, 16], Africa [17], and Oceania [18]. Younger-age onset diabetes appears to be more familial than diabetes that occurs at older ages [19–22]. Among the Pima Indians, the prevalence of diabetes is highest amongst individuals whose parents both developed the disease before the age of 45, while it is lowest in people with non-diabetic parents [22]. Similar findings have been reported in southern Asian Indians by Viswanathan *et al.* [16]. The lifetime risk of developing type 2 diabetes is approximately 40% in offspring of diabetic parents [23]. If both parents had type 2 diabetes, the risk may be as high as 70% [23]. In the Framingham Offspring Study, the risk of type 2 diabetes in the offspring was 3.4- to 3.5-fold higher if either parent had diabetes, but was 6.1-fold if both parents were affected [24].

13.1.3
Parent-specific Effects

Parent-specific effects provide further evidence for genetic contributions. It has been demonstrated that the risk of diabetes associated with low birth weight is strongly related to paternal diabetes [25]. Generally, however, there is an excess of maternal transmission of diabetes, although in most populations the difference is small [9, 21, 26, 27]. It is not known to what extent this excess maternal transmission is due to mitochondrial variants or genomic imprinting, and to what extent it can be explained by non-genetic factors such as *in utero* exposure to maternal diabetes.

Mitochondria are subcellular organelles that provide energy in the form of ATP. They are inherited solely from the mother and have a semi-autonomous genetic system which is partly under the control of the cell nucleus. Variation in the mitochondrial DNA (mtDNA) has been identified as the cause of several disorders including rare syndromic forms of diabetes, often associated with hearing loss [28–31]. Type 2 diabetes has been linked to over 40 mitochondrial variants and mutations, including a common T16189C variant [32], an adenine-to-guanine mutation in the gene coding for the leucine transfer RNA [33], and a reduction in mtDNA copy number [33].

However, not all transmission of diabetes from mother to child is by genetic means. Environmental exposure to diabetes *in utero* is known to increase the risk of developing diabetes. Among Pima Indians the prevalence of diabetes is higher in individuals whose mothers had diabetes during the pregnancy than in those whose mothers developed diabetes after giving birth [34], and individuals born after the mother developed diabetes were at higher risk of developing the disease themselves than the siblings who were born before the mother developed diabetes. The same effect is not seen with the onset of paternal diabetes [35].

Matters are further complicated by survival bias. This is due to the fact that women, on average, live longer than men, and so are more likely at some point in their lives to develop a late-onset disease such as type 2 diabetes. Questioning adults about whether their parents ever developed diabetes in order to determine disease inheritance patterns may therefore result in a natural excess of maternal diabetes.

Table 13.1. Studies on the concordance of type 2 diabetes in twins

Reference	Country	Ascertainment	MZ concordance	DZ concordance	P-value
Barnett *et al.*, 1981 [40]	UK	Proband	0.91	–	–
Newman *et al.*, 1987[a] [41]	USA	Population-based	0.41	0.10	<0.01
Japan Diabetes Society, 1988 [42]	Japan	Proband	0.83	0.40	<0.01
Kaprio *et al.*, 1992 [43]	Finland	Population-based	0.20	0.09	<0.01
Matsuda and Kuzuya 1994 [44]	Japan	Proband	0.87	0.43	<0.01
Poulsen *et al.*, 1999 [45]	Denmark	Population-based	0.33	0.23	0.40

P-values are for the null hypothesis that MZ and DZ concordance rates
are equal using Fisher's exact test.
[a] Concordance rates for a second examination were reported.

13.1.4
Twin Studies

Although there is substantial evidence of aggregation of type 2 diabetes in families, this alone is not conclusive evidence that there is a genetic component to the disease. Families share many other factors including environments, culture, and habits. In twin studies, the extent to which familial aggregation of disease can be accounted for by inherited genetic factors can be assessed. In these studies the concordance rates for the disease under investigation are compared in monozygotic (MZ) and dizygotic (DZ) twins. MZ twins are genetically identical while DZ twins share on average only half their genes. As both types of twins have shared environmental factors, increased concordance rates for disease in MZ twins compared with DZ twins are indicative of genetic factors. Results from twin studies in type 2 diabetes are summarized in Table 13.1. Concordance rates were found to be higher in MZ twins than in DZ, and this difference is statistically significant in all studies except for the Danish study. This study was too small, however, to allow for precise estimation of genetic predisposition.

Despite the consistency of the results, there may be valid additional interpretations. The higher concordance observed in MZ twins could also reflect a correlation of intrauterine environment rather than just increased sharing of genotypes, as MZ twins frequently share a common placenta while DZ twins do not [36]. It might also be influenced by increased sharing of environmental risk factors postnatally, since identical twins behave much more similarly than non-identical twins [37]. Whether this concordance in behavior is itself genetically determined, or is due to factors such as gender and outward appearance, still remains to be investigated.

Type 2 diabetes is a disease with variable age of onset but most study designs have not taken this into consideration. It is possible that many discordant twin pairs would be concordant for disease later on in life. In one study 76% of MZ twins that were initially selected as being discordant for disease became concordant after a 15-year follow-up [38]. Ghosh and Schork suggest that the age-adjusted concordance rate in MZ twins may be as high as 70–80% [39].

13.2
Approaches to Gene Identification

There are two main approaches used to identify disease-causing genes: linkage analysis and association studies. Linkage analysis identifies regions of the genome inherited with the disease by mapping the inheritance patterns of genetic markers in families. Until recently this approach was the main route to the discovery of new genes that could plausibly be linked to disease. Association studies quantify the relationship of variants with disease by investigating whether specific genetic variants are more common in people with the disease than in those without. These studies tend to be carried out on unrelated populations, but are sometimes undertaken within families. Most association studies so far have focused on candidate genes, where involvement in the disease process is hypothesized from prior knowledge, but there is currently a shift to broaden this approach into whole genome association studies. These studies test variants throughout the genome, regardless of biological plausibility. This change in focus is largely being led by advances in technology, as previously it was simply not feasible to type a sufficiently large number of variants in substantial populations.

13.2.1
Linkage Analysis

Linkage studies take advantage of recombination – the exchange of segments of DNA between homologous chromosomes during gamete production [46, 47]. This process means that the further apart two loci are on a chromosome, the less likely they are to be inherited together. Family members are typed for many different genetic markers, usually highly polymorphic microsatellites, which are simply used to identify physical locations in the genome without any assumption of involvement in the disease process. The inheritance patterns of the markers are compared with that of the disease. This is quantified by calculating a score for each marker that describes the likelihood that it is co-inherited with the disease locus, the logarithm of odds (LOD) score. A LOD greater than 2.2 is considered suggestive evidence for linkage, while 3.6 is currently recognized as the threshold for significance [48].

Early studies focused on regions identified as likely candidates for diabetogenic loci, but most groups now use markers spanning the entire genome at regular intervals (usually 5–10 cM). Once a given chromosomal interval has been identified as being inherited with the disease, linkage disequilibrium studies or fine mapping must follow. These techniques are required to narrow the region in order to identify the underlying gene and mutation as the initial linkage interval is usually large and will contain many genes.

One of the advantages of this type of study is that they are hypothesis-free and allow for the discovery of new, previously unthought-of, disease genes, and so can lead to the identification of novel pathways and disease mechanisms. The only requirement is that the genetic locus has a sufficient impact on the disease susceptibility to be detectable through an observed increase in segregation of the chromosomal region among family members with type 2 diabetes relative to that expected from chance alone.

There are two types of linkage study: parametric and non-parametric.

13.2.1.1 Parametric Studies

Parametric studies build a theoretical model of how the disease segregates from estimates of factors – or parameters – which affect the inheritance, such as the mode of inheritance, allele frequencies, and penetrance. The pedigree under scrutiny is then analyzed to see whether any of the polymorphic markers are inherited in the pattern described by the model. If they do, this is taken as evidence that the disease-causing gene is physically close to that marker. This has historically worked well for genetically simple diseases such as Huntingdon's, but is not a robust method for analysing complex diseases as inaccuracies in estimating parameters can considerably affect the results. In the case of type 2 diabetes, these factors are almost impossible to measure precisely.

Early linkage studies in type 2 diabetes used the parametric methods and few genetic markers available at the time. These studies used large multigenerational families and were based on the improbable assumption that all illness was caused by one or a very small number of genes, each with large effects. Problems with a parametric analysis of diabetes have been largely caused by the following factors:

- Lack of a Mendelian inheritance pattern. This means that at the outset it is impossible to estimate the parameters required for this type of linkage analysis, as discussed above.
- The high mean age at diagnosis of 60 years. It is very difficult to ascertain multigenerational families for study as often parents are no longer available for study, while offspring may not yet have developed the disease.
- Only affected subjects can be used for linkage studies. This is due to the variable age of onset and age-related reduced penetrance, and so unaffected subjects in a family provide only limited information since their disease status cannot be clearly known.
- The presence of many phenocopies. Mutations or polymorphisms in any one of several genes may result in an identical phenotype and any given contributing chromosomal region may co-segregate with the disease in only a subset of the

families being studied. This reduces the signal from linkage and reduces power to find the predisposing gene locations. In addition, phenocopies due to environmental causes are not uncommon, further complicating the analysis.

For these reasons, the development of non-parametric methods was a crucial breakthrough in the genetic analysis of diabetes.

13.2.1.2 **Non-parametric Studies**

Most of the more recent genome scans in type 2 diabetes have tended to employ non-parametric, or "model-free" methods, which do not require any knowledge of inheritance mechanisms. The most common non-parametric method used in mapping studies is the affected sib-pair analysis (ASP), as it only requires pairs of affected siblings [49–52]. The basis of ASP analysis is the theory that siblings concordant for the disease should share marker alleles that are linked to the disease more often than expected by chance. If the marker is not linked to the disease locus, then it would be expected to be shared on average 50% of the time. Polymorphic markers are typed throughout the genome as described above, and allele sharing between affected siblings is analyzed at each locus using software packages such as GENEHUNTER. Above-average sharing of a marker allele should, in theory, be an indication of linkage to a disease locus.

13.2.1.3 **Results of Linkage Studies**

Many genome scans for type 2 diabetes have been carried out in a range of ethnic groups (recently reviewed in Refs [53, 54]), and numerous chromosome regions have demonstrated evidence suggestive of linkage. However, only a few regions have reached the classical threshold for significance, and fewer still have been replicated in multiple studies. Linkage studies in type 2 diabetes have been plagued by the same problems encountered in other complex diseases, namely a lack of replication of peaks of linkage, and difficulty in identifying the underlying genes. This is likely to be largely due to differences in the etiology of diabetes between sampled groups, further complicated by variations in study design and analytical technique.

Despite these challenges, progress has been made in identifying promising regions of linkage. Those that are most relevant are discussed below, and a summary of current results is given in Table 13.2.

Chromosome 2q: calpain 10   The identification of calpain 10 as a susceptibility gene for polygenic type 2 diabetes has been hailed as proof-of-concept for the linkage analysis approach. The locus on chromosome 2q was mapped by a genome scan in a Mexican-American population and labelled NIDDM1 by Hanis *et al.* in 1996 [55]. Four years later, Horikawa *et al.* identified the source of the linkage as the gene calpain 10 (*CAPN10*), a ubiquitously expressed cysteine protease, and an entirely novel candidate gene for diabetes [56]. Subsequent studies in Old Order Amish [57], French [58], and Indo-Mauritian [59] populations have provided further evidence of linkage at the locus. A meta-analysis supported these findings

Table 13.2. Summary of results from whole-genome scans for type 2 diabetes and abnormal glucose homeostasis

Chromosome region	LOD/P-value	Population	Analysis method	Marker with highest LOD score	Flanking marker (left)	Flanking marker (right)	Reference
1p36-32	1.53	Japanese	Affected siblings	D1S199			83
1q21-23	4.3	Caucasian American	Parametric		CRP	APOA2/D1S196	67
1q21-23	2.5	Pima Indian	Haseman-Elston	D1S1677	D1S2125	D1S2127	64
1q21-24	3.04	Caucasian French	Non-parametric	APOA2/D1S484	D1S2635 & D1S398	D1S2768 & D1S2844	69
1q24-25	1.98	Caucasian UK	Non-parametric		D1S2799	D1S452	68
1q42.2	2.38	Caucasian Finn and Swedish	Affected siblings	D1S3462			84
1q44	2.14	Indo-Mauritian	Affected siblings	D1S2836			59
2p11	2.0	Caucasian Finn	Non-parametric	D2S1777			85
2p25	2.02	Caucasian French	Non-parametric	D2S319			69
2p21-16	2.27	Caucasian French	Non-parametric	D2S2259/D2S391			69
2q24-23	1.22	Caucasian French	Non-parametric	D2S2330			69
2q24.3-31.1	2.97	Indigenous Australian	Parametric	D2S2345	D2S2330	D2S335	86
2q34	1.68	Japanese	Affected siblings	D2S325			83
2q36	1.68	Japanese	Affected siblings	D2S396			83
2q37	4.3	Mexican American	Affected siblings	D2S125	D2S2285	D2S125	55

Location	LOD	Population	Method	Marker 1	Marker 2	Marker 3	%
3p24.1	3.91	GENNID (Mexican-American subgroup)	Non-parametric	D3S2432			75
3p24–22	2.2	Caucasian Finn	Non-parametric		D3S1561	D3S1768	85
3p14	1.73	Japanese	Affected siblings	D3S1566			83
3q22	2.06	Indo-Mauritian	Affected siblings	D3S1292			59
3q26–28	1.38	Japanese	Affected siblings	D3S1565			83
3q27	2.13	Indo-Mauritian	Affected siblings		D3S1571	D3S3686	59
3q27–qter	4.67	Caucasian French	Non-parametric	D3S1580	D3S1262	D3S1601	69
3q27–qter	3.3	Caucasian French	Non-parametric		D3S1565	D3S1262	69
3q29	1.8	Indigenous Australian	Parametric	D3S1311			86
4p16	1.35	Caucasian French	Non-parametric	D4S2935/D4S412			69
4q21	1.4	Caucasian Finnish and Swedish	Affected siblings	D4S2361			84
4q31	2.09	Caucasian French	Non-parametric	D4S1539			69
4q31	1.63	Japanese	Affected siblings	D4S415			83
4q32–33	2.5	Caucasian Finn	Non-parametric		D4S3015	D4S2951	85
4q34.1–34.3	2.41	Ashkenazi Jew	Non-parametric	D4S1501	D4S1539	D4S2967	81
5q13	1.22	Caucasian UK	Non-parametric	D5S647			68
5q13.3	2.8	GENNID (Caucasian subgroup)	Non-parametric	D5S1404			75
5q31.1	2.41	Caucasian Finn and Swedish	Affected siblings	D5S816			84
5q32	1.22	Caucasian UK	Non-parametric	D5S436			68

Table 13.2 *(continued)*

Chromosome region	LOD/P-value	Population	Analysis method	Marker with highest LOD score	Flanking marker (left)	Flanking marker (right)	Reference
5q31–33	1.67	Caucasian French	Non-parametric	D5S410/D5S436			69
5q34–35.2	2.9	Icelandic	Non-parametric	D5S625			87
6p23	1.52	Japanese	Affected siblings	D6S289			83
6q23.3–24.2	1.39	Pima Indian	Variance components		D6S1009	D6S1003	64
6q24.1	1.8	Icelandic	Non-parametric	D6S1569			87
7p22–21	3.51	Japanese	Affected siblings	D7S517			83
7p15.3	1.31	Caucasian UK	Non-parametric	D7S493			68
7q11	1.68	Caucasian French	Non-parametric	D7S669			69
7q22	1.8	Pima Indians	Affected siblings	D7S1799	D7S479	D7S1822	64
8p22	1.77	Indigenous Australian	Parametric	D8S549			86
8p22–21	2.55	Caucasian UK	Non-parametric	D8S258	D8S549	D8S258	68
8q12.1	1.66	Ashkenazi Jew	Non-parametric	D8S593			81
8q23	2.55	Indo-Mauritian	Affected siblings	D8S1784			59
8q24.2	1.41	Caucasian UK	Non-parametric		D8S284	D8S272	68
9p21	3.29	Chinese Han	Non-parametric	D9S171			82

9p13–21	3.9	Caucasian Finn	Non-parametric	D9S166/D9S301	D9S1874	D9S153	85
9q21.13	2.94	Chinese Han	Non-parametric	D9S175			82
9q31.2	1.22	Pima Indian	Variance components		D9S299	D9S2026	64
9q34	1.22	Caucasian French	Non-parametric	D9S158			69
10p13	2.39	GENNID (African-American subgroup)	Non-parametric	D10S1412			75
10q23.3	1.99	Caucasian UK	Non-parametric	D10S1765	D10S1765	D10S185	68
10q25.3	3.8	Mexican American	Variance components	D10S587	D10S187	D10S1223	88
10q25.3	1.69	Icelandic	Non-parametric	D10S1773			87
10q26	1.24	Caucasian French	Non-parametric	D10S1655			69
10q26	1.69	Caucasian French	Non-parametric	D10S212			69
10q26.2	$P = 0.0074$	Caucasian Finn	Affected siblings	D10S217			89
11p15.1	1.3	Icelandic	Non-parametric	D11S928			87
11p15	1.25	Caucasian French	Non-parametric	D11S1338			69
11p13–12	3.08	Japanese	Affected siblings	D11S935	D11S904	D11S905	83
11p–11q	$P = 0.0013$	Caucasian American	Parametric & non-parametric	D11S935	D11S935	D11S901	90
11q21	1.75	Caucasian Finn	Affected siblings	D11S901	D11S1314	D11S1311	89
11q21	$P = 0.0042$	Caucasian Finn	Affected siblings	D11S1314			89

Table 13.2 *(continued)*

Chromosome region	LOD/P-value	Population	Analysis method	Marker with highest LOD score	Flanking marker (left)	Flanking marker (right)	Reference
12q14.3–21.2	6.2	Caucasian, Hispanic and African American	Parametric & non-parametric	D12S375 & D12S1052	D12S1702	D12S1684	91
12q15	3.1	Caucasian American & Canadian	Multipoint parametric		D12S1693	D12S326	91
12q21.32	2.7	Caucasian American	Affected relatives	D12S853			92
12q21.32	2.81	GENNID (Caucasian subgroup)	Non-parametric	D12S853			75
12q23	1.05	Icelandic	Non-parametric	D12S346			87
12q23.3–24.33	3.9	Caucasian Finn	Non-parametric	D12S1349/D12S366	D12S338	D12S97	72
12q24	2.1	Caucasian Finn	Non-parametric	D12S366	D12S304	D12S1614	85
12q24.23	1.44	Icelandic	Non-parametric	D12S366			87
12q24.23–24.31	3.65	Pacific Islander ancestry	Parametric	D12S321	D12S86	D12S807	74
14q11–13	2.37	Japanese	Affected siblings	D14S275			83
14q32.12–32.2	1.69/1.90	Ashkenazi Jew	Non-parametric	D14S749/D14S605			81
15q13–21	3.91/2.44	Japanese	Affected siblings	D15S994			83
16p12–11	1.7	Caucasian Finn	Non-parametric	D16S420			85
17p12–11	1.9	Caucasian Finn	Non-parametric	D17S953			85
17q25.1	1.29	Caucasian Finn and Swedish	Affected siblings	D17S1301			84

Location	LOD/P	Population	Method	Marker	Marker	Marker	No.
18p11.3	1.31	Icelandic	Non-parametric	D18S63			87
18p11.3	2.4	Caucasian American	Affected relatives	D18S59			67
18q23	1.52	Japanese	Affected siblings	D18S70			83
19p13	1.26	Caucasian French	Non-parametric	D19S221			69
19q13	1.69	Japanese	Affected siblings	D19S571			83
20p13	1.8	Ashkenazi Jew	Non-parametric	D20S103			81
20p13–12.3	3.1	Caucasian Finn	Affected siblings		D20S482	D20S115	79
20q11.22–12	2.74	Caucasian Finn	Affected siblings		D20S909	D20S170	79
20q12–13	2.32	Japanese	Affected siblings	D20S119			83
20q13	1.81	Caucasian French	Affected siblings	RPNII	D20S112	ADA/D20S119	78
20q13	2.74	Caucasian French	Affected siblings	7 cM proximal to PCK1	D20S176	PCK1	78
20q13.12	2.05	Ashkenazi Jew	Non-parametric	D20S195	D20S106	D20S481	81
20q13.12–13.13	2.1	Caucasian Finn	Affected siblings	D20S119	D20S119	D20S866	79
20q13.12–13.13	2.7	Caucasian Finn	Affected siblings	D20S892	D20S481 & D20S119	D20S866	89
20q13.12–13.13	3.3	Caucasian American	Non-parametric & TDT	D20S178/D20S197/D20S176	ADA8PR	D20S196	77
20q13.12	2.37	Caucasian	Parametric & non-parametric	D20S197			67
20q13.1–13.2	$P = 1.8 \times 10^{-5}$	Caucasian	Affected relatives	D20S196	D20S196	D20S428	80
20q13.13	2.92	Chinese Han	Non-parametric	D20S196	D20S196		82
22q13.1	$P = 0.00007$	Caucasian Finn	Affected siblings	D22S423			89
Xq23	2.99	GENNID (Caucasian subgroup)	Non-parametric	GATA172D05			75

with an overall odds ratio of 1.2 (1.1–1.4) for the variant identified as the likely cause of the linkage (SNP44) [60]. However, other studies in similar populations have failed to replicate the findings [61–63], and those that have replicated the area did so using different analytical techniques and subgroups of the population. The inconsistent results of attempts to replicate the original findings with *CAPN10* in other populations is a striking example of the daunting nature of the task of identifying true diabetes polygenes with confidence.

Chromosome 1q21–24 Genome scans in at least seven populations including Pima Indians [64], Chinese [65, 66], and several distinct Caucasian populations [57, 67–69] have all identified chromosome 1q21–24 as a likely diabetogenic region. Support also comes from localization of familial combined hyperlipidemia to the same area [70] and identification of a diabetes locus in the GK rat in an equivalent position in the rat genome [71].

Chromosome 12q21–25 Initial evidence for linkage on chromosome 12q21–25 came from a study by Mahtani *et al.* [72]. Detection of suggestive linkage at the same locus by another three groups [73–75] has maintained interest in the region, supported by recent evidence for linkage of age-of-onset effects at the same locus [76].

Chromosome 20q11–13 Three reports published in 1997 provided evidence for linkage in chromosome 20q in Caucasian populations [73, 77, 78]. Stronger evidence followed from studies in two more Caucasian populations [79, 80] and Ashkenazi Jews [81], and an international collaboration carried out a combined analysis of genotypic data from nearly 2000 families, which provided additional support for an association. There has since been further replication from Chinese [82] and Japanese [83] groups.

13.2.1.4 Linkage Analysis – Where Next?

Currently many collaborative groups are undertaking fine mapping and linkage disequilibrium (LD) studies in the most promising regions identified above. These techniques aim to narrow the linkage region to a number of genes that can feasibly be investigated with association studies, to identify the cause of the linkage.

One additional approach for uncovering the genetic etiology of complex diseases is to study the intermediate phenotypes that characterize the early disease processes rather than studying the final disease endpoint itself. This approach has led to the mapping of many additional loci for quantitative, continuous traits that relate to type 2 diabetes in families with high levels of diabetes [93–96]. Examples of traits investigated are fasting and post-glucose challenge insulin, glucose and triglyceride levels. With few exceptions, these traits have not mapped to the same locations as the putative type 2 diabetes loci. It remains to be seen whether the genes affecting intermediate traits are genuinely different to those underlying diabetes.

Attempts have also been made to evaluate gene–gene interactions through linkage results. It is known that these effects play an important role in diabetes pathology, and several groups have attempted to analyze joint effects of loci mapping to different chromosome intervals [75, 89, 97].

13.2.2
Association Studies

Association studies aim to determine whether a specific genetic variant is more common in people with the disorder than in those without. Most association studies employ the traditional case–control design in which the prevalence of the putative disease variant among people with the disorder (cases) is compared with that in people without the disorder (controls), although more recently developed methods are described below. This type of study may either be population or family based.

13.2.2.1 Population-based Studies
Population-based association studies compare unrelated cases and controls, and although they can be quite powerful for detection of genetic effects, they can lead to false-positive results due to population stratification which can be difficult to detect. However, if the cases and controls are adequately matched for potential confounders such as age, gender, ethnicity, or geographical origin, as well as other disease-specific confounders, then it is reasonable to assume that population stratification will not be sufficient to explain the observed associations [98]. In addition, new methods that control for population stratification may be used [99, 100].

13.2.2.2 Family-based Studies
Several family-based methods have been put forward as alternative ways to control for population stratification. Some groups have proposed that unaffected siblings be used as controls despite the problems with variable age of onset of diabetes discussed earlier [101–103]. Another method is to use the transmission distortion/disequilibrium test (TDT), in which control genotypes are constructed from parental alleles that are not transmitted to affected offspring [104]. This has been further developed into a test requiring only siblings, the "sib TDT", to overcome the problem that parents are frequently not available for analysis in late-onset diseases such as type 2 diabetes [103].

These family-based analyses only produce positive results if the marker is both linked and associated with the disease. This robustness comes at additional financial cost compared to population-based case–control studies. In order to obtain similar power to detect an association, approximately the same number of trios as the number of cases in a population-based case control study (assuming a 1:1 ratio of cases to controls) are required. This carries a higher expenditure given the requirement to genotype three rather than two individuals for the same power [105], and the increased difficulty in recruiting families for studying complex diseases.

13.2.2.3 **Results from Association Studies**

Genes are selected for association studies either through positional cloning, as discussed above, or because prior knowledge suggests that it may be a good candidate. This can either be from an understanding of how the function of the gene may impact on the molecular pathogenesis of the disease, or from the gene being implicated in related disorders. In type 2 diabetes most association studies so far have been of the candidate gene type, and given that mechanisms influencing insulin secretion and action both predate and predict the development of type 2 diabetes, most groups have focused on these pathways. In particular, evidence has been found for associations with diabetes in genes that encode proteins involved in insulin secretion, insulin-induced glucose uptake in muscle and fat, and insulin regulation of gluconeogenic pathways. The genes described below are those which currently have the most consistent evidence for association with type 2 diabetes.

KCNJ11 **and** *ABCC8*   Glucose-stimulated insulin release by pancreatic β-cells is controlled by ATP-sensitive potassium channels. These consist of two subunits, Kir6.2 and SUR1, encoded by genes *KCNJ11* and *ABCC8* respectively. Their status as candidate genes does not only come from their central role in insulin secretion, but is supported by the fact that the SUR1 protein is the target for the sulfonylurea family of diabetes drugs – evidence that modifying the channel function does indeed impact on glycemic levels. Polymorphisms in both genes have been investigated for evidence of association with diabetes. The evidence has been conflicting, but this may have largely been due to the small sizes of the early studies. Analysis of these two genes is complicated by the fact that they lie adjacent to one another on chromosome 11 in a region of high linkage disequilibrium, so identifying which variant is the cause of the association is not easy [106].

Recent large studies have highlighted the E23K polymorphism in Kir6.2 as the most likely functional variant [107–110], and a recent meta-analysis calculated an odds ratio for a recessive effect of 1.5 (95% confidence interval (CI) 1.2–1.8) [110]. As 10–20% of people in Caucasian populations have the risk-carrying KK genotype [111], this translates into a significant contribution to population risk. The E23K K allele has also been linked to the diabetes-related traits of obesity [110] and reduced insulin secretion [112], although both these observations have been contradicted [113].

Peroxisome proliferator-activated receptor gamma   Peroxisome proliferator-activated receptor gamma (PPARγ) is one of the PPAR family of hormone nuclear receptors that is important in adipose tissue development and function, and is the receptor for the thiazolidinedione group of antidiabetes drugs. Several variants within the gene have been investigated for association with diabetes, although only the Pro12Ala variant has been replicated in a number of studies (see also chapter 11). There is now strong evidence that the common Pro12 allele is associated with an increased risk of type 2 diabetes [114–116]. A meta-analysis of over 3000 individuals suggested that many of the conflicting negative results may have been due to underpowered studies, and estimated that the rare Ala12 allele provides a protective effect with an odds ratio for developing diabetes of 0.8 (95% CI 0.7–0.9) [114]. As the Pro allele is very common (83–87% in Caucasian populations) this small effect at the individual level will have a significant impact on population risk [114]. The Ala12 allele has been linked to beneficial intermediate metabolic traits such as higher high-density lipoprotein (HDL)-cholesterol [117, 118] and lower triglycerides [118, 119], although also, surprisingly, obesity [120, 121]. However, the evidence for these associations is conflicting [114, 115].

Insulin receptor substrate 1   Insulin receptor substrate 1 (IRS1) mediates downstream signalling from the insulin receptor. As such, it is a highly plausible candidate gene, and several studies have found associations with the Gly972Arg polymorphism (Table 13.3). Evaluation of the combined data through meta-analysis supported the link, with a summary odds ratio of 1.3 (95% CI 1.1–1.5) [122], although this has subsequently been contradicted by a large follow-up study in 9000 people which failed to find any association [123]. This polymorphism has been implicated in the development of insulin resistance in obesity [124, 125], and *in vitro* work indicates that the mutation interferes with insulin-stimulated signalling [126]. The rare allele frequency is 8–11%, but due to the recessive nature of this variant the contribution to population risk is quite small (2%) [122].

13.2.2.4 Association Studies – Where Next?

Many of the early studies were fraught with problems related to suboptimal study designs, and this continues to be an important issue. Common problems include poor matching of cases and controls and a wide usage of convenience samples, the testing of a very limited number of markers per gene, and small sizes of studies with consequently little power to detect genuine effects. As a result, few associations have been replicated in additional populations. Currently the focus is on better and larger study designs with well-defined and matched samples, and it is hoped that this will lead to greater consistency in results [188, 294]. In addition, the use of meta-analysis to accurately combine results from diverse studies has become more prominent and has successfully identified diabetes risk alleles [60, 107, 109, 110, 114].

It is clear that the size of gene effects for complex diseases will be small, and much larger and better-characterized populations will be required to detect the associations [188, 295]. Many population samples with several thousands of participants are currently being collected for this purpose. It is expected that these, in conjunction with a reduction in the price of genotyping, will lead to whole-genome association studies becoming a reality [296]. It is hoped that this new generation of studies will identify associations with higher confidence and will additionally be able to conclusively rule out irrelevant genes. This shift in focus will bring new challenges, not least in the arena of data analysis. Better statistical methods will be required to appropriately correct for the multiple non-independent tests that will be done.

The feasibility of whole-genome association studies has generated much controversy in the past [297–301]. However, the number of single-nucleotide polymor-

Table 13.3. Population studies of genetic polymorphisms associated with type 2 diabetes

Gene	Common names	Polymorphism	Population with evidence for association	References supporting association	Population with evidence of no association	References not supporting association
ABCC8	SUR1; sulfonylurea receptor	Exon 22 Thr761Thr ACC → ACT	Caucasian	127	Caucasian	114
		Arg1273Arg AGG → AGA	Caucasian	128–130		
		Exon 7 Ser1370Ala			Japanese Caucasian	127, 131
ADRB2	β_2-Adrenergic receptor	Gln27Glu	Japanese Caucasian	132, 133	Caucasian Korean Taiwanese	114, 134, 135
		Arg16Gly	Taiwanese	135	Korean	134
ADRB3	β_3-Adrenergic receptor	Trp64Arg	Caucasian Japanese	136–138	Caucasian Pima Indian Tamil Indian Indian	114, 133, 139–145
APM1	Adiponectin	−11426 A → G	Caucasian	146	Caucasian	147
		−11391	Caucasian	147		
		+45 T → G	Japanese	147, 148	Caucasian Pima Indian	149, 150
		+276 G → T	Caucasian			
		−11377 G → C	Caucasian Japanese	146, 147, 151		

Gene	Protein	Variant	Population	Ref.	Population	Ref.
CAPN10	Calpain-10	SNP19	Mexican-American	56	Caucasian, Japanese, Chinese, Samoan	61, 63, 152–154
					West African	155
		SNP56			Japanese	156, 157
		Indel-19			Mexican	
		7920in/del32 bp			Japanese	158
		SNP44	Mexican, Caucasian, Japanese	60, 152, 157		
		SNP63	Mexican-American	56	Mexican, Caucasian, Japanese, Chinese, Samoan	61, 63, 152–154, 157
		SNP110	Mexican	157	Japanese, West African, Mexican, Caucasian, Chinese, Pima Indian, Samoan	61, 63, 152–158, 160
		SNP43	Mexican-American, African-American	56, 159		
		4852 G → A				
ENPP1	PC-1	Lys121Gln	Caucasian, Dominican Republic	161, 162	Caucasian	163, 164
FABP2	Liver fatty acid-binding protein	Ala54Thr			Tongan, Chinese, Japanese	165–169
GC	DBP; vitamin D-binding protein	Asp416Glu	Japanese	170	Caucasian	171–173
		Thr420Lys	Japanese	170	Caucasian	171–173

Table 13.3 (continued)

Gene	Common names	Polymorphism	Population with evidence for association	References supporting association	Population with evidence of no association	References not supporting association
GCGR	Glucagon receptor	Gly40Ser	Caucasian	174, 175	Chinese Brazilian Taiwanese Russian Caucasian Japanese	176–185
GCK	Glucokinase	Dinucleotide repeat	African-American Mauritian	186, 187		
		IVS6 + 38			Caucasian	188
GYS1	Glycogen synthase	Met416Val	Pima Indian	189	Japanese Caucasian Russian	190–193
		XbaI	Caucasian	194		
		Tandem repeat	Japanese	195	Japanese	196
HNF1A	Hepatocyte nuclear factor 1 alpha	Gly319Ser	Oji-Cree	197		
		Gly191Asp			Japanese	198
		Ile27Leu and others			Caucasian	199
HNF1B	Hepatocyte nuclear factor 1 beta	Ser465Arg	Japanese	200		
		Arg177X			Japanese	201

Gene	Gene name	Variant	Population	References	Population	References
HNF4A	Hepatocyte nuclear factor 4 alpha	Thr130Ile	Japanese	202, 203	Caucasian	79, 204
		rs1800961	Caucasian		Japanese	205
		Gln268X			Caucasian	206
		−79 C → T			Caucasian	206
		−276 G → T			Caucasian	204, 208
		rs3818247	Ashkenazi Jew	207		
		rs2425640	Amish	209		
		rs2144908	Ashkenazi Jew / Caucasian	207, 210, 211		
		rs1884614	Ashkenazi Jew / Caucasian	207, 208		
		rs4810424	Caucasian	211		
		Haplotype effects	Caucasian	188, 212		
IAPP	Insulin amyloid protein	Ser20Gly	Japanese / Chinese	213–215	Japanese	216
		Gln10Arg			Maori	217
		−132 G → A	Caucasian	218	Maori	217, 219
		−215 T → G	Maori	217	Caucasian	
IGF1	Insulin-like growth factor 1	5′ CA microsatellite	Caucasian	220	Caucasian	221
INS	Insulin	−23 A → T (HphI)	Indian / Caucasian	222–225	Caucasian / Pima Indian / Japanese / African-American	226–230
		VNTR				
INSR	Insulin receptor	Val985Met	Caucasian	231, 232	Tamil Indian / Caucasian	144, 233

Table 13.3 *(continued)*

Gene	Common names	Polymorphism	Population with evidence for association	References supporting association	Population with evidence of no association	References not supporting association
IPF1	Insulin promoter factor 1	C18R	Caucasian	234	Taiwanese	235
		R197H	Caucasian	234	Taiwanese	235
		Q59L	Caucasian	236	Taiwanese	235
		D76N	Caucasian	234, 236	Taiwanese	235
IRS1	Insulin receptor substrate 1	Gly972Arg	Caucasian	237, 238	Mexican Parakana Brazilian Indian Caucasian Mexican-American	125, 232, 239–245
IRS2	Insulin receptor substrate 2	Gly1057Asp	Pima Indian Caucasian	246, 247	Caucasian	248, 249
KCNJ11	KIR6.2	Glu23Lys (E23K)	Caucasian	106–110, 128, 250, 251	Caucasian	114, 252, 253
NOS3	eNOS	−786 T → C	Japanese	254		
		Glu298Asp	Caucasian	255		
		IVS18 + 27C	Caucasian	255		
PIK3R1	Phosphoinositide-3-kinase regulatory subunit p85 alpha	Met326Ile	Pima Indian	256	Caucasian	233, 257
		IVS4 + 82	Caucasian	188	Japanese	

Gene	Description	Variant	Population		Population	
PPARG	Peroxisome proliferator-activated receptor gamma	Pro12Ala	Caucasian, Japanese, Japanese-Americans	114–116, 119, 258–261	Caucasian, Taiwanese	262–267
PPARGC1	PGC1; peroxisome proliferator-activated receptor gamma co-activator 1	Gly482Ser	Caucasian, Japanese, Caucasian	261, 268, 269	Caucasian, Pima Indian	270, 271
		Haplotype effect		272		
PPP1R3A	Protein phosphatase 1, regulatory (inhibitor) subunit 3A	Asp905Lys	Chinese, Pima Indian	273, 274	Chinese, Oji-Cree, Japanese, Caucasian	275–278
		3′ UTR ins/del	Pima Indian	274	Chinese	279
RRAD	Ras-related associated with diabetes	4681 C → T	Chinese, Caucasian, Japanese	280	Caucasian	283
		RAD1 repeat		281, 282		
SLC2A2	GLUT2; glucose transporter 2	Thr110Ile	Caucasian	188	Japanese	284
		IVS5–15	Caucasian	188		
		T198	Caucasian	188		
		T110I	Caucasian	188	Caucasian	286
		−44 A → T	Korean	285		
TNF	Tumor necrosis factor	−238 A → G	Caucasian	287	Taiwanese, Pima Indian, Caucasian	288, 289
		−308 G → A	Caucasian	290	Taiwanese, Chinese, Caucasian, Pima Indian	288, 289, 291
UCP3	Uncoupling protein 3	−55 C → T	Chinese, Caucasian	292, 293		

phisms (SNPs) required to perform such studies has been drastically reduced by the HapMap project (www.hapmap.org), which is working to identify the most common haplotypes (clusters of alleles inherited together) in four distinct populations. At the time of writing, the first genome-wide haplotype map has just been completed, and work has begun on a more detailed second phase. The first whole-genome association studies have been published, and a number of companies now have access to their own haplotype tagging SNPs [302]. Early results are showing promise in the identification of novel genes involved in complex disease predisposition [303–305].

13.3
The Success Story of Maturity-onset Diabetes of the Young

Despite the almost overwhelming problems discussed above stemming from inconsistent results and lack of replication, the world of diabetes genetics has had some notable successes. In the 1960s an unusual familial form of type 2 diabetes was recognized, with onset usually occurring before the age of 25 [312]. In the 1970s it was documented that familial transmission of the disease followed autosomal dominant inheritance [313, 314]. This clinical group of heterogeneous disorders was called maturity-onset diabetes of the young (MODY) and was characterized by early age of onset, autosomal dominant inheritance and non-ketotic diabetes with a primary defect in the β-cells of the pancreas. Genetic linkage studies successfully identified loci on chromosome 20q (MODY1) in a portion of a single large American pedigree [315], on chromosome 7p (MODY2) near the glucokinase gene (*GCK*) [316] in French families and on chromosome 12q (MODY3) [317]. Glucokinase functions as the glucose sensor in the pancreatic β-cell, controlling the rate of entry of glucose into the glycolytic pathway. In the liver, it plays a role in the storage of glucose as glycogen. It was an obvious candidate gene to study, and mutations in this gene were shown to be responsible for the linkage on chromosome 7. Further work on chromosome 12 led to the identification of missense and nonsense mutations in hepatocyte nuclear factor 1 alpha (*HNF1A*), a transcription factor that regulates many liver-specific genes but which had not previously been considered a candidate gene for diabetes. This led to a very similar gene, *HNF4A*, to be considered a candidate gene for MODY1. Mutations in *HNF4A* were subsequently shown to cause MODY1. To date, six genes responsible for MODY have been described (Table 13.4).

All of these genes are expressed in the pancreatic β-cells, and mutation in any of them leads to β-cell dysfunction and diabetes mellitus.

The identification of the MODY genes rapidly stimulated research into links between them and late-onset type 2 diabetes, but the early results suggested that there was no connection [199, 205, 318–320]. Later studies, however, have demonstrated significant associations (Table 13.3). Although the precise role of these

Table 13.4. Summary of mature-onset diabetes of the young (MODY) genes

MODY	Location	Gene	Reference
MODY1	20q	*HNF4A*	306
MODY2	7p	*GCK*	307
MODY3	12q	*HNF1A*	308
MODY4	13q	*IPF1*	309
MODY5	17q	*HNF1B*	310
MODY6	2q	*NEUROD1*	311

genes in forms of diabetes other than MODY is still open to debate, work on MODY has already improved our understanding of the genetic causes of β-cell dysfunction and genetic pathways in glucose homeostasis.

13.4
The Impact of Identifying Diabetes Polygenes

It is not possible to predict fully how clinical practice will be affected by the understanding of how genes control type 2 diabetes, as it will depend on the structural protein class encoded by each gene and the impact of each on disease risk. The finding of a diabetes gene with a "major" effect would be tremendously valuable, but is, perhaps, unlikely. If such a gene existed it would probably already be known. Nevertheless, even if each of the genes identified imparts only a small increase in risk, the information could be used to improve disease prognosis and treatment, and prevention of the serious consequences of diabetes.

The identification of new targets for drugs is a high priority, and there is the potential that a number of novel type 2 diabetes genes will code for proteins suitable for drug development. It is also possible that even if new genes are not tractable targets in themselves, they will identify new potential pathways for therapeutic intervention. Even if only a subset of the patients carry the predisposing genotype or if the risk imparted by that genotype is low, pharmacological intervention in the function of that gene may be beneficial for most patients, including those that do not carry the risk variant.

One of the most promising areas of progress is in understanding the variation in responses to drug therapies. There are currently several diabetes drugs on the market, but few methods for predicting which will be the most successful for a specific individual. An understanding of the genetic causes of the disease and of the variation in response to treatment may allow clinicians to identify the optimal treatment for each patient from the outset. There are already clear examples of this in sulfonylurea treatment of diabetes: people with MODY caused by mutations in

HNF1α respond better to sulfonylureas than other treatments [321], while those carrying the IRS1 variant Arg972 have twice the normal risk of failing to respond to sulfonylurea treatment [322].

It may be that diabetes from certain genetic causes has a different prognosis. A greater understanding of this would allow health care to be focused on those with the greatest treatment requirement and response, and the identification of those most at risk from specific environmental risk factors. A good example of this is women who develop diabetes during pregnancy; diabetes in pregnancy due to *GCK* mutations is stable and does not require further treatment after delivery, while non-*GCK* gestational diabetes commonly progresses to full-blown type 2 diabetes over the following 10–20 years [323].

Regardless of the ultimate characteristics of the collective set of diabetes genes, their identification will play an important role in the development of new therapies and treatment regimes to reduce the impact of this serious disease.

13.5
Gene–Environment Interactions

When analyzing the genetics of a disease, gene–environment interactions are usually left unstudied until the disease-causing/predisposing loci have been identified. However, when working with a disorder such as diabetes, the effect of some important genetic variants may only be observed under specific environmental exposures. Ignoring the effects of environment can result in failure to identify important genetic contributions [324]. Precise measurement of exposure to lifestyle factors such as diet and exercise is challenging, but precision is crucial to allow the detection of the small effects involved.

Unsurprisingly, progress in unravelling this problem has been slow, but evidence is beginning to accumulate. Work in this field has focused on interactions between specific genes and environmental exposures and their effects on intermediate traits, rather than tackling head-on the enormous and complex problem of the contribution of gene–environment interactions to diabetes. Diet and genes have been shown to measurably affect each other: changes in diet have been shown to modify gene expression [325], and gene variants have been linked to differences in response to diet. For example, plasma low-density lipoprotein (LDL)-cholesterol concentration responds to changes in dietary fat, and this change in concentration is modified by polymorphisms in the genes coding for apolipoproteins (a family of proteins important in fat absorption) [326–328]. Interactions between exercise and genetic factors have also been identified; physical activity modifies the effect of polymorphisms in the β-adrenergic receptor on the concentration of non-esterified fatty acids in the blood [329]. The effects of diet and exercise are additive, as shown by Franks *et al.* [330] in a study that demonstrated interactions between the ratio of polyunsaturated:saturated fats in the diet, physical activity, and PPARγ Pro12Ala genotype on fasting insulin levels [330].

Although work on gene–environment interactions suffers from the same problems seen in association studies, such as lack of replication, it is promising that techniques are being developed which are capable of measuring these elusive effects.

13.6
Conclusion

Type 2 diabetes is a serious illness, both for the individual and the population. A better understanding of the mechanisms of the disease will be required to produce novel preventions and cures, and due to the complex nature of the disorder this will be a challenging task. This will require the cooperation of researchers from many different fields, and it is only with the combined evidence from many different study designs that progress will be made.

References

1 KING, H., REWERS, M. Global estimates for prevalence of diabetes mellitus and impaired glucose tolerance in adults. WHO Ad Hoc Diabetes Reporting Group. *Diabetes Care* **1993**, *16*, 157–77.

2 KNOWLER, W.C. *et al.* Gm3;5,13,14 and type 2 diabetes mellitus: an association in American Indians with genetic admixture. *Am J Hum Genet* **1988**, *43*, 520–6.

3 SERJEANTSON, S.W. *et al.* Genetics of diabetes in Nauru: effects of foreign admixture, HLA antigens and the insulin-gene-linked polymorphism. *Diabetologia* **1983**, *25*, 13–7.

4 PICKUP, J., WILLIAMS, G. *Textbook of Diabetes*, 2nd edn. **1997**, Oxford: Blackwell Science Ltd.

5 NEEL, J.V. Diabetes mellitus: a "thrifty" genotype rendered detrimental by "progress"? *Am J Hum Genet* **1962**, *14*, 353–62.

6 DIAMOND, J. The double puzzle of diabetes. *Nature* **2003**, *423*, 599–602.

7 RICH, S.S. Mapping genes in diabetes – genetic epidemiologic perspective. *Diabetes* **1990**, *39*, 1315–9.

8 HARRIS, M.I. Epidemiological correlates of NIDDM in Hispanics, whites, and blacks in the U.S.

population. *Diabetes Care* **1991**, *14*, 639–48.

9 MITCHELL, B.D. *et al.* Is there an excess in maternal transmission of NIDDM? *Diabetologia* **1995**, *38*, 314–17.

10 KNOWLER, W.C. *et al.* Diabetes incidence in Pima indians: contributions of obesity and parental diabetes. *Am J Epidemiol* **1981**, *113*, 144–56.

11 LEE, E.T. *et al.* Diabetes and impaired glucose tolerance in three American Indian populations aged 45–74 years. The Strong Heart Study. *Diabetes Care* **1995**, *18*, 599–610.

12 OLIVEIRA, J.E., MILECH, A., FRANCO, L.J. The prevalence of diabetes in Rio de Janeiro, Brazil. The Cooperative Group for the Study of Diabetes Prevalence in Rio De Janeiro. *Diabetes Care* **1996**, *19*, 663–6.

13 MYKKANEN, L. *et al.* Prevalence of diabetes and impaired glucose tolerance in elderly subjects and their association with obesity and family history of diabetes. *Diabetes Care* **1990**, *13*, 1099–105.

14 CHARLES, M.A. *et al.* Risk factors for NIDDM in white population. Paris prospective study. *Diabetes* **1991**, *40*, 796–9.

15 Pan, X.R. *et al.* Prevalence of diabetes and its risk factors in China, 1994. National Diabetes Prevention and Control Cooperative Group. *Diabetes Care* **1997**, *20*, 1664–9.

16 Viswanathan, M. *et al.* High prevalence of type 2 (non-insulin-dependent) diabetes among the offspring of conjugal type 2 diabetic parents in India. *Diabetologia* **1985**, *28*, 907–10.

17 Mengesha, B., Abdulkadir, J. Heritability of diabetes mellitus in Ethiopian diabetics. *East Afr Med J* **1997**, *74*, 37–40.

18 Zimmet, P.Z., McCarty, D.J., de Courten, M.P. The global epidemiology of non-insulin-dependent diabetes mellitus and the metabolic syndrome. *J Diabetes Complications* **1997**, *11*, 60–8.

19 Kobberling, J. Studies on the genetic heterogeneity of diabetes mellitus. *Diabetologia* **1971**, *7*, 46–9.

20 Simpson, N.E. Multifactorial inheritance: a possible hypothesis for diabetes. *Diabetes* **1964**, *13*, 462–71.

21 Thomas, F. *et al.* Maternal effect and familial aggregation in NIDDM. The CODIAB Study. CODIAB-INSERM-ZENECA Study Group. *Diabetes* **1994**, *43*, 63–7.

22 Knowler, W.C., Pettitt, D.J., and Lillioja, S. *et al.* Genetic and environmental factors in the development of diabetes mellitus in Pima Indians. In: *Genetic Susceptibility to Environmental Factors – a Challenge for Public Intervention*, U. Smith, S. Eriksson, F. Lindgarde, editors, **1998**, Stockholm: Almqvist and Wiksell International, pp. 67–76.

23 Groop, L.C., Tuomi, T. Non-insulin-dependent diabetes mellitus – a collision between thrifty genes and an affluent society. *Ann Med* **1997**, *29*, 37–53.

24 Meigs, J.B., Cupples, L.A., Wilson, P.W. Parental transmission of type 2 diabetes: the Framingham Offspring Study. *Diabetes* **2000**, *49*, 2201–7.

25 Lindsay, R.S. *et al.* Type 2 diabetes and low birth weight: the role of paternal inheritance in the association of low birth weight and diabetes. *Diabetes* **2000**, *49*, 445–9.

26 Alcolado, J.C., Alcolado, R. Importance of maternal history of non-insulin dependent diabetic patients. *BMJ* **1991**, *302*, 1178–80.

27 Karter, A.J. *et al.* Excess maternal transmission of type 2 diabetes. The Northern California Kaiser Permanente Diabetes Registry. *Diabetes Care* **1999**, *22*, 938–43.

28 Ballinger, S.W. *et al.* Maternally transmitted diabetes and deafness associated with a 10.4 kb mitochondrial DNA deletion. *Nat Genet* **1992**, *1*, 11–15.

29 Reardon, W. *et al.* Diabetes mellitus associated with a pathogenic point mutation in mitochondrial DNA. *Lancet* **1992**, *340*, 1376–9.

30 Kadowaki, T. *et al.* A subtype of diabetes mellitus associated with a mutation of mitochondrial DNA. *N Engl J Med* **1994**, *330*, 962–8.

31 van den Ouweland, J.M. *et al.* Mutation in mitochondrial tRNA(Leu)(UUR) gene in a large pedigree with maternally transmitted type II diabetes mellitus and deafness. *Nat Genet* **1992**, *1*, 368–71.

32 Poulton, J. *et al.* Type 2 diabetes is associated with a common mito-chondrial variant: evidence from a population-based case-control study. *Hum Mol Genet* **2002**, *11*, 1581–3.

33 Lamson, D.W., Plaza, S.M. Mitochondrial factors in the pathogenesis of diabetes: a hypothesis for treatment. *Altern Med Rev* **2002**, *7*, 94–111.

34 Pettitt, D.J., Knowler, W.C. Long-term effects of the intrauterine environment, birth weight, and breast-feeding in Pima Indians. *Diabetes Care* **1998**, *21* (Suppl 2), B138–41.

35 Dabelea, D. *et al.* Intrauterine exposure to diabetes conveys risk for diabetes and obesity in offspring above that attributable to genetics. *Diabetes* **1999**, *48*, (Suppl 1), A52.

36 Hales, C.N., Desai, M., Ozanne, S.E. The Thrifty Phenotype hypothesis: how does it look after 5 years? *Diabet Med* **1997**, *14*, 189–95.

37 BOOMSMA, D., BUSJAHN, A., PELTONEN, L. Classical twin studies and beyond. *Nat Rev Genet* **2002**, *3*, 872–82.

38 MEDICI, F. *et al.* Concordance rate for type II diabetes mellitus in monozygotic twins: actuarial analysis. *Diabetologia* **1999**, *42*, 146–50.

39 GHOSH, S., SCHORK, N.J. Genetic analysis of NIDDM. The study of quantitative traits. *Diabetes* **1996**, *45*, 1–14.

40 BARNETT, A.H. *et al.* Diabetes in identical twins. A study of 200 pairs. *Diabetologia* **1981**, *20*, 87–93.

41 NEWMAN, B. *et al.* Concordance for type 2 (non-insulin-dependent) diabetes mellitus in male twins. *Diabetologia* **1987**, *30*, 763–8.

42 Diabetes mellitus in twins: a cooperative study in Japan. Committee on Diabetic Twins, Japan Diabetes Society. *Diabetes Res Clin Pract* **1988**, *5*, 271–80.

43 KAPRIO, J. *et al.* Concordance for type 1 (insulin-dependent) and type 2 (non-insulin-dependent) diabetes mellitus in a population-based cohort of twins in Finland. *Diabetologia* **1992**, *35*, 1060–7.

44 MATSUDA, A., KUZUYA, T. Diabetic twins in Japan. *Diabetes Res Clin Pract* **1994**, *24* (Suppl.), S63–7.

45 POULSEN, P., VAAG, A., BECK-NIELSEN, H. Does zygosity influence the metabolic profile of twins? A population based cross sectional study. *BMJ* **1999**, *319*, 151–4.

46 ALBERTS, B. *et al. Molecular Biology of the Cell*, 4th edn. **2002**, London: Garland Science.

47 NUSSBAUM, R.L., MCINNES, R.R., WILLARD, H.F. *Genetics in Medicine*, 6th edn. **2004**, Philadelphia: Saunders.

48 LANDER, E., KRUGLYAK, L. Genetic dissection of complex traits – guidelines for interpreting and reporting linkage results. *Nat Genet* **1995**, *11*, 241–7.

49 LANDER, E.S., SCHORK, N.J. Genetic dissection of complex traits. *Science* **1994**, *265*, 2037–48.

50 KRUGLYAK, L., LANDER, E.S. Complete multipoint sib-pair analysis of qualitative and quantitative traits. *Am J Hum Genet* **1995**, *57*, 439–54.

51 WEEKS, D.E. *et al.* Computer programs for multilocus haplotyping of general pedigrees. *Am J Hum Genet* **1995**, *56*, 1506–7.

52 BLACKWELDER, W.C., ELSTON, R.C. A comparison of sib-pair linkage tests for disease susceptibility loci. *Genet Epidemiol* **1985**, *2*, 85–97.

53 MCCARTHY, M.I. Growing evidence for diabetes susceptibility genes from genome scan data. *Curr Diab Rep* **2003**, *3*, 159–67.

54 FLOREZ, J.C., HIRSCHHORN, J., ALTSHULER, D. The inherited basis of diabetes mellitus: implications for the genetic analysis of complex traits. *Annu Rev Genomics Hum Genet* **2003**, *4*, 257–91.

55 HANIS, C.L. *et al.* A genome-wide search for human non-insulin-dependent (type 2) diabetes genes reveals a major susceptibility locus on chromosome 2. *Nat Genet* **1996**, *13*, 161–6.

56 HORIKAWA, Y. *et al.* Genetic variation in the gene encoding calpain-10 is associated with type 2 diabetes mellitus. *Nat Genet* **2000**, *26*, 163–75.

57 HSUEH, W.C. *et al.* Genome-wide and fine-mapping linkage studies of type 2 diabetes and glucose traits in the Old Order Amish: evidence for a new diabetes locus on chromosome 14q11 and confirmation of a locus on chromosome 1q21-q24. *Diabetes* **2003**, *52*, 550–7.

58 HANI, E.H. *et al.* Mapping NIDDM susceptibility loci in French families: studies with markers in the region of NIDDM1 on chromosome 2q. *Diabetes* **1997**, *46*, 1225–6.

59 FRANCKE, S. *et al.* A genome-wide scan for coronary heart disease suggests in Indo-Mauritians a susceptibility locus on chromosome 16p13 and replicates linkage with the metabolic syndrome on 3q27. *Hum Mol Genet* **2001**, *10*, 2751–65.

60 WEEDON, M.N. *et al.* Meta-analysis and a large association study confirm a role for calpain-10 variation in type 2

diabetes susceptibility. *Am J Hum Genet* **2003**, *73*, 1208–12.

61 HORIKAWA, Y. *et al.* Genetic variations in calpain-10 gene are not a major factor in the occurrence of type 2 diabetes in Japanese. *J Clin Endocrinol Metab* **2003**, *88*, 244–7.

62 RASMUSSEN, S.K. *et al.* Variants within the calpain-10 gene on chromosome 2q37 (NIDDM1) and relationships to type 2 diabetes, insulin resistance, and impaired acute insulin secretion among Scandinavian Caucasians. *Diabetes* **2002**, *51*, 3561–7.

63 SUN, H.X. *et al.* Single nucleotide polymorphisms in CAPN10 gene of Chinese people and its correlation with type 2 diabetes mellitus in Han people of northern China. *Biomed Environ Sci* **2002**, *15*, 75–82.

64 HANSON, R.L. *et al.* An autosomal genomic scan for loci linked to type II diabetes mellitus and body-mass index in Pima Indians. *Am J Hum Genet* **1998**, *63*, 1130–8.

65 NG, M. *et al.* Identification of a susceptibility locus on chromosome 1q for metabolic syndrome traits in Hong Kong Chinese diabetic families. *Diabetes* **2003**, *52*, A35.

66 XIANG, K.S. *et al.* Genome-wide search for type 2 diabetes/impaired glucose homeostasis susceptibility genes in the Chinese – Significant linkage to chromosome 6q2l-q23 and chromosome 1q21-q24. *Diabetes* **2004**, *53*, 228–234.

67 ELBEIN, S.C. *et al.* A genome-wide search for type 2 diabetes susceptibility genes in Utah Caucasians. *Diabetes* **1999**, *48*, 1175–82.

68 WILTSHIRE, S. *et al.* A genomewide scan for loci predisposing to type 2 diabetes in a U.K. population (the Diabetes UK Warren 2 Repository): analysis of 573 pedigrees provides independent replication of a susceptibility locus on chromosome 1q. *Am J Hum Genet* **2001**, *69*, 553–69.

69 VIONNET, N. *et al.* Genomewide search for type 2 diabetes-susceptibility genes in French whites: evidence for a novel susceptibility locus for early-onset diabetes on chromosome 3q27-qter

and independent replication of a type 2-diabetes locus on chromosome 1q21-q24. *Am J Hum Genet* **2000**, *67*, 1470–80.

70 PAJUKANTA, P. *et al.* Linkage of familial combined hyperlipidaemia to chromosome 1q21-q23. *Nat Genet* **1998**, *18*, 369–73.

71 GAUGUIER, D. *et al.* Chromosomal mapping of genetic loci associated with non-insulin dependent diabetes in the GK rat. *Nat Genet* **1996**, *12*, 38–43.

72 MAHTANI, M.M. *et al.* Mapping of a gene for type 2 diabetes associated with an insulin secretion defect by a genome scan in Finnish families. *Nat Genet* **1996**, *14*, 90–4.

73 BOWDEN, D.W. *et al.* Linkage of genetic markers on human chromosomes 20 and 12 to NIDDM in Caucasian sib pairs with a history of diabetic nephropathy. *Diabetes* **1997**, *46*, 882–6.

74 SHAW, J.T. *et al.* Novel susceptibility gene for late-onset NIDDM is localized to human chromosome 12q. *Diabetes* **1998**, *47*, 1793–6.

75 EHM, M.G. *et al.* Genomewide search for type 2 diabetes susceptibility genes in four American populations. *Am J Hum Genet* **2000**, *66*, 1871–81.

76 WILTSHIRE, S. *et al.* Evidence from a large UK family collection that genes influencing age of onset of type 2 diabetes map to chromosome 12p and to ghe MODY3/NIDDM2 locus on 12q24. *Diabetes* **2004**, *53*, 855–60.

77 JI, L. *et al.* New susceptibility locus for NIDDM is localized to human chromosome 20q. *Diabetes* **1997**, *46*, 876–81.

78 ZOUALI, H. *et al.* A susceptibility locus for early-onset non-insulin dependent (type 2) diabetes mellitus maps to chromosome 20q, proximal to the phosphoenolpyruvate carboxykinase gene. *Hum Mol Genet* **1997**, *6*, 1401–8.

79 GHOSH, S. *et al.* Type 2 diabetes: evidence for linkage on chromosome 20 in 716 Finnish affected sib pairs. *Proc Natl Acad Sci USA* **1999**, *96*, 2198–203.

80 KLUPA, T. *et al.* Further evidence for a

susceptibility locus for type 2 diabetes on chromosome 20q13.1-q13.2. *Diabetes* **2000**, *49*, 2212–16.

81 PERMUTT, M.A. *et al.* A genome scan for type 2 diabetes susceptibility loci in a genetically isolated population. *Diabetes* **2001**, *50*, 681–5.

82 LUO, T.H. *et al.* A genome-wide search for type II diabetes susceptibility genes in Chinese Hans. *Diabetologia* **2001**, *44*, 501–6.

83 MORI, Y. *et al.* Genome-wide search for type 2 diabetes in Japanese affected sib-pairs confirms susceptibility genes on 3q, 15q, and 20q and identifies two new candidate Loci on 7p and 11p. *Diabetes* **2002**, *51*, 1247–55.

84 PARKER, A. *et al.* A gene conferring susceptibility to type 2 diabetes in conjunction with obesity is located on chromosome 18p11. *Diabetes* **2001**, *50*, 675–80.

85 LINDGREN, C.M. *et al.* Genomewide search for type 2 diabetes mellitus susceptibility loci in Finnish families: the Botnia study. *Am J Hum Genet* **2002**, *70*, 509–16.

86 BUSFIELD, F. *et al.* A genomewide search for type 2 diabetes-susceptibility genes in indigenous Australians. *Am J Hum Genet* **2002**, *70*, 349–57.

87 REYNISDOTTIR, I. *et al.* Localization of a susceptibility gene for type 2 diabetes to chromosome 5q34-q35.2. *Am J Hum Genet* **2003**, *73*, 323–35.

88 DUGGIRALA, R. *et al.* Linkage of type 2 diabetes mellitus and of age at onset to a genetic location on chromosome 10q in Mexican Americans. *Am J Hum Genet* **1999**, *64*, 1127–40.

89 GHOSH, S. *et al.* The Finland-United States investigation of non-insulin-dependent diabetes mellitus genetics (FUSION) study. I. An autosomal genome scan for genes that predispose to type 2 diabetes. *Am J Hum Genet* **2000**, *67*, 1174–85.

90 ELBEIN, S.C. *et al.* Linkage studies of NIDDM with 23 chromosome 11 markers in a sample of whites of northern European descent. *Diabetes* **1996**, *45*, 370–5.

91 BEKTAS, A. *et al.* Evidence of a novel type 2 diabetes locus 50 cM centromeric to NIDDM2 on chromosome 12q. *Diabetes* **1999**, *48*, 2246–51.

92 EHM, M.G., KARNOUB, M.C., ST JEAN, P. Search for susceptibility genes, gene × gene interactions, and gene × environment interactions utilizing nonparametric linkage analysis. *Genet Epidemiol* **1999**, *17* (Suppl 1), S539–43.

93 STERN, M.P. *et al.* Evidence for linkage of regions on chromosomes 6 and 11 to plasma glucose concentrations in Mexican Americans. *Genome Res* **1996**, *6*, 724–34.

94 WATANABE, R.M. *et al.* The Finland-United States investigation of non-insulin-dependent diabetes mellitus genetics (FUSION) study. II. An autosomal genome scan for diabetes-related quantitative-trait loci. *Am J Hum Genet* **2000**, *67*, 1186–200.

95 PRATLEY, R.E. *et al.* An autosomal genomic scan for loci linked to prediabetic phenotypes in Pima Indians. *J Clin Invest* **1998**, *101*, 1757–64.

96 ELBEIN, S.C., HASSTEDT, S.J. Genome-wide scan for prediabetic traits in Caucasian familial type 2 diabetes kindreds. *Diabetes* **2000**, *49* (Suppl 1), A200.

97 COX, N.J. *et al.* Loci on chromosomes 2 (NIDDM1) and 15 interact to increase susceptibility to diabetes in Mexican Americans. *Nat Genet* **1999**, *21*, 213–15.

98 ARDLIE, K.G., LUNETTA, K.L., SEIELSTAD, M. Testing for population subdivision and association in four case-control studies. *Am J Hum Genet* **2002**, *71*, 304–11.

99 PRITCHARD, J.K., ROSENBERG, N.A. Use of unlinked genetic markers to detect population stratification in association studies. *Am J Hum Genet* **1999**, *65*, 220–8.

100 SHMULEWITZ, D., ZHANG, J., GREENBERG, D.A. Case-control association studies in mixed populations: correcting using genomic control. *Hum Hered* **2004**, *58*, 145–53.

101 CURTIS, D. Use of siblings as controls in case-control association studies.

Ann Hum Genet **1997**, *61* (Pt 4), 319–33.

102 HORVATH, S., LAIRD, N.M. A discordant-sibship test for disequilibrium and linkage: no need for parental data. *Am J Hum Genet* **1998**, *63*, 1886–97.

103 SPIELMAN, R.S., EWENS, W.J. A sibship test for linkage in the presence of association: the sib transmission/disequilibrium test. *Am J Hum Genet* **1998**, *62*, 450–8.

104 SPIELMAN, R.S., MCGINNIS, R.E., EWENS, W.J. Transmission test for linkage disequilibrium: the insulin gene region and insulin-dependent diabetes mellitus (IDDM). *Am J Hum Genet* **1993**, *52*, 506–16.

105 MCGINNIS, R., SHIFMAN, S., and DARVASI, A. Power and efficiency of the TDT and case-control design for association scans. *Behav Genet* **2002**, *32*, 135–44.

106 FLOREZ, J.C. *et al.* Haplotype structure and genotype-phenotype correlations of the sulfonylurea receptor and the islet ATP-sensitive potassium channel gene region. *Diabetes* **2004**, *53*, 1360–8.

107 HANI, E.H. *et al.* Missense mutations in the pancreatic islet beta cell inwardly rectifying K^+ channel gene (KIR6.2/BIR): a meta-analysis suggests a role in the polygenic basis of type II diabetes mellitus in Caucasians. *Diabetologia* **1998**, *41*, 1511–15.

108 LOVE-GREGORY, L. *et al.* An E23K single nucleotide polymorphism in the islet ATP-sensitive potassium channel gene (Kir6.2) contributes as much to the risk of Type II diabetes in Caucasians as the PPAR Pro12Ala variant. *Diabetologia* **2003**, *46*, 136–7.

109 GLOYN, A.L. *et al.* Large-scale association studies of variants in genes encoding the pancreatic beta-cell K(ATP) channel subunits Kir6.2 (KCNJ11) and SUR1 (ABCC8) confirm that the KCNJ11 E23K variant is associated with type 2 diabetes. *Diabetes* **2003**, *52*, 568–72.

110 NIELSEN, E.M. *et al.* The E23K variant of Kir6.2 associates with impaired post-OGTT serum insulin response and increased risk of type 2 diabetes. *Diabetes* **2003**, *52*, 573–7.

111 RIEDEL, M.J., STECKLEY, D.C., and LIGHT, P.E. Current status of the E23K Kir6.2 polymorphism: implications for type-2 diabetes. *Hum Genet* **2005**, *116*, 133–45.

112 'T HART, L.M. *et al.* Variations in insulin secretion in carriers of the E23K variant in the KIR6.2 subunit of the ATP-sensitive K(+) channel in the beta-cell. *Diabetes* **2002**, *51*, 3135–8.

113 HANSEN, L. *et al.* Amino acid polymorphisms in the ATP-regulatable inward rectifier Kir6.2 and their relationships to glucose- and tolbutamide-induced insulin secretion, the insulin sensitivity index, and NIDDM. *Diabetes* **1997**, *46*, 508–12.

114 ALTSHULER, D. *et al.* The common PPARgamma Pro12Ala polymorphism is associated with decreased risk of type 2 diabetes. *Nat Genet* **2000**, *26*, 76–80.

115 DEEB, S.S. *et al.* A Pro12Ala substitution in PPARgamma2 associated with decreased receptor activity, lower body mass index and improved insulin sensitivity. *Nat Genet* **1998**, *20*, 284–7.

116 HARA, K. *et al.* The Pro12Ala polymorphism in PPAR gamma2 may confer resistance to type 2 diabetes. *Biochem Biophys Res Commun* **2000**, *271*, 212–16.

117 KIM, K.S. *et al.* Effects of peroxisome proliferator-activated receptor-gamma 2 Pro12Ala polymorphism on body fat distribution in female Korean subjects. *Metabolism* **2004**, *53*, 1538–43.

118 PIHLAJAMAKI, J. *et al.* The Pro12Ala substitution in the peroxisome proliferator activated receptor gamma 2 is associated with an insulin-sensitive phenotype in families with familial combined hyperlipidemia and in nondiabetic elderly subjects with dyslipidemia. *Atherosclerosis* **2000**, *151*, 567–74.

119 DOUGLAS, J.A. *et al.* The peroxisome proliferator-activated receptor-gamma2 Pro12Ala variant: association with type

2 diabetes and trait differences. *Diabetes* **2001**, *50*, 886–90.

120 BEAMER, B.A. *et al.* Association of the Pro12Ala variant in the peroxisome proliferator-activated receptor-gamma2 gene with obesity in two Caucasian populations. *Diabetes* **1998**, *47*, 1806–8.

121 HSUEH, W.C. *et al.* Interactions between variants in the beta3-adrenergic receptor and peroxisome proliferator-activated receptor-gamma2 genes and obesity. *Diabetes Care* **2001**, *24*, 672–7.

122 JELLEMA, A. *et al.* Gly972Arg variant in the insulin receptor substrate-1 gene and association with Type 2 diabetes: a meta-analysis of 27 studies. *Diabetologia* **2003**, *46*, 990–5.

123 FLOREZ, J.C. *et al.* Association testing in 9,000 people fails to confirm the association of the insulin receptor substrate-1 G972R polymorphism with type 2 diabetes. *Diabetes* **2004**, *53*, 3313–18.

124 BARONI, M.G. *et al.* The G972R variant of the insulin receptor substrate-1 (IRS-1) gene is associated with insulin resistance in "uncomplicated" obese subjects evaluated by hyperinsulinemic-euglycemic clamp. *J Endocrinol Invest* **2004**, *27*, 754–9.

125 ZHANG, Y. *et al.* UKPDS 19: heterogeneity in NIDDM: separate contributions of IRS-1 and beta 3-adrenergic-receptor mutations to insulin resistance and obesity respectively with no evidence for glycogen synthase gene mutations. UK Prospective Diabetes Study. *Diabetologia* **1996**, *39*, 1505–11.

126 ALMIND, K. *et al.* A common amino acid polymorphism in insulin receptor substrate-1 causes impaired insulin signaling. Evidence from transfection studies. *J Clin Invest* **1996**, *97*, 2569–75.

127 INOUE, H. *et al.* Sequence variants in the sulfonylurea receptor (SUR) gene are associated with NIDDM in Caucasians. *Diabetes* **1996**, *45*, 825–31.

128 LAUKKANEN, O. *et al.* Polymorphisms of the SUR1 (ABCC8) and Kir6.2 (KCNJ11) genes predict the conversion from impaired glucose tolerance to type 2 diabetes. The Finnish Diabetes Prevention Study. *J Clin Endocrinol Metab* **2004**, *89*, 6286–90.

129 REIS, A.F. *et al.* Association of a variant in exon 31 of the sulfonylurea receptor 1 (SUR1) gene with type 2 diabetes mellitus in French Caucasians. *Hum Genet* **2000**, *107*, 138–44.

130 RISSANEN, J. *et al.* Sulfonylurea receptor 1 gene variants are associated with gestational diabetes and type 2 diabetes but not with altered secretion of insulin. *Diabetes Care* **2000**, *23*, 70–3.

131 OHTA, Y. *et al.* Identification and functional analysis of sulfonylurea receptor 1 variants in Japanese patients with NIDDM. *Diabetes* **1998**, *47*, 476–81.

132 ISHIYAMA-SHIGEMOTO, S. *et al.* Association of polymorphisms in the beta2-adrenergic receptor gene with obesity, hypertriglyceridaemia, and diabetes mellitus. *Diabetologia* **1999**, *42*, 98–101.

133 CARLSSON, M. *et al.* Common variants in the beta2-(Gln27Glu) and beta3-(Trp64Arg)-adrenoceptor genes are associated with elevated serum NEFA concentrations and type II diabetes. *Diabetologia* **2001**, *44*, 629–36.

134 KIM, S.H. *et al.* Significance of beta2-adrenergic receptor gene polymorphism in obesity and type 2 diabetes mellitus in Korean subjects. *Metabolism* **2002**, *51*, 833–7.

135 CHANG, T.J. *et al.* The Arg16Gly polymorphism of human beta2-adrenoreceptor is associated with type 2 diabetes in Taiwanese people. *Clin Endocrinol (Oxford)* **2002**, *57*, 685–90.

136 FUJISAWA, T. *et al.* Association of Trp64Arg mutation of the beta3-adrenergic-receptor with NIDDM and body weight gain. *Diabetologia* **1996**, *39*, 349–52.

137 OIZUMI, T. *et al.* Genotype Arg/Arg, but not Trp/Arg, of the Trp64Arg polymorphism of the beta(3)-adrenergic receptor is associated with type 2 diabetes and obesity in a large

Japanese sample. *Diabetes Care* **2001**, *24*, 1579–83.

138 WIDEN, E. *et al.* Association of a polymorphism in the beta 3-adrenergic-receptor gene with features of the insulin resistance syndrome in Finns. *N Engl J Med* **1995**, *333*, 348–51.

139 HEGELE, R.A. *et al.* Absence of association between genetic variation of the beta 3-adrenergic receptor and metabolic phenotypes in Oji-Cree. *Diabetes Care* **1998**, *21*, 851–4.

140 OEVEREN VAN-DYBICZ, A.M. *et al.* Beta 3-adrenergic receptor gene polymorphism and type 2 diabetes in a Caucasian population. *Diabetes Obes Metab* **2001**, *3*, 47–51.

141 BOULLU-SANCHIS, S. *et al.* Type 2 diabetes mellitus: association study of five candidate genes in an Indian population of Guadeloupe, genetic contribution of FABP2 polymorphism. *Diabetes Metab* **1999**, *25*, 150–6.

142 VENDRELL, J. *et al.* Beta 3-adrenoreceptor gene polymorphism and leptin. Lack of relationship in type 2 diabetic patients. *Clin Endocrinol (Oxford)* **1998**, *49*, 679–83.

143 BUETTNER, R. *et al.* The Trp64Arg polymorphism of the beta 3-adrenergic receptor gene is not associated with obesity or type 2 diabetes mellitus in a large population-based Caucasian cohort. *J Clin Endocrinol Metab* **1998**, *83*, 2892–7.

144 LEPRETRE, F. *et al.* Genetic studies of polymorphisms in ten non-insulin-dependent diabetes mellitus candidate genes in Tamil Indians from Pondichery. *Diabetes Metab* **1998**, *24*, 244–50.

145 RISSANEN, J. *et al.* The Trp64Arg polymorphism of the beta 3-adrenergic receptor gene. Lack of association with NIDDM and features of insulin resistance syndrome. *Diabetes Care* **1997**, *20*, 1319–23.

146 GU, H.F. *et al.* Single nucleotide polymorphisms in the proximal promoter region of the adiponectin (APM1) gene are associated with type 2 diabetes in Swedish caucasians. *Diabetes* **2004**, *53* (Suppl 1), S31–5.

147 VASSEUR, F. *et al.* Single-nucleotide polymorphism haplotypes in the both proximal promoter and exon 3 of the APM1 gene modulate adipocyte-secreted adiponectin hormone levels and contribute to the genetic risk for type 2 diabetes in French Caucasians. *Hum Mol Genet* **2002**, *11*, 2607–14.

148 HARA, K. *et al.* Genetic variation in the gene encoding adiponectin is associated with an increased risk of type 2 diabetes in the Japanese population. *Diabetes* **2002**, *51*, 536–40.

149 MENZAGHI, C. *et al.* A haplotype at the adiponectin locus is associated with obesity and other features of the insulin resistance syndrome. *Diabetes* **2002**, *51*, 2306–12.

150 VOZAROVA DE COURTEN, B. *et al.* Common polymorphisms in the adiponectin gene ACDC are not associated with diabetes in Pima Indians. *Diabetes* **2005**, *54*, 284–9.

151 POPULAIRE, C. *et al.* Does the −11377 promoter variant of APM1 gene contribute to the genetic risk for Type 2 diabetes mellitus in Japanese families? *Diabetologia* **2003**, *46*, 443–5.

152 EVANS, J.C. *et al.* Studies of association between the gene for calpain-10 and type 2 diabetes mellitus in the United Kingdom. *Am J Hum Genet* **2001**, *69*, 544–52.

153 ELBEIN, S.C. *et al.* Role of calpain-10 gene variants in familial type 2 diabetes in Caucasians. *J Clin Endocrinol Metab* **2002**, *87*, 650–4.

154 TSAI, H.J. *et al.* Type 2 diabetes and three calpain-10 gene polymorphisms in Samoans: no evidence of association. *Am J Hum Genet* **2001**, *69*, 1236–44.

155 CHEN, Y. *et al.* Calpain-10 gene polymorphisms and type 2 diabetes in West Africans: the Africa America Diabetes Mellitus (AADM) Study. *Ann Epidemiol* **2005**, *15*, 153–9.

156 IWASAKI, N. *et al.* Genetic variants in the calpain-10 gene and the development of type 2 diabetes in the Japanese population. *J Hum Genet* **2005**, *50*, 92–8.

157 DEL BOSQUE-PLATA, L. *et al.* Association of the calpain-10 gene with type 2 diabetes mellitus in a Mexican population. *Mol Genet Metab* **2004**, *81*, 122–6.

158 DAIMON, M. *et al.* Calpain 10 gene polymorphisms are related, not to type 2 diabetes, but to increased serum cholesterol in Japanese. *Diabetes Res Clin Pract* **2002**, *56*, 147–52.

159 GARANT, M.J. *et al.* SNP43 of CAPN10 and the risk of type 2 diabetes in African-Americans: the Atherosclerosis Risk in Communities Study. *Diabetes* **2002**, *51*, 231–7.

160 HEGELE, R.A. *et al.* Absence of association of type 2 diabetes with CAPN10 and PC-1 polymorphisms in Oji-Cree. *Diabetes Care* **2001**, *24*, 1498–9.

161 KUBASZEK, A. *et al.* The association of the K121Q polymorphism of the plasma cell glycoprotein-1 gene with type 2 diabetes and hypertension depends on size at birth. *J Clin Endocrinol Metab* **2004**, *89*, 2044–7.

162 HAMAGUCHI, K. *et al.* The PC-1 Q121 allele is exceptionally prevalent in the Dominican Republic and is associated with type 2 diabetes. *J Clin Endocrinol Metab* **2004**, *89*, 1359–64.

163 LAUKKANEN, O. *et al.* Common polymorphisms in the genes regulating the early insulin signalling pathway: effects on weight change and the conversion from impaired glucose tolerance to Type 2 diabetes. The Finnish Diabetes Prevention Study. *Diabetologia* **2004**, *47*, 871–7.

164 RASMUSSEN, S.K. *et al.* The K121Q variant of the human PC-1 gene is not associated with insulin resistance or type 2 diabetes among Danish Caucasians. *Diabetes* **2000**, *49*, 1608–11.

165 DUARTE, N.L. *et al.* Obesity, type II diabetes and the Ala54Thr polymorphism of fatty acid binding protein 2 in the Tongan population. *Mol Genet Metab* **2003**, *79*, 183–8.

166 XIANG, K. *et al.* The impact of codon 54 variation in intestinal fatty acid binding protein gene on the pathogenesis of diabetes mellitus in Chinese. *Chin Med J (Engl)* **1999**, *112*, 99–102.

167 HAYAKAWA, T. *et al.* Variation of the fatty acid binding protein 2 gene is not associated with obesity and insulin resistance in Japanese subjects. *Metabolism* **1999**, *48*, 655–7.

168 ITO, K. *et al.* Codon 54 polymorphism of the fatty acid binding protein gene and insulin resistance in the Japanese population. *Diabet Med* **1999**, *16*, 119–24.

169 YAGI, T. *et al.* A population association study of four candidate genes (hexokinase II, glucagon-like peptide-1 receptor, fatty acid binding protein-2, and apolipoprotein C-II) with type 2 diabetes and impaired glucose tolerance in Japanese subjects. *Diabet Med* **1996**, *13*, 902–7.

170 HIRAI, M. *et al.* Group specific component protein genotype is associated with NIDDM in Japan. *Diabetologia* **1998**, *41*, 742–3.

171 MALECKI, M.T. *et al.* Vitamin D binding protein gene and genetic susceptibility to type 2 diabetes mellitus in a Polish population. *Diabetes Res Clin Pract* **2002**, *57*, 99–104.

172 YE, W.Z. *et al.* Variations in the vitamin D-binding protein (Gc locus) and risk of type 2 diabetes mellitus in French Caucasians. *Metabolism* **2001**, *50*, 366–9.

173 KLUPA, T. *et al.* Amino acid variants of the vitamin D-binding protein and risk of diabetes in white Americans of European origin. *Eur J Endocrinol* **1999**, *141*, 490–3.

174 GOUGH, S.C. *et al.* Mutation of the glucagon receptor gene and diabetes mellitus in the UK: association or founder effect? *Hum Mol Genet* **1995**, *4*, 1609–12.

175 HAGER, J. *et al.* A missense mutation in the glucagon receptor gene is associated with non-insulin-dependent diabetes mellitus. *Nat Genet* **1995**, *9*, 299–304.

176 DENG, H., TANG, W.L., PAN, Q. [Gly40Ser mutation of glucagon receptor gene and NIDDM in Han nationality]. *Hunan Yi Ke Da Xue Xue Bao* **2001**, *26*, 291–3.

177 SHIOTA, D. *et al.* Role of the Gly40Ser mutation in the glucagon receptor gene in Brazilian patients with type 2 diabetes mellitus. *Pancreas* **2002**, *24*, 386–90.

178 HUANG, C.N. *et al.* Screening for the Gly40Ser mutation in the glucagon receptor gene among patients with type 2 diabetes or essential hypertension in Taiwan. *Pancreas* **1999**, *18*, 151–5.

179 BABADJANOVA, G. *et al.* Polymorphism of the glucagon receptor gene and non-insulin-dependent diabetes mellitus in the Russian population. *Exp Clin Endocrinol Diabetes* **1997**, *105*, 225–6.

180 TONOLO, G. *et al.* Physiological and genetic characterization of the Gly40Ser mutation in the glucagon receptor gene in the Sardinian population. The Sardinian Diabetes Genetic Study Group. *Diabetologia* **1997**, *40*, 89–94.

181 ODAWARA, M., TACHI, Y., and YAMASHITA, K. Absence of association between the Gly40 → Ser mutation in the human glucagon receptor and Japanese patients with non-insulin-dependent diabetes mellitus or impaired glucose tolerance. *Hum Genet* **1996**, *98*, 636–9.

182 OGATA, M. *et al.* Absence of the Gly40-ser mutation in the glucagon receptor gene in Japanese subjects with NIDDM. *Diabetes Res Clin Pract* **1996**, *33*, 71–4.

183 RISTOW, M. *et al.* Restricted geographical extension of the association of a glucagon receptor gene mutation (Gly40Ser) with non-insulin-dependent diabetes mellitus. *Diabetes Res Clin Pract* **1996**, *32*, 183–5.

184 HUANG, X. *et al.* Lack of association between the Gly40Ser polymorphism in the glucagon receptor gene and NIDDM in Finland. *Diabetologia* **1995**, *38*, 1246–8.

185 FUJISAWA, T. *et al.* A mutation in the glucagon receptor gene (Gly40Ser): heterogeneity in the association with

diabetes mellitus. *Diabetologia* **1995**, *38*, 983–5.

186 CHIU, K.C., PROVINCE, M.A., and PERMUTT, M.A. Glucokinase gene is genetic marker for NIDDM in American blacks. *Diabetes* **1992**, *41*, 843–9.

187 CHIU, K.C. *et al.* A genetic marker at the glucokinase gene locus for type 2 (non-insulin-dependent) diabetes mellitus in Mauritian Creoles. *Diabetologia* **1992**, *35*, 632–8.

188 BARROSO, I. *et al.* Candidate gene association study in type 2 diabetes indicates a role for genes involved in beta-cell function as well as insulin action. *PLoS Biol* **2003**, *1*, 41.

189 MAJER, M. *et al.* Association of the glycogen synthase locus on 19q13 with NIDDM in Pima Indians. *Diabetologia* **1996**, *39*, 314–21.

190 The Met416 → Val variant in the glycogen synthase gene: the prevalence and the association with diabetes in a large number of Japanese individuals. Study Group for the Identification of Type 2 Diabetes Genes in Japanese. *Diabetes Care* **2000**, *23*, 1709–10.

191 RISSANEN, J. *et al.* New variants in the glycogen synthase gene (Gln71His, Met416Val) in patients with NIDDM from eastern Finland. *Diabetologia* **1997**, *40*, 1313–19.

192 SHIMOMURA, H. *et al.* A missense mutation of the muscle glycogen synthase gene (M416V) is associated with insulin resistance in the Japanese population. *Diabetologia* **1997**, *40*, 947–52.

193 BABADJANOVA, G. *et al.* Polymorphism of the glycogen synthase gene and non-insulin-dependent diabetes mellitus in the Russian population. *Metabolism* **1997**, *46*, 121–2.

194 GROOP, L.C. *et al.* Association between polymorphism of the glycogen synthase gene and non-insulin-dependent diabetes mellitus. *N Engl J Med* **1993**, *328*, 10–14.

195 KUROYAMA, H. *et al.* Simple tandem repeat DNA polymorphism in the human glycogen synthase gene is associated with NIDDM in Japanese

References 259

subjects. *Diabetologia* **1994**, *37*, 536–9.

196 SHIMOKAWA, K. *et al.* Molecular scanning of the glycogen synthase and insulin receptor substrate-1 genes in Japanese subjects with non-insulin-dependent diabetes mellitus. *Biochem Biophys Res Commun* **1994**, *202*, 463–9.

197 HEGELE, R.A. *et al.* Hepatocyte nuclear factor-1 alpha G319S. A private mutation in Oji-Cree associated with type 2 diabetes. *Diabetes Care* **1999**, *22*, 524.

198 BABAYA, N. *et al.* Hepatocyte nuclear factor-1alpha gene and non-insulin-dependent diabetes mellitus in the Japanese population. *Acta Diabetol* **1998**, *35*, 150–3.

199 URHAMMER, S.A. *et al.* Genetic variation in the hepatocyte nuclear factor-1 alpha gene in Danish Caucasians with late-onset NIDDM. *Diabetologia* **1997**, *40*, 473–5.

200 FURUTA, H. *et al.* Nonsense and missense mutations in the human hepatocyte nuclear factor-1 beta gene (TCF2) and their relation to type 2 diabetes in Japanese. *J Clin Endocrinol Metab* **2002**, *87*, 3859–63.

201 BABAYA, N. *et al.* Lack of association between hepatocyte nuclear factor-1beta gene and common forms of type 2 diabetes in the Japanese population. *Diabetes Nutr Metab* **2001**, *14*, 220–4.

202 ZHU, Q. *et al.* T130I mutation in HNF-4alpha gene is a loss-of-function mutation in hepatocytes and is associated with late-onset Type 2 diabetes mellitus in Japanese subjects. *Diabetologia* **2003**, *46*, 567–73.

203 EK, J. *et al.* The functional Thr130Ile and Val255Met polymorphisms of the hepatocyte nuclear factor-4 α (HNF-4A) gene associate with type 2 diabetes and altered β-cell function among Danes. *J Clin Endocrinol Metab* **2005**, *90*, 3054–9.

204 WINCKLER, W. *et al.* Association testing of variants in the hepatocyte nuclear factor 4alpha gene with risk of type 2 diabetes in 7,883 people. *Diabetes* **2005**, *54*, 886–92.

205 NAKAJIMA, H. *et al.* Hepatocyte nuclear factor-4 alpha gene mutations in Japanese non-insulin dependent diabetes mellitus (NIDDM) patients. *Res Commun Mol Pathol Pharmacol* **1996**, *94*, 327–30.

206 MITCHELL, S.M. *et al.* Rare variants identified in the HNF-4 alpha beta-cell-specific promoter and alternative exon 1 lack biological significance in maturity onset diabetes of the young and young onset type II diabetes. *Diabetologia* **2002**, *45*, 1344–8.

207 LOVE-GREGORY, L.D. *et al.* A common polymorphism in the upstream promoter region of the hepatocyte nuclear factor-4 alpha gene on chromosome 20q is associated with type 2 diabetes and appears to contribute to the evidence for linkage in an ashkenazi jewish population. *Diabetes* **2004**, *53*, 1134–40.

208 HANSEN, S.K. *et al.* Variation near the hepatocyte nuclear factor (HNF)-4alpha gene associates with type 2 diabetes in the Danish population. *Diabetologia* **2005**, *48*, 452–8.

209 DAMCOTT, C.M. *et al.* Polymorphisms in both promoters of hepatocyte nuclear factor 4-alpha are associated with type 2 diabetes in the Amish. *Diabetes* **2004**, *53*, 3337–41.

210 SILANDER, K. *et al.* Genetic variation near the hepatocyte nuclear factor-4 alpha gene predicts susceptibility to type 2 diabetes. *Diabetes* **2004**, *53*, 1141–9.

211 WEEDON, M.N. *et al.* Common variants of the hepatocyte nuclear factor-4alpha P2 promoter are associated with type 2 diabetes in the U.K. population. *Diabetes* **2004**, *53*, 3002–6.

212 BAGWELL, A.M. *et al.* Genetic analysis of HNF4A polymorphisms in Caucasian-American type 2 diabetes. *Diabetes* **2005**, *54*, 1185–90.

213 SEINO, S. S20G mutation of the amylin gene is associated with type II diabetes in Japanese. Study Group of Comprehensive Analysis of Genetic Factors in Diabetes Mellitus. *Diabetologia* **2001**, *44*, 906–9.

214 LEE, S.C. *et al.* The islet amyloid polypeptide (amylin) gene S20G mutation in Chinese subjects: evidence for associations with type 2 diabetes and cholesterol levels. *Clin Endocrinol (Oxford)* **2001**, *54*, 541–6.

215 SAKAGASHIRA, S. *et al.* Missense mutation of amylin gene (S20G) in Japanese NIDDM patients. *Diabetes* **1996**, *45*, 1279–81.

216 HAYAKAWA, T. *et al.* S20G mutation of the amylin gene in Japanese patients with type 2 diabetes. *Diabetes Res Clin Pract* **2000**, *49*, 195–7.

217 POA, N.R., COOPER, G.J., EDGAR, P.F. Amylin gene promoter mutations predispose to type 2 diabetes in New Zealand Maori. *Diabetologia* **2003**, *46*, 574–8.

218 NOVIALS, A. *et al.* A novel mutation in islet amyloid polypeptide (IAPP) gene promoter is associated with type II diabetes mellitus. *Diabetologia* **2001**, *44*, 1064–5.

219 PILDAL, J. *et al.* Studies of variability in the islet amyloid polypeptide gene in relation to type 2 diabetes. *Diabet Med* **2003**, *20*, 491–4.

220 VAESSEN, N. *et al.* A polymorphism in the gene for IGF-I: functional properties and risk for type 2 diabetes and myocardial infarction. *Diabetes* **2001**, *50*, 637–42.

221 FRAYLING, T.M. *et al.* A putative functional polymorphism in the IGF-I gene: association studies with type 2 diabetes, adult height, glucose tolerance, and fetal growth in U.K. populations. *Diabetes* **2002**, *51*, 2313–16.

222 KAMBO, P.K. The genetic predisposition to fibrocalculous pancreatic diabetes. *Diabetologia* **1989**, *32*, 45–51.

223 ONG, K. The insulin gene VNTR, type 2 diabetes and birth weight genetics. *Nature* **1999**, *21*, 262–263.

224 MEIGS, J.B. *et al.* The insulin gene variable number tandem repeat and risk of type 2 diabetes in a population-based sample of families and unrelated men and women. *J Clin Endocrinol Metab* **2005**, *90*, 1137–43.

225 JOWETT, N.I. Diabetic hypertriglyceridaemia and related 5′ flanking polymorphism of the human insulin gene. *BMJ* **1984**, *288*, 96–99.

226 HANSEN, S.K. *et al.* Large-scale studies of the HphI insulin gene variable-number-of-tandem-repeats polymorphism in relation to type 2 diabetes mellitus and insulin release. *Diabetologia* **2004**, *47*, 1079–87.

227 LINDSAY, R.S. *et al.* The insulin gene variable number tandem repeat class I/III polymorphism is in linkage disequilibrium with birth weight but not type 2 diabetes in the Pima population. *Diabetes* **2003**, *52*, 187–93.

228 AOYAMA, N. *et al.* Low frequency of 5′-flanking insertion of human insulin gene in Japanese non-insulin-dependent diabetic subjects. *Diabetes Care* **1986**, *9*, 365–9.

229 ELBEIN, S.C. Lack of association of the polymorphic locus in the 5′-flanking region of the human insulin gene and diabetes in american blacks. *Diabetes* **1985**, *34*, 433–9.

230 MORGAN, R. Allelic variants at insulin-receptor and insulin gene loci and susceptibility in NIDDM in Welsh population. *Diabetes* **1990**, *39*, 1479–84.

231 HART, L.M. *et al.* Association of the insulin-receptor variant Met-985 with hyperglycemia and non-insulin-dependent diabetes mellitus in the Netherlands: a population-based study. *Am J Hum Genet* **1996**, *59*, 1119–25.

232 HART, L.M. *et al.* Prevalence of variants in candidate genes for type 2 diabetes mellitus in The Netherlands: the Rotterdam study and the Hoorn study. *J Clin Endocrinol Metab* **1999**, *84*, 1002–6.

233 HANSEN, L. *et al.* The Val985Met insulin-receptor variant in the Danish Caucasian population: lack of associations with non-insulin-dependent diabetes mellitus or insulin resistance. *Am J Hum Genet* **1997**, *60*, 1532–5.

234 MACFARLANE, W.M. *et al.* Missense mutations in the insulin promoter factor-1 gene predispose to type 2 diabetes. *J Clin Invest* **1999**, *104*, R33–9.

235 SHIAU, M.Y. *et al.* Missense mutations in the human insulin promoter factor-1 gene are not a common cause of

type 2 diabetes mellitus in Taiwan. *J Endocrinol Invest* **2004**, *27*, 1076–80.

236 HANI, E.H. *et al.* Defective mutations in the insulin promoter factor-1 (IPF-1) gene in late-onset type 2 diabetes mellitus. *J Clin Invest* **1999**, *104*, R41–8.

237 ALMIND, K. *et al.* Amino acid polymorphisms of insulin receptor substrate-1 in non-insulin-dependent diabetes mellitus. *Lancet* **1993**, *342*, 828–32.

238 SIGAL, R.J. *et al.* Codon 972 polymorphism in the insulin receptor substrate-1 gene, obesity, and risk of noninsulin-dependent diabetes mellitus. *J Clin Endocrinol Metab* **1996**, *81*, 1657–9.

239 FLORES-MARTINEZ, S.E. *et al.* DNA polymorphism analysis of candidate genes for type 2 diabetes mellitus in a Mexican ethnic group. *Ann Genet* **2004**, *47*, 339–48.

240 BEZERRA, R.M. *et al.* Lack of Arg972 polymorphism in the IRS1 gene in Parakana Brazilian Indians. *Hum Biol* **2004**, *76*, 147–51.

241 VAN DAM, R.M. *et al.* The insulin receptor substrate-1 Gly972Arg polymorphism is not associated with type 2 diabetes mellitus in two population-based studies. *Diabet Med* **2004**, *21*, 752–8.

242 CELI, F.S. *et al.* Molecular scanning for mutations in the insulin receptor substrate-1 (IRS-1) gene in Mexican Americans with type 2 diabetes mellitus. *Diabetes Metab Res Rev* **2000**, *16*, 370–7.

243 LAAKSO, M. *et al.* Insulin receptor substrate-1 variants in non-insulin-dependent diabetes. *J Clin Invest* **1994**, *94*, 1141–6.

244 HITMAN, G.A. *et al.* Insulin receptor substrate-1 gene mutations in NIDDM; implications for the study of polygenic disease. *Diabetologia* **1995**, *38*, 481–6.

245 YAMADA, K. *et al.* Codon 972 polymorphism of the insulin receptor substrate-1 gene in impaired glucose tolerance and late-onset NIDDM. *Diabetes Care* **1998**, *21*, 753–6.

246 STEFAN, N. *et al.* Metabolic effects of the Gly1057Asp polymorphism in IRS-2 and interactions with obesity. *Diabetes* **2003**, *52*, 1544–50.

247 MAMMARELLA, S. *et al.* Interaction between the G1057D variant of IRS-2 and overweight in the pathogenesis of type 2 diabetes. *Hum Mol Genet* **2000**, *9*, 2517–21.

248 D'ALFONSO, R. *et al.* Polymorphisms of the insulin receptor substrate-2 in patients with type 2 diabetes. *J Clin Endocrinol Metab* **2003**, *88*, 317–22.

249 BERNAL, D. *et al.* Insulin receptor substrate-2 amino acid polymorphisms are not associated with random type 2 diabetes among Caucasians. *Diabetes* **1998**, *47*, 976–9.

250 GLOYN, A.L. Association studies of variants in promoter and coding regions of beta-cell ATP-sensitive K-channel genes SUR1 and Kir6.2 with type 2 diabetes mellitus (UKPDS 53). *Diabetic Med* **2001**, *18*, 206–12.

251 SCHWANSTECHER, C., SCHWANSTECHER, M. Nucleotide sensitivity of pancreatic ATP-sensitive potassium channels and type 2 diabetes. *Diabetes* **2002**, *51* (Suppl 3), S358–62.

252 INOUE, H. *et al.* Sequence variants in the pancreatic islet beta-cell inwardly rectifying K+ channel Kir6.2 (Bir) gene: identification and lack of role in Caucasian patients with NIDDM. *Diabetes* **1997**, *46*, 502–7.

253 SAKURA, H. *et al.* Sequence variations in the human Kir6.2 gene, a subunit of the beta-cell ATP-sensitive K-channel: no association with NIDDM in while Caucasian subjects or evidence of abnormal function when expressed in vitro. *Diabetologia* **1996**, *39*, 1233–6.

254 OHTOSHI, K. *et al.* Association of (−)786T-C mutation of endothelial nitric oxide synthase gene with insulin resistance. *Diabetologia* **2002**, *45*, 1594–601.

255 MONTI, L.D. *et al.* Endothelial nitric oxide synthase polymorphisms are associated with type 2 diabetes and the insulin resistance syndrome. *Diabetes* **2003**, *52*, 1270–5.

256 BAIER, L.J. *et al.* Variant in the regulatory subunit of phosphatidylinositol 3-kinase (p85alpha): preliminary evidence indicates a potential role of this variant in the acute insulin response and type 2 diabetes in Pima women. *Diabetes* **1998**, *47*, 973–5.

257 KAWANISHI, M. *et al.* Prevalence of a polymorphism of the phosphatidylinositol 3-kinase p85 alpha regulatory subunit (codon 326 Met → Ile) in Japanese NIDDM patients. *Diabetes Care* **1997**, *20*, 1043.

258 DONEY, A.S. *et al.* Association of the Pro12Ala and C1431T variants of PPARG and their haplotypes with susceptibility to Type 2 diabetes. *Diabetologia* **2004**, *47*, 555–8.

259 EVANS, D. *et al.* Association between the P12A and c1431t polymorphisms in the peroxisome proliferator activated receptor gamma (PPAR gamma) gene and type 2 diabetes. *Exp Clin Endocrinol Diabetes* **2001**, *109*, 151–4.

260 LI, W.D., LEE, J.H., PRICE, R.A. The peroxisome proliferator-activated receptor gamma 2 Pro12Ala mutation is associated with early onset extreme obesity and reduced fasting glucose. *Mol Genet Metab* **2000**, *70*, 159–61.

261 ANDRULIONYTE, L. *et al.* Common polymorphisms of the PPAR-gamma2 (Pro12Ala) and PGC-1alpha (Gly482Ser) genes are associated with the conversion from impaired glucose tolerance to type 2 diabetes in the STOP-NIDDM trial. *Diabetologia* **2004**, *47*, 2176–84.

262 MALECKI, M.T. *et al.* The Pro12Ala polymorphism of PPARgamma2 gene and susceptibility to type 2 diabetes mellitus in a Polish population. *Diabetes Res Clin Pract* **2003**, *62*, 105–11.

263 LEI, H.H. *et al.* Peroxisome proliferator-activated receptor gamma 2 Pro12Ala gene variant is strongly associated with larger body mass in the Taiwanese. *Metabolism* **2000**, *49*, 1267–70.

264 CLEMENT, K. *et al.* The Pro115Gln and Pro12Ala PPAR gamma gene mutations in obesity and type 2 diabetes. *Int J Obes Relat Metab Disord* **2000**, *24*, 391–3.

265 MEIRHAEGHE, A. *et al.* Impact of the peroxisome proliferator activated receptor gamma2 Pro12Ala polymorphism on adiposity, lipids and non-insulin-dependent diabetes mellitus. *Int J Obes Relat Metab Disord* **2000**, *24*, 195–9.

266 MANCINI, F.P. *et al.* Pro12Ala substitution in the peroxisome proliferator-activated receptor-gamma2 is not associated with type 2 diabetes. *Diabetes* **1999**, *48*, 1466–8.

267 RINGEL, J. *et al.* Pro12Ala missense mutation of the peroxisome proliferator activated receptor gamma and diabetes mellitus. *Biochem Biophys Res Commun* **1999**, *254*, 450–3.

268 EK, J. *et al.* Mutation analysis of peroxisome proliferator-activated receptor-gamma coactivator-1 (PGC-1) and relationships of identified amino acid polymorphisms to type II diabetes mellitus. *Diabetologia* **2001**, *44*, 2220–6.

269 HARA, K. *et al.* A genetic variation in the PGC-1 gene could confer insulin resistance and susceptibility to type II diabetes. *Diabetologia* **2002**, *45*, 740–3.

270 LACQUEMANT, C. *et al.* No association between the G482S polymorphism of the proliferator-activated receptor-gamma coactivator-1 (PGC-1) gene and type II diabetes in French Caucasians. *Diabetologia* **2002**, *45*, 602–3; author reply 604.

271 MULLER, Y.L. *et al.* A Gly482Ser missense mutation in the peroxisome proliferator-activated receptor gamma coactivator-1 is associated with altered lipid oxidation and early insulin secretion in Pima Indians. *Diabetes* **2003**, *52*, 895–8.

272 OBERKOFLER, H. *et al.* Complex haplotypes of the PGC-1alpha gene are associated with carbohydrate metabolism and type 2 diabetes. *Diabetes* **2004**, *53*, 1385–93.

273 WANG, G. *et al.* The association between PPP1R3 gene polymorphisms and type 2 diabetes mellitus. *Chin Med J (Engl)* **2001**, *114*, 1258–62.

274 XIA, J. *et al.* A common variant in

PPP1R3 associated with insulin resistance and type 2 diabetes. *Diabetes* **1998**, *47*, 1519–24.

275 CHEN, M.W. *et al.* [Study on the association of PPP1R3 gene polymorphism with type 2 diabetes in Han population of Anhui province]. *Zhonghua Liu Xing Bing Xue Za Zhi* **2004**, *25*, 534–6.

276 HANSEN, L. *et al.* Polymorphism in the glycogen-associated regulatory subunit of type 1 protein phosphatase (PPP1R3) gene and insulin sensitivity. *Diabetes* **2000**, *49*, 298–301.

277 SHEN, G.Q. *et al.* Asp905Tyr polymorphism of the gene for the skeletal muscle-specific glycogen-targeting subunit of protein phosphatase 1 in NIDDM. *Diabetes Care* **1998**, *21*, 1086–9.

278 HANSEN, L. *et al.* A widespread amino acid polymorphism at codon 905 of the glycogen-associated regulatory subunit of protein phosphatase-1 is associated with insulin resistance and hypersecretion of insulin. *Hum Mol Genet* **1995**, *4*, 1313–20.

279 CHEN, M.W. *et al.* [Study on association of PPP1R3 gene 5 bp deletion/insertion within 3′-untranslated region polymorphism with type 2 diabetes]. *Zhonghua Yi Xue Yi Chuan Xue Za Zhi* **2004**, *21*, 29–31.

280 WANG, G.Y. *et al.* A novel Rad gene polymorphism combined with obesity increases risk for type 2 diabetes mellitus. *Chin Med J (Engl)* **2004**, *117*, 770–1.

281 DORIA, A. *et al.* Trinucleotide repeats at the rad locus. Allele distributions in NIDDM and mapping to a 3-cM region on chromosome 16q. *Diabetes* **1995**, *44*, 243–7.

282 YUAN, X. *et al.* Analysis of trinucleotide-repeat combination polymorphism at the rad gene in patients with type 2 diabetes mellitus. *Metabolism* **1999**, *48*, 173–5.

283 ORHO, M. *et al.* Polymorphism at the rad gene is not associated with NIDDM in Finns. *Diabetes* **1996**, *45*, 429–33.

284 MATSUBARA, A. *et al.* Sequence variations of the pancreatic islet/liver glucose transporter (GLUT2) gene in

Japanese subjects with noninsulin dependent diabetes mellitus. *J Clin Endocrinol Metab* **1995**, *80*, 3131–5.

285 CHA, J.Y. *et al.* Analysis of polymorphism of the GLUT2 promoter in NIDDM patients and its functional consequence to the promoter activity. *Ann Clin Lab Sci* **2002**, *32*, 114–22.

286 MOLLER, A.M. *et al.* Studies of genetic variability of the glucose transporter 2 promoter in patients with type 2 diabetes mellitus. *J Clin Endocrinol Metab* **2001**, *86*, 2181–6.

287 DAY, C.P. *et al.* Tumour necrosis factor-alpha gene promoter polymorphism and decreased insulin resistance. *Diabetologia* **1998**, *41*, 430–4.

288 SHIAU, M.Y. *et al.* TNF-alpha polymorphisms and type 2 diabetes mellitus in Taiwanese patients. *Tissue Antigens* **2003**, *61*, 393–7.

289 HAMANN, A. *et al.* Genetic variability in the TNF-alpha promoter is not associated with type II diabetes mellitus (NIDDM). *Biochem Biophys Res Commun* **1995**, *211*, 833–9.

290 KUBASZEK, A. *et al.* Promoter polymorphisms of the TNF-alpha (G-308A) and IL-6 (C-174G) genes predict the conversion from impaired glucose tolerance to type 2 diabetes: the Finnish Diabetes Prevention Study. *Diabetes* **2003**, *52*, 1872–6.

291 KO, G.T. *et al.* Tumour necrosis factor-alpha promoter gene polymorphism at −308 (genotype AA) in Chinese subjects with Type 2 diabetes. *Diabet Med* **2003**, *20*, 167–8.

292 SHEN, H., XIANG, K., and JIA, W. [Effects of uncoupling protein 3 gene −55 C → T variant on lipid metabolism, body fat, its distribution and non-insulin-dependent diabetes mellitus in Chinese]. *Zhonghua Yi Xue Yi Chuan Xue Za Zhi* **2002**, *19*, 317–21.

293 MEIRHAEGHE, A. *et al.* An uncoupling protein 3 gene polymorphism associated with a lower risk of developing Type II diabetes and with atherogenic lipid profile in a French cohort. *Diabetologia* **2000**, *43*, 1424–8.

294 CARDON, L.R., PALMER, L.J. Population

stratification and spurious allelic association. *Lancet* **2003**, *361*, 598–604.

295 IOANNIDIS, J.P. *et al.* Genetic associations in large versus small studies: an empirical assessment. *Lancet* **2003**, *361*, 567–71.

296 HIRSCHHORN, J.N., DALY, M.J. Genome-wide association studies for common diseases and complex traits. *Nat Rev Genet* **2005**, *6*, 95–108.

297 DALY, M.J. *et al.* High-resolution haplotype structure in the human genome. *Nat Genet* **2001**, *29*, 229–32.

298 GABRIEL, S.B. *et al.* The structure of haplotype blocks in the human genome. *Science* **2002**, *296*, 2225–9.

299 REICH, D.E. *et al.* Linkage disequilibrium in the human genome. *Nature* **2001**, *411*, 199–204.

300 TERWILLIGER, J.D. *et al.* A biased assessment of the use of SNPs in human complex traits. *Curr Opin Genet Dev* **2002**, *12*, 726–34.

301 TIRET, L. *et al.* Heterogeneity of linkage disequilibrium in human genes has implications for association studies of common diseases. *Hum Mol Genet* **2002**, *11*, 419–29.

302 PATIL, N. *et al.* Blocks of limited haplotype diversity revealed by high-resolution scanning of human chromosome 21. *Science* **2001**, *294*, 1719–23.

303 KAMMERER, S. *et al.* Amino acid variant in the kinase binding domain of dual-specific A kinase-anchoring protein 2: a disease susceptibility polymorphism. *Proc Natl Acad Sci USA* **2003**, *100*, 4066–71.

304 JONASDOTTIR, A. *et al.* A whole genome association study in Icelandic multiple sclerosis patients with 4804 markers. *J Neuroimmunol* **2003**, *143*, 88–92.

305 LAAKSONEN, M. *et al.* A whole genome association study in Finnish multiple sclerosis patients with 3669 markers. *J Neuroimmunol* **2003**, *143*, 70–3.

306 YAMAGATA, K. *et al.* Mutations in the hepatocyte nuclear factor-4alpha gene in maturity-onset diabetes of the young (MODY1). *Nature* **1996**, *384*, 458–60.

307 VIONNET, N. *et al.* Nonsense mutation in the glucokinase gene causes early-onset non-insulin-dependent diabetes mellitus. *Nature* **1992**, *356*, 721–2.

308 YAMAGATA, K. *et al.* Mutations in the hepatocyte nuclear factor-1alpha gene in maturity-onset diabetes of the young (MODY3). *Nature* **1996**, *384*, 455–8.

309 STOFFERS, D.A. *et al.* Early-onset type-II diabetes mellitus (MODY4) linked to IPF1. *Nat Genet* **1997**, *17*, 138–9.

310 HORIKAWA, Y. *et al.* Mutation in hepatocyte nuclear factor-1 beta gene (TCF2) associated with MODY. *Nat Genet* **1997**, *17*, 384–5.

311 MALECKI, M.T. *et al.* Mutations in NEUROD1 are associated with the development of type 2 diabetes mellitus. *Nat Genet* **1999**, *23*, 323–8.

312 FAJANS, S.S., CONN, J.W. Prediabetes, subclinical diabetes and latent clinical diabetes: interpretation, diagnosis and treatment. In: *On the Nature and Treatment of Diabetes*, D.S. LEIBEL and G.S. WRENSHALL, editors. **1965**, Amsterdam: Excerpta Medica, pp. 641–656.

313 TATTERSALL, R.B. Mild familial diabetes with dominant inheritance. *Q J Med* **1974**, *43*, 339–57.

314 TATTERSALL, R.B., FAJANS, S.S. A difference between the inheritance of classical juvenile-onset and maturity-onset type diabetes of young people. *Diabetes* **1975**, *24*, 44–53.

315 BELL, G.I. *et al.* Gene for non-insulin-dependent diabetes mellitus (maturity-onset diabetes of the young subtype) is linked to DNA polymorphism on human chromosome 20q. *Proc Natl Acad Sci USA* **1991**, *88*, 1484–8.

316 FROGUEL, P. *et al.* Close linkage of glucokinase locus on chromosome 7p to early-onset non-insulin-dependent diabetes mellitus. *Nature* **1992**, *356*, 162–4.

317 VAXILLAIRE, M. *et al.* A gene for maturity onset diabetes of the young (MODY) maps to chromosome 12q. *Nat Genet* **1995**, *9*, 418–23.

318 FROGUEL, P. *et al.* Familial hyper-glycemia due to mutations in gluco-kinase. Definition of a subtype of diabetes mellitus. *N Engl J Med* **1993**, *328*, 697–702.

319 MOLLER, A.M. *et al.* Studies of the genetic variability of the coding region of the hepatocyte nuclear factor-4alpha in Caucasians with maturity onset NIDDM. *Diabetologia* **1997**, *40*, 980–3.

320 NISHI, S. *et al.* Mutations in the glucokinase gene are not a major cause of late-onset type 2 (non-insulin-dependent) diabetes mellitus in Japanese subjects. *Diabet Med* **1994**, *11*, 193–7.

321 PEARSON, E.R. *et al.* Genetic cause of hyperglycaemia and response to treatment in diabetes. *Lancet* **2003**, *362*, 1275–81.

322 SESTI, G. *et al.* The Arg972 variant in insulin receptor substrate-1 is associated with an increased risk of secondary failure to sulfonylurea in patients with type 2 diabetes. *Diabetes Care* **2004**, *27*, 1394–8.

323 STRIDE, A., HATTERSLEY, A.T. Different genes, different diabetes: lessons from maturity-onset diabetes of the young. *Ann Med* **2002**, *34*, 207–16.

324 WAREHAM, N.J., FRANKS, P.W., HARDING, A.H. Establishing the role of gene-environment interactions in the etiology of type 2 diabetes. *Endocrinol Metab Clin North Am* **2002**, *31*, 553–66.

325 KAPUT, J. Diet-disease gene interactions. *Nutrition* **2004**, *20*, 26–31.

326 VINCENT, S. *et al.* Genetic polymorphisms and lipoprotein responses to diets. *Proc Nutr Soc* **2002**, *61*, 427–34.

327 RIBALTA, J. *et al.* Apolipoprotein and apolipoprotein receptor genes, blood lipids and disease. *Curr Opin Clin Nutr Metab Care* **2003**, *6*, 177–87.

328 RANTALA, M. *et al.* Apolipoprotein B gene polymorphisms and serum lipids: meta-analysis of the role of genetic variation in responsiveness to diet. *Am J Clin Nutr* **2000**, *71*, 713–24.

329 MEIRHAEGHE, A. *et al.* The effect of the Gly16Arg polymorphism of the beta-adrenergic receptor gene on plasma free fatty acid levels is modulated by physical activity. *J Clin Endocrinol Metab* **2001**, *86*, 5881–7.

330 FRANKS, P.W. *et al.* Does peroxisome proliferator-activated receptor gamma genotype (Pro12ala) modify the association of physical activity and dietary fat with fasting insulin level? *Metabolism* **2004**, *53*, 11–16.

14
Gene Variants and Obesity

Günter Brönner, A. Hinney, K. Reichwald, A.-K. Wermter,
A. Scherag, S. Friedel, and Johannes Hebebrand

The molecular genetic basis of body weight regulation has been thoroughly analyzed in the past decade. Genetic studies were mainly triggered by the identification of the leptin gene in 1994. Candidate genes for obesity can be derived from different sources: (a) molecular genetic studies in humans, including syndromal forms of obesity; (b) animal models; (c) physiological considerations; and (d) genome-wide scans.

Single (extremely) rare monogenic forms of obesity have been elucidated; these forms are additionally associated with distinct endocrinological abnormalities. Obesity due to a mutation in the melanocortin-4 receptor gene (*MC4R*) represents an exception; functionally relevant mutations in *MC4R* occur in approximately 2–4% of extremely obese individuals who do not show additional phenotypic alterations ("common" obesity). This gene is currently regarded as the most relevant "obesity gene." The number of positive association studies pertaining to obesity has increased steadily; unfortunately only a few of these results have been confirmed in independent studies or remained positive in meta-analyses. Hence, it remains largely unclear which of the studies represent true positive findings. More than 30 genome-wide scans pertaining to obesity and associated phenotypes have been performed; few linkage peaks at specific chromosomal regions have repeatedly been observed. Recently, two underlying candidate genes had been described. Whereas independent confirmation was shown for one of these genes non-confirmation was recently described for the other gene. Hopefully, the recent completion of the Human Genome Project and the ever improving large-scale chip technology will accelerate the search for obesity genes.

14.1
Introduction

Despite intensive efforts biomedical research has not been able to exactly pinpoint the reasons underlying the obesity epidemic. An altered environment is commonly

Nutritional Genomics. Edited by Regina Brigelius-Flohé and Hans-Georg Joost
Copyright © 2006 WILEY-VCH Verlag GmbH & Co. KGaA, Weinheim
ISBN: 3-527-31294-3

assumed to underlie the secular trend; both an increased energy intake and a reduced energy expenditure figure as the two prominent factors in this discussion [1]. The reasons inherent to the difficulties in pinpointing the exact factors are presumably of physiological nature. Thus, energy intake needs to only minimally exceed energy expenditure on a daily basis; such a slight degree of overeating cannot be detected using available technology. In addition, several factors work in concert, each factor in itself making only a minute contribution. Palatability, availability, and low price of food, portion size, TV and media consumption, and decreased levels of manual work and physical activity are relevant in this context.

An attempt to explain a genetic predisposition for obesity was made in the "thrifty genotype hypothesis" originally postulated by Neel in 1962 for diabetes mellitus (see Ref. [2]). According to this hypothesis genetic variations favoring energy storage have been fixed during evolution, thus increasing the probability of survival during periods of famine. Accordingly, both in humans and animals alleles predisposing to an elevated body weight and thus obesity should be more common than alleles that do not favor energy storage. Whereas this hypothesis intuitively makes sense, it has not been convincingly demonstrated to apply to obesity.

14.1.1
Heritability Estimates

Twin studies have produced the most consistent and highest heritability estimates with values in the range of 0.6–0.9 for body mass index (BMI; kg/m^2). Heritability estimates of this magnitude indicate that the genetic component for body weight is almost as high as that for height (0.8). These high estimates apply to twins reared both together and apart. Except for the newborn period, age does not affect heritability estimates to a substantial degree. Heritability of BMI is maximal (up to 0.9) during late childhood and adolescence. In comparison to twin studies adoption and family studies have mostly resulted in lower heritability estimates. However, large and more recent family studies have also come up with heritability estimates of approx. 0.7. The genes relevant for weight regulation in childhood presumably only partially overlap with those operative in adulthood [3].

14.1.2
Syndromal Obesity

Obesity is a frequent feature in many syndromes. Pre-eminent among these is the Prader–Labhardt–Willi syndrome (PWS) characterized by neonatal hypotonia and feeding problems in infancy, short stature, a reduced intelligence quotient, behavior problems, and hyperphagia leading to severe obesity (incidence: 1/10 000). PWS is a complex genetic disorder caused by the loss of function of imprinted genes on chromosome 15q11–q13 [4] as the result of one of three genetic lesions: paternal deletion, maternal uniparental disomy, and imprinting defect. Although all the genes within the critical region are known, it is unclear which of them are causally involved in hyperphagia. A number of patients with features partly mimicking PWS have chromosomal aberrations that do not involve the critical PWS region (for example, del (6)(q22.2–q23.1, dup (6)(q24–q27), del 10 (q26.3), del X (pter–q26.1), dup X (q23–q25)). Other syndromes associated with obesity include Cohen syndrome, Bardet–Biedl syndrome, Ohdo-like blepharophimosis, and Albright syndrome. Among these the Bardet–Biedl syndrome has been studied intensively; triallelic inheritance has been reported in some cases [5, 6].

14.1.3
Association Studies

A widely used approach to find genes involved in "common" obesity are association studies. A vast number of such studies have been performed for obesity-related traits (Human Obesity Gene Map; http://www.obesity.chair.ulaval.ca/genes.html). Whereas many associations have been reported, it is largely unclear which of these represent true positive findings. False-positive findings often result from not correcting for multiple testing, particularly if multiple (endo)phenotypes are analyzed [3]. On the other hand, many of these studies are underpowered for the detection of minor gene effects. Based on large-scale studies incorporated into a meta-analysis encompassing more than 7000 individuals, Geller *et al.* identified for the first time a polygenic influence in body weight regulation. Thus, the V103I polymorphism in *MC4R* is negatively associated with obesity (odds ratio 0.69); a finding that has recently been confirmed in approximately 8000 individuals from an epidemiologic study group [7].

In this chapter we present an overview describing different modules accounting for the current research in the genetics of obesity. Completion of the genome sequences of human and many animal species was a prerequisite for many novel technological advances accelerating and transforming genetic research. Genetic and physiological studies in animal models form the basis for many discoveries that are later applied to the human situation. We describe some of the known human obesity genes and recent discoveries on gene variants that influence body weight in the common population. Finally, we focus on approaches used to map complex traits and give an outlook on emerging novel strategies.

14.2
The Human Genome Project

As outlined above, elucidation of genetic components that underlie complex phenotypes such as obesity is one of the greatest challenges in current biomedical research. The recently completed sequence of the human genome [8] and large efforts to identify the full set of human genes will provide one means to tackle the task. This data, complemented with information about genetic variation, regulatory segments, epigenetic modifications, protein–protein interactions, etc., will continually improve our understanding of how genes and environment interact and cumulate in a phenotype. Eventually, this knowledge will lead to the identification of the large number of genetic loci likely to be involved in common obesity. In this respect, we will briefly outline the history, outcome, and benefits of the Human Genome Project (HGP).

The project was launched in the late 1980s in the United States [9] and in 1990 became an international initiative with the joining in of the UK, France, and Japan; in 1995 Germany and in 1998 China followed. The coordinating scientific body of the HGP was provided by the Human Genome Organization (HUGO, www.hugo-international.org). The project was run in three stages: a pilot (1996–1999), data production (1999–2000), and finishing phase (2001–2004). Preceding work focused on establishing genetic and physical maps of both the human and mouse genomes. Large-insert clones formed the basis of these maps and, once ordered along chromosomes, also defined the basis of the hierarchical shotgun sequencing strategy [10] adopted by the HGP. These clone maps were also essential for positional gene cloning and the study of inherited disease. For example, both the human and murine leptin genes were identified by a positional cloning approach [11] and subsequently rare genetic variations causing extreme obesity in humans were detected [12, 13].

In the pilot phase, HGP researchers committed themselves to making sequence data immediately and freely available to the public [14, 15] and the International Human Genome Sequencing Consortium (IHGSC) was formed by researchers of 20 groups in six countries (USA, UK, Japan, France, Germany, and China). Pilot sequencing was completed in 1999 with 15% of the human genome provided as finished sequence of highest quality (99.99% accuracy, gap-free). Fueled by competition from the private sector, the idea in the following production phase was to generate a draft sequence covering about 90% of the genome with less accuracy (99.9%). This way a genome sequence could be generated more rapidly and would, although not complete, allow researchers to extract main information as well as gain insights on global genome features. Quality and genome coverage of the draft sequences published in 2001 by the IHGSC [10] and the private company Celera Genomics [16] granted identification of genes, regulatory regions, repetitive elements, and other genomic features on a global scale. Most striking was the small number of genes identified in the draft sequence: 30 000–40 000 [10]. Also, it became apparent that no more than ~1.5% of the human genome contains coding information. About ~50% is composed of repetitive elements. Conceivably, human complexity, in addition to culturally derived influences, is built on diversity and finely tuned interaction of gene products such as RNA and proteins rather than gene numbers. Consistent with this, ~50% of human protein coding genes exhibit alternative splicing [17–19] creating a proteome of more than 90 000 proteins [20], and gene expression is regulated by the complex interaction of a wide variety of transcription factors [21, 22].

With respect to identifying disease genes, a most important achievement potentiated by the HGP was the construction of a (first) map of naturally occurring polymorphisms in which ~2–4 unique single-nucleotide polymorphisms (SNPs) were placed per human gene [23]. The data set allowed linkage disequilibrium mapping

(LD: the co-occurrence of a specific DNA marker and a second marker relevant for disease at a higher frequency than would be predicted by random chance). LD extends on average 60 kb from common alleles, and although this figure varies considerably between loci and human populations, the number of genotyped markers needed to map a disease gene allele might potentially be drastically reduced [24]. Furthermore, identification of haplotype blocks (sizeable regions over which there is little evidence for historical recombination and within which only a few common haplotypes are observed) became feasible [25, 26]. Both availability of SNPs and information on haplotype structures facilitates comprehensive genetic studies of human disease [27]. Presently, oligonucleotide microarrays containing ~10 000 SNPs evenly spaced across the genome are widely used for SNP-based linkage analysis to identify disease gene loci [28, 29]. To perform genome-wide association studies, an even higher SNP coverage (100 000–500 000 SNPs) is essential.

In the last phase of the HGP, IHGSC worked on converting the draft into a highly accurate and complete human genome sequence. In April 2003, in the 50th anniversary year of the discovery of the double-helical structure of DNA, the work was completed. The finished sequence represents ~99% of the euchromatic portion of the human genome (2.85 Gb) with 99,999% accuracy, that is one error per 100 000 nucleotides [8]. While there were ~150 000 gaps in the draft version, only 341 gaps are left in the finished sequence, of which 308 (euchromatic gaps) are associated with segmental duplications which cannot be resolved with current technology.

Detailed analyses requiring highest sequence accuracy and completion are now possible. Of major interest is the identification of all genes and a comprehensive genome annotation. Currently, 24 194 genes (including 1978 pseudogenes) are listed in the human gene catalogue (Ensemble Release 29.35b). The total number of protein coding genes is now estimated at 20 000–25 000 which is consistent with data from cross-species comparison [30, 31]. Interspecies comparison will also be essential to identify regulatory regions and functional motifs, and so sequencing of many prokaryotic as well as eukaryotic organisms including mammals (mouse, rat, cat, chimpanzee, cow, dog) is completed or well under way.

Coordinated efforts are neccessary and already put into effect to meet the challenge "to translate genomic information into health benefits" [32]. For example, to systematically identify all genetic variations in the human population, the International HapMap Project was initiated in 2002 (www.hapmap.org/index.html.en). Populations with African, Asian, and European ancestry are studied to identify and catalogue genetic similarities and differences in humans. Currently, there are ~8 million SNPs known, placing on average one SNP per 0.5 kb. Genotyping a SNP subset in the four populations will provide an invaluable resource for discovering genes related to complex disease. Another international project aimed at better understanding disease state is the Human Epigenome Project (www.epigenome.org) in which genome-wide DNA methylation patterns are studied. DNA methylation is a natural modification of the nucleotide cytosine by which gene expression is controlled. It is tissue specific, changes over time, and in response to environmental factors. In this respect, DNA methylation represents a direct link between environment and an individual's state of health. There are many more projects, such as the international Human Brain Proteome Project which is concerned with the brain proteome in health, aging, and neurodiseases (www.hbpp.org/5602.html) and the ENCODE project aimed at identifying all functional elements in the human genome (www.genome.gov/1), both launched in 2003. Together with the initiatives emerged and more to foresee, the HGP demonstrates the immense power lying in coordinated efforts to provide a foundation of biological and biomedical research at a new level. For identifying and mapping of candidate genes for complex disorders, which in the case of obesity likely exert a truly small effect, this way of combining resources and data (as has recently been done through meta-analysis, e.g. Ref. [33]) across laboratories will be crucial.

14.3
Genetic Studies in Different Animal Species

Before the Human Genome Project was completed, the full sequences of some animal genomes had became available. This increased the attractiveness of the respective animal models considerably. Animals offer unique research possibilities either by their technical or procedural amenability or by economy of time and cost. Furthermore, animals can be reared under constant environmental conditions from birth on. This is important since human obesity is a complex multifactorial condition. Many interacting factors such as ethnicity, food intake, food preference, energy expenditure, age, gender, educational level, smoking status, etc. likely contribute to obesity and may make the identification of the impact of a single gene variation very difficult. The possibility of limiting confounding environmental parameters in laboratory animals will in many cases increase the likelihood of detecting weak genetic inputs but may also lead to an overestimation of the importance of a genetic contribution. Another caveat upon the use of animal models is the risk that evolutionary differences between species narrow the transferability of findings to the human situation. Therefore, it is important to carefully select the animal model considering both its strengths and weaknesses.

Here we present some examples of animal models that are used for metabolic research. We give a short summary of their respective characteristics that we deem important.

14.3.1
Caenorhabditis elegans

Caenorhabditis elegans is a small (about 1 mm long) soil nematode. In the 1960s Sydney Brenner began using it to study the genetics of development and neurobiology. Adult *C. elegans* are usually self-fertilizing hermaphrodites, thus it is easy to generate homozygous mutant stocks. *C. elegans* is diploid and has five pairs of autosomal chromosomes (named I, II, III, IV, and V) and a pair of sex chromosomes (X). It was the first animal with a completely sequenced genome [34]. The genome comprises approximately 100 megabases (Mb), containing 18 000–20 000 protein-coding genes; in comparison, the human genome comprises approximately 3300 Mb, and the gene count is currently 20 000–25 000 genes. The *C. elegans* genome data are collated into the databases ACeDB (http://nema.cap.ed.ac.uk/Caenorhabditis/C_elegans_genome/acedb.html) and Wormbase (www.wormbase.org).

An advantage of *C. elegans* is that it is possible to freeze worms and recover living animals years and even decades later. This facilitates the systematic generation of knockout lines (e.g. in the *C. elegans* Gene Knockout Consortium), which can be stored frozen and distributed to interested scientists. Another feature of worms is the very good response to gene silencing by RNA interference [35]. The small size of the worms and availability of RNAi libraries representing most of the worm genes, allows for high-throughput silencing experiments (e.g. in microtiter plates) and the study of the phenotypic consequences.

This approach has been used to identify genes that affect fat storage in worms [36]. Lipid droplets accumulate in intestinal cells, the principal site of fat storage in *C. elegans*. Feeding worms with the lipophilic dye Nile Red stains the lipid droplets that can be observed in the transparent animals. After silencing of genes by RNAi treatment (by feeding RNAi-expressing bacteria), the size and distribution of stained droplets can be estimated to judge a gene's effect on fat storage. Of 16 757 tested genes, 0.7% (112 genes) resulted in increased fat storage or droplet size, and 1.8% (305 genes) caused reduced fat or distorted droplet deposition pattern. Despite some concerns about technical aspects (e.g. effectiveness of RNAi silencing, stability of Nile Red staining, and indirect effects on fat storage by a potential dysfunction of intestinal cells, which are the prime target tissue of RNAi application), the work shows that a relatively high number of genes may have an impact on fat storage. Even if these experiments cannot answer the question of how "polygenic" human obesity really is, these results should be kept in mind.

14.3.2
Drosophila melanogaster

The fruitfly *Drosophila melanogaster* is a small insect (about 3 mm long) and belongs to the family of flies that like to accumulate around rotting fruit. *D. melanogaster* is the classical geneticist's pet. A hundred years of research has accumulated extensive knowledge about its biology and numerous genetic research tools. Mutant lines with defects in thousands of genes are available from the stock centers. *D. melanogaster* has four pairs of chromosomes: the X/Y sex chromosomes and the autosomes 2, 3, and 4. The size of the genome is about 160 Mb and contains an estimated 14 000 protein-coding genes. The genome was (almost) completely sequenced in 2000. Genome data are collected in the databases BDGP (www.fruitfly.org) and Flybase (www.flybase.org).

The prime site of fat storage in *Drosophila* is called the fatbody. Probably the first mutation describing an obesity-like phenotype with increased lipid storage in the fatbody was *adipose* (*adp*), derived from a natural population in Nigeria [37]. Well-fed *adp* mutant flies can store more than twice as much triglycerides as wild-type flies and outlive the leaner flies under starvation conditions. The occurrence of the mutation in a natural population suggests that mutant animals may have a survival advantage under certain environmental conditions, outbalancing disadvantages of the mutation like reduced physical fitness and reduced fecundity [38]. Encouraged by the interesting phenotype of the *adp* mutation and the high degree of homology between the *Drosophila* and human *adp* homologs, scientists at the biotech company DeveloGen (www.develogen.com) started a systematic mutation screen for metabolic phenotypes including obesity. A transposon was used as mutagenic agent, which allows for fast identification of the mutated gene by inverse polymerase chain reaction (PCR). In addition, the transposon was engineered to facilitate overexpression of neighboring genes in a tissue of choice (e.g. fatbody or nervous system) [39]. That way loss-of-function mutations (by integration of the transposon into essential regions of a gene) or gain-of-function mutations (by activation of the overexpression system) were induced and the resulting phenotype (e.g. triglyceride storage) recorded. No detailed list of the identified genes is currently publicly available. However, again a relatively high percentage of the mutated genes produced a fat storage phenotype, comparable to the percentage obtained in the *C. elegans* screen [40]. The considerable degree of homology of metabolic pathways in *Drosophila* and humans (www.genome.jp/kegg/pathway) and the existence of many "proof of concept" genes with similar metabolic functions in both species [41] suggest that many of the genes identified in the mutation screen will have a relevant function in the control of fat storage in humans as well.

14.3.3
Rodents

Several rodent species, such as mouse, rat, Siberian hamster, and Israeli sand rat (*Psammomys obesus*), have been used for obesity research. In the eyes of the geneticist, *Mus musculus* is certainly the most interesting rodent model system. The mouse genome is approximately 3000 Mb in size, consisting of 20 pairs of autosomes and the X/Y sex chromosomes. The gene number is quite similar to the human gene number. Mice reach sexual maturity only four weeks after birth, making genetic experiments spanning several generations feasible in this mammal. An important advantage of mouse genetics is the availability of a vast array of techniques. For example, generation of transgenic animals, knockout animals, or conditional knockouts is possible.

Historically, the domain of mice in obesity research has been the field of monogenic obesity models. Several single-gene mutations were discovered, leading to severe obesity in certain mouse strains. While monogenic causes of obesity are rare both in mice and in the human population, these animals are nevertheless of outstanding value for unravelling genetic pathways and physiologic mechanisms controlling energy homeostasis. The most noted mutant is the *obese* (*ob*) mouse lacking the product of the hormone/cytokine-encoding gene *leptin* (*Lep^{ob}*). This mutant was discovered at the Jackson Laboratories in 1949 [42]. Another spontaneous mutation discovered at Jackson Labs is the diabetes mutation (*db*) lacking the long form of the leptin receptor (*Lepr*^{db}) [43]. Parabiosis experiments with joined circulatory systems of *Lep*^{ob} and *Lepr*^{db} mice suggested that *ob* mice lack a secreted factor, while in *db* mice the receptor is missing [44]. In 1994 and 1995 the corresponding genes were cloned, confirming this assumption [11, 45, 46]. Complete failure of leptin signaling results in severe obesity. In humans, rare recessive mutations in the *LEP* gene have been discovered, resulting in extreme early-onset obesity [12, 47] that can be cured by leptin treatment [13, 48, 49]. This approach marks the first pharmacological treatment of a monogenic form of obesity in humans based on knowledge of the underlying biological mechanism.

Another obese mouse model is the agouti Yellow (*A*^y) mouse [50]. *Yellow* mice have been known for 200 years; they show a yellow coat color and obesity. Overexpression of the agouti peptide in the skin blocks α-melanocyte-stimulating hormone (α-MSH) signaling at melanocortin-1 receptors in the hair follicle, resulting in the production of yellow pigments (pheomelanin). In the *A*^y mouse agouti protein is ectopically expressed in all somatic cells [51]. Therefore, overexpression of agouti in the brain also antagonizes the anorectic action of α-MSH signaling at the melanocortin-4 receptor (MC4R), causing hyperphagia and subsequently obesity. α-MSH is produced by proteolytic cleavage of a prohormone by carboxypeptidase E (CPE) [52]. Mutated CPE leads to the *fat* mouse, another obese mouse model discovered at Jackson Labs [53].

In addition to the aforementioned spontaneous mutations, knockout and transgenic rodent models have proved successful in revealing and resolving gene function. They are the most straightforward way of doing reverse genetics (i.e. going from gene to phenotype). Several thousand knockout and transgenic models have been generated and many of them show evidence that the targeted gene is involved in processes of body weight regulation. Currently (version 11), the Obesity Gene Map Database (www.obesity.chair.ulaval.ca/genes.html) lists a selection of 164 knockouts and transgenics with obesity related phenotypes. An example that will be discussed later in this chapter is the MC4R-knockout mouse [54] (see also Fig. 14.1). Inactivation of this receptor results in mice that develop maturity-onset obesity. Heterozygous mice show an intermediate phenotype, suggesting that MC4R gene dosage is important. Not surprisingly pro-opiomelanocortin (POMC)-knockout mice that lack the MC4R agonist α-MSH are obese as well [55]. Accordingly, transgenic overexpression of the MC4R antagonist agouti-related protein (AGRP) causes obesity in mice [56]. Analyses of such mouse models generated not only a better understanding of the function of the leptinergic–melanocortinergic system in energy homeostasis, but also ignited progress in analyses of genetic mechanisms in human obesity. For example, studies of the human *MC4R* gene revealed that functionally relevant mutations are currently the most prevalent cause for monogenic human obesity [57].

14.3.3.1 Rodent Models of Polygenic Obesity

Mice (and other animal models) can also be used to investigate polygenic control of energy homeostasis directly. Strain-related differences in fat mass, diet-induced obesity, or other obesity-related phenotypes can be mapped by phenotyping the F2 offspring of interstrain crosses. An example is a back-cross model of New Zealand obese (NZO) mouse with the lean Swiss Jackson Laboratory (SJL) strain [58, 59]. In such quantitative trait loci (QTL) mapping experiments, two parental lines differing for the obese phenotype are crossed, and the F1 generation is intercrossed or backcrossed to produce an F2 or a backcross generation. These offspring populations are analyzed using molecular markers across the genome to search for obesity QTL. For example, more than 100 obesity-related QTL have been mapped in mice [60, 61]. Usually, the loci identified in different strain pairs show little overlap. This indicates that mapping experiments for obesity QTL are far from saturation and, presumably, a great many genes can account for differences in body weight between strains. Many of the identified loci correspond to syntenic human chromosomal regions, which have also been identified in obesity-related linkage studies. Such QTL indicate regions which appear most interesting in terms of the search for candidate genes.

14.3.4
Livestock: Pig as a Valuable Animal Model

The pig (*Sus scrofa*) is one of the most important livestock worldwide. Therefore, there is a great interest in elucidating the genetic basis of economically important performance traits such as backfat thickness, fat content of the muscle, growth rate, and food intake and utilization. In addition, the pig is an important model organism for health issues such as obesity, cardiovascular disease, and organ transplantation [62]. The anatomical, physiological, and genetic similarities between humans and pigs are especially pronounced and exceed those of the species discussed above. Comparative genetic studies have found that there is more similarity between the organization of the genomes of humans and pigs than there is between either humans and mice or pigs and mice (comparative map: www.toulouse. inra.fr/lgc/pig/compare/compare.html) [63]. The medium length of conserved syntenic regions between humans and pigs is approximately twice as long as the corresponding regions between mice and humans [64, 65]. The pig genome comprises 18 autosomes, with X and Y sex chromosomes. The genome size is similar to that of humans and is estimated at 2700 Mb.

In the last decade efforts to unravel the pig genome have progressed quickly (currently there are about 1600 genes and about 2500 markers in the public database (www.thearkdb.org/browser?species=pig) of the Roslin Institute, Scotland) [66, 62]. A large number of genome-wide QTL scans for traits of economic interest have been published [60]. The pig QTL database (pig quantitative trait loci (QTL) database; PigQTLdb; http://www.animalgenome.org/QTLdb/) comprises all pig QTL data published during the last 10 years or so. To date, 990 QTL affecting 233 different traits were fed into the database. Most QTL ($n = 219$) were identified for the trait type "fatness." For the trait "average backfat thickness" 48 QTL were found, and those explaining most of the phenotypic variance were located on the porcine chromosomes 1, 2, 4, 7, and X [67]. Obesity-relevant QTL were identified mainly on the porcine chromosomes 1, 2, 4–7, 13, 14, and X [68–93].

Several positional candidate genes have been examined [66, 94, 95]. As QTL experiments are expanded and the comparative map improves, it is likely that additional positional candidates will be identified and the causative quantitative trait nucleotide (QTN) discovered [66]. However, there are several difficulties. Only a few significant associations for positional candidate genes have been published [67]. The still incomplete sequence of the pig genome complicates the search for candidate genes in QTL areas. Additionally, unravelling the genes that underlie detected QTLs for multifactorial traits appears to be a great challenge [96]. One of the complicating factors, reviewed by Andersson and Georges [96] is that individual QTL only account for part of the phenotypic variance, whereas the rest is due to environmental factors as well as other QTL. The authors note further complicating factors such as mild phenotypic effects and the possibility of underlying regulatory mutations that are poorly detectable at present. Nonetheless, there are a few examples for which mutations that underlie mapped QTL have been identified in domestic animals, such as for *RYR1* (ryanodine receptor 1; SSC6; HSA19q13) [97], *PRKAG3* (protein kinase, AMP-activated, gamma 3 non-catalytic subunit) [83], and *IGF2* (insulin-like growth factor 2; SSC2; HSA11p15.5) [98]. In the future, additional methods like high-resolution identical by descent (IBD) mapping and functional genomic analyses will accelerate the (more accurate) detection of further positional and physiological candidate genes [96, 62].

14.3.5
Concluding Remarks on Animal Models

Ideally, an animal model should resemble the human situation in as many behavioral and physical features as possible, for example, in neuroendocrine mechanisms, adipocyte physiology, structure of the alimentary tract, or social aspects of food intake. Therefore, it seems that primates would be the ideal animal model [99, 100]. However, other aspects such as ethical concerns, economic constraints, and practical reasons make "lower" animal species in many cases more attractive. Therefore, it is important to have profound knowledge of the characteristics of an animal model and to be aware of possible pitfalls in order to plan experiments properly. For example, somebody interested in modeling vomiting would not consider mice or rats as their animal model. On the other hand, energy homeostasis is an essential feature of all multicellular, free-living animals. It is likely that fundamental mechanisms controlling energy homeostasis are conserved even in lower animal species. Not all factors and mechanisms important for humans may be present in animals. But biological functions discovered in animals have a good chance of being relevant in humans, too.

14.4
Candidate Gene Analyses in Humans

Two major avenues for a candidate gene approach can be envisaged in humans, both of which can be applied at the same time: (1) Genome-wide scans depict chromosomal regions harboring candidate genes for obesity and related disorders; fine mapping leads to a narrowed region, allowing candidate gene analyses. These techniques will be outlined in the next section of this chapter. (2) Some genes will be considered because they are involved in relevant central or peripheral pathways as shown in animal models or via other evidence. This approach should not be viewed as an alternative to the identification of the genes contributing to linkage peaks. Instead such studies are complementary because linkage studies cannot readily lead to the detection of minor gene effects or infrequent major gene effects.

The conservation of hypothalamic and metabolic pathways between rodents and humans has led to successful candidate gene analyses [11, 12, 47, 54, 101–103]. Most human candidate genes have been derived from animal models. All spontaneous mutations leading to obesity in mice [11, 46, 104–106] either led to the identification of mutations within the same genes in humans [12, 107] or have led to the identification of a system/pathway in which other genes were found to be mutated in humans (e.g. carboxypeptidase mutations in mice [52] and human prohormone convertase 1 gene mutations [108]). Most of the mutations in humans have been detected via specific endocrinologic findings such as elevated pro-insulin levels [108] or hypoleptinemia that led to the screen of a specific gene in thus identified individuals. These analyses were hypothesis driven. Some of the candidate gene screens will be depicted in more detail:

14.4.1
Leptinergic–Melanocortinergic Pathway

One of the pathways that are currently regarded as the most relevant central circuits involved in body weight regulation is the leptinergic–melanocortinergic pathway (Fig. 14.1). Initially, the cloning of the five spontaneous mutations that led to the monogenic forms of obesity in rodents (for review see Ref. [60]) aided the discovery of critical checkpoints in this feedback mechanism. Mutations in the homologous genes in humans showed that the same mechanisms seemingly work here as well. An autosomal recessive mode of inheritance has been described for obesity due to mutations in the *leptin* gene in single individuals worldwide [12, 48, 109] who all stem from inbred families. Their detection was made possible by the groundbreaking cloning of the *leptin* gene via a positional cloning approach based on the *obese* mouse [11]. The cardinal features of the *obese* mouse encompass hyperphagia, extreme obesity, hypothermia, and infertility. The symptoms in leptin-deficient humans are similar for the core phenotype except that they have a normal body temperature. Treatment of affected children and adults with recombinant leptin initially resulted in a striking and rapid normalization of their eating behavior [48, 49, 110]. Long-term treatment has led to a near normal body weight in both young and older patients [49, 109]. It has been shown that leptin can have substantial effects on tissue composition in the human brain. The potential spectrum of the role of leptin seems to broaden beyond feeding behavior and endocrine function [111].

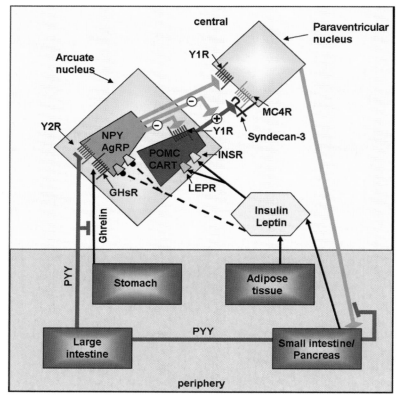

Fig. 14.1. Interactions among hormonal and neural pathways that regulate food intake and body fat mass (adapted from Ref. [179]). In this schematic diagram, the dashed lines indicate hormonal inhibitory effects and the solid lines indicate stimulatory effects. The paraventricular and arcuate nuclei each contain neurons that are capable of stimulating or inhibiting food intake. Y1R, neuropeptide Y (NPY) receptor 1; Y2R, NPY receptor 2; MC4R, melanocortin-4 receptor; PYY, peptide YY$_{3-36}$; GHsR, growth hormone secretagogue receptor; AgRP, agouti-related protein; POMC, proopiomelanocortin; α-MSH, α-melanocyte-stimulating protein; LEPR, leptin receptor; INSR, insulin receptor; BDNF, brain-derived neurotrophic factor; NTRK2, neurotrophic tyrosine kinase receptor type 2.

The leptin signal is centrally received by the leptin receptor. Recessive mutations in the leptin receptor gene in humans were shown to lead to extreme early-onset obesity [107]. Neurons expressing the leptin receptor project to neurons that express the melanocortin-4 receptor (MC4R). α-Melanocyte-stimulating hormone (α-MSH, a cleavage product of pro-opiomelanocortin (POMC)) is a potent agonist at the MC4R and induces satiety. Agouti-related protein (AgRP) is an inverse agonist at the MC4R [112, 113]. A single study showed that a polymorphism in the human *AGRP* was involved in the etiology of anorexia nervosa [114]; recently the same polymorphism was also associated with lower body weight in humans [115].

A mouse model showed that transgenic expression of syndecan-1, a cell surface heparan sulfate proteoglycan (HSPG) and modulator of ligand–receptor interactions, in the hypothalamus produces mice with hyperphagia and maturity-onset obesity thereby resembling mice with reduced action of α-MSH [116]. Syndecans presumably potentiate the action of AgRP and agouti-signaling protein. In wild-type mice the predominantly neural syndecan (syndecan-3) is ectopically expressed in hypothalamic regions that control energy balance. Syndecan-3-null mice respond to food deprivation with markedly reduced reflex hyperphagia [116]. It was proposed that oscillation of hypothalamic syndecan-3 levels physiologically modulates feeding behavior [116]. Surprisingly, a mutation screen of the *syndecan 3* homolog in humans has not been reported yet.

The *POMC* gene is located within a chromosomal region for which linkage of serum leptin levels, fat mass, and saturated fat intake have been shown [117–119]. Fine mapping of the region increased the LOD scores considerably [118]. Hence, the *POMC* gene has been regarded as the most obvious candidate gene underlying the linkage peak. However, although more than 8 years of hard molecular genetic work have passed since the original publication [117], evidence is lacking so far that variation(s) in the *POMC* gene can readily explain the linkage peak. Variation of leptin levels has been associated with *POMC* polymorphisms [118]. Mutations in the *POMC* gene have been identified in patients with early-onset obesity, adrenal insufficiency due to adrenocorticotropic hormone (ACTH) deficiency (ACTH is also a cleavage product of POMC), and red hair [120, 121]. Relevant *POMC* mutations have been detected only very infrequently; the syndrome is due to either compound heterozygosity or homozygosity. Hyperphagia and obesity result from a lack of α-MSH. A mutation at the C-terminus was reported that affects the dibasic cleavage site necessary for β-MSH generation. A dominant-negative effect, rather than a loss of function resulting in haplo-insufficiency, was implied, as the resulting fusion peptide showed preserved MC4R binding but loss of signal transduction [122].

The transference of results obtained in animal models to humans has not only proven to be successful for rare syndromal monogenic forms (e.g. pertaining to mutations in the *POMC, leptin,* and *leptin receptor*) of obesity. Mutation screens in large samples are required to identify alleles involved in obesity not readily associated with a specific endocrinologic or behavioral phenotype. The occurrence of obesity in *Mc4r$^{-/-}$* mice [54] led to the detection of mutations in the human *MC4R* marking the first time a major gene effect has been detected in non-syndromal obesity. Obesity due to functionally relevant *MC4R* mutations is not readily distinguishable from "common" obesity; the mode of inheritance is semi-dominant. Originally, Yeo *et al.* [102] and Vaisse *et al.* [101] each detected a single mutation segregating with obesity in small pedigrees. Shortly thereafter, Hinney *et al.* [103] described a total of nine mutations in 306 obese children and adolescents including frameshift, nonsense, and missense mutations. Individuals with *MC4R* mutations have exert for obesity no readily recognizable phenotype.

14.4.2
Evaluation of Mutations/Polymorphisms with a Known Effect on Body Weight

Until now approximately 70 different infrequent missense, nonsense, and frame-shift mutations in the *MC4R* have been described in (extremely) obese individuals (see Ref. [123]). Most of the mutations were shown to lead to total or partial loss of function in *in vitro* assays. Combined frequencies for all functionally relevant mutations typically range from 2–3% in extremely obese individuals [124, 123]. The quantitative effect of functionally relevant human *MC4R* mutations on body weight was recently determined in a family-based setting [124]. Carriers of these mutations had a significantly higher current BMI than their wild-type relatives. The observed effect was about twice as strong in women as in men. BMI differences between mutation carriers and their family members harboring the wild-type genotype were approximately 2.5 and 1.3 standard deviations, equivalent to 9.5 and 4 kg/m^2 in middle-aged women and men, respectively. The fact that relatives of extremely obese mutation carriers that harbor wild-type alleles are often also obese further complicated the assessment of the effect size of *MC4R* mutations. Seemingly additional genetic and/or environmental factors are operating in these families, which accordingly could also contribute to the obesity of the index cases [124].

14.4.3
V103I Polymorphism in the Melanocortin-4 Receptor Gene

The *MC4R* polymorphism V103I (rs2229616) has been detected by several groups (see Refs [103, 57]). Because both association and functional studies had been negative, this polymorphism had been considered as irrelevant for body weight regulation. However, a transmission disequilibrium test (TDT) revealed transmission disequilibrium for the I103 allele in 520 trios ascertained via an obese offspring (10 transmissions, 25 non-transmissions; *P*-value = 0.02). This finding led to a meta-analysis. Most groups had reported a higher frequency of the I103 allele in controls in comparison with obese cases. A meta-analysis considering newly generated data (including 2334 subjects of the epidemiologic KORA S2000 sample, which is representative of the Augsburg region) and all previously published reports (7937 probands in total) provided clear evidence for a negative association of the I103 allele with obesity (odds ratio 0.69; 95% confidence interval 0.59–0.99). In light of the previous negative functional findings the two receptor variants were again compared in *in vitro* assays. Again, no differences were found although minor functional differences cannot be excluded via these assays. Additionally, the 5′ and 3′ regions of the *MC4R* were sequenced and four other SNPs were detected, three of which were in total linkage disequilibrium with the original polymorphism. None of these readily explained the negative association with obesity in functional terms [33]. However, the negative association with obesity was recently reconfirmed in a large epidemiologic study group (extension of the original KORA S2000 cohort) [7].

14.4.4
Signals Downstream of the Melanocortin-4 Receptor

There are molecules downstream of the MC4R that are also involved in body weight regulation. Several lines of evidence indicate an involvement of brain-derived neurotrophic factor (BDNF) in body weight regulation and activity: (1) heterozygous *Bdnf*-knockout mice (*Bdnf*$^{+/-}$) are hyperphagic, obese, and hyperactive; (2) central infusion of BDNF leads to severe, dose-dependent appetite suppression and weight loss in rats; (3) BDNF infusion into the brain suppresses the hyperphagia and excessive weight gain observed on higher fat diets in mice with deficient MC4R signaling. These results showed that MC4R signaling controls BDNF expression in the ventro-medial hypothalamus and support the hypothesis that BDNF is an important effector through which MC4R signaling controls energy balance [125]. However, a recent mutation screen did not detect an association between variations in *BDNF* (p.V66M, c.−46C → T) and obesity in humans [126].

The BDNF receptor (TRKB) is also a relevant candidate for body weight regulation. Mouse mutants expressing TrkB at a severely reduced amount showed hyperphagia and excessive weight gain on higher fat diet [127]. Recently, an 8-year-old boy with severe obesity and a complex developmental syndrome was shown to be heterozygous for a *de novo* missense mutation resulting in a Y722C substitution in the neurotrophin receptor TRKB. The mutated TRKB led to markedly impaired receptor autophosphorylation and to reduced signaling to MAP kinase. Mutations of *NTRK2*, the gene encoding TRKB, seem to result in a unique human syndrome including hyperphagic obesity. The associated impairment in memory, learning, and nociception seen in the proband reflects the crucial role of TRKB in the human nervous system [128]. Again, mutations in this gene seem to be (very) rare and cannot explain a substantial proportion of the obesity epidemic.

14.4.5
Validation of an Obesity Gene

The *MC4R* V103I polymorphism suggests that it will be especially difficult to pinpoint the effect of gene variants that exert only a minor influence on body weight. In association studies large case numbers will be required to firmly establish a role of the respective variants; stratification according to weight class appears reasonable. Probands could hence be categorized as underweight, normal weight, overweight, and obese; allele frequencies would accordingly be expected to systematically increase. However, as gene variations predisposing to leanness might be non-allelic to variations predisposing to overweight/obesity this systematic increase is probable but not a prerequisite. For every particular candidate gene it needs to be devised how to confirm that a specific allele indeed predisposes to obesity irrespective of the approach leading to the gene identification [129]. A vast amount of positive association studies has been published so far; only some of these have been followed up [60]. One has to bear in mind that most of the positive studies have resulted after multiple tests; hence the "significant" finding has to be viewed critically and should not be taken for granted until confirmed independently in a sufficiently powered study. On the other hand it is crucial that negative findings are also published to avoid publication bias. Not to discourage other scientists from conducting studies, high standards should be set for publication of such negative findings. To allow a better interpretation of negative findings, the power of the study for a given (previously reported) effect should be stated. To be sure whether or not an identified allele/gene is involved in the phenotype, positive findings need to be followed up in a systematic fashion so that the scientific community can get a balanced opinion. To pursue this purpose, defined (epidemiologic) population samples could be referred to in addition to large trio samples to allow for TDTs [130]; as soon as a sufficient number of studies become available meta-analyses are possible and should prove helpful. A decision needs to be reached at some point as to whether current evidence is sufficient to unequivocally conclude that a particular allele(s) is relevant for the phenotype. The decision should be best based on epidemiologic studies and on an appropriate meta-analysis of all available studies.

As a general rule medical, epidemiologic and functional studies only appear warranted if an initial association finding can be confirmed. Otherwise the risk of pursuing false-positive findings is too large.

14.5
Current Approaches for the Genetic Mapping of Complex Traits in Humans

As depicted above, for most candidate genes it turned out to be difficult or impossible to pinpoint a clear function in common human obesity. Complex traits such as obesity do not conform to Mendelian patterns of segregation and are thought to result from the combined effects of single genes (oligogenic) to many genes (polygenic). It is assumed that each gene has only a small effect on the development of the respective phenotype/disorder. Here we provide a current overview of the two major approaches (linkage analysis and association/linkage disequilibrium tests) that are used to identify genes/alleles that contribute to complex traits. Novel developments, such as high-density oligonucleotide chips and adapted statistical methods will increase the possibility of dissecting complex traits.

14.5.1
Genetic Linkage Analysis

Segments of DNA closely positioned along a chromosome segregate together and are therefore inherited together. Genetic linkage studies take advantage of this fact. The closer the loci, the lower the probability that they will be separated at meiosis, and hence the greater the probability that they will be inherited together. By analyzing meiotic recombination frequencies between specific loci, genetic linkage analysis can be used to localize susceptibility genes within a framework map of genetic markers with known positions in the genome. Instead of focusing on large, multiply affected families, in more recent linkage studies sibling pairs and small nuclear families have been analyzed [131]. Pairs of siblings affected with the same disorder or concordant for the same phenotype are likely to share susceptibility genes inherited from the same parent. Where affected siblings share parental alleles more often than expected by chance, this indicates linkage between a susceptibility gene locus and the analyzed phenotype. As no assumptions are required about the underlying mode of inheritance these studies are often termed model-free.

14.5.2
Linkage Disequilibrium Analysis

Linkage disequilibrium (LD) or population allelic association describes the phenomenon whereby two alleles at different loci (haplotypes) co-occur more frequently than can be expected from their allele frequencies. Alleles associated in such a way may reflect fragments of ancestral chromosomes that did not undergo recombination despite many meiotic events over multiple generations and therefore appear to be associated even in individuals from different families. Risch and Merikangas [132] showed that this characteristic can be used for study designs that are more powerful for the detection of small genetic effects than genetic linkage designs.

14.5.3
Genome Scans

To date more than 30 genome scans pertaining to obesity and related phenotypes have been performed (Fig. 14.2). Genome scans published for obesity, related traits, and BMI [133–163] have resulted in some coinciding regions. Analyses at

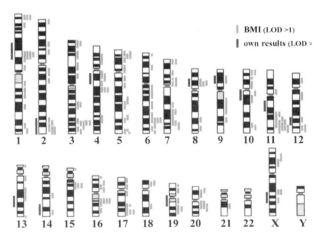

Fig. 14.2. Human obesity gene map. Chromosomal regions that were identified by linkage studies for the phenotype BMI (status: 2/2005). Each bar represents one positive linkage finding. Own results [158].

human chromosomes 1p, 3q, 6q, 11q, and 16q especially have shown repeated evidence of linkage.

The upcoming years will reveal to what extent single mutations or SNPs and haplotypes underlie these peaks. Furthermore, it seems possible that some of the peaks actually represent the combined effect of SNPs or haplotypes at more than one locus. In two very recent studies genes potentially underlying linkage peaks have for the first time been identified via fine mapping and the candidate gene approach. A *glutamate decarboxylase 2 gene* (*GAD2*) haplotype comprising three SNPs located in the linkage region on chromosome 10p was found to predispose to obesity [164]. However, this finding could not be confirmed in large study groups comprising children, adolescents, and adults [180]. In contrast, a second locus could be confirmed. A SNP in the gene encoding amino acid transporter solute carrier family 6 (neurotransmitter transporter) member 14 (*SLC6A14*) located within a peak region on chromosome Xp was associated and linked with obesity [165]. Durand *et al.* [166] were able to confirm the association of *SLC6A14* with obesity.

14.5.4
Current Developments in Biostatistical/Biometrical Analyses

In most cases, association approaches have been used for fine mapping or candidate genes analyses. Compared with genetic linkage studies, genome-wide association approaches are still rarely applied (e.g. Ref. [167]) as until recently technological constraints posed a major limitation. Genome-wide association studies based on linkage disequilibrium mapping with SNP markers are expected to allow for a more extensive and faster identification of genes that underlie complex disorders. Since linkage disequilibrium is known, for example, to decrease quickly with physical distance [168], hundreds of thousands of these markers will be required for comprehensive genome-wide association studies. As a consequence, both the genotyping strategy as well as the development of new, refined statistical methods becomes more and more important. Concepts like DNA pooling, haplotype-tagging SNPs, the identification of regions with reduced LD, haplotype analyses, staged or data adaptive study designs – to name a few – have been suggested to improve the efficiency of these studies (e.g. Refs [169, 170, 171]). Excellent overviews of important issues in the development of statistical methods are given by Terwilliger and Göring [172] or more recently by Freimer and Sabatti [173]. Articles by Schadt *et al.* [174]; or van Steen *et al.* [175] may serve as examples of the latest suggestions for dealing with the problem of multiple testing, for example. Finally, as association studies have unfortunately yielded a number of contradictory results, attention has to be paid to issues like differences in phenotype definition, study populations, or inadequate sample sizes for the assumed small effects (e.g. Ref. [176]). In our view, reliable result can only be achieved if investigators cooperate in networks which operate on the basis of appropriate and explicit decision rules. In addition, these networks should commit themselves to the idea of confirmation and replication [177, 178] instead of publishing premature findings.

Acknowledgments

Our research on genetic mechanisms of body weight regulation is supported by the Deutsche Forschungsgemeinschaft, the Bundesministerium für Bildung und Forschung (German Human Genome Project and National Genome Network 1 and 2; 01GS0482) and the European Union (Framework V and VI; LSHM-CT2003-503041).

References

1 TAUBES, G. (**1998**) As obesity rates rise, experts struggle to explain why. *Science* 280, 1367–1368.

2 NEEL, J.V., A.B. WEDER, S. JULIUS (**1998**) Type II diabetes, essential hypertension, obesity as "syndromes of impaired genetic homeostasis": the "thrifty genotype" hypothesis enters the 21st century. *Perspect Biol Med* 42, 44–74.

3 HEBEBRAND, J., S. FRIEDEL, N. SCHAUBLE, F. GELLER, A. HINNEY (**2003**) Perspectives: molecular genetic research in human obesity. *Obes Rev* 4, 139–146.

4 HORSTHEMKE, B., H. NAZLICAN, J. HUSING, L. KLEIN-HITPASS, U. CLAUSSEN, S. MICHEL, C. LICH, G. GILLESSEN-KAESBACH, K. BUITING (**2003**) Somatic mosaicism for maternal uniparental disomy 15 in a girl with Prader-Willi syndrome: confirmation by cell cloning and identification of candidate downstream genes. *Hum Mol Genet* 12, 2723–2732.

5 KATSANIS, N., S.J. ANSLEY, J.L. BADANO, E.R. EICHERS, R.A. LEWIS, B.E. HOSKINS, P.J. SCAMBLER, W.S. DAVIDSON, P.L. BEALES, J.R. LUPSKI (**2001**) Triallelic inheritance in Bardet-Biedl syndrome, a Mendelian recessive disorder. *Science* 293, 2256–2259.

6 EICHERS, E.R., R.A. LEWIS, N. KATSANIS, J.R. LUPSKI (**2004**) Triallelic inheritance: a bridge between Mendelian and multifactorial traits. *Ann Med* 36, 262–272.

7 HEID, I.M., C. VOLLMERT, A. HINNEY, A. DORING, F. GELLER, H. LOWEL, H.E. WICHMANN, T. ILLIG, J. HEBEBRAND, F. KRONENBERG (**2005**) Association of the 103I MC4R allele with decreased body mass in 7937 participants of two population based surveys. *J Med Genet* 42, e21.

8 ANON (**2004**) Finishing the euchromatic sequence of the human genome. *Nature* 431, 931–945.

9 SHORT, E.M. (**1988**) Proposed ASHG position on mapping/sequencing the human genome. *Am J Hum Genet* 43, 101–102.

10 LANDER, E.S., L.M. LINTON, B. BIRREN, C. NUSBAUM, M.C. ZODY, J. BALDWIN, K. DEVON, K. DEWAR, M. DOYLE, W. FITZHUGH, R. FUNKE, D. GAGE, K. HARRIS, A. HEAFORD, J. HOWLAND, L. KANN, J. LEHOCZKY, R. LEVINE, P. MCEWAN, K. MCKERNAN, J. MELDRIM, J.P. MESIROV, C. MIRANDA, W. MORRIS, J. NAYLOR, C. RAYMOND, M. ROSETTI, R. SANTOS, A. SHERIDAN, C. SOUGNEZ, N. STANGE-THOMANN, N. STOJANOVIC, A. SUBRAMANIAN, D. WYMAN, J. ROGERS, J. SULSTON, R. AINSCOUGH, S. BECK, D. BENTLEY, J. BURTON, C. CLEE, N. CARTER, A. COULSON, R. DEADMAN, P. DELOUKAS, A. DUNHAM, I. DUNHAM, R. DURBIN, L. FRENCH, D. GRAFHAM, S. GREGORY, T. HUBBARD, S. HUMPHRAY, A. HUNT, M. JONES, C. LLOYD, A. MCMURRAY, L. MATTHEWS, S. MERCER, S. MILNE, J.C. MULLIKIN, A. MUNGALL, R. PLUMB, M. ROSS, R. SHOWNKEEN, S. SIMS, R.H. WATERSTON, R.K. WILSON, L.W. HILLIER, J.D. MCPHERSON, M.A. MARRA, E.R. MARDIS, L.A. FULTON, A.T. CHINWALLA, K.H. PEPIN, W.R.

Gish, S.L. Chissoe, M.C. Wendl, K.D. Delehaunty, T.L. Miner, A. Delehaunty, J.B. Kramer, L.L. Cook, R.S. Fulton, D.L. Johnson, P.J. Minx, S.W. Clifton, T. Hawkins, E. Branscomb, P. Predki, P. Richardson, S. Wenning, T. Slezak, N. Doggett, J.F. Cheng, A. Olsen, S. Lucas, C. Elkin, E. Uberbacher, M. Frazier, R.A. Gibbs, D.M. Muzny, S.E. Scherer, J.B. Bouck, E.J. Sodergren, K.C. Worley, C.M. Rives, J.H. Gorrell, M.L. Metzker, S.L. Naylor, R.S. Kucherlapati, D.L. Nelson, G.M. Weinstock, Y. Sakaki, A. Fujiyama, M. Hattori, T. Yada, A. Toyoda, T. Itoh, C. Kawagoe, H. Watanabe, Y. Totoki, T. Taylor, J. Weissenbach, R. Heilig, W. Saurin, F. Artiguenave, P. Brottier, T. Bruls, E. Pelletier, C. Robert, P. Wincker, D.R. Smith, L. Doucette-Stamm, M. Rubenfield, K. Weinstock, H.M. Lee, J. Dubois, A. Rosenthal, M. Platzer, G. Nyakatura, S. Taudien, A. Rump, H. Yang, J. Yu, J. Wang, G. Huang, J. Gu, L. Hood, L. Rowen, A. Madan, S. Qin, R.W. Davis, N.A. Federspiel, A.P. Abola, M.J. Proctor, R.M. Myers, J. Schmutz, M. Dickson, J. Grimwood, D.R. Cox, M.V. Olson, R. Kaul, C. Raymond, N. Shimizu, K. Kawasaki, S. Minoshima, G.A. Evans, M. Athanasiou, R. Schultz, B.A. Roe, F. Chen, H. Pan, J. Ramser, H. Lehrach, R. Reinhardt, W.R. McCombie, M. de la Bastide, N. Dedhia, H. Blocker, K. Hornischer, G. Nordsiek, R. Agarwala, L. Aravind, J.A. Bailey, A. Bateman, S. Batzoglou, E. Birney, P. Bork, D.G. Brown, C.B. Burge, L. Cerutti, H.C. Chen, D. Church, M. Clamp, R.R. Copley, T. Doerks, S.R. Eddy, E.E. Eichler, T.S. Furey, J. Galagan, J.G. Gilbert, C. Harmon, Y. Hayashizaki, D. Haussler, H. Hermjakob, K. Hokamp, W. Jang, L.S. Johnson, T.A. Jones, S. Kasif, A. Kaspryzk, S. Kennedy, W.J. Kent, P. Kitts, E.V. Koonin, I. Korf, D. Kulp, D. Lancet, T.M. Lowe, A. McLysaght, T. Mikkelsen, J.V. Moran, N. Mulder, V.J. Pollara, C.P. Ponting, G. Schuler, J. Schultz, G. Slater, A.F. Smit, E. Stupka, J. Szustakowski, D. Thierry-Mieg, J. Thierry-Mieg, L. Wagner, J. Wallis, R. Wheeler, A. Williams, Y.I. Wolf, K.H. Wolfe, S.P. Yang, R.F. Yeh, F. Collins, M.S. Guyer, J. Peterson, A. Felsenfeld, K.A. Wetterstrand, A. Patrinos, M.J. Morgan, P. de Jong, J.J. Catanese, K. Osoegawa, H. Shizuya, S. Choi, Y.J. Chen (2001) Initial sequencing and analysis of the human genome. *Nature* 409, 860–921.

11 Zhang, Y., R. Proenca, M. Maffei, M. Barone, L. Leopold, J.M. Friedman (1994) Positional cloning of the mouse obese gene and its human homologue. *Nature* 372, 425–432.

12 Montague, C.T., I.S. Farooqi, J.P. Whitehead, M.A. Soos, H. Rau, N.J. Wareham, C.P. Sewter, J.E. Digby, S.N. Mohammed, J.A. Hurst, C.H. Cheetham, A.R. Earley, A.H. Barnett, J.B. Prins, S. O'Rahilly (1997) Congenital leptin deficiency is associated with severe early-onset obesity in humans. *Nature* 387, 903–908.

13 Farooqi, I.S., J.M. Keogh, S. Kamath, S. Jones, W.T. Gibson, R. Trussell, S.A. Jebb, G.Y. Lip, S. O'Rahilly (2001) Partial leptin deficiency and human adiposity. *Nature* 414, 34–35.

14 Bentley, D.R. (1996) Genomic sequence information should be released immediately and freely in the public domain. *Science* 274, 533–534.

15 Marshall, E. (2001) Bermuda rules: community spirit, with teeth. *Science* 291, 1192.

16 Venter, J.C., M.D. Adams, E.W. Myers, P.W. Li, R.J. Mural, G.G. Sutton, H.O. Smith, M. Yandell, C.A. Evans, R.A. Holt, J.D. Gocayne, P. Amanatides, R.M. Ballew, D.H. Huson, J.R. Wortman, Q. Zhang, C.D. Kodira, X.H. Zheng, L. Chen, M. Skupski, G. Subramanian, P.D. Thomas, J. Zhang, G.L. Gabor Miklos, C. Nelson, S. Broder, A.G. Clark, J. Nadeau, V.A. McKusick, N. Zinder, A.J. Levine, R.J. Roberts,

M. Simon, C. Slayman, M. Hunkapiller, R. Bolanos, A. Delcher, I. Dew, D. Fasulo, M. Flanigan, L. Florea, A. Halpern, S. Hannenhalli, S. Kravitz, S. Levy, C. Mobarry, K. Reinert, K. Remington, J. Abu-Threideh, E. Beasley, K. Biddick, V. Bonazzi, R. Brandon, M. Cargill, I. Chandramouliswaran, R. Charlab, K. Chaturvedi, Z. Deng, V. Di Francesco, P. Dunn, K. Eilbeck, C. Evangelista, A.E. Gabrielian, W. Gan, W. Ge, F. Gong, Z. Gu, P. Guan, T.J. Heiman, M.E. Higgins, R.R. Ji, Z. Ke, K.A. Ketchum, Z. Lai, Y. Lei, Z. Li, J. Li, Y. Liang, X. Lin, F. Lu, G.V. Merkulov, N. Milshina, H.M. Moore, A.K. Naik, V.A. Narayan, B. Neelam, D. Nusskern, D.B. Rusch, S. Salzberg, W. Shao, B. Shue, J. Sun, Z. Wang, A. Wang, X. Wang, J. Wang, M. Wei, R. Wides, C. Xiao, C. Yan, A. Yao, J. Ye, M. Zhan, W. Zhang, H. Zhang, Q. Zhao, L. Zheng, F. Zhong, W. Zhong, S. Zhu, S. Zhao, D. Gilbert, S. Baumhueter, G. Spier, C. Carter, A. Cravchik, T. Woodage, F. Ali, H. An, A. Awe, D. Baldwin, H. Baden, M. Barnstead, I. Barrow, K. Beeson, D. Busam, A. Carver, A. Center, M.L. Cheng, L. Curry, S. Danaher, L. Davenport, R. Desilets, S. Dietz, K. Dodson, L. Doup, S. Ferriera, N. Garg, A. Gluecksmann, B. Hart, J. Haynes, C. Haynes, C. Heiner, S. Hladun, D. Hostin, J. Houck, T. Howland, C. Ibegwam, J. Johnson, F. Kalush, L. Kline, S. Koduru, A. Love, F. Mann, D. May, S. McCawley, T. McIntosh, I. McMullen, M. Moy, L. Moy, B. Murphy, K. Nelson, C. Pfannkoch, E. Pratts, V. Puri, H. Qureshi, M. Reardon, R. Rodriguez, Y.H. Rogers, D. Romblad, B. Ruhfel, R. Scott, C. Sitter, M. Smallwood, E. Stewart, R. Strong, E. Suh, R. Thomas, N.N. Tint, S. Tse, C. Vech, G. Wang, J. Wetter, S. Williams, M. Williams, S. Windsor, E. Winn-Deen, K. Wolfe, J. Zaveri, K. Zaveri, J.F. Abril, R. Guigo, M.J. Campbell, K.V. Sjolander, B. Karlak, A. Kejariwal, H. Mi, B. Lazareva, T. Hatton, A. Narechania, K. Diemer, A. Muruganujan, N. Guo, S. Sato, V. Bafna, S. Istrail, R. Lippert, R. Schwartz, B. Walenz, S. Yooseph, D. Allen, A. Basu, J. Baxendale, L. Blick, M. Caminha, J. Carnes-Stine, P. Caulk, Y.H. Chiang, M. Coyne, C. Dahlke, A. Mays, M. Dombroski, M. Donnelly, D. Ely, S. Esparham, C. Fosler, H. Gire, S. Glanowski, K. Glasser, A. Glodek, M. Gorokhov, K. Graham, B. Gropman, M. Harris, J. Heil, S. Henderson, J. Hoover, D. Jennings, C. Jordan, J. Jordan, J. Kasha, L. Kagan, C. Kraft, A. Levitsky, M. Lewis, X. Liu, J. Lopez, D. Ma, W. Majoros, J. McDaniel, S. Murphy, M. Newman, T. Nguyen, N. Nguyen, M. Nodell, S. Pan, J. Peck, M. Peterson, W. Rowe, R. Sanders, J. Scott, M. Simpson, T. Smith, A. Sprague, T. Stockwell, R. Turner, E. Venter, M. Wang, M. Wen, D. Wu, M. Wu, A. Xia, A. Zandieh, X. Zhu (**2001**) The sequence of the human genome. *Science* 291, 1304–1351.

17 Mironov, A.A., J.W. Fickett, M.S. Gelfand (**1999**) Frequent alternative splicing of human genes. *Genome Res* 9, 1288–1293.

18 Brett, D., J. Hanke, G. Lehmann, S. Haase, S. Delbruck, S. Krueger, J. Reich, P. Bork (**2000**) EST comparison indicates 38% of human mRNAs contain possible alternative splice forms. *FEBS Lett* 474, 83–86.

19 Modrek, B., C. Lee (**2002**) A genomic view of alternative splicing. *Nat Genet* 30, 13–19.

20 Harrison, P.M., A. Kumar, N. Lang, M. Snyder, M. Gerstein (**2002**) A question of size: the eukaryotic proteome and the problems in defining it. *Nucleic Acids Res* 30, 1083–1090.

21 Fickett, J.W., W.W. Wasserman (**2000**) Discovery and modeling of transcriptional regulatory regions. *Curr Opin Biotechnol* 11, 19–24.

22 Stamm, S. (**2002**) Signals and their transduction pathways regulating

alternative splicing: a new dimension of the human genome. *Hum Mol Genet* 11, 2409–2416.

23 SACHIDANANDAM, R., D. WEISSMAN, S.C. SCHMIDT, J.M. KAKOL, L.D. STEIN, G. MARTH, S. SHERRY, J.C. MULLIKIN, B.J. MORTIMORE, D.L. WILLEY, S.E. HUNT, C.G. COLE, P.C. COGGILL, C.M. RICE, Z. NING, J. ROGERS, D.R. BENTLEY, P.Y. KWOK, E.R. MARDIS, R.T. YEH, B. SCHULTZ, L. COOK, R. DAVENPORT, M. DANTE, L. FULTON, L. HILLIER, R.H. WATERSTON, J.D. MCPHERSON, B. GILMAN, S. SCHAFFNER, W.J. VAN ETTEN, D. REICH, J. HIGGINS, M.J. DALY, B. BLUMENSTIEL, J. BALDWIN, N. STANGE-THOMANN, M.C. ZODY, L. LINTON, E.S. LANDER, D. ALTSHULER (**2001**) A map of human genome sequence variation containing 1.42 million single nucleotide polymorphisms. *Nature* 409, 928–933.

24 REICH, D.E., M. CARGILL, S. BOLK, J. IRELAND, P.C. SABETI, D.J. RICHTER, T. LAVERY, R. KOUYOUMJIAN, S.F. FARHADIAN, R. WARD, E.S. LANDER (**2001**) Linkage disequilibrium in the human genome. *Nature* 411, 199–204.

25 GABRIEL, S.B., S.F. SCHAFFNER, H. NGUYEN, J.M. MOORE, J. ROY, B. BLUMENSTIEL, J. HIGGINS, M. DEFELICE, A. LOCHNER, M. FAGGART, S.N. LIU-CORDERO, C. ROTIMI, A. ADEYEMO, R. COOPER, R. WARD, E.S. LANDER, M.J. DALY, D. ALTSHULER (**2002**) The structure of haplotype blocks in the human genome. *Science* 296, 2225–2229.

26 GOODSTADT, L., C.P. PONTING (**2001**) Sequence variation and disease in the wake of the draft human genome. *Hum Mol Genet* 10, 2209–2214.

27 DALY, M.J., J.D. RIOUX, S.F. SCHAFFNER, T.J. HUDSON, E.S. LANDER (**2001**) High-resolution haplotype structure in the human genome. *Nat Genet* 29, 229–232.

28 SELLICK, G.S., C. GARRETT, R.S. HOULSTON (**2003**) A novel gene for neonatal diabetes maps to chromo-some 10p12.1-p13. *Diabetes* 52, 2636–2638.

29 JOHN, S., N. SHEPHARD, G. LIU, E. ZEGGINI, M. CAO, W. CHEN, N. VASAVDA, T. MILLS, A. BARTON, A. HINKS, S. EYRE, K.W. JONES, W. OLLIER, A. SILMAN, N. GIBSON, J. WORTHINGTON, G.C. KENNEDY (**2004**) Whole-genome scan, in a complex disease, using 11,245 single-nucleotide polymorphisms: comparison with microsatellites. *Am J Hum Genet* 75, 54–64.

30 WATERSTON, R.H., K. LINDBLAD-TOH, E. BIRNEY, J. ROGERS, J.F. ABRIL, P. AGARWAL, R. AGARWALA, R. AINSCOUGH, M. ALEXANDERSSON, P. AN, S.E. ANTONARAKIS, J. ATTWOOD, R. BAERTSCH, J. BAILEY, K. BARLOW, S. BECK, E. BERRY, B. BIRREN, T. BLOOM, P. BORK, M. BOTCHERBY, N. BRAY, M.R. BRENT, D.G. BROWN, S.D. BROWN, C. BULT, J. BURTON, J. BUTLER, R.D. CAMPBELL, P. CARNINCI, S. CAWLEY, F. CHIAROMONTE, A.T. CHINWALLA, D.M. CHURCH, M. CLAMP, C. CLEE, F.S. COLLINS, L.L. COOK, R.R. COPLEY, A. COULSON, O. COURONNE, J. CUFF, V. CURWEN, T. CUTTS, M. DALY, R. DAVID, J. DAVIES, K.D. DELEHAUNTY, J. DERI, E.T. DERMITZAKIS, C. DEWEY, N.J. DICKENS, M. DIEKHANS, S. DODGE, I. DUBCHAK, D.M. DUNN, S.R. EDDY, L. ELNITSKI, R.D. EMES, P. ESWARA, E. EYRAS, A. FELSENFELD, G.A. FEWELL, P. FLICEK, K. FOLEY, W.N. FRANKEL, L.A. FULTON, R.S. FULTON, T.S. FUREY, D. GAGE, R.A. GIBBS, G. GLUSMAN, S. GNERRE, N. GOLDMAN, L. GOODSTADT, D. GRAFHAM, T.A. GRAVES, E.D. GREEN, S. GREGORY, R. GUIGO, M. GUYER, R.C. HARDISON, D. HAUSSLER, Y. HAYASHIZAKI, L.W. HILLIER, A. HINRICHS, W. HLAVINA, T. HOLZER, F. HSU, A. HUA, T. HUBBARD, A. HUNT, I. JACKSON, D.B. JAFFE, L.S. JOHNSON, M. JONES, T.A. JONES, A. JOY, M. KAMAL, E.K. KARLSSON, D. KAROLCHIK, A. KASPRZYK, J. KAWAI, E. KEIBLER, C. KELLS, W.J. KENT, A. KIRBY, D.L. KOLBE, I. KORF, R.S. KUCHERLAPATI, E.J. KULBOKAS, D. KULP, T. LANDERS, J.P. LEGER, S. LEONARD, I. LETUNIC, R. LEVINE, J. LI, M. LI, C. LLOYD, S. LUCAS, B. MA, D.R. MAGLOTT, E.R. MARDIS, L. MATTHEWS,

E. Mauceli, J.H. Mayer, M. McCarthy, W.R. McCombie, S. McLaren, K. McLay, J.D. McPherson, J. Meldrim, B. Meredith, J.P. Mesirov, W. Miller, T.L. Miner, E. Mongin, K.T. Montgomery, M. Morgan, R. Mott, J.C. Mullikin, D.M. Muzny, W.E. Nash, J.O. Nelson, M.N. Nhan, R. Nicol, Z. Ning, C. Nusbaum, M.J. O'Connor, Y. Okazaki, K. Oliver, E. Overton-Larty, L. Pachter, G. Parra, K.H. Pepin, J. Peterson, P. Pevzner, R. Plumb, C.S. Pohl, A. Poliakov, T.C. Ponce, C.P. Ponting, S. Potter, A. Quail, A. Reymond, B.A. Roe, K.M. Roskin, E.M. Rubin, A.G. Rust, R. Santos, V. Sapojnikov, B. Schultz, J. Schultz, M.S. Schwartz, S. Schwartz, C. Scott, S. Seaman, S. Searle, T. Sharpe, A. Sheridan, R. Shownkeen, S. Sims, J.B. Singer, G. Slater, A. Smit, D.R. Smith, B. Spencer, A. Stabenau, N. Stange-Thomann, C. Sugnet, M. Suyama, G. Tesler, J. Thompson, D. Torrents, E. Trevaskis, J. Tromp, C. Ucla, A. Ureta-Vidal, J.P. Vinson, A.C. Von Niederhausern, C.M. Wade, M. Wall, R.J. Weber, R.B. Weiss, M.C. Wendl, A.P. West, K. Wetterstrand, R. Wheeler, S. Whelan, J. Wierzbowski, D. Willey, S. Williams, R.K. Wilson, E. Winter, K.C. Worley, D. Wyman, S. Yang, S.P. Yang, E.M. Zdobnov, M.C. Zody, E.S. Lander (**2002**) Initial sequencing and comparative analysis of the mouse genome. *Nature* 420, 520–562.

31 Roest Crollius, H., O. Jaillon, A. Bernot, C. Dasilva, L. Bouneau, C. Fischer, C. Fizames, P. Wincker, P. Brottier, F. Quetier, W. Saurin, J. Weissenbach (**2000**) Estimate of human gene number provided by genome-wide analysis using Tetraodon nigroviridis DNA sequence. *Nat Genet* 25, 235–238.

32 Collins, F.S., E.D. Green, A.E. Guttmacher, M.S. Guyer (**2003**) A vision for the future of genomics research. *Nature* 422, 835–847.

33 Geller, F., K. Reichwald, A.

Dempfle, T. Illig, C. Vollmert, S. Herpertz, W. Siffert, M. Platzer, C. Hess, T. Gudermann, H. Biebermann, H.E. Wichmann, H. Schafer, A. Hinney, J. Hebebrand (**2004**) Melanocortin-4 receptor gene variant I103 is negatively associated with obesity. *Am J Hum Genet* 74, 572–581.

34 (**1998**) Genome sequence of the nematode *C. elegans*: a platform for investigating biology. *Science* 282, 2012–2018.

35 Tabara, H., A. Grishok, C.C. Mello (**1998**) RNAi in *C. elegans*: soaking in the genome sequence. *Science* 282, 430–431.

36 Ashrafi, K., F.Y. Chang, J.L. Watts, A.G. Fraser, R.S. Kamath, J. Ahringer, G. Ruvkun (**2003**) Genome-wide RNAi analysis of *Caenorhabditis elegans* fat regulatory genes. *Nature* 421, 268–272.

37 Doane, W.W. (**1960**) Developmental physiology of the mutant female sterile(2)adipose of *Drosophila melanogaster*. I. Adult morphology, longevity, egg production, egg lethality. *J Exp Zool* 145, 1–21.

38 Doane, W.W. (**1980**) Selection for amylase allozymes in *Drosophila melanogaster*: some questions. *Evolution* 34, 868–874.

39 Brand, A.H., N. Perrimon (**1993**) Targeted gene expression as a means of altering cell fates and generating dominant phenotypes. *Development* 118, 401–415.

40 Dohrmann, C.E. (**2004**) Target discovery in metabolic disease. *Drug Discov Today* 9, 785–794.

41 Gronke, S., M. Beller, S. Fellert, H. Ramakrishnan, H. Jackle, R.P. Kuhnlein (**2003**) Control of fat storage by a *Drosophila* PAT domain protein. *Curr Biol* 13, 603–606.

42 Ingalls, A.M., M.M. Dickie, G.D. Snell (**1950**) Obese, a new mutation in the house mouse. *J Hered* 41, 317–318.

43 Hummel, K.P., M.M. Dickie, D.L. Coleman (**1966**) Diabetes, a new mutation in the mouse. *Science* 153, 1127–1128.

44 COLEMAN, D.L. (**1973**). Effects of parabiosis of obese with diabetes and normal mice. *Diabetologia* 9, 294–298.

45 CHUA, S.C., JR., W.K. CHUNG, X.S. WU-PENG, Y. ZHANG, S.M. LIU, L. TARTAGLIA, R.L. LEIBEL (**1996**) Phenotypes of mouse diabetes and rat fatty due to mutations in the OB (leptin) receptor. *Science* 271, 994–996.

46 TARTAGLIA, L.A., M. DEMBSKI, X. WENG, N. DENG, J. CULPEPPER, R. DEVOS, G.J. RICHARDS, L.A. CAMPFIELD, F.T. CLARK, J. DEEDS, C. MUIR, S. SANKER, A. MORIARTY, K.J. MOORE, J.S. SMUTKO, G.G. MAYS, E.A. WOOL, C.A. MONROE, R.I. TEPPER (**1995**) Identification and expression cloning of a leptin receptor, OB-R. *Cell* 83, 1263–1271.

47 STROBEL, A., T. ISSAD, L. CAMOIN, M. OZATA, A.D. STROSBERG (**1998**) A leptin missense mutation associated with hypogonadism and morbid obesity. *Nat Genet* 18, 213–215.

48 FAROOQI, I.S., S.A. JEBB, G. LANGMACK, E. LAWRENCE, C.H. CHEETHAM, A.M. PRENTICE, I.A. HUGHES, M.A. MCCAMISH, S. O'RAHILLY (**1999**) Effects of recombinant leptin therapy in a child with congenital leptin deficiency. *N Engl J Med* 341, 879–884.

49 LICINIO, J., S. CAGLAYAN, M. OZATA, B.O. YILDIZ, P.B. DE MIRANDA, F. O'KIRWAN, R. WHITBY, L. LIANG, P. COHEN, S. BHASIN, R.M. KRAUSS, J.D. VELDHUIS, A.J. WAGNER, A.M. DEPAOLI, S.M. MCCANN, M.L. WONG (**2004**) Phenotypic effects of leptin replacement on morbid obesity, diabetes mellitus, hypogonadism, behavior in leptin-deficient adults. *Proc Natl Acad Sci USA* 101, 4531–4536.

50 BULTMAN, S.J., E.J. MICHAUD, R.P. WOYCHIK (**1992**) Molecular characterization of the mouse agouti locus. *Cell* 71, 1195–1204.

51 MILLER, M.W., D.M. DUHL, H. VRIELING, S.P. CORDES, M.M. OLLMANN, B.M. WINKES, G.S. BARSH (**1993**) Cloning of the mouse agouti gene predicts a secreted protein ubiquitously expressed in mice

carrying the lethal yellow mutation. *Genes Dev* 7, 454–467.

52 NAGGERT, J.K., L.D. FRICKER, O. VARLAMOV, P.M. NISHINA, Y. ROUILLE, D.F. STEINER, R.J. CARROLL, B.J. PAIGEN, E.H. LEITER (**1995**) Hyper-proinsulinaemia in obese fat/fat mice associated with a carboxypeptidase E mutation which reduces enzyme activity. *Nat Genet* 10, 135–142.

53 COLEMAN, D.L., E.M. EICHER (**1990**) Fat (fat) and tubby (tub): two autosomal recessive mutations causing obesity syndromes in the mouse. *J Hered* 81, 424–427.

54 HUSZAR, D., C.A. LYNCH, V. FAIRCHILD-HUNTRESS, J.H. DUNMORE, Q. FANG, L.R. BERKEMEIER, W. GU, R.A. KESTERSON, B.A. BOSTON, R.D. CONE, F.J. SMITH, L.A. CAMPFIELD, P. BURN, F. LEE (**1997**) Targeted disruption of the melanocortin-4 receptor results in obesity in mice. *Cell* 88, 131–141.

55 YASWEN, L., N. DIEHL, M.B. BRENNAN, U. HOCHGESCHWENDER (**1999**) Obesity in the mouse model of pro-opiomelanocortin deficiency responds to peripheral melanocortin. *Nat Med* 5, 1066–1070.

56 OLLMANN, M.M., B.D. WILSON, Y.K. YANG, J.A. KERNS, Y. CHEN, I. GANTZ, G.S. BARSH (**1997**) Antagonism of central melanocortin receptors in vitro and in vivo by agouti-related protein. *Science* 278, 135–138.

57 HINNEY, A., S. HOHMANN, F. GELLER, C. VOGEL, C. HESS, A.K. WERMTER, B. BROKAMP, H. GOLDSCHMIDT, W. SIEGFRIED, H. REMSCHMIDT, H. SCHAFER, T. GUDERMANN, J. HEBEBRAND (**2003**) Melanocortin-4 receptor gene: case-control study and transmission disequilibrium test confirm that functionally relevant mutations are compatible with a major gene effect for extreme obesity. *J Clin Endocrinol Metab* 88, 4258–4267.

58 ORTLEPP, J.R., R. KLUGE, K. GIESEN, L. PLUM, P. RADKE, P. HANRATH, H.G. JOOST (**2000**) A metabolic syndrome of hypertension, hyperinsulinaemia and hypercholesterolaemia in the New

Zealand obese mouse. *Eur J Clin Invest* 30, 195–202.

59 GIESEN, K., L. PLUM, R. KLUGE, J. ORTLEPP, H.G. JOOST (**2003**) Diet-dependent obesity and hypercholesterolemia in the New Zealand obese mouse: identification of a quantitative trait locus for elevated serum cholesterol on the distal mouse chromosome 5. *Biochem Biophys Res Commun* 304, 812–817.

60 PERUSSE, L., T. RANKINEN, A. ZUBERI, Y.C. CHAGNON, S.J. WEISNAGEL, G. ARGYROPOULOS, B. WALTS, E.E. SNYDER, C. BOUCHARD (**2005**) The human obesity gene map: the (**2004**) update. *Obes Res* 13, 381–490.

61 SNYDER, E.E., B. WALTS, L. PERUSSE, Y.C. CHAGNON, S.J. WEISNAGEL, T. RANKINEN, C. BOUCHARD (**2004**) The human obesity gene map: the (**2003**) update. *Obes Res* 12, 369–439.

62 ROTHSCHILD, M.F. (**2004**) Porcine genomics delivers new tools and results: this little piggy did more than just go to market. *Genet Res* 83, 1–6.

63 ROHRER, G.A. (**2003**) Porcine genomic sequencing initiative. http://www.genome.iastate.edu/community/PigWhitePaper/PigWhitePaper.html

64 ELLEGREN, H., B.P. CHOWDHARY, M. JOHANSSON, L. MARKLUND, M. FREDHOLM, I. GUSTAVSSON, L. ANDERSSON (**1994**) A primary linkage map of the porcine genome reveals a low rate of genetic recombination. *Genetics* 137, 1089–1100.

65 RETTENBERGER, G., C. KLETT, U. ZECHNER, J. KUNZ, W. VOGEL, H. HAMEISTER (**1995**) Visualization of the conservation of synteny between humans and pigs by heterologous chromosomal painting. *Genomics* 26, 372–378.

66 ROTHSCHILD, M.F. (**2003**) From a sow's ear to a silk purse: real progress in porcine genomics. *Cytogenet Genome Res* 102, 95–99.

67 BIDANEL, J.P., M.F. ROTHSCHILD (**2002**) Current status of quantitative trait locus mapping in pigs. *Pig News and Info* 23, 39–54.

68 ANDERSSON, L., C.S. HALEY, H. ELLEGREN, S.A. KNOTT, M. JOHANSSON, K. ANDERSSON, L. ANDERSSON-EKLUND, I. EDFORS-LILJA, M. FREDHOLM, I. HANSSON *et al.* (**1994**) Genetic mapping of quantitative trait loci for growth and fatness in pigs. *Science* 263, 1771–1774.

69 KNOTT, S.A., L. MARKLUND, C.S. HALEY, K. ANDERSSON, W. DAVIES, H. ELLEGREN, M. FREDHOLM, I. HANSSON, B. HOYHEIM, K. LUNDSTROM, M. MOLLER, L. ANDERSSON (**1998**) Multiple marker mapping of quantitative trait loci in a cross between outbred wild boar and large white pigs. *Genetics* 149, 1069–1080.

70 ROHRER, G.A., J.W. KEELE (**1998**) Identification of quantitative trait loci affecting carcass composition in swine: I. Fat deposition traits. *J Anim Sci* 76, 2247–2254.

71 WALLING, G.A., A.L. ARCHIBALD, J.A. CATTERMOLE, A.C. DOWNING, H.A. FINLAYSON, D. NICHOLSON, P.M. VISSCHER, C.A. WALKER, C.S. HALEY (**1998**) Mapping of quantitative trait loci on porcine chromosome 4. *Anim Genet* 29, 415–424.

72 WALLING, G.A., A.L. ARCHIBALD, P.M. VISSCHER, C.S. HALEY (**1998**) Mapping genes for growth rate and fatness in a Large White x Meishan F_2 pig population. *Proceedings of the British Society of Animal Science* 7.

73 WANG, L., T.P. YU, C.K. TUGGLE, H.C. LIU, M.F. ROTHSCHILD (**1998**) A directed search for quantitative trait loci on chromosomes 4 and 7 in pigs. *J Anim Sci* 76, 2560–2567.

74 DE KONING, D.J., L.L. JANSS, A.P. RATTINK, P.A. VAN OERS, B.J. DE VRIES, M.A. GROENEN, J.J. VAN DER POEL, P.N. DE GROOT, E.W. BRASCAMP, J.A. VAN ARENDONK (**1999**) Detection of quantitative trait loci for backfat thickness and intramuscular fat content in pigs (*Sus scrofa*). *Genetics* 152, 1679–1690.

75 DE KONING, D.J., B. HARLIZIUS, A.P. RATTINK, M.A. GROENEN, E.W. BRASCAMP, J.A. VAN ARENDONK (**2001**) Detection and characterization of quantitative trait loci for meat quality traits in pigs. *J Anim Sci* 79, 2812–2819.

76 Perez-Enciso, M., A. Clop, J.L. Noguera, C. Ovilo, A. Coll, J.M. Folch, D. Babot, J. Estany, M.A. Oliver, I. Diaz, A. Sanchez (**2000**) A QTL on pig chromosome 4 affects fatty acid metabolism: evidence from an Iberian by Landrace intercross. *J Anim Sci* 78, 2525–2531.

77 Harlizius, B., A.P. Rattink, D.J. de Koning, M. Faivre, R.G. Joosten, J.A. van Arendonk, M.A. Groenen (**2000**) The X chromosome harbors quantitative trait loci for backfat thickness and intramuscular fat content in pigs. *Mamm Genome* 11, 800–802.

78 Rohrer, G.A. (**2000**) Identification of quantitative trait loci affecting birth characters and accumulation of backfat and weight in a Meishan-White Composite resource population. *J Anim Sci* 78, 2547–2553.

79 Wada, Y., T. Akita, T. Awata, T. Furukawa, N. Sugai, Y. Inage, K. Ishii, Y. Ito, E. Kobayashi, H. Kusumoto, T. Matsumoto, S. Mikawa, M. Miyake, A. Murase, S. Shimanuki, T. Sugiyama, Y. Uchida, S. Yanai, H. Yasue (**2000**) Quantitative trait loci (QTL) analysis in a Meishan × Gottingen cross population. *Anim Genet* 31, 376–384.

80 Bidanel, J.P., D. Milan, N. Iannuccelli, Y. Amigues, M.Y. Boscher, F. Bourgeois, J.C. Caritez, J. Gruand, P. Le Roy, H. Lagant, R. Quintanilla, C. Renard, J. Gellin, L. Ollivier, C. Chevalet (**2001**) Detection of quantitative trait loci for growth and fatness in pigs. *Genet Sel Evol* 33, 289–309.

81 Grindflek, E., J. Szyda, Z. Liu, S. Lien (**2001**) Detection of quantitative trait loci for meat quality in a commercial slaughter pig cross. *Mamm Genome* 12, 299–304.

82 Malek, M., J.C. Dekkers, H.K. Lee, T.J. Baas, M.F. Rothschild (**2001**) A molecular genome scan analysis to identify chromosomal regions influencing economic traits in the pig. I. Growth and body composition. *Mamm Genome* 12, 630–636.

83 Milan, D., J.P. Bidanel, N. Iannuccelli, J. Riquet, Y. Amigues, J. Gruand, P. Le Roy, C. Renard, C. Chevalet (**2002**) Detection of quantitative trait loci for carcass composition traits in pigs. *Genet Sel Evol* 34, 705–728.

84 Geldermann, H., E. Müller, G. Moser, G. Reiner, H. Bartenschlager, S. Cepica, A. Straitl, J. Kuryl, C. Moran, R. Davoli, C. Brunsch (**2003**) Genome-wide linkage and QTL mapping in porcine F2 families generated from Peitrain, Meishan and Wild Boar Crosses. *J Anim Breed Genet* 120, 363–393.

85 Nagamine, Y., C.S. Haley, A. Sewalem, P.M. Visscher (**2003**) Quantitative trait loci variation for growth and obesity between and within lines of pigs (*Sus scrofa*). *Genetics* 164, 629–635.

86 Vidal, O., J.L. Noguera, M. Amills, L. Varona, M. Gil, N. Jimenez, G. Davalos, J.M. Folch, A. Sanchez (**2005**) Identification of carcass and meat quality quantitative trait loci in a Landrace pig population selected for growth and leanness. *J Anim Sci* 83, 293–300.

87 Sato, S., Y. Oyamada, K. Atsuji, T. Nade, S. Sato, E. Kobayashi, T. Mitsuhashi, K. Nirasawa, A. Komatsuda, Y. Saito, S. Terai, T. Hayashi, Y. Sugimoto (**2003**) Quantitative trait loci analysis for growth and carcass traits in a Meishan × Duroc F2 resource population. *J Anim Sci* 81, 2938–2949.

88 Jeon, J.T., O. Carlborg, A. Tornsten, E. Giuffra, V. Amarger, P. Chardon, L. Andersson-Eklund, K. Andersson, I. Hansson, K. Lundstrom, L. Andersson (**1999**) A paternally expressed QTL affecting skeletal and cardiac muscle mass in pigs maps to the IGF2 locus. *Nat Genet* 21, 157–158.

89 Nezer, C., L. Moreau, B. Brouwers, W. Coppieters, J. Detilleux, R. Hanset, L. Karim, A. Kvasz, P. Leroy, M. Georges (**1999**) An imprinted QTL with major effect on muscle mass and fat deposition maps to the IGF2 locus in pigs. *Nat Genet* 21, 155–156.

90 DE KONING, D.J., A.P. RATTINK, B. HARLIZIUS, J.A. VAN ARENDONK, E.W. BRASCAMP, M.A. GROENEN (**2000**) Genome-wide scan for body composition in pigs reveals important role of imprinting. *Proc Natl Acad Sci USA* 97, 7947–7950.

91 RATTINK, A.P., D.J. DE KONING, M. FAIVRE, B. HARLIZIUS, J.A. VAN ARENDONK, M.A. GROENEN (**2000**) Fine mapping and imprinting analysis for fatness trait QTLs in pigs. *Mamm Genome* 11, 656–661.

92 THOMSEN, H., H.K. LEE, M.F. ROTHSCHILD, M. MALEK, J.C. DEKKERS (**2004**) Characterization of quantitative trait loci for growth and meat quality in a cross between commercial breeds of swine. *J Anim Sci* 82, 2213–2228.

93 MARKLUND, L., P.E. NYSTROM, S. STERN, L. ANDERSSON-EKLUND, L. ANDERSSON (**1999**) Confirmed quantitative trait loci for fatness and growth on pig chromosome 4. *Heredity* 82 (Pt 2):134–141.

94 KIM, K.S., N. LARSEN, T. SHORT, G. PLASTOW, M.F. ROTHSCHILD (**2000**) A missense variant of the porcine melanocortin-4 receptor (MC4R) gene is associated with fatness, growth, feed intake traits. *Mamm Genome* 11, 131–135.

95 KIM, K.S., H. THOMSEN, J. BASTIAANSEN, N.T. NGUYEN, J.C. DEKKERS, G.S. PLASTOW, M.F. ROTHSCHILD (**2004**) Investigation of obesity candidate genes on porcine fat deposition quantitative trait loci regions. *Obes Res* 12, (**1981**)–(**1994**)

96 ANDERSSON, L., M. GEORGES (**2004**) Domestic-animal genomics: deciphering the genetics of complex traits. *Nat Rev Genet* 5, 202–212.

97 FUJII, J., K. OTSU, F. ZORZATO, S. DE LEON, V.K. KHANNA, J.E. WEILER, P.J. O'BRIEN, D.H. MACLENNAN (**1991**) Identification of a mutation in porcine ryanodine receptor associated with malignant hyperthermia. *Science* 253, 448–451.

98 GEORGES, M., C. CHARLIER, N. COCKETT (**2003**) The callipyge locus:

evidence for the trans interaction of reciprocally imprinted genes. *Trends Genet* 19, 248–252.

99 COMUZZIE, A.G., S.A. COLE, L. MARTIN, K.D. CAREY, M.C. MAHANEY, J. BLANGERO, J.L. VANDEBERG (**2003**) The baboon as a nonhuman primate model for the study of the genetics of obesity. *Obes Res* 11, 75–80.

100 HANSEN, B.C. (**1999**) The metabolic syndrome X. *Ann NY Acad Sci* 892, 1–24.

101 VAISSE, C., K. CLEMENT, B. GUY-GRAND, P. FROGUEL (**1998**) A frameshift mutation in human MC4R is associated with a dominant form of obesity. *Nat Genet* 20, 113–114.

102 YEO, G.S., I.S. FAROOQI, S. AMINIAN, D.J. HALSALL, R.G. STANHOPE, S. O'RAHILLY (**1998**) A frameshift mutation in MC4R associated with dominantly inherited human obesity. *Nat Genet* 20, 111–112.

103 HINNEY, A., A. SCHMIDT, K. NOTTEBOM, O. HEIBULT, I. BECKER, A. ZIEGLER, G. GERBER, M. SINA, T. GORG, H. MAYER, W. SIEGFRIED, M. FICHTER, H. REMSCHMIDT, J. HEBEBRAND (**1999**) Several mutations in the melanocortin-4 receptor gene including a nonsense and a frameshift mutation associated with dominantly inherited obesity in humans. *J Clin Endocrinol Metab* 84, 1483–1486.

104 CHEN, Y., D.M. DUHL, G.S. BARSH (**1996**) Opposite orientations of an inverted duplication and allelic variation at the mouse agouti locus. *Genetics* 144, 265–277.

105 KLEYN, P.W., W. FAN, S.G. KOVATS, J.J. LEE, J.C. PULIDO, Y. WU, L.R. BERKEMEIER, D.J. MISUMI, L. HOLMGREN, O. CHARLAT, E.A. WOOLF, O. TAYBER, T. BRODY, P. SHU, F. HAWKINS, B. KENNEDY, L. BALDINI, C. EBELING, G.D. ALPERIN, J. DEEDS, N.D. LAKEY, J. CULPEPPER, H. CHEN, M.A. GLUCKSMANN-KUIS, G.A. CARLSON, G.M. DUYK, K.J. MOORE (**1996**) Identification and characterization of the mouse obesity gene tubby: a member of a novel gene family. *Cell* 85, 281–290.

106 NOBEN-TRAUTH, K., J.K. NAGGERT,

M.A. NORTH, P.M. NISHINA (**1996**)
A candidate gene for the mouse
mutation tubby. *Nature* 380, 534–538.

107 CLEMENT, K., C. VAISSE, N. LAHLOU, S.
CABROL, V. PELLOUX, D. CASSUTO, M.
GOURMELEN, C. DINA, J. CHAMBAZ,
J.M. LACORTE, A. BASDEVANT, P.
BOUGNERES, Y. LEBOUC, P. FROGUEL,
B. GUY-GRAND (**1998**) A mutation in
the human leptin receptor gene
causes obesity and pituitary
dysfunction. *Nature* 392, 398–401.

108 JACKSON, R.S., J.W. CREEMERS, S.
OHAGI, M.L. RAFFIN-SANSON, L.
SANDERS, C.T. MONTAGUE, J.C.
HUTTON, S. O'RAHILLY (**1997**) Obesity
and impaired prohormone processing
associated with mutations in the
human prohormone convertase 1
gene. *Nat Genet* 16, 303–306.

109 FAROOQI, I.S., G. MATARESE, G.M.
LORD, J.M. KEOGH, E. LAWRENCE, C.
AGWU, V. SANNA, S.A. JEBB, F. PERNA,
S. FONTANA, R.I. LECHLER, A.M.
DEPAOLI, S. O'RAHILLY (**2002**)
Beneficial effects of leptin on obesity,
T cell hyporesponsiveness,
neuroendocrine/metabolic dysfunction
of human congenital leptin deficiency.
J Clin Invest 110, 1093–1103.

110 WILLIAMSON, D.A., E. RAVUSSIN, M.L.
WONG, A. WAGNER, A. DIPAOLI, S.
CAGLAYAN, M. OZATA, C. MARTIN, H.
WALDEN, C. ARNETT, J. LICINIO (**2005**)
Microanalysis of eating behavior of
three leptin deficient adults treated
with leptin therapy. *Appetite* 45, 75–80.

111 MATOCHIK, J.A., E.D. LONDON, B.O.
YILDIZ, M. OZATA, S. CAGLAYAN, A.M.
DEPAOLI, M.L. WONG, J. LICINIO
(**2005**) Effect of leptin replacement on
brain structure in genetically leptin-
deficient adults. *J Clin Endocrinol
Metab* 90, 2851–2854.

112 HASKELL-LUEVANO, C., E.K. MONCK
(**2001**) Agouti-related protein functions
as an inverse agonist at a constitutively
active brain melanocortin-4 receptor.
Regul Pept 99, 1–7.

113 NIJENHUIS, W.A., J. OOSTEROM, R.A.
ADAN (**2001**) AgRP(83–132) acts
as an inverse agonist on the human-
melanocortin-4 receptor. *Mol
Endocrinol* 15, 164–171.

114 VINK, T., A. HINNEY, A.A. VAN
ELBURG, S.H. VAN GOOZEN, L.A.
SANDKUIJL, R.J. SINKE, B.M.
HERPERTZ-DAHLMANN, J. HEBEBRAND,
H. REMSCHMIDT, H. VAN ENGELAND,
R.A. ADAN (**2001**) Association between
an agouti-related protein gene
polymorphism and anorexia nervosa.
Mol Psychiatry 6, 325–328.

115 MARKS, D.L., N. BOUCHER, C.M.
LANOUETTE, L. PERUSSE, G.
BROOKHART, A.G. COMUZZIE, Y.C.
CHAGNON, R.D. CONE (**2004**)
Ala67Thr polymorphism in the
Agouti-related peptide gene is
associated with inherited leanness in
humans. *Am J Med Genet* A 126, 267–
271.

116 REIZES, O., J. LINCECUM, Z. WANG,
O. GOLDBERGER, L. HUANG, M.
KAKSONEN, R. AHIMA, M.T. HINKES,
G.S. BARSH, H. RAUVALA, M.
BERNFIELD (**2001**) Transgenic
expression of syndecan-1 uncovers a
physiological control of feeding
behavior by syndecan-3. *Cell* 106, 105–
116.

117 COMUZZIE, A.G., J.E. HIXSON, L.
ALMASY, B.D. MITCHELL, M.C.
MAHANEY, T.D. DYER, M.P. STERN,
J.W. MACCLUER, J. BLANGERO (**1997**)
A major quantitative trait locus
determining serum leptin levels and
fat mass is located on human
chromosome 2. *Nat Genet* 15, 273–
276.

118 HIXSON, J.E., L. ALMASY, S. COLE,
S. BIRNBAUM, B.D. MITCHELL, M.C.
MAHANEY, M.P. STERN, J.W.
MACCLUER, J. BLANGERO, A.G.
COMUZZIE (**1999**) Normal variation in
leptin levels in associated with poly-
morphisms in the proopiomelano-
cortin gene, POMC. *J Clin Endocrinol
Metab* 84, 3187–3191.

119 CAI, G., S.A. COLE, R.A.
BASTARRACHEA-SOSA, J.W. MACCLUER,
J. BLANGERO, A.G. COMUZZIE (**2004**)
Quantitative trait locus determining
dietary macronutrient intakes is
located on human chromosome 2p22.
Am J Clin Nutr 80, 1410–1414.

120 KRUDE, H., H. BIEBERMANN, W. LUCK,
R. HORN, G. BRABANT, A. GRUTERS

(**1998**) Severe early-onset obesity, adrenal insufficiency and red hair pigmentation caused by POMC mutations in humans. *Nat Genet* 19, 155–157.

121 KRUDE, H., H. BIEBERMANN, D. SCHNABEL, M.Z. TANSEK, P. THEUNISSEN, P.E. MULLIS, A. GRUTERS (**2003**) Obesity due to proopiomelanocortin deficiency: three new cases and treatment trials with thyroid hormone and ACTH4–10. *J Clin Endocrinol Metab* 88, 4633–4640.

122 CHALLIS, B.G., L.E. PRITCHARD, J.W. CREEMERS, J. DELPLANQUE, J.M. KEOGH, J. LUAN, N.J. WAREHAM, G.S. YEO, S. BHATTACHARYYA, P. FROGUEL, A. WHITE, I.S. FAROOQI, S. O'RAHILLY (**2002**) A missense mutation disrupting a dibasic prohormone processing site in pro-opiomelanocortin (POMC) increases susceptibility to early-onset obesity through a novel molecular mechanism. *Hum Mol Genet* 11, 1997–2004.

123 TAO, Y.X. (**2005**) Molecular mechanisms of the neural melanocortin receptor dysfunction in severe early onset obesity. *Mol Cell Endocrinol* 239, 1–14.

124 DEMPFLE, A., A. HINNEY, M. HEINZEL-GUTENBRUNNER, M. RAAB, F. GELLER, T. GUDERMANN, H. SCHAFER, J. HEBEBRAND (**2004**) Large quantitative effect of melanocortin-4 receptor gene mutations on body mass index. *J Med Genet* 41, 795–800.

125 XU, H., J. STEVEN RICHARDSON, X.M. LI (**2003**) Dose-related effects of chronic antidepressants on neuroprotective proteins BDNF, Bcl-2 and Cu/Zn-SOD in rat hippocampus. *Neuropsychopharmacology* 28, 53–62.

126 FRIEDEL, S., F.F. HORRO, A.K. WERMTER, F. GELLER, A. DEMPFLE, K. REICHWALD, J. SMIDT, G. BRONNER, K. KONRAD, B. HERPERTZ-DAHLMANN, A. WARNKE, U. HEMMINGER, M. LINDER, H. KIEFL, H.P. GOLDSCHMIDT, W. SIEGFRIED, H. REMSCHMIDT, A. HINNEY, J. HEBEBRAND (**2005**) Mutation screen of the brain derived neurotrophic factor gene (BDNF): identification of several genetic

variants and association studies in patients with obesity, eating disorders, attention-deficit/hyperactivity disorder. *Am J Med Genet B Neuropsychiatr Genet* 132, 96–99.

127 KERNIE, S.G., D.J. LIEBL, L.F. PARADA (**2000**) BDNF regulates eating behavior and locomotor activity in mice. *EMBO J* 19, 1290–1300.

128 YEO, G.S., C.C. CONNIE HUNG, J. ROCHFORD, J. KEOGH, J. GRAY, S. SIVARAMAKRISHNAN, S. O'RAHILLY, I.S. FAROOQI (**2004**) A de novo mutation affecting human TrkB associated with severe obesity and developmental delay. *Nat Neurosci* 7, 1187–1189.

129 CAMPBELL, H., I. RUDAN (**2002**) Interpretation of genetic association studies in complex disease. *Pharmacogenomics J* 2, 349–360.

130 SPIELMAN, R.S., R.E. MCGINNIS, W.J. EWENS (**1993**) Transmission test for linkage disequilibrium: the insulin gene region and insulin-dependent diabetes mellitus (IDDM). *Am J Hum Genet* 52, 506–516.

131 RISCH, N. (**2000**) Searching for genes in complex diseases: lessons from systemic lupus erythematosus. *J Clin Invest* 105, 1503–1506.

132 RISCH, N., K. MERIKANGAS (**1996**) The future of genetic studies of complex human diseases. *Science* 273, 1516–1517.

133 ADEYEMO, A., A. LUKE, R. COOPER, X. WU, B. TAYO, X. ZHU, C. ROTIMI, N. BOUZEKRI, R. WARD (**2003**) A genome-wide scan for body mass index among Nigerian families. *Obes Res* 11, 266–273.

134 ATWOOD, L.D., N.L. HEARD-COSTA, L.A. CUPPLES, C.E. JAQUISH, P.W. WILSON, R.B. D'AGOSTINO (**2002**) Genomewide linkage analysis of body mass index across 28 years of the Framingham Heart Study. *Am J Hum Genet* 71, 1044–1050.

135 BELL, C.G., M. BENZINOU, A. SIDDIQ, C. LECOEUR, C. DINA, A. LEMAINQUE, K. CLEMENT, A. BASDEVANT, B. GUY-GRAND, C.A. MEIN, D. MEYRE, P. FROGUEL (**2004**) Genome-wide linkage analysis for severe obesity in french

caucasians finds significant susceptibility locus on chromosome 19q. *Diabetes* 53, 1857–1865.

136 CHEN, W., S. LI, N.R. COOK, B.A. ROSNER, S.R. SRINIVASAN, E. BOERWINKLE, G.S. BERENSON (**2004**) An autosomal genome scan for loci influencing longitudinal burden of body mass index from childhood to young adulthood in white sibships: The Bogalusa Heart Study. *Int J Obes Relat Metab Disord* 28, 462–469.

137 CHEN, G., A.A. ADEYEMO, T. JOHNSON, J. ZHOU, A. AMOAH, S. OWUSU, J. ACHEAMPONG, K. AGYENIM-BOATENG, B.A. EGHAN, J. OLI, G. OKAFOR, F. ABBIYESUKU, G.M. DUNSTON, Y. CHEN, F. COLLINS, C. ROTIMI (**2005**) A genome-wide scan for quantitative trait loci linked to obesity phenotypes among West Africans. *Int J Obes Relat Metab Disord* 29, 255–259.

138 DENG, H.W., H. DENG, Y.J. LIU, Y.Z. LIU, F.H. XU, H. SHEN, T. CONWAY, J.L. LI, Q.Y. HUANG, K.M. DAVIES, R.R. RECKER (**2002**) A genomewide linkage scan for quantitative-trait loci for obesity phenotypes. *Am J Hum Genet* 70, 1138–1151.

139 FEITOSA, M.F., I.B. BORECKI, S.S. RICH, D.K. ARNETT, P. SHOLINSKY, R.H. MYERS, M. LEPPERT, M.A. PROVINCE (**2002**) Quantitative-trait loci influencing body-mass index reside on chromosomes 7 and 13: the National Heart, Lung, Blood Institute Family Heart Study. *Am J Hum Genet* 70, 72–82.

140 HAGER, J., C. DINA, S. FRANCKE, S. DUBOIS, M. HOUARI, V. VATIN, E. VAILLANT, N. LORENTZ, A. BASDEVANT, K. CLEMENT, B. GUY-GRAND, P. FROGUEL (**1998**) A genome-wide scan for human obesity genes reveals a major susceptibility locus on chromosome 10. *Nat Genet* 20, 304–308.

141 HANSON, R.L., M.G. EHM, D.J. PETTITT, M. PROCHAZKA, D.B. THOMPSON, D. TIMBERLAKE, T. FOROUD, S. KOBES, L. BAIER, D.K. BURNS, L. ALMASY, J. BLANGERO, W.T. GARVEY, P.H. BENNETT, W.C. KNOWLER (**1998**) An autosomal genomic scan for loci linked to type II diabetes mellitus and body-mass index in Pima Indians. *Am J Hum Genet* 63, 1130–1138.

142 HSUEH, W.C., B.D. MITCHELL, J.L. SCHNEIDER, P.L. ST JEAN, T.I. POLLIN, M.G. EHM, M.J. WAGNER, D.K. BURNS, H. SAKUL, C.J. BELL, A.R. SHULDINER (**2001**) Genome-wide scan of obesity in the Old Order Amish. *J Clin Endocrinol Metab* 86, 1199–1205.

143 HUNT, S.C., V. ABKEVICH, C.H. HENSEL, A. GUTIN, C.D. NEFF, D.L. RUSSELL, T. TRAN, X. HONG, S. JAMMULAPATI, R. RILEY, J. WEAVER-FELDHAUS, T. MACALMA, M.M. RICHARDS, R. GRESS, M. FRANCIS, A. THOMAS, G.C. FRECH, T.D. ADAMS, D. SHATTUCK, S. STONE (**2001**) Linkage of body mass index to chromosome 20 in Utah pedigrees. *Hum Genet* 109, 279–285.

144 IWASAKI, N., N.J. COX, Y.Q. WANG, P.E. SCHWARZ, G.I. BELL, M. HONDA, M. IMURA, M. OGATA, M. SAITO, N. KAMATANI, Y. IWAMOTO (**2003**) Mapping genes influencing type 2 diabetes risk and BMI in Japanese subjects. *Diabetes* 52, 209–213.

145 KISSEBAH, A.H., G.E. SONNENBERG, J. MYKLEBUST, M. GOLDSTEIN, K. BROMAN, R.G. JAMES, J.A. MARKS, G.R. KRAKOWER, H.J. JACOB, J. WEBER, L. MARTIN, J. BLANGERO, A.G. COMUZZIE (**2000**) Quantitative trait loci on chromosomes 3 and 17 influence phenotypes of the metabolic syndrome. *Proc Natl Acad Sci USA* 97, 14478–14483.

146 LEE, J.H., D.R. REED, W.D. LI, W. XU, E.J. JOO, R.L. KILKER, E. NANTHAKU-MAR, M. NORTH, H. SAKUL, C. BELL, R.A. PRICE (**1999**) Genome scan for human obesity and linkage to markers in 20q13. *Am J Hum Genet* 64, 196–209.

147 LEMBERTAS, A.V., L. PERUSSE, Y.C. CHAGNON, J.S. FISLER, C.H. WARDEN, D.A. PURCELL-HUYNH, F.T. DIONNE, J. GAGNON, A. NADEAU, A.J. LUSIS, C. BOUCHARD (**1997**) Identification of an obesity quantitative trait locus on mouse chromosome 2 and evidence of

linkage to body fat and insulin on the human homologous region 20q. *J Clin Invest* 100, 1240–1247.

148 LINDSAY, R.S., S. KOBES, W.C. KNOWLER, P.H. BENNETT, R.L. HANSON (**2001**) Genome-wide linkage analysis assessing parent-of-origin effects in the inheritance of type 2 diabetes and BMI in Pima Indians. *Diabetes* 50, 2850–2857.

149 MEYRE, D., C. LECOEUR, J. DELPLANQUE, S. FRANCKE, V. VATIN, E. DURAND, J. WEILL, C. DINA, P. FROGUEL (**2004**) A genome-wide scan for childhood obesity-associated traits in French families shows significant linkage on chromosome 6q22.31-q23.2. *Diabetes* 53, 803–811.

150 MOSLEHI, R., A.M. GOLDSTEIN, M. BEERMAN, L. GOLDIN, A.W. BERGEN (**2003**) A genome-wide linkage scan for body mass index on Framingham Heart Study families. *BMC Genet* 4 (Suppl 1), S97.

151 NORRIS, J.M., C.D. LANGEFELD, A.L. SCHERZINGER, S.S. RICH, E. BOOKMAN, S.R. BECK, M.F. SAAD, S.M. HAFFNER, R.N. BERGMAN, D.W. BOWDEN, L.E. WAGENKNECHT (**2005**) Quantitative trait loci for abdominal fat and BMI in Hispanic-Americans and African-Americans: the IRAS Family study. *Int J Obes Relat Metab Disord* 29, 67–77.

152 OHMAN, M., L. OKSANEN, J. KAPRIO, M. KOSKENVUO, P. MUSTAJOKI, A. RISSANEN, J. SALMI, K. KONTULA, L. PELTONEN (**2000**) Genome-wide scan of obesity in Finnish sibpairs reveals linkage to chromosome Xq24. *J Clin Endocrinol Metab* 85, 3183–3190.

153 PALMER, L.J., S.G. BUXBAUM, E. LARKIN, S.R. PATEL, R.C. ELSTON, P.V. TISHLER, S. REDLINE (**2003**) A whole-genome scan for obstructive sleep apnea and obesity. *Am J Hum Genet* 72, 340–350.

154 PEROLA, M., M. OHMAN, T. HIEKKALINNA, J. LEPPAVUORI, P. PAJUKANTA, M. WESSMAN, M. KOSKENVUO, A. PALOTIE, K. LANGE, J. KAPRIO, L. PELTONEN (**2001**) Quantitative-trait-locus analysis of body-mass index and of stature, by

combined analysis of genome scans of five Finnish study groups. *Am J Hum Genet* 69, 117–123.

155 PLATTE, P., G.J. PAPANICOLAOU, J. JOHNSTON, C.M. KLEIN, K.F. DOHENY, E.W. PUGH, M.H. ROY-GAGNON, A.J. STUNKARD, C.A. FRANCOMANO, A.F. WILSON (**2003**) A study of linkage and association of body mass index in the Old Order Amish. *Am J Med Genet C Semin Med Genet* 121, 71–80.

156 PRICE, R.A., W.D. LI, R. KILKER (**2002**) An X-chromosome scan reveals a locus for fat distribution in chromosome region Xp21–22. *Diabetes* 51, 1989–1991.

157 REED, D.R., Y. DING, W. XU, C. CATHER, E.D. GREEN, R.A. PRICE (**1996**) Extreme obesity may be linked to markers flanking the human OB gene. *Diabetes* 45, 691–694.

158 SAAR, K., F. GELLER, F. RUSCHENDORF, A. REIS, S. FRIEDEL, N. SCHAUBLE, P. NURNBERG, W. SIEGFRIED, H.P. GOLDSCHMIDT, H. SCHAFER, A. ZIEGLER, H. REMSCHMIDT, A. HINNEY, J. HEBEBRAND (**2003**) Genome scan for childhood and adolescent obesity in German families. *Pediatrics* 111, 321–327.

159 STONE, S., V. ABKEVICH, S.C. HUNT, A. GUTIN, D.L. RUSSELL, C.D. NEFF, R. RILEY, G.C. FRECH, C.H. HENSEL, S. JAMMULAPATI, J. POTTER, D. SEXTON, T. TRAN, D. GIBBS, D. ILIEV, R. GRESS, B. BLOOMQUIST, J. AMATRUDA, P.M. RAE, T.D. ADAMS, M.H. SKOLNICK, D. SHATTUCK (**2002**) A major predisposition locus for severe obesity, at 4p15-p14. *Am J Hum Genet* 70, 1459–1468.

160 VAN DER KALLEN, C.J., R.M. CANTOR, M.M. VAN GREEVENBROEK, J.M. GEURTS, F.G. BOUWMAN, B.E. AOUIZERAT, H. ALLAYEE, W.A. BUURMAN, A.J. LUSIS, J.I. ROTTER, T.W. DE BRUIN (**2000**) Genome scan for adiposity in Dutch dyslipidemic families reveals novel quantitative trait loci for leptin, body mass index and soluble tumor necrosis factor receptor superfamily 1A. *Int J Obes Relat Metab Disord* 24, 1381–1391.

161 WATANABE, R.M., S. GHOSH, C.D.

LANGEFELD, T.T. VALLE, E.R. HAUSER, V.L. MAGNUSON, K.L. MOHLKE, K. SILANDER, D.S. ALLY, P. CHINES, J. BLASCHAK-HARVAN, J.A. DOUGLAS, W.L. DUREN, M.P. EPSTEIN, T.E. FINGERLIN, H.S. KALETA, E.M. LANGE, C. LI, R.C. MCEACHIN, H.M. STRINGHAM, E. TRAGER, P.P. WHITE, J. BALOW, JR, G. BIRZNIEKS, J. CHANG, W. ELDRIDGE (**2000**) The Finland-United States investigation of non-insulin-dependent diabetes mellitus genetics (FUSION) study. II. An autosomal genome scan for diabetes-related quantitative-trait loci. *Am J Hum Genet* 67, 1186–1200.

162 WU, X., R.S. COOPER, I. BORECKI, C. HANIS, M. BRAY, C.E. LEWIS, X. ZHU, D. KAN, A. LUKE, D. CURB (**2002**) A combined analysis of genomewide linkage scans for body mass index from the National Heart, Lung, Blood Institute Family Blood Pressure Program. *Am J Hum Genet* 70, 1247–1256.

163 ZHU, X., R.S. COOPER, A. LUKE, G. CHEN, X. WU, D. KAN, A. CHAKRAVARTI, A. WEDER (**2002**) A genome-wide scan for obesity in African-Americans. *Diabetes* 51, 541–544.

164 BOUTIN, P., C. DINA, F. VASSEUR, S. DUBOIS, L. CORSET, K. SERON, L. BEKRIS, J. CABELLON, B. NEVE, V. VASSEUR-DELANNOY, M. CHIKRI, M.A. CHARLES, K. CLEMENT, A. LERNMARK, P. FROGUEL (**2003**) GAD2 on chromosome 10p12 is a candidate gene for human obesity. *PLoS Biol* 1, E68.

165 SUVIOLAHTI, E., L.J. OKSANEN, M. OHMAN, R.M. CANTOR, M. RIDDERSTRALE, T. TUOMI, J. KAPRIO, A. RISSANEN, P. MUSTAJOKI, P. JOUSILAHTI, E. VARTIAINEN, K. SILANDER, R. KILPIKARI, V. SALOMAA, L. GROOP, K. KONTULA, L. PELTONEN, P. PAJUKANTA (**2003**) The SLC6A14 gene shows evidence of association with obesity. *J Clin Invest* 112, 1762–1772.

166 DURAND, E., P. BOUTIN, D. MEYRE, M.A. CHARLES, K. CLEMENT, C. DINA, P. FROGUEL (**2004**) Polymorphisms in the amino acid transporter solute carrier family 6 (neurotransmitter transporter) member 14 gene contribute to polygenic obesity in French Caucasians. *Diabetes* 53, 2483–2486.

167 KLEIN, R.J., C. ZEISS, E.Y. CHEW, J.Y. TSAI, R.S. SACKLER, C. HAYNES, A.K. HENNING, J.P. SANGIOVANNI, S.M. MANE, S.T. MAYNE, M.B. BRACKEN, F.L. FERRIS, J. OTT, C. BARNSTABLE, J. HOH (**2005**) Complement factor H polymorphism in age-related macular degeneration. *Science* 308, 385–389.

168 ABECASIS, G.R., D. GHOSH, T.E. NICHOLS (**2005**) Linkage disequilibrium: ancient history drives the new genetics. *Hum Hered* 59, 118–124.

169 SHAM, P., J.S. BADER, I. CRAIG, M. O'DONOVAN, M. OWEN (**2002**) DNA Pooling: a tool for large-scale association studies. *Nat Rev Genet* 3, 862–871.

170 SCHAID, D.J. (**2004**) Genetic epidemiology and haplotypes. *Genet Epidemiol* 27, 317–320.

171 GU, C., D.C. RAO (**2001**) Optimum study designs. *Adv Genet* 42, 439–457.

172 TERWILLIGER, J.D., H.H. GORING (**2000**) Gene mapping in the 20th and 21st centuries: statistical methods, data analysis, experimental design. *Hum Biol* 72, 63–132.

173 FREIMER, N., C. SABATTI (**2004**) The use of pedigree, sib-pair and association studies of common diseases for genetic mapping and epidemiology. *Nat Genet* 36, 1045–1051.

174 SCHADT, E.E., J. LAMB, X. YANG, J. ZHU, S. EDWARDS, D. GUHATHAKURTA, S.K. SIEBERTS, S. MONKS, M. REITMAN, C. ZHANG, P.Y. LUM, A. LEONARDSON, R. THIERINGER, J.M. METZGER, L. YANG, J. CASTLE, H. ZHU, S.F. KASH, T.A. DRAKE, A. SACHS, A.J. LUSIS (**2005**) An integrative genomics approach to infer causal associations between gene expression and disease. *Nat Genet* 37, 710–717.

175 VAN STEEN, K., M.B. MCQUEEN, A. HERBERT, B. RABY, H. LYON, D.L. DEMEO, A. MURPHY, J. SU, S. DATTA, C. ROSENOW, M. CHRISTMAN, E.K.

SILVERMAN, N.M. LAIRD, S.T. WEISS, C. LANGE (**2005**) Genomic screening and replication using the same data set in family-based association testing. *Nat Genet* 37, 683–691.

176 IOANNIDIS, J.P., T.A. TRIKALINOS, E.E. NTZANI, D.G. CONTOPOULOS-IOANNIDIS (**2003**) Genetic associations in large versus small studies: an empirical assessment. *Lancet* 361, 567–571.

177 VIELAND, V.J. (**2001**) The replication requirement. *Nat Genet* 29, 244–245.

178 NEALE, B.M., P.C. SHAM (**2004**) The future of association studies: gene-based analysis and replication. *Am J Hum Genet* 75, 353–362.

179 KORNER, J., R.L. LEIBEL (**2003**). To eat or not to eat – how the gut talks to the brain. *N Engl J Med* 349, 926–928.

180 SWARBRICK, M.M., B. WALDENMAIER, L.A. PENNACCHIO, D.L. LIND, M.M. CAVAZOS, F. GELLER, R. MERRIMAN, A. USTASZEWSKA, M. MALLOY, A. SCHERAG, W.C. HSUEH, W. RIEF, F. MAUVAIS-JARVIS, C.R. PULLINGER, J.P. KANE, R. DENT, R. MCPHERSON, P.Y. KWOK, A. HINNEY, J. HEBEBRAND, C. VAISSE (**2005**) Lack of support for the association between GAD2 polymorphisms and severe human obesity. *PLOS Biol.*

15
Gene Polymorphisms, Nutrition, and the Inflammatory Response

Robert F. Grimble

The study of the influence of genomic factors on the responsiveness to nutrient intake is a newly developed field of research. There is growing interest in determining the interactions between nutrition, inflammation, and aging and the possible impact of these factors, and the interaction between them, on lifespan and disease development.

Inflammation adversely affects health in many diseases that have an overt inflammatory basis, such as infection, and in those that have a covert inflammatory basis, such as atherosclerosis and type 2 diabetes mellitus. Furthermore new insights into obesity indicate that adipose tissue exerts an inflammatory influence on the body. It has been shown that inflammation also plays a part in the pathology of obesity.

The metabolic effects of inflammation are mediated by proinflammatory cytokines. Metabolic effects include insulin insensitivity, hyperlipidemia, muscle protein loss, hepatic acute phase protein synthesis, and oxidant stress. It is becoming increasingly apparent that inflammation, changes in the mass of adipose tissue and in blood lipids are interlinked. Furthermore dietary manipulation may have linked influences on all three components of this interaction. Aging is also characterized by an increase in inflammatory stress and contains some of the hallmarks of inflammatory disease. It is also noted that inflammatory diseases rise in incidence with increasing age.

Evidence is accumulating that the individual level of production of cytokines and other molecules intimately linked to the inflammatory process is influenced by single-nucleotide polymorphisms (SNPs). The combination of SNPs in an individual might control the relative level of inflammatory stress in response to inflammatory stimuli and disease. These genomic characteristics might therefore influence lifespan, morbidity, and mortality in diseases with an infectious or inflammatory basis and the course of chronic inflammatory disease.

Certain nutrients, such as polyunsaturated fatty acids and antioxidants, decrease the strength of the inflammatory response. Recent studies show that the extent of the anti-inflammatory effects of these nutrients is modulated by SNPs. In addi-

Nutritional Genomics. Edited by Regina Brigelius-Flohé and Hans-Georg Joost
Copyright © 2006 WILEY-VCH Verlag GmbH & Co. KGaA, Weinheim
ISBN: 3-527-31294-3

tion, the effects of polyunsaturated fats on plasma lipids are also modulated by genotype.

A better understanding of this aspect of nutrient–gene interactions and of the genomic factors that influence the intensity of inflammation in diseases will help in the targeting of nutritional therapy.

15.1
Introduction

In 2001 the details of the structure of the human genome became available to the scientific community. These findings from the study of the human genome shed new light on the interaction of nutrients with the immune system, and health and disease processes within individuals. They have indicated the direction in which clinical nutrition and public health strategy might profitably travel if immune function is to be optimized, chronic inflammatory disease successfully treated, and healthy lifespan maximized.

The Human Genome Project indicated that small base changes, termed single-nucleotide polymorphisms (SNPs), occur in genes. These differences in genotype alter either the function or the amount of product produced following activation of the gene. The parts of the genome that control the functioning of the immune system are particularly rich in SNPs [1].

An increasing body of research clearly demonstrates that SNPs not only influence the level of cytokine production in an individual but are closely associated with mortality and morbidity in a wide range of diseases [2].

15.2
Involvement of Inflammation in Life and Disease Processes

Inflammation is an essential part of the response of the body to infection, surgery, and injury. It fulfils an important role in killing pathogens. This outcome is achieved by the creation of a hostile tissue environment through production of oxidant molecules and activation of T and B lymphocytes. From the evolutionary perspective it is the most primitive of the responses of complex organisms to invasion by potential pathogens. Unlike the components of the mammalian immune response that belong to the category of acquired immunity, inflammation is associated with wide-ranging and profound metabolic changes that are triggered by four proinflammatory cytokines: interleukins 1α and 1β (IL-1), interleukin 6 (IL-6), and tumor necrosis factor alpha (TNFα) [3]. These cytokines have hormone-like properties. Although produced by a wide range of cells, including fixed and circulating phagocytes, lymphocytes, T helper cells, mast cells, and fibroblasts, in evolutionary terms cytokine production is a key property of phagocytic cells. Examples of cytokine production by phagocytic cells can be found in a number of species. Pro-

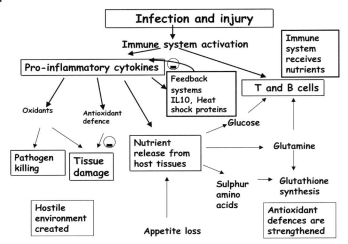

Fig. 15.1. Key aspects of the cytokine-mediated and coordinated response to infection and injury.

inflammatory interleukins have been cloned in birds, fish, and amphibians, but not reptiles; monoclonal antibody studies also suggest that an IL-6-like factor exists in starfish [4].

Of prime importance among the actions of proinflammatory cytokines is their ability to make substrate available from endogenous sources to support the activity of T and B lymphocytes in destruction of invading microorganisms. The nutrients also enhance antioxidant defenses, so that healthy tissue may be protected from the potent mediators released during inflammation [3] (Fig. 15.1). Under the influence of cytokines, blood lipids are raised, gluconeogenesis is enhanced, muscle protein is lost, catabolic hormones are produced in increased amounts, and decreased insulin sensitivity occurs [3, 5]. Evidence for the production of proinflammatory cytokines can be found in the response of the body to infection and injury. Likewise, cytokine production occurs in diseases that have an overt inflammatory basis, such as rheumatoid arthritis and Crohn's disease, as well as in diseases with a covert inflammatory basis, such as atherosclerosis, type 2 diabetes mellitus, and obesity [5] (Fig. 15.2). In the diseases with an inflammatory basis, the changes do not have a purposeful nature and contribute to pathology. From the evolutionary perspective the inflammatory response is designed to ensure the survival of the mammalian species rather than survival of individual members of the species.

Proinflammatory cytokine production is also associated with a number of life processes. Cytokine production occurs at ovulation, in response to vigorous prolonged exercise, and during the process of aging [6–8]. Brain function is influenced by proinflammatory cytokine production. Not only do cytokines have somnambulistic properties but sleep deprivation will increase plasma IL-6 production and the ability of peripheral blood mononuclear cells (PBMCs) to produce IL-1

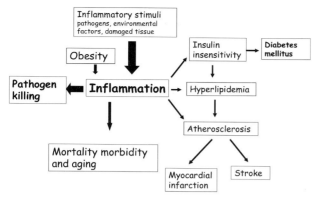

Fig. 15.2. Linkage between cytokine-mediated mechanisms for pathogen killing and diseases with an overt or covert inflammatory basis.

when stimulated with endotoxin. In patients with obstructive sleep apnea there was a significant, positive correlation between blood levels of IL-6 and TNFα and the percentage of time of apnea and hypopnea [9]. It is well known that sleep deprivation lowers the resistance of the body to infections but it is unclear how the increased level of inflammatory stress associated with lack of sleep contributes to this phenomenon.

15.3
Genetic Effects on the Intensity of the Inflammatory Process

One of the earliest indications that mediators of inflammation came under genomic influence arose in *in vitro* studies on the production of TNFα, by PBMCs, from healthy and diseased subjects. The cells were stimulated with inflammatory agents and showed a remarkable constancy in TNFα production in male and post-menopausal female subjects [10]. This individual constancy suggested that genetic factors exert a strong influence. Over the last 15 years the complexity of the interaction between genotype and inflammation has become increasingly apparent. SNPs in the genes responsible for producing the molecules involved in the inflammatory process exert a modulatory effect on the intensity of inflammation. Polymorphisms occur in genes that control events at all levels in the inflammatory process, influencing the receptors that bind bacterial endotoxin, pro- and anti-inflammatory cytokine production, transcription factor activation, and eicosanoid production.

The largest amount of information is available for genomic influences on cytokine production. The earliest studies showed that SNPs in the promoter regions for the TNF and lymphotoxin α (LT-α) genes are associated with differential TNF production [11–13]. The SNP site located at base 308 of the promoter, TNFα−308,

Table 15.1. Single-nucleotide polymorphisms (SNPs) in cytokine genes and genotypes associated with altered levels of cytokine production. The location of the polymorphism is indicated by the nucleotide position in the promoter region

Gene and location of polymorphism in promoter region	Base change involved in creation of variant allele	Resultant genotype associated with raised cytokine production and/or altered clinical outcome to inflammation[a]
Proinflammatory cytokines		
TNFα−308	G → A (TNF1 → TNF2)	A (TNF2) allele
LT-α+252	G → A (TNFB1 → TNFB2)	AA (TNFB2:2)
IL-1β−511	C → T	CT or TT
IL-6−174	C → G	G allele
Anti-inflammatory cytokines		
IL-10−1082†	A → G	GG
TGF-1β+915 (Arg25-Pro)	G → C	GG

[a] In the case of proinflammatory cytokines outcome worsens, in the case of anti-inflammatory cytokines outcome improves with the genotype indicated in the right-hand column.
TNF, tumor necrosis factor; LT, lymphotoxin; IL, interleukin; TGF-1, transforming growth factor 1; C, cytosine; G, guanosine; T, thymidine.

has been shown to influence TNFα expression in a number of studies [12–14]. In addition, a polymorphism in the first intron of the LT-α gene (*LTA*+252AA) is closely linked to TNFα production, and found in linkage disequilibrium with HLA-A1, B8, DR3 [11, 12]. This genotype has also been reported to define a TNF "high expresser" haplotype [15], in addition to modifying expression of LT-α itself [11]. Subsequently a large body of research indicated that SNPs occur in the upstream regulatory (promoter) regions of many cytokine genes [16, 17]. Many of these genetic variations influence the level of expression of genes. Both pro- and anti-inflammatory cytokines are influenced by the differences in genotype [2]. A number of SNPs that have been implicated in the outcome to inflammatory stress are shown in Table 15.1.

Endotoxin (lipopolysaccharide (LPS)) from the bacterial cell wall is a major stimulator of the inflammatory process that follows invasion of the body by microorganisms. Upstream of cytokine production, endotoxin binds to LPS-binding protein, which interacts with receptors in the cell membrane of macrophages and other immune cells. There are polymorphisms in these receptors (Toll-like receptors (TLR)), which influence the intensity and outcome of the inflammatory process. TLR4 is an important receptor in triggering an inflammatory response to bacteria [18]. A polymorphism occurs at position 299, resulting in a substitution of

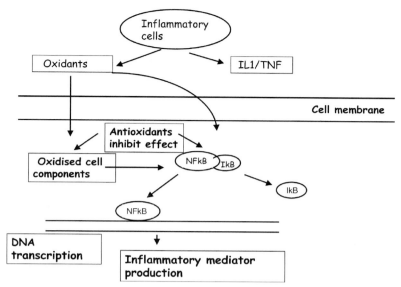

Fig. 15.3. The interaction between oxidant stress, antioxidant defenses, nuclear factor kappa-B activation during the inflammatory response to infection, injury and during chronic inflammatory disease.

glycine for aspartic acid and a decrease in the intensity of inflammation. This change is associated with a reduced risk of carotid artery atherosclerosis and acute coronary events, lower plasma concentrations of IL-6 and an increased risk of Gram-negative infections [18].

An increase in the production of oxidant molecules follows from activation of the immune system. In addition to being part of the inflammatory response, the oxidants may upregulate inflammation. The nuclear factor kappa-B (NFκB) is activated by oxidants and switches on many of the genes involved in the inflammatory response (cytokines, adhesion molecules, and acute phase proteins). Enhancement of antioxidant defenses is important in protecting healthy tissues and in preventing excessive activation of NFκB by the oxidative cellular environment during inflammation [19] (Fig. 15.3). The upregulation of inflammation, caused by activation of NFκB, increases the risk of host damage [20]. Genetic factors influence the propensity of individuals to produce oxidant molecules and thereby influence NFκB activation. Natural resistance-associated macrophage protein 1 (NRAMP1) modulates many macrophage functions, including TNFα production and activation of inducible nitric oxide synthase (iNOS), which occurs by cooperation between the NRAMP1, TNFα, and LT-α genes [21, 22].

There are four variations in the NRAMP1 gene, resulting in different basal levels of activity and differential sensitivity to stimulation by inflammatory agents. Alleles 1, 2, and 4 are poor promoters, while allele 3 causes high gene expression. Hyperactivity of macrophages, associated with allele 3, is linked to autoimmune disease susceptibility and high resistance to infection, while allele 2 increases susceptibility to infection and protects against autoimmune disease [22].

Many symptoms of inflammation (fever, muscle proteolysis, chemotaxis) are mediated by eicosanoids (prostaglandins (PGs) and leukotrienes (LTs)). These are formed, together with many other types of lipid mediator (e.g. platelet-activating factor), after the liberation of long-chain polyunsaturated fatty acids from phospholipids in cell membranes by phospholipases. The latter enzymes are activated by proinflammatory cytokines. Recent studies have shown that there are a number of variants in the promoter of the gene for 5-lipoxygenase (a key enzyme in LT formation), which alter the intensity of the atherosclerotic process [23]. The variants showed differences in the number of tandem Sp1-binding motifs (5'GGGCGG3') in the promoter region of the gene for the enzyme. Middle-aged men and women carrying the variant gene have a greater degree of atherosclerosis (assessed by measuring intima-media thickness) and a higher level of chronic inflammation (assessed by plasma C-reactive protein (CRP) concentrations) than individuals carrying the wild type of the gene.

A number of molecules suppress production of proinflammatory cytokines and exert an anti-inflammatory influence. These include antioxidant defenses (via modulation of NFκB activation) and interleukin 10 (IL-10) and transforming growth factor 1β (TGFβ) [24, 25]. There are at least three polymorphic sites (-1082, -819, -592) in the IL-10 promoter that influence production [26]. SNPs also occur in genes encoding enzymatic components of antioxidant defenses, such as catalase, superoxide dismutase (SOD), and glutathione peroxidase-1, which influence levels of activity [27–29].

Despite the complexity of the interactions between genotype and the inflammatory process a biological framework for these interactions is starting to emerge. Each individual possesses combinations of SNPs in the genes associated with inflammation, resulting in an "inflammatory drive" within the body. Individual genotype will determine the level of the inflammatory drive. Currently the critical genes which fully define the strength of the inflammatory drive remain to be fully defined but certain SNPs appear to exert a strong influence on the prognosis of inflammation. At an individual level the inflammatory drive may express itself as differing degrees of morbidity and mortality (Fig. 15.4). The strength of the genomic influence on the inflammatory process may affect the chances of an individual developing inflammatory disease, or succumbing to the potentially adverse effects of the inflammatory process on host tissues, particularly if their antioxidant defenses are poor.

15.4
Sex-linked Differences in the Interaction of Genotype with Inflammatory Processes

In general, men are more sensitive to genomic influences on the strength of the inflammatory process than women. In a study on LT genotype and mortality from

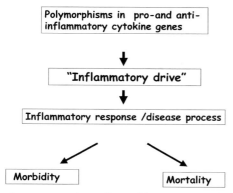

Fig. 15.4. Resultant effects of the interaction of single-nucleotide polymorphisms in genes controlling production of inflammatory mediators.

sepsis it was found that men with a *TNFB22* (*LTA*+252AA) genotype had a 72% mortality rate compared with men who were *TNFB11* (*LTA*+252GG), who had a 42% mortality rate. In female patients the mortalities for the two genotypes were 53% and 33% respectively [30]. In a study on patients undergoing surgery for gastrointestinal cancer it was found that postoperative C-reactive protein and IL-6 concentrations were higher in men than in women and that in multivariate analysis, males possessing the *TNF2* allele had greater responses than men without it. The genomic influence was not seen in women (Table 15.2) [31]. In a study on hospitalized geriatric care patients, men possessing the *TNFB22* or *IL1*−511CT or TT genotype had lower three-year survival rates than men possessing the *TNFB11* or *B12*, or *IL1*−511CC genotype. Furthermore possession of the *IL1*−511T allele was associated with a 48% greater length of stay in hospital in men (Table 15.3) [32, 33]. In women, proinflammatory genotypes (*IL6*−174GG and *TNF*−308A allele) were associated with longer survival times. Thus genotypes associated with enhanced cytokine production adversely effect men but offer a "survival advantage" to women. The mechanisms underlying these gender differences in genomic effects on inflammation are yet to be investigated.

15.5
Genotype, Gender, Inflammation, and Longevity

15.5.1
Longevity and the Metabolic Effects of Aging

Aging is associated with an increase in the incidence of both overt and covert inflammatory diseases. A further feature of the aging process is a decrease in insulin sensitivity and increased incidence in obesity. Aging also contains a number of fea-

Table 15.2. Influence of TNFα−308 polymorphism and gender on the inflammatory response to surgery in gastrointestinal cancer patients

	Males ($n = 65$)	Females ($n = 56$)
Duration of operation (min)	214 ± 125 (65)	172 ± 76 (56)
Blood loss (ml)	473 ± 521 (65)	258 ± 348 (54)
Peak CRP concentration (mg/ml)[a]	150 ± 81(45)	126 ± 48 (38)
TNFα−308 without 2 allele	132 ± 46 (33)	128 ± 57 (25)
with 2 allele	193 ± 116(12)‡	121 ± 37 (13)
Peak IL-6 concentration (pg/ml)[b]	467 ± 411 (31)	342 ± 310 (20)
TNFα−308 without 2 allele	439 ± 402 (24)	362 ± 376 (15)
with 2 allele	676 ± 544 (7)‡	315 ± 147 (5)

[a] Two days postoperatively; [b] one day postoperatively.
‡ Significantly different from females with same genotype by multivariate analysis allowing for longer operation time and greater blood loss $P = 0.013$ and $P = 0.027$ for CRP and IL-6, respectively.
Means ± SD, values in brackets are the number of assays.
TNF, tumor necrosis factor. IL, interleukin. CRP, C-reactive protein.

Table 15.3. Influence of genotype and gender on hospital length of stay and survival in geriatric care patients (mean age 83 ± 7 years). The location of the polymorphism is indicated by the nucleotide position in the IL-1β and LT-α genes, *TNFB11*, *TNFB12*, *TNFB22*

	Males ($n = 50$)	Females ($n = 39$)
Hospital length of stay (days)		
Patients with IL-1β−511 CC genotype	9 ± 11 (9)	15 ± 7 (26)
Patients with IL-1β−511 CT or TT genotype	14 ± 6 (13)‡	14 ± 12 (28)
Survival post-hospitalization (months)		
Patients with TNFB11 or 12 genotype	21 ± 12 (11)	21 ± 15 (19)
Patients with TNFB22 genotype	10 ± 12 (10)‡	22 ± 15 (28)
Patients with IL-1β−511 CC genotype	27 ± 13 (9)	19 ± 15 (26)
Patients with IL-1β−511 CT or TT genotype	14 ± 13 (16)‡	25 ± 14 (28)

TNF, tumor necrosis factor; IL, interleukin. C, cytosine, T, thymidine.
‡ Significantly different from value for same sex possessing the other genotype.
$P < 0.05$ Mann–Whitney test.
Means ± SD, values in brackets are the number of patients.

tures of the chronic inflammatory response, such as a decrease in muscle protein and a rise in plasma acute phase protein and cytokine concentrations [34, 35]. Antioxidant status may decline with age [36] and may thus be linked to increased TNFα production [37, 38]. Chronic inflammatory diseases, such as rheumatoid arthritis are associated with increased oxidant stress [38].

15.5.2
Sex-linked Differences the Impact of Inflammatory Stress on Longevity

Evidence is accumulating that the intensity of the reaction between an individual's immune system and microbes, together with childhood exposure to infections, may have a long-term impact upon longevity [39].

In a review on the different impact of genetic factors on the probability of reaching old age, Franceschi *et al.* [40] concluded from studies conducted in Italy, that (regarding mitochondrial DNA haplogroups, tyrosine hydroxylase, and IL-6 genes) female longevity is less dependent on genetics than male longevity, and that female centenarians are more likely to have had a healthier lifestyle and more favorable environmental conditions than men because of gender-specific cultural and anthropological characteristics.

Nonetheless, human longevity may be directly correlated with optimal functioning of the immune system. It is likely therefore that one of the genetic determinants of longevity resides in the influence of genes controlling the activity of the immune system. A number of cross-sectional studies have examined the role of HLA genes on human longevity by comparing HLA antigen frequencies between groups of young and elderly people. Conflicting findings have been obtained. However a review of this issue [41] concluded that in humans there is possibly an association between longevity and some HLA-DR alleles, or the HLA-B8, DR3 haplotype. These genotypes are involved in the antigen non-specific control of the immune response. Indeed, polymorphisms in cytokine genes may play a role in longevity by modulating an individual's responses to inflammatory stimuli. It has been estimated that there are up to 7000 variations in the genome that contribute to lifespan [42]. Loss of muscle and bone mass result in poorer survival following infection and injury. It is thus unsurprising that genes associated with lifespan that contribute to loss of muscle and bone mass during aging are related to the inflammatory process and include pro- and anti-inflammatory cytokines and their receptors.

Possession of SNPs in certain pro- and anti-inflammatory cytokine genes influences lifespan. When 700 individuals between 60 and 110 years of age were studied it was noted that not only was plasma IL-6 concentration positively related to age but individuals with a SNP in the promoter region of the IL-6 gene that predisposes to high levels of production of the cytokine ($-174GG$) decreased in frequency with age. The effect was seen in men but not in women [39]. While men with the SNP made up 58% of the 60- to 80-year-old age group, the percentage fell to 38% in subjects <99 years of age. Conversely, one of three SNPs in the

IL-10 gene (−1082GG) that is closely linked to higher production of the anti-inflammatory cytokine IL-10 [17, 44] was found in higher proportions in male centenarians than in younger controls (58% versus 34%). In females this genotype exerted no effect upon longevity [45]. Thus it would appear that genetic characteristics that might influence the balance between pro- and anti-inflammatory cytokines influences mortality in men but not women [43]. A study on SNPs (e.g. +874 T → A) that influence interferon γ (IFNγ) production further reinforces the concept that possession of a genotype which predisposes to a raised proinflammatory status adversely influences lifespan [46]. In women, possession of the A allele, which is known to be associated with low production of IFNγ, significantly increased the probability of reaching old age. It might be concluded that possession of "high-producing" alleles of IL-10 is universally protective against morbidity and mortality. Indeed possession of a genotype that results in low levels of IL-10 production (−1082AA) increases the risk of developing inflammatory diseases [47–49]. However in a large survey of hospital admission in the Netherlands patients with raised IL-10 to IL-6 ratios had higher mortality in a wide range of disease, indicating that a "low inflammatory drive" is not always protective from adverse life event [50].

The implication that apparent "protective" or "non-protective" effects of cytokine genotypes against adverse life events exist, however, is still open to question since not all studies implicate cytokine gene SNP in longevity. When cytokine gene SNPs in the genes producing IL-1α, IL-1β, IL-1Ra, IL-6, IL-10, and TNFα were characterized in 250 Finnish nonagenarians (52 men and 198 women) and in 400 healthy blood donors (18–60 years) as controls no statistically significant differences were found in the distribution of genotype, allelic frequencies, and A2+ carrier status between nonagenarians and younger controls [51].

15.6
Genotype, Insulin Sensitivity, Obesity, and Cardiovascular Disease

Glucose is utilized by a wide range of tissues and cell types in the body. It is of prime importance as a metabolic fuel in the immune system. Paradoxically, insulin insensitivity may initially exert a beneficial effect during the response to infection and injury, but have an adverse influence in chronic disease processes [52]. Studies on rats given endotoxin and observations on patients with sepsis show that gluconeogenesis is enhanced under the influence of proinflammatory cytokines. Large increases in glucose utilization by immune cells occur, gluconeogenesis increases, and hypertriglyceridemia develops [53, 54]. An insulin-insensitive state will reduce glucose uptake by tissues in which the process is insulin dependent (muscle), thereby increasing availability for tissues in which the process is not insulin dependent (immune tissue). During inflammation, secretion of catabolic hormones, which enhances muscle protein breakdown and glutamine release, will oppose insulin action.

There are close links between obesity, oxidant stress, and inflammation [52]. The

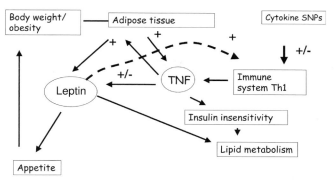

Fig. 15.5. Interaction of leptin and tumor necrosis factor (TNF), with adipose tissue mass, lipid metabolism, and inflammation. TNF and leptin stimulate the immune system and adipose tissue, respectively. Adipose tissue exerts a proinflammatory influence. The interrelationship influences lipid metabolism and plasma triglyceride concentrations.

links lie in the ability of adipose tissue to produce proinflammatory cytokines [55, 56]. There is a positive relationship between adiposity and the strength of the inflammatory response. A positive correlation between serum TNFα production and body mass index (BMI) has been noted in non-insulin-dependent diabetes mellitus (NIDDM) patients, obese subjects, and healthy women [57–59]. Abdominal subcutaneous adipose tissue was found to produce 7.5-fold higher amounts of TNFα than adipose tissue of lean subjects. Furthermore insulin sensitivity was inversely correlated with both TNFα and IL-6 production [60]. Genetic factors may play a part in the interaction between inflammation, oxidant stress, and insulin insensitivity. Single nucleotide polymorphisms in the NFκB gene were investigated in a group of type 1 diabetic patients and compared with normal controls. It was found that there was a higher frequency of allele 138 bp (highly bioactive) and lower frequency of allele 146 bp (less bioactive) in diabetics than in controls [61].

Leptin production is stimulated by proinflammatory cytokines (Fig. 15.5). The influence is complex, involving initially a stimulation followed by inhibition, of leptin by cytokines [56]. Nonetheless a positive relationship between TNFα and leptin production in adipose tissue has been observed [62]. Thus plasma triglycerides, body fat mass and inflammation may be loosely associated because of these endocrine relationships. In an investigation of cytokine production in healthy male subjects we found that while there were no statistically significant relationships between BMI, plasma fasting triglycerides, and the ability of PBMCs to produce TNFα in the study population as a whole, individuals with the LTA+252AA genotype (associated with raised TNF production) showed significant relationships between TNF production and BMI and fasting triglycerides (Fig. 15.6) [63]. Thus despite the study population being made up of healthy subjects, within that population were individuals with a genotype that resulted in an "aged" phenotype as far as plasma lipids, BMI, and inflammation were concerned.

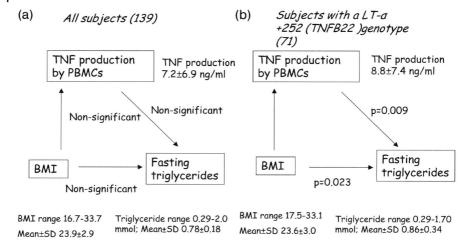

Fig. 15.6. Relationships between TNF production, BMI and fasting triglycerides and the ability of PBMCs to produce TNFα in healthy men. Data for the study population as (a) a whole, as well as (b) in individuals with a *LTA*+252AA (TNF22 genotype). NS, non-significant; BMI, body mass index; PBMC, peripheral blood mononuclear cell. The number of subjects is shown in brackets, the correlations were examined by Spearman's rank correlation.

There is increasing recognition that atherosclerosis has a strong underlying inflammatory component [64–66]. Observations from epidemiological studies on diet and heart disease showed associations between factors (such as "reduced serum albumin, periodontitis, and *Helicobacter pylori, Chlamydia pneumoniae* infections") which did not have an immediately apparent association with plasma cholesterol or dietary saturated fat intake (risk factors for cardiovascular disease). Atheromatous plaques are commonly found in the arteries of healthy adults. Proinflammatory cytokines, such as TNFα, are produced by cells found in the atheromatous artery wall. These cells include macrophages and foam cells derived from them, T lymphocytes, and smooth muscle cells (particularly of the invasive secretory phenotype) [67]. Migration of leukocytes from the circulation into the atherosclerotic lesion is believed to play an important part in plaque development [67]. Thus from these studies and the research findings on obesity and inflammation cited earlier, and insights gained from the modulatory influence of genotype, it can be seen that with the current rise in the proportion of the population who are obese, individuals with certain genotypes (e.g. *LTA*+252AA) may be at increased risk from atherosclerosis. This latter genotype is found in approximately 40% of the population. Thus it is important to gain insights into the way that genotype influences responsiveness to dietary intervention.

15.7
Influence of Genotype on the Interactions Between Changes in Nutrient Intake and Inflammation

The intensity of the inflammatory process is influenced by changes in the pattern of nutrient intake. In general, nutrients exert an effect over a period of weeks, rather than hours and, depending on the mode of action of the nutrient, act at a number of levels in the sequence of events between the initial stimulus and the final inflammatory process.

Nutrients may modulate transcription of inflammatory mediators directly, as in the case of 1,25-dihydroxyvitamin D and retinoic acid, or indirectly as is the case for antioxidants such as vitamins C and E by influencing NFκB activation. Omega-3 and omega-6 polyunsaturated fatty acids (PUFAs) also exert indirect effects on inflammation by modulating cytokine production and altering the potency of inflammatory mediators such as the eicosanoids and platelet activating factor. In addition, these PUFAs may act via stimulation of peroxisomal proliferator-activated receptors [68]. Clinical trials have been conducted to reduce the level of inflammation by nutritional means in a number of diseases. These have targeted in particular proinflammatory cytokine production and NFκB activation. This strategy has met with mixed levels of success.

An important variable in whether statistically demonstrable changes in inflammation can be obtained is the genotype of the individual. Pharmacologists have for a number of years studied the issue of genotype and responsiveness to drugs (e.g. Ref. [69]). More recently the question of whether the responsiveness of inflammation to nutritional modulation is, like the response to drugs, independent of genotype or whether, in any population there will be individual "non-responders" due to genotype. These latter individuals might require larger changes in nutrient intake to achieve an anti-inflammatory effect than other individuals, or might be inherently "non-responsive."

It has been clearly demonstrated that SNPs influence the responsiveness of individuals to changes in nutrient intake. For example SNPs may be responsible for variations in the lipemic response to dietary lipids [70] and influence the interrelationship between plasma vitamin B_{12}, folate, and homocysteine [71].

Currently studies examining the interaction between genotype, nutrients, and inflammation have focused on the suppression of proinflammatory cytokine production or alteration in the strength of antioxidant defenses. Fish oil has been employed in the first type of study and vitamin E is in the second.

15.8
The Influence of Genotype on the Modulatory Effects of Changes in Dietary Fat Intake

Inflammation and lipid metabolism are intimately linked. Lipid mediators are the products and controllers of the inflammatory process. Furthermore inflammation alters fat metabolism, as is evident from the hyperlipidemia that occurs during the course of an inflammatory response. Thus alterations in dietary fat intake can impact upon all aspects of this complex interrelationship. Plasma triglycerides and cholesterol are not only related to obesity and risk of cardiovascular disease but are raised as a part of the collection of metabolic changes that occur during the inflammatory response. The mechanisms for the rise in plasma lipids are complex but are due in part to the ability of proinflammatory cytokines to increase very low-density lipoprotein (VLDL) secretion as a result of increased adipose tissue lipolysis, to increase *de novo* hepatic fatty acid synthesis, and to suppress fatty acid oxidation. However in severe infection, when cytokines are grossly elevated, VLDL clearance decreases secondary to decreased lipoprotein lipase and apolipoprotein E (apoE) in VLDL. In rodents, hypercholesterolemia occurs due to increased hepatic cholesterol synthesis and decreased low-density lipoprotein (LDL) clearance, conversion of cholesterol to bile acids, and secretion of cholesterol into the bile. Marked alterations in proteins important in high-density lipoprotein (HDL) metabolism lead to decreased reverse cholesterol transport and increased cholesterol delivery to immune cells. Oxidation of LDL and VLDL increases during inflammation and HDL becomes a proinflammatory molecule. Lipoproteins become enriched in ceramide, glucosylceramide, and sphingomyelin, leading to an increase in uptake by macrophages. These latter changes may exert a pro-atherogenic influence [72].

Fish oil supplementation is not universally efficacious in the treatment of inflammatory disease. Rheumatoid arthritis and inflammatory bowel disease have been the most successfully treated of all inflammatory diseases [73]. The anti-inflammatory mechanism may be through suppression of TNF production. Endres *et al.* [74] showed that large doses (15 g/day for 6 weeks) of oil in nine healthy volunteers resulted in a small but statistically significant reduction in TNFα and interleukin 1β (IL-1β) production from PBMCs. Subsequently less than half of 11 similar small intervention studies were unable to demonstrate a statistically significant reduction in cytokine production. To understand the differences in response more closely, we conducted a study on 111 young men fed 6 g fish oil/day for 12 weeks and measured TNFα production by PBMCs before and after supplementation, in relation to the SNP at −308 in the TNF, and at +252 in the LT genes. No significant effect of fish oil on cytokine production was noted in the group as a whole. However when the results were examined according to tertile of TNFα production prior to supplementation, homozygosity for the LT-α A allele (*TNFB2*) was 2.5-times more frequent in the highest than in the lowest tertile of production. The percentage of individuals in whom fish oil suppressed production was lowest (22%) in the lowest tertile and doubled with each ascending tertile. In the highest tertile mean values were decreased by 43%. In the lowest tertile mean values were increased by 62%. Individuals who were homozygotes for the A allele were strongly represented among unresponsive individuals. In the lowest tertile of TNFα production only heterozygous subjects were responsive to suppressive effects of fish oil. In the medium tertile this genotype was six times more frequent than other LT-α genotypes among responsive individuals. No relationship between possession of *TNFA*−308 genotype and responsiveness to fish oil was found. Clearly, while the level of inflammation determines whether fish oil will exert an anti-inflammatory influence or not, and is influenced by the *LTA*−252A allele, the precise genomic mechanism for an anti-inflammatory effect is unclear at present [75].

In studies on atherosclerosis it was found that the increase in intima-media thickness associated with variants in the number of tandem Sp1-binding motifs in the promoter region of the 5-lipoxygenase gene was blunted by a raised omega-3 PUFA intake and worsened by high levels of omega-6 PUFA intake [23].

Inflammation promotes the development of atherosclerotic plaques and influences their stability. Fats rich in omega-3 and omega-6 PUFAs are considered to prevent cardiovascular disease. Omega *n*-3 PUFAs, in addition to reducing pro-inflammatory cytokine production, decrease the incidence of re-infarction in heart disease patients. In the GISSI study a daily supplement of 1 g of omega-3 PUFA decreased the rate of re-infarction in by 10–15%. We examined whether polymorphisms in pro- and anti-inflammatory cytokine genes are associated with the response of atherosclerotic plaques to dietary *n*-3 and *n*-6 PUFA intake. Supplements (6 g/day) of fish or sunflower or placebo oil were given to patients with severe atherosclerosis prior to surgical removal of carotid plaques. Patients were genotyped for SNPs in *TNFA, LTA, IL1B, IL6, IL10*, and *TGF1B* genes and plaque structure was examined by histomorphometry. It was found that an AA genotype for the SNP at +252 in the *LTA* gene was associated with significantly higher plaque soft lipid and lower fibrous tissue content in patients given sunflower oil supplements. A raised PUFA intake in individuals with a lower capacity for IL-10 production (possessing the A allele of the SNP at −1082) was associated with a higher proportion of foamy macrophages. In patients who possessed an A allele in the SNP at −308 in the *TNFA* gene or CC genotype for the *IL1B*−511 SNP, fish oil supplementation was associated with improved plaque pathology as evident from the a lower modified American Heart Association grading [76]. There is therefore the possibility that a proinflammatory genotype may have an adverse influence on plaque structure but bestow benefit when fish oil supplementation is administered. It is interesting to note that in a study on the ability of fish oil to reduce fasting plasma triglyceride concentrations in healthy young men we noted that fish oil appeared to reduce concentrations only in all individuals with a BMI > 25 and in particular in individuals with an *LTA*+252AA genotype [63, 77]. Thus it would appear that in healthy populations possession of the A allele in the *LTA*+252 SNP may bestow risk during the response to inflammatory stimuli but an advantage in terms of responsiveness to the lipid-lowering properties of fish oil (Fig. 15.7).

Population studies have revealed further interesting insights into nutrient–gene interactions in the area of inflammation, dietary fats, and lipid metabolism. Pima Indians have been studied for many years in relation to obesity and its related metabolic problems. There is an association between high IL-6 levels and insulin resistance in both Pima Indian and White populations. When investigators studied the *IL6*−174G/C polymorphism in Native American and Spanish White populations they found that among the Spanish White subjects there was a significant difference in genotypic distribution between diabetic and non-diabetic subjects; the GG

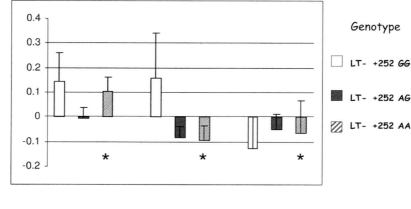

Fig. 15.7. The influence of body mass index and *LTA+252* genotype on the ability of a dietary supplement of 6 g/day fish oil for 12 weeks to lower fasting plasma triglyceride concentrations in healthy men.

genotype was more common in diabetic (0.40) than in non-diabetic (0.29) subjects. The G allele was much more frequent in the Native Americans. When healthy Native Americans were compared with those who had type 2 diabetes mellitus, the GG genotype was significantly more common in diabetic subjects. When this sample population was stratified according to ethnic heritage, all subjects who were of full Pima Indian heritage had the GG genotype [78].

Studies on the health of the population of Framingham in Massachusetts provided seminal work on the understanding of dietary risk factors for cardiovascular disease. Responsiveness of plasma lipids to dietary manipulations is highly variable, with diet–gene interactions thought to explain, in part, interindividual responses. In addition, polymorphisms at specific gene loci are thought to be significant determinants of fasting lipid levels and may also explain the highly heterogeneous nature of individuals' postprandial responses to a standard fat load.

Recently genomic investigations have revealed some interesting insights into the interactions between dietary fat intake and plasma cholesterol concentration in the population. Cholesterol is transported by lipoproteins. Polymorphisms influence synthesis of these molecules. ApoE is a functional and structural component of several classes of lipoproteins, including chylomicrons, VLDLs, and their remnants, and has a major influence on the metabolism and clearance of these particles by the liver. Three common isoforms of the ApoE gene loci exist, yielding apoE2, apoE3, and apoE4, with between 55% and 60% of the population homozygous for the E3 allele [76]. A study assessed the efficacy of fish oil supplementation in counteracting the classic dyslipidemia of the atherogenic lipoprotein phenotype (ALP) in 55 men. In addition, the impact of the common apoE polymorphism on the fasting and postprandial lipid profile and on responsiveness to the dietary intervention was measured. The subjects completed a randomized placebo-controlled crossover trial of fish oil (3.0 g eicosapentaenoic acid/docosahexaenoic acid per day) and placebo (olive oil) capsules. The six-week treatment arms were separated by a 12-week washout period. Fish oil supplementation resulted in a reduction in fasting plasma triglyceride (TG) concentrations of 35%, a decrease in the postprandial TG response of 26%, and a fall in LDL-3 of 26%. However, no change in concentrations was evident. Baseline values of HDL-cholesterol were significantly lower in apoE4 carriers. The apoE genotype also had a striking impact on lipid responses to fish oil intervention. Individuals with an apoE2 allele displayed a marked reduction in the extent of the postprandial TG increase. In apoE4 individuals, a significant increase in total cholesterol and a trend toward a reduction in HDL-cholesterol relative to the common homozygous E3/E3 profile was evident. Thus the study demonstrates that the efficacy of fish oil fatty acids in counteracting the pro-atherogenic lipid profile in individuals with an atherogenic lipid profile is influenced by the apoE genotype, individuals with an apoE2 genotype being more likely to benefit from the anti-atherogenic influence of fish oil on postprandial lipemia. It could be hypothesized that individuals with raised plasma lipids with the E4 genotype are at greatest risk of atherosclerosis [79].

An SNP at −2854 in the gene for apolipoprotein B100, the *APOC3*−2854T > G polymorphism, lies in the *APOC3*−A4 intergenic region. In healthy adults, this polymorphism was associated with plasma triglyceride concentration. Individuals with a homozygous T allele (wild-type) had a 55% lower concentration compared with individuals who are homozygous for the rarer G allele. Age and gender had a significant impact on genotype–triglyceride interactions [80].

In the Framingham study population some interesting insights were gained into nutrient–gene interactions between PUFA intake and HDL-cholesterol metabolism. *APOA1* is the primary protein constituent of HDL. An SNP at −75 in the *APOA1* gene involving a G to A substitution influences the effects of PUFA intake on HDL concentration. A positive relationship between PUFA intake and HDL cholesterol concentration existed in individuals with a GG genotype. HDL increased by more than 40% over a range of PUFA intake from <4 to >8% of dietary energy. In individuals who were homozygous for the A allele HDL-cholesterol was not influenced by the amount of PUFA consumed. Heterozygotes showed a modest rise in HDL-cholesterol over the range of PUFA intake. These findings have important public health implications. The anti-atherogenic effects of PUFAs associated with HDL-cholesterol are only possible in individuals with a G allele for the SNP [81].

The association between the APOE genotype and carotid atherosclerosis, defined as carotid artery intima-media thickness (IMT) and stenosis >25%, was studied to assess if other cardiovascular risk factors modified any association. A total of 1315 men and 1408 women from the Framingham Offspring Study underwent carotid ultrasound and were genotyped for APOE variants. The data were analyzed according to three genotypes: *APOE2* (including E2/E2, E3/E2 genotypes), *APOE3* (E3/E3), and *APOE4* (including E4/E3, E4/E4 genotypes). There were gender differences in the interaction between genotype and pathology. In women, the *APOE2* group was associated with lower carotid IMT (0.67 versus 0.73 mm) and lower prevalence of stenosis compared with the *APOE3* group. In men, APOE genotype was not associated with carotid IMT or stenosis in the group as a whole; however, diabetes modified the association between APOE genotype and carotid IMT. Among diabetic men, the *APOE4* group was associated with a higher internal carotid artery IMT (1.22 mm) than the *APOE3* group (0.90 mm) or the *APOE2* group (0.84 mm). The E2 allele was associated with lower levels of carotid atherosclerosis in women, and the E4 allele was associated with higher internal carotid IMT in diabetic men [82].

15.9
Influence of Antioxidants and Genotype on Inflammation

Paraoxonase-1 is hypothesized to protect serum lipoproteins from oxidative stress in humans. Decreased serum activity of paraoxonase-1 in animal models is associated with an increased risk of vascular disease and has been linked to the antioxidant capacity of the enzyme. An SNP in the human paraoxonase-1 gene strongly influences serum concentrations of the enzyme. The hypothesis that SNPs in this gene might be genetic risk factors for vascular disease in humans was examined by Leviev *et al.* [83]. In an earlier study the same group tested the hypothesis that promoter polymorphism T(−107)C of the human paraoxonase gene (*PON1*) is associated with risk of coronary disease [83]. Genotypes arising from the promoter C(−907)G polymorphism were analyzed in a the "Etude Cas-Temoins de l'Infarctus du Myocard" population. The global odds ratio for myocardial infarction, comparing the high expresser GG genotype to other genotypes, was 0.77. The association with the SNP for the paraoxonase-1 gene was more pronounced in the youngest age group (odds ratio 0.52) and was progressively lost with age. The study supports the idea that the protective, antioxidant capacity of HDLs is at least in part genetically determined and that genotype acting via this route would reduce inflammatory stress and its consequences.

Proteomic studies have shown that inducible nitric oxide synthase (iNOS) and superoxide dismutase (SOD) are both influenced by the natural resistance-associated macrophage protein (NRAMP) 1 gene [84]. The molecule controls a wide range of macrophage activities. The production of oxidant molecules, enhancing proinflammatory cytokine production via high levels of NFκB activation may thus be under a genomic influence due to the aforementioned variations in the NRAMP1 gene [85].

Antioxidant intake also modifies cytokine production. In a study on healthy men and women and smokers, dietary supplementation with 600 IU/day α-tocopherol for one month, suppressed the ability of PBMCs to produce TNFα. Production was reduced by 22 and 33% in non-smokers and smokers respectively [86]. In a similar dietary intervention study on normolipemic and hypertriglyceridemic subjects receiving α-tocopherol at 600 IU/day for six weeks, TNFα, IL-1β, and IL-8 production by LPS-stimulated blood mononuclear cells was reduced [87]. A similar effect of α-tocopherol was noted in a study on normal subjects and type 2 diabetics [88]. However, there were large standard deviations in the data from these studies, indicating major intra-individual variability in the ability of vitamin E to suppress production of the cytokine. While a number of studies have shown that α-tocopherol suppresses superoxide production, the situation with regards to nitric oxide is less clear [86, 87]. The α-tocopherol derivative pentamethylhydroxychromane inhibited LPS-stimulated NFκB and iNOS activation in cultured J774 macrophages [89].

At present it is unknown whether antioxidants interact with SNP in the genes associated with oxidant stress and inflammation in a differential manner as may occur with the other anti-inflammatory nutrient *n*-3 PUFAs. This topic is currently an area of active research. It is likely that genomic differences in the response to antioxidant therapy will be found because recent studies of the effects on polymorphisms in genes that influence endogenous components of antioxidant defense have been found. Selenium (Se) is an essential micronutrient for human health. The essential nature of this micronutrient is attributed to its presence in a range of 20–30 selenoproteins including the cytosolic and phospholipid hydroperoxide glutathione peroxidases (GPX1 and GPX4) (see also chapter 8). In eukaryotes selenium is incorporated into selenoproteins as the amino acid selenocysteine in a process requiring a stem–loop within the 3′ untranslated region (3′ UTR) of the mRNA. GPX4 may play a role in regulation of leukotriene biosynthesis and thus inflammation. In a study on healthy subjects the region of the GPX4 gene corresponding to the 3′ UTR was scanned for mutations. The data revealed a T/C variant at position 718. The distribution of this SNP in the study population was 34% CC, 25% TT and 41% TC. Individuals of different genotypes exhibited significant differences in the levels of lymphocyte 5-lipoxygenase total products formed when the cells were stimulated. Individuals with the C allele at 718 showed increased levels of those products compared with T718 and T/C718 (36% and 44% increases, respectively). The data suggest that the SNP718 has functional effects and support the hypothesis that GPX4 plays a regulatory role in leukotriene biosynthesis [90] (see chapter 8).

It is unclear at present what role the SNP might play in susceptibility to asthma since leukotrienes play an important role in hypersensitivity of the lungs. A better understanding of this interaction and of the interaction of *n*-3 PUFAs with genotype may allow a better design of nutrient products for the treatment of inflammatory disease.

15.10
Conclusion

Inflammation is both an essential part of the ability to survive the continual attack of pathogens on the human body and a process that can seriously debilitate or kill the individual. In addition, inflammation plays a further disadvantageous role in a wide range of diseases having a covert inflammatory basis. This duality in effects on human health and survival arises from the actions of cytokines, which are key modulators of the response. The pro- and anti-inflammatory cytokines, nuclear transcription factors, and antioxidant defenses influence the intensity of the response. The recent insights from the human genome have revealed individual differences in the degree to which key proteins are expressed. This variability, induced by single-nucleotide polymorphisms in the genes associated with the inflammatory process, is being shown to be an important determinant of the strength of the inflammatory process. Inflammation has complex and interrelated effects on metabolism. It is intimately linked with lipid metabolism. The recent findings that adipose tissue is an organ capable of exerting an inflammatory influence on the body has further strengthened these two areas of human biology which impact upon health. It is becoming increasingly apparent that the outcome from dietary intervention is influenced by genotype. Although studies of nutrient–gene–inflammation interactions are still in their infancy, further research into the interaction between SNPs and nutrients will permit nutritional therapy and health promotion to be more effectively tailored at an individual level. With greater insight into gene–nutrient interactions, alterations in diet and single nutrient interventions may be better able to protect against cancer, decrease the occurrence of cardiovascular and other chronic diseases, and perhaps increase human longevity.

Acknowledgments

The author is grateful to the BBSRC and the Food Standards Agency for funding which has facilitated a number of studies reported in this chapter.

References

1 Howell, W.M., P.C. Calder, R.F. Grimble (2002) Gene polymorphisms, inflammatory diseases and cancer. *Proc. Nutr. Soc.* 4, 447–456.

2 Paoloni-Giacobino, A., R. Grimble, C. Pichard (2003) Genomic interactions with disease and nutrition. *Clin. Nutr.* 22, 507–514.

3 Grimble, R. (2001) Nutritional modulation of immune function. *Proc. Nutr. Soc.* 60, 389–397.

4 Kaiser, P., L. Rothwell, S. Avery, S. Balu (2004) Evolution of the interleukins. *Dev. Comp. Immunol.* 28, 375–394.

5 Grimble, R.F. (2002) Inflammatory status and insulin resistance. *Curr. Opin. Clin. Nutr. Metab. Care* 5, 551–559.

6 Willis, C., J.M. Morris, V. Danis, E.D. Gallery (2003) Cytokine production by peripheral blood monocytes during the normal human ovulatory menstrual cycle. *Hum. Reprod.* 18, 1173–1178.

7 Pedersen, B.K. (2000) Special feature for the Olympics: effects of exercise on the immune system: exercise and cytokines. *Immunol. Cell. Biol.* 78, 532–535.

8 GRIMBLE, R.F. (**2003**) Inflammatory response in the elderly. *Curr. Opin. Clin. Nutr. Metab. Care* 6, 21–29.

9 HATIPOGLU, U., I. RUBINSTEIN (**2003**) Inflammation and obstructive sleep apnea syndrome pathogenesis: a working hypothesis. *Respiration* 70, 665–671.

10 JACOB, C., Z. FRONEK, G. LEWIS, M. KOO, J. HANSEN, H. McDEVITT (**1990**) Heritable major histocompatibility complex class II-associated differences in production of tumor necrosis factor alpha: relevance to genetic predisposition to systemic lupus erythematosus. *Proc. Natl Acad. Sci. USA* 87, 1233–1237.

11 MESSER, G., U. SPENGLER, M. JUNG, G. HONOLD, K. BLOMER, G. PAPE, G. RIETHMULLER (**1991**) Polymorphic structure of the tumour necrosis factor (TNF) locus: an Nco I polymorphism in the first intron of the human TNF-b gene correlates with a variant amino acid in position 26 and a reduced level of TNFa production. *J. Exp. Med.* 173, 209–219.

12 WILSON, A., N. DE VRIES, F. POCIOT, F. DI GIOVINE, L. VAN DER PUTTE, G. DUFF (**1993**) An allelic polymorphism within the human tumor necrosis factor alpha promoter region is strongly associated with HLA-A1, B8, and DR3 alleles. *J. Exp. Med.* 177, 557–560.

13 ALLEN, R. (**1999**) Polymorphism of the human TNFα promoter – random variation or functional diversity? *Mol. Immunol.* 36, 1017–1027.

14 KROEGER, K., J. STEER, D. JOYCE, L. ABRAHAM (**2000**) Effects of stimulus and cell type on the expression of the −308 tumour necrosis factor promoter polymorphism. *Cytokine* 12, 110–119.

15 WARZOCHA, K., P. RIBEIRO, J. BIENVENU, P. ROY, C. CHARLOT, D. RIGAL, B. COIFFIER, G. SALLES (**1998**) Genetic polymorphisms in the tumor necrosis factor locus influence non-Hodgkin's lymphoma outcome. *Blood* 91, 3574–3581.

16 BIDWELL, J., L. KEEN, G. GALLAGHER, R. KIMBERLY, T. HUIZINGA, M. McDERMOTT, J. OKSENBERG, J. McNICHOLL, F. POCIOT, C. HARDT, S. D'ALFONSO (**1999**) Cytokine gene polymorphisms in human disease: on-line databases. *Genes. Immun.* 1, 3–19.

17 TURNER, D., D. WILLIAMS, D. SANKARAN, M. LAZARUS, H. SINNOTT IV (**1997**) An investigation of polymorphisms in the interleukin-10 gene promoter. *Eur. J. Immunogenet.* 24, 108–112.

18 COOK, D.N., D.S. PISETSKY, D.A. SCHWARTZ (**2004**) Toll-like receptors in the pathogenesis of human disease. *Nat. Immunol.* 5, 975–979.

19 SCHRECK, R., P. RIEBER, P. BAEUERLE (**1991**) Reactive oxygen intermediates as apparently widely used messengers in the activation of the NF-kappa B transcription factor and HIV-1. *EMBO J.* 10, 2247–2258.

20 JERSMANN, H., C. HII, J. FERRANTE, A. FERRANTE (**2001**) Bacterial lipopolysaccharide and tumor necrosis factor alpha synergistically increase expression of human endothelial adhesion molecules through activation of NF-kappaB and p38 mitogen-activated protein kinase signaling pathways. *Infect. Immun.* 69, 1273–1279.

21 ABLES, G., D. TAKAMATSU, H. NOMA, S. EL-SHAZLY, H. JIN, T. TANIGUCHI, K. SEKIKAWA, T. WATANABE (**2001**) The roles of Nramp1 and Tnfa genes in nitric oxide production and their effect on the growth of *Salmonella typhimurium* in macrophages from Nramp1 congenic and tumor necrosis factor-alpha-/-mice. *J. Interferon Cytokine Res.* 21, 53–62.

22 SEARLE, S., J. BLACKWELL (**1999**) Evidence for a functional repeat polymorphism in the promoter of the human NRAMP1 gene that correlates with autoimmune versus infectious disease susceptibility. *J. Med. Genet.* 36, 295–299.

23 DWYER, J.H., H. ALLAYEE, K.M. DWYER, J. FAN, H. WU, R. MAR, A.J. LUSIS, M. MEHRABIAN (**2004**) Arachidonate 5-lipoxygenase promoter genotype, dietary arachidonic acid, and atherosclerosis. *N. Engl. J. Med.* 350, 29–37.

24 Espevik, T., I. Figari, M. Shalaby, G. Lackides, G. Lewis, H. Shepard, M.J. Palladino (**1987**) Inhibition of cytokine production by cyclosporin A and transforming growth factor beta. *J. Exp. Med.* 166, 571–576.

25 Chernoff, A., E. Granowitz, L. Shapiro, E. Vannier, G. Lonnemann, J. Angel, J. Kennedy, A. Rabson, S. Wolff, C.A. Dinarello (**1995**) A randomized, controlled trial of IL-10 in humans. Inhibition of inflammatory cytokine production and immune responses. *J. Immunol.* 154, 5492–5499.

26 Perrey, C., V. Pravica, P. Sinnott, I. Hutchinson (**1998**) Genotyping for polymorphisms in interferon-gamma, interleukin-10, transforming growth factor-beta 1 and tumour necrosis factor-alpha genes: a technical report. *Transplant Immunol.* 6, 193–197.

27 Forsberg, L., L. Lyrenas, U. de Faire, R.A. Morgenstern (**2001**) A common functional C-T substitution polymorphism in the promoter region of the human catalase gene influences transcription factor binding, reporter gene transcription and is correlated to blood catalase levels. *Free Radic. Biol. Med.* 30, 500–505.

28 Mitrunen, K., P. Sillanpaa, V. Kataja, M. Eskelinen, V. Kosma, S. Benhamou, M. Uusitupa, A. Hirvonen (**2001**) Association between manganese superoxide dismutase (MnSOD) gene polymorphism and breast cancer risk. *Carcinogenesis* 22, 827–829.

29 Chorazy, P., H.J. Schumacher, T. Edlind (**1992**) Role of glutathione peroxidase in rheumatoid arthritis: analysis of enzyme activity and DNA polymorphism. *DNA. Cell. Biol.* 11, 221–225.

30 Schroder, J., V. Kahlke, M. Book, F. Stuber (**2000**) Gender differences in sepsis: genetically determined? *Shock* 14, 307–310.

31 Thorell, A., J. Nygren, O. Ljungqvist, N. Barber, S. Grant, J. Madden, R. Grimble (**2003**) Cytokine genotype and gender influence the inflammatory response to surgery. *Clin. Nutr.* 22, S45.

32 Grimble, R., P. Andersson, J. Madden, J. Palmblad, M. Persson, I. Vedin, T. Cederholm (**2003**) Gene:gene interactions influence the outcome in elderly patients. *Clin. Nutr.* 22, S39.

33 Persson, M., K. Brismar, K. Katzarski, J. Nordenstrom, T. Cederholm (**2002**) Nutritional status using mini nutritional assessment and subjective global assessment predict mortality in geriatric patients. *J. Am. Geriatr. Soc.* 50, 1996–2002.

34 Grimble, R. (**2003**) Inflammatory response in the elderly. *Curr. Opin. Clin. Nutr. Metab. Care* 6, 21–29.

35 Nuttall, S.L., M.J. Kendall, U. Martin (**1999**) Age-independent oxidative stress in elderly patients with non-insulin-dependent diabetes mellitus. *Q. J. Med.* 92, 33–38.

36 Rink, L., I. Cakman, H. Kirchner (**1998**) Altered cytokine production in the elderly. *Mech. Ageing Dev.* 102, 199–209.

37 Kudoh, A., H. Katagai, T. Takazawa, A. Matsuki (**2001**) Plasma pro-inflammatory cytokine response to surgical stress in elderly patients. *Cytokine* 15, 270–273.

38 Ozturk, H., M. Cimen, O. Cimen, M. I. Kacmaz (**1999**) Oxidant/antioxidant status of plasma samples from patients with rheumatoid arthritis. *Rheumatol. Int.* 19, 35–37.

39 Finch, C.E., and E.M. Crimmins (**2004**) Inflammatory exposure and historical changes in human life-spans. *Science* 305, 1736–1739.

40 Franceschi, C., L. Motta, S. Valensin, R. Rapisarda, A. Franzone, M. Berardelli, M. Motta, D. Monti, M. Bonafe, L. Ferrucci, L. Deiana, G. Pes, C. Carru, M. Desole, C. Barbi, G. Sartoni, C. Gemelli, F. Lescai, F. Olivieri, F. Marchegiani, M. Cardelli, L. Cavallone, P. Gueresi, A. Cossarizza, L. Troiano, G. Pini, P. Sansoni, G. Passeri, R. Lisa, L. Spazzafumo, L. Amadio, S. Giunta, R. Stecconi, R. Morresi, C.

VITICCHI, R. MATTACE, G. DE
BENEDICTIS, G. BAGGIO (**2002**) Do
men and women follow different
trajectories to reach extreme longevity?
*Italian Multicenter Study on
Centenarians (IMUSCE) Aging
(Milano)* 12, 77–84.

41 CARUSO, C., G. CANDORE, G. ROMANO,
D. LIO, M. BONAFE, S. VALENSIN, C.
FRANCESCHI (**2001**) Immunogenetics
of longevity. Is major histocompatibility complex polymorphism relevant
to the control of human longevity? A
review of literature data. *Mech. Ageing
Dev.* 122, 445–462.

42 MARTIN, G. (**1997**) Genetics and the
pathobiology of ageing. *Philos. Trans.
R. Soc. Lond. B Biol. Sci.* 352, 1773–
1780.

43 BONAFE, M., F. MARCHEGIANI, M.
CARDELLI, F. OLIVIERI, L. CAVALLONE,
S. GIOVAGNETTI, C. PIERI, M. MARRA,
R. ANTONICELLI, L. TROIANO, P.
GUERESI, G. PASSERI, M. BERARDELLI,
G. PAOLISSO, M. BARBIERI, S. TESEI,
R. LISA, G. DE BENEDICTIS, C.A.
FRANCESCHI (**2001**) A gender-
dependent genetic predisposition to
produce high levels of IL-6 is
detrimental to longevity. *Eur. J.
Immunol.* 31, 2357–2361.

44 HUTCHINSON, IV P.V.A., P.J. HAJEER,
(**1999**) Identification of high and low
responders to allographs. *Rev.
Immunogenet.* 1, 323–333.

45 LIO, D., L. SCOLA, A. CRIVELLO,
G. COLONNA-ROMANO, G. CANDORE,
M. BONAFE, L. CAVALLONE, C.
FRANCESCHI, C. CARUSO (**2002**)
Gender specific association between –
1082 IL-10 promoter polymorphism
and longevity. *Genes Immun.* 3, 30–33.

46 LIO, D., L. SCOLA, A. CRIVELLO, M.
BONAFE, C. FRANCESCHI, F. OLIVIERI,
G. COLONNA ROMANO, G. CANDORE,
C. CARUSO (**2002**) Allele frequencies
of +874T → A single nucleotide
polymorphism at the first intron of
interferon-gamma gene in a group of
Italian centenarians. *Exp. Gerontol.* 37,
315–319.

47 HAJEER, A., M. LAZARUS, D. TURNER,
R. MAGEED, J. VENCOVSKY, P.
SINNOTT, I. HUTCHINSON, W. OLLIER

(**1998**) IL-10 gene promoter
polymorphisms in rheumatoid
arthritis. *Scand. J. Rheumatol.* 27, 142–
145.

48 TAGORE, A., W. GONSALKORALE, V.
PRAVICA, A. HAJEER, R. MCMAHON,
P. WHORWELL, P. SINNOTT, I.
HUTCHINSON (**1999**) Interleukin 10
(IL-10) genotypes in inflammatory
bowel disease. *Tissues Antigens* 54,
386–390.

49 T. HUIZINGA, V. KEIJSERS, G. YANNI,
M. HALL, W. RAMAGE, J. LANCHBURY,
C. PITZALIS, W. DROSSAERS-BAKKER,
R. WESTENDORP, F. BREEDVELD, G.
PANAYI, C. VERWEIJ (**2002**) Are
differences in interleukin 10
production associated with joint
damage? *Rheumatology* 39, 1180–1188.

50 VAN DISSEL, J., P. VAN LANGEVELDE, R.
WESTENDORP, K. KWAPPENBERG, M.
FROLICH (**1998**) Anti-inflammatory
cytokine profile and mortality in
febrile patients. *Lancet* 351, 950–953.

51 WANG, X., M. HURME, M.A.H. JYLHA
(**2001**) Lack of association between
human longevity and polymorphisms
of IL-1 cluster, IL-6, IL-10 and TNF-
alpha genes in Finnish nonagenarians.
Mech. Ageing Dev. 123, 29–38.

52 GRIMBLE, R. (**2002**) Inflammatory
status and insulin resistance. *Curr.
Opin. Clin. Nutr. Metab. Care* 5, 551–
559.

53 SPITZER, J., G. BAGBY, K. MESZAROS,
C. LANG (**1988**) Alterations in lipid
and carbohydrate metabolism in
sepsis. *J. Parent. Enteral. Nutr.* 12
(Suppl), 53S–58S.

54 SPITZER, J., G. BAGBY, K. MESZAROS,
C. LANG (**1989**) Altered control of
carbohydrate metabolism in
endotoxemia. *Prog. Clin. Biol. Res.* 286,
145–165

55 TRAYHURN, P., I.S. WOOD (**2004**)
Adipokines: inflammation and the
pleiotropic role of white adipose
tissue. *Br. J. Nutr.* 92, 347–355.

56 COPPACK, S.W. (**2001**) Proinflammatory cytokines and adipose
tissue. *Proc. Nutr. Soc.* 60, 349–356.

57 NILSSON, J., S. JOVINGE, A. NIEMANN,
R. RENELAND, H. LITHELL (**1998**)
Relation between plasma tumor

necrosis factor-alpha and insulin sensitivity in elderly men with non-insulin-dependent diabetes mellitus. *Arterioscler. Thromb. Vasc. Biol.* 18, 1199–1202.

58 YAQOOB, P., E. NEWSHOLME, P.C. CALDER (**1999**) Comparison of cytokine production in cultures of whole blood and peripheral blood mononuclear cells. *Cytokine* 11, 600–605.

59 HOTAMISLIGIL, G.S., P. ARNER, J.F. CARO, R.L. ATKINSON, B.M. SPIEGELMAN (**1995**) Increased adipose tissue expression of tumor necrosis factor-alpha in human obesity and insulin resistance. *J. Clin. Invest.* 95, 2409–2415.

60 KERN, P.A., S. RANGANATHAN, C. LI, L. WOOD, G. RANGANATHAN (**2001**) Adipose tissue tumor necrosis factor and interleukin-6 expression in human obesity and insulin resistance. *Am. J. Physiol. Endocrinol. Metab.* 280, E745–51.

61 HEGAZY, D., D. O'REILLY, B. YANG, A. HODGKINSON, B. MILLWARD, A. DEMAINE (**2001**) NFkappaB polymorphisms and susceptibility to type 1 diabetes. *Genes Immun.* 2, 304–308.

62 M. BULLO, P. GARCIA-LORDA, J. PEINADO-ONSURBE, M. HERNANDEZ, D. DEL CASTILLO, J.M. ARGILES, J. SALAS-SALVADO (**2002**) TNFalpha expression of subcutaneous adipose tissue in obese and morbid obese females: relationship to adipocyte LPL activity and leptin synthesis. *Int. J. Obes. Relat. Metab. Disord.* 26, 652–658.

63 MARKOVIC, O., G. O'REILLY, H.M. FUSSELL, S.J. TURNER, P.C. CALDER, W.M. HOWELL, R.F. GRIMBLE (**2004**) Role of single nucleotide polymorphisms of pro-inflammatory cytokine genes in the relationship between serum lipids and inflammatory parameters, and the lipid-lowering effect of fish oil in healthy males. *Clin. Nutr.* 23, 1084–1095.

64 PHILLIPS, A., A.G. SHAPER, P.H. WHINCUP (**1989**) Association between serum albumin and mortality from cardiovascular disease, cancer and other causes. *Lancet* 2, 1434–1437.

65 GRIMBLE, R. (**1990**) Serum albumin and mortality. *Lancet* 1, 348.

66 FLEET, J.C., S.K. CLINTON, R.N. SALOMAN (**1992**) Atherogenic diets enhance endotoxin-stimulated interleukin-1 and tumor necrosis factor gene expression in rabbit aorta. *J. Nutr.* 12, 294–298.

67 ROSS, R. (**1993**) The pathogenesis of atherosclerosis: a perspective for the 1990s. *Nature* 362, 801–809.

68 STULNIG, T.M. (**2003**) Immunomodulation by polyunsaturated fatty acids: mechanisms and effects. *Int. Arch. Allergy. Immunol.* 132, 310–321.

69 INGELMAN-SUNDBERG, M. (**2004**) Genetic polymorphisms of cytochrome P450 2D6 (CYP2D6): clinical consequences, evolutionary aspects and functional diversity. *Pharmacogenomics J.* Oct 19.

70 MINIHANE, A., S. KHAN, E. LEIGH-FIRBANK, P. TALMUD, J. WRIGHT, M. MURPHY, B. GRIFFIN, C. WILLIAMS (**2000**) ApoE polymorphism and fish oil supplementation in subjects with an atherogenic lipoprotein phenotype. *Arterioscler. Thromb. Vasc. Biol.* 20, 1990–1997.

71 M. ANDREASSI, N. BOTTO, F. COCCI, D. BATTAGLIA, E. ANTONIOLI, S. MASETTI, S. MANFREDI, M. COLOMBO, A. BIAGINI, A. CLERICO (**2003**) Methylenetetrahydrofolate reductase gene C677T polymorphism, homocysteine, vitamin B12, DNA damage in coronary artery disease. *Hum. Genet.* 112, 171–177.

72 KHOVIDHUNKIT, W., M.S. KIM, R.A. MEMON, J.K. SHIGENAGA, A.H. MOSER, K.R. FEINGOLD, C. GRUNFELD (**2004**) Effects of infection and inflammation on lipid and lipoprotein metabolism: mechanisms and consequences to the host. *J. Lipid Res.* 45, 1169–1196.

73 CALDER, P. (**2001**) Polyunsaturated fatty acids, inflammation and immunity. *Lipids* 36, 1007–1024.

74 ENDRES, S., R. GHORBANI, V. KELLEY, K. GEORGILIS, G. LONNEMANN, J. VAN

DER MEER, J. CANNON, T. ROGERS, M. KLEMPNER, P. WEBER, E. SCHAEFFER, S. WOLFF, C. DINARELLO (1989) The effect of dietary supplementation with n-3 polyunsaturated fatty acids on the synthesis of interleukin-1 and tumor necrosis factor by mononuclear cells. *N. Engl. J. Med.* 320, 265–271.

75 GRIMBLE, R., W. HOWELL, G. O'REILLY, S. TURNER, O. MARKOVIC, S. HIRRELL, M. EAST, P. CALDER (2002) The ability of fish oil to suppress tumor necrosis factor-alpha production by peripheral blood mononuclear cells in healthy men is associated with polymorphisms in genes which influence TNF-alpha production. *Am. J. Clin. Nutr.* 76, 454–459.

76 GRIMBLE, R.F., P.C. CALDER, P.J. GALLAGHER, J. GARRY, W.M. HOWELL, K. RERKASEM, C.P. SHEARMAN, F. THIES, J. WILLIAMS (2004) Association of pro- and anti-inflammatory cytokine genotypes with the response of advanced atherosclerotic plaque composition to dietary n-3 and n-6 polyunsaturated fatty acid supplementation. In: *Omega-3 Fatty acids: New Research*, F. COLUMBUS, editor, New York: Nova Science.

77 GRIMBLE, R.F., O. MARKOVIC, G. O'REILLY, H. FUSSELL, S. TURNER, P.C. CALDER, W.M. HOWELL (2004) Polymorphism in LT-α gene and body mass index influence the lipid lowering properties of fish oil in healthy men. *Clin. Nutr.* 23, 832.

78 VOZAROVA, B., J.M. FERNANDEZ-REAL, W.C. KNOWLER, L. GALLART, R.L. HANSON, J.D. GRUBER, W. RICART, J. VENDRELL, C. RICHART, P.A. TATARANNI, J.K. WOLFORD (2003) The interleukin-6 (−174) G/C promoter polymorphism is associated with type-2 diabetes mellitus in Native Americans and Caucasians. *Hum. Genet.* 112, 409–413.

79 MINIHANE, A.M., S. KHAN, E.C. LEIGH-FIRBANK, P. TALMUD, J.W. WRIGHT, M.C. MURPHY, B.A. GRIFFIN, C.M. WILLIAMS (2000) ApoE polymorphism and fish oil supplementation in subjects with an atherogenic lipoprotein phenotype. *Arterioscler. Thromb. Vasc. Biol.* 20, 1990–1997.

80 MINIHANE, A.M., Y.E. FINNEGAN, P. TALMUD, E.C. LEIGH-FIRBANK, C.M. WILLIAMS (2002) Influence of the APOC3−2854T > G polymorphism on plasma lipid levels: effect of age and gender. *Biochim. Biophys. Acta* 1583, 311–314.

81 ORDOVAS, J.M., D. CORELLA, L.A. CUPPLES, S. DEMISSIE, A. KELLEHER, O. COLTELL, P.W. WILSON, E.J. SCHAEFER, K. TUCKER (2002) Polyunsaturated fatty acids modulate the effects of the APOA1 G-A polymorphism on HDL-cholesterol concentrations in a sex-specific manner: the Framingham Study. *Am. J. Clin. Nutr.* 75, 38–46.

82 ELOSUA, R., J.M. ORDOVAS, L.A. CUPPLES, C.S. FOX, J.F. POLAK, P.A. WOLF, R.A. D'AGOSTINO SR, C.J. O'DONNELL (2004) Association of APOE genotype with carotid atherosclerosis in men and women: the Framingham Heart Study. *J. Lipid Res.* 45, 1868–1875.

83 LEVIEV, I., O. POIRIER, V. NICAUD, A. EVANS, F. KEE, D. ARVEILER, C. MORRISSON, F. CAMBIEN, R. JAMES (2002) High expressor paraoxonase PON1 gene promoter polymorphisms are associated with reduced risk of vascular disease in younger coronary patients. *Atherosclerosis* 161, 463–467.

84 KOVAROVA, H., R. NECASOVA, S. PORKERTOVA, D. RADZIOCH, A. MACELA (2001) Natural resistance to intracellular pathogens: modulation of macrophage signal transduction related to the expression of the Bcg locus. *Proteomics* 1, 587–596.

85 FORMICA, S., T. ROACH, J. BLACKWELL (1994) Interaction with extracellular matrix proteins influences Lsh/Ity/Bcg (candidate Nramp) gene regulation of macrophage priming/activation for tumour necrosis factor-alpha and nitrite release. *Immunology* 82, 42–50.

86 MOL, M., Y. DE RIJKE, P. DEMACKER, A. STALENHOEF (1997) Plasma levels of lipid and cholesterol oxidation products and cytokines in diabetes

mellitus and cigarette smoking: effects of vitamin E treatment. *Atherosclerosis* 129, 169–176.

87 van Tits, L., P. Demacker, J. de Graaf, H. Hak-Lemmers, A. Stalenhoef (**2000**) Alpha-tocopherol supplementation decreases production of superoxide and cytokines by leukocytes ex vivo in both normolipidemic and hypertriglyceridemic individuals. *Am. J. Clin. Nutr.* 71, 458–464.

88 Devaraj, S., I. Jialal (**2000**) Low-density lipoprotein postsecretory modification, monocyte function, and circulating adhesion molecules in type 2 diabetic patients with and without macrovascular complications: the

effect of alpha-tocopherol supplementation. *Circulation* 102, 191–196.

89 Hattori, S., Y. Hattori, N. Banba, K. Kasai, S. Shimoda (**1995**) Pentamethyl-hydroxychromane, vitamin E derivative, inhibits induction of nitric oxide synthase by bacterial lipopolysaccharide. *Biochem. Mol. Biol. Int.* 35, 177–183.

90 Villette, S., J.A. Kyle, K.M. Brown, K. Pickard, J.S. Milne, F. Nicol, J.R. Arthur, J.E. Hesketh (**2002**) A novel single nucleotide polymorphism in the 3′ untranslated region of human glutathione peroxidase 4 influences lipoxygenase metabolism. *Blood Cells Mol. Dis.* 29, 174–178.

16
Gene Variants, Nutritional Parameters, and Hypertension

Maolian Gong and Norbert Hübner

Summary

Essential hypertension (EH) is a major public health problem due to its high prevalence and its association with coronary heart disease, stroke, renal disease, peripheral vascular disease, and other disorders in many countries. Dietary salt is the major cause of the rise in blood pressure with age and the development of hypertension, while dietary potassium lowers the risk to develop hypertension and stroke. Epidemiological and clinical data concluded that the reduction of daily sodium intake, through salt restricted diets, lowers BP effectively. It was estimated that a universal reduction in dietary intake of salt by 50 mmol/d would lead to a 50% reduction in the number of people requiring anti-hypertensive therapy, a 22% reduction in number of deaths due to strokes and a 16% reduction in number of death from coronary heart disease (CHD). An increase in dietary intake of potassium, from approximately 60–80 mmol/d was shown to be inversely and significantly related to the incidence of stroke mortality in women. However, the mechanisms whereby high salt intake raises the blood pressure remain unclear. Existing concepts focus on the tendency for an increase in extracellular fluid volume (ECV), but an increased salt intake also induces a small elevation in plasma sodium, which shifts fluid from the intracellular to the extacellular space, and stimulates the thirst centre. Accordingly, the rise in plasma sodium is responsible for the tendency for an increase in ECV. Although the change in ECV may have a pressor effect, the associated rise in plasma sodium itself may also cause the blood pressure to rise. The heritability of hypertension has been demonstrated in several studies. The genetic contribution to blood pressure variability ranges from 20 to 50%, according to different studies. Therefore, identifying hypertension susceptibility genes will help understanding the pathophysiology of the disease. Due to the fact that blood pressure is controlled by cardiac output and total peripheral resistance, many molecular pathways are believed to be involved in the disease. Several strategies and methods have been used to identify hypertension susceptibility genes. In this review, recent genetic studies investigating the molecular basis of hypertension including different molecular pathways will be highlighted.

Nutritional Genomics. Edited by Regina Brigelius-Flohé and Hans-Georg Joost
Copyright © 2006 WILEY-VCH Verlag GmbH & Co. KGaA, Weinheim
ISBN: 3-527-31294-3

16.1
Introduction

Essential hypertension (EH) is a major public health problem due to its high prevalence and its association with coronary heart disease, stroke, renal disease, peripheral vascular disease, and other disorders in many countries [1]. EH is regarded as a multifactorial condition, the onset and severity of which are influenced by both genetic and environmental factors. The role of genetic factors in the etiology of hypertension is supported by cross-sectional studies that document familial aggregation of the disorder despite the different environment factors [2, 3]. Twins and adoption studies indicated a greater degree of trait concordance among identical as compared with dizygotic twins [4] and among natural as compared with adoptive siblings [5], which also stresses the importance of genetic factors [6]. The exact form of the underlying genetic mechanism remains unanswered. The estimates on genetic variance range from 20% to 50% [6–8].

The identification of variant genes that contribute to the development of hypertension is complicated by the fact that the two entities that determine blood pressure (BP) – cardiac output and total peripheral resistance – are themselves controlled by other intermediary phenotypes, including the autonomic nervous system, vasopressor/vasodepresser hormones, the structure of the cardiovascular system, body fluid volume and renal function, and many others (Fig. 16.1). The identification of genes underlying blood pressure variation has the capacity to define primary physiologic mechanisms causing this trait, thereby clarifying disease pathogenesis, establishing molecular diagnostics and developing a novel therapy of hypertension [9]. Substantial progress has been made in the last decade towards detection of genes underpinning several Mendelian forms of hypertension traits, which may present early in life with distinct phenotypes [9, 10]. Currently, most published data on human EH arises from association-based studies and genome-wide screens [11].

16.2
Blood Pressure Regulation is Related to Sodium and Potassium Intake

Dietary salt is a major cause for elevating BP levels across populations, while dietary potassium lowers the risk of developing hypertension and stroke. Comprehensive epidemiologic evidence was provided by the INTERSALT Study [12–14], which investigated the relationship of 24-h urinary electrolyte excretion to BP in 52 population groups across 32 countries with 10 074 subjects, following standardized protocols. In adults aged 20–59 years there was a significant positive correlation between urinary sodium excretion and elevated BP. The relationship between urinary sodium excretion and elevated BP was stronger in older than in younger individuals. With adjustment for confounding variables, potassium excretion was significantly negatively correlated with BP. Again these relationships tended to be more marked at older ages. Further, it was also observed that in four of these pop-

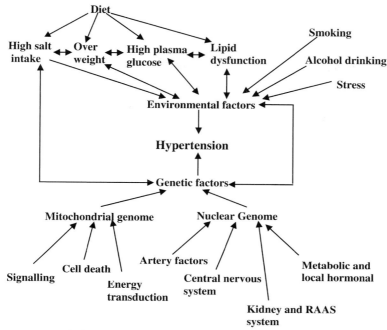

Fig. 16.1. Multiple interactions between genetics and environmental factors leading to hypertension.

ulations in whom the mean 24-h urinary sodium excretion was lower than 100 mmol/day (median salt intake ranged from under 1 g to 3 g daily versus more than 9 g in the rest of the INTERSALT populations), systolic blood pressure (SBP) did not rise with age [15].

An overview of observational data in populations suggested that a difference in sodium intake of 100 mmol/day could be associated with average difference in SBP of 5 mmHg at age 15–19 years and 10 mmHg at age 60–69 years. It was estimated that a universal reduction in dietary intake of salt by 50 mmol/day would lead to a 50% reduction in the number of people requiring antihypertensive therapy, a 22% reduction in number of deaths due to strokes, and a 16% reduction in number of death from coronary heart disease (CHD) [16]. Meta-analysis of randomized controlled clinical trials confirmed that potassium supplementation was associated with a significant reduction in mean (95% confidence interval, CI) systolic and diastolic BP of −4.44 (range −2.53 to −6.36) and −2.45 (range −0.74 to −4.16) mmHg, respectively. After exclusion of a trial with extreme results, potassium supplementation was still associated with a significant reduction in mean (95% CI) systolic and diastolic BP of −3.11 (range −1.91 to −4.31) and −1.97 (range −0.52 to −3.42) mmHg, respectively. The BP effects of potassium administration appeared to be enhanced in studies where participants were concurrently exposed to a high intake of sodium [17].

The results of a low-sodium diet in the Dietary Approaches to Stop Hypertension (DASH) trial [18] further strengthened the conclusion that reduction of daily sodium intake through salt-restricted diets lowers BP effectively and is additive to the benefit conferred by the DASH diet. The effects of composite dietary interventions on BP levels in normotensive and hypertensive individuals were studied in well-designed clinical trails [18, 19]. The initial dietary intervention used in the DASH trial involved a diet that emphasized fruits, vegetables, and low-fat dairy products and included whole grains, poultry, fish, and nuts while reducing the amount of red meat, sweets, and sugar-containing beverages. The DASH diet was more effective in substantially reducing systolic and diastolic BP, both in hypertensive and normotensive subjects compared with the "typical" low-fat and low-cholesterol diets in the United States, even though the latter diet intervention also lowers BP. The DASH was also demonstrated to be effective as first-line therapy in individuals with stage I isolated systolic hypertension, with 78% of the people on the DASH diet reducing their SBP to below 140 mmHg, in comparison to 24% in the control group.

The DASH trial was followed by a well-designed factorial trial combining the diet with high, intermediate, and low levels of sodium consumption and measuring the effects on BP in comparison with a control diet typical of the United States, administrated with similar graded variations in the sodium content [18]. Within each assigned group (DASH versus typical), participants ate foods with high, intermediate, and low levels of sodium for 30 consecutive days. Reduction in sodium intake at each level resulted in significant lowering of systolic and diastolic BP in both DASH and control groups. The decline in BP was maximum, however, when the DASH diet was modified to reduce the sodium content. As compared with the control diet with a high sodium level, the DASH diet with a low sodium level led to mean systolic BP 7.1 mmHg lower in normotensive participants and 11.5 mmHg lower in hypertensive participants. There was also a 4.5 mmHg decrease in the mean diastolic BP level between the low-sodium DASH diet phase and the high-sodium control diet phase of the trial [18].

The effects of the low-sodium intake and potassium-supplementation diet has great potential for application in both population-based and individual-focused strategies for prevention and control of high blood pressure (HBP) and associated cardiovascular disease (CVD). Adoption of the low-sodium and potassium-supplementation diet by large populations is likely to be safe and beneficial in shifting the population distributions of BP towards lower levels of cumulative risk of CVD in those populations. This diet will also provide an effective non-pharmacological therapeutic intervention in the clinical management of individuals identified to be at an increased risk of CVD because of HBP and associated risk factors.

16.2.1
Plasma Sodium, Mechanisms and Hypertension

There is evidence that those who develop HBP have an underlying defect in the ability of the kidney to excrete salt, and that the greater compensatory response

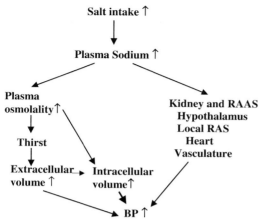

Fig. 16.2. Plasma sodium and blood pressure. BP, blood pressure; RAAS, renin–angiotensin–aldosterone system.

required to restore sodium balance is responsible for the increase in BP. Sodium balance is controlled entirely by the kidney's ability to vary the urinary excretion of sodium. The immediate effects of dietary sodium are to alter plasma sodium and, consequently, the extracellular fluid volume (ECV) (Fig. 16.2). The rise in pressure is associated with the rise in cardiac output. These changes must therefore be primarily responsible for the subsequent alterations that affect the BP [20].

The impairment of the kidney's ability to excrete sodium is either genetic, as in human essential hypertension (EH) and a rat model of human EH, the spontaneously hypertensive rat (SHR), or it can be superimposed as in obesity, primary hyperaldosteronism, or renal disease [20]. The primacy of the kidney's control of sodium excretion in the regulation of BP has been confirmed by the finding of 20 genes so far identified to be associated with EH or responsible for rare Mendelian forms of high and low BP; all are involved in the regulation of sodium handling by the kidney (see below) [9].

High salt intake has also been demonstrated to have direct harmful effects on the cardiovascular system independent of and additive to the effect of salt on BP (Fig. 16.2) [21]. There is direct evidence that a rise in plasma sodium *in vivo* can induce changes in the arterial which could contribute to a rise in BP, and that, *in vitro*, a rise in sodium concentration can cause intracellular changes in the vessels and the heart that are similar to those found in vascular tissue from hypertensive individuals or animals. The intracellular sodium concentration was directly related to the systolic and diastolic BP. A rise in intracellular sodium increases muscle tone due to the resultant increase in intracellular Ca^{2+} concentration. The results demonstrated that *in vivo*, an acute increase in plasma sodium not accompanied by a rise in ECV can raise BP, and that such a rise in BP is associated with an increase in the sodium concentration of smooth muscle. *In vitro*, a change in sodium con-

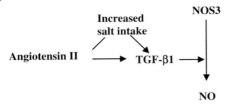

Fig. 16.3. The relationship between angiotensin II and transforming growth factor-*β*1. TGF*β*1, transforming growth factor-*β*1; NOS3, endothelial isoform of nitric oxide synthase; NO, nitric oxide.

centration in cultures of myocardial myoblasts and vascular muscle cells resulted in an increased cell diameter, volume, and protein content of these cells.

Other experiments showed similarly that increasing the sodium concentration by just 2 mmol/l above normal can result in an increase in the cellular protein content of cultured coronary artery smooth muscle cells by 84.5%. Similar results were obtained in cultured umbilical vein endothelial cells. An increase in the concentration of NaCL by 10 mmol/l in thoracic aorta and cultured rat vascular muscle cells caused a time-dependent rise in AT_1 receptor mRNA levels that appeared at 12 h, and was sustained for 48 h when the experiment was ended. The above results directly demonstrate that a rise in plasma sodium induces multiple changes in vascular tissue similar to those in hypertension [20].

Epidemiologically, as the BP rises with age there is an inverse association between plasma renin activity and systolic BP, yet the administration of angiotensin inhibitors to EH patients lowers their BP, which suggests that such patients may show an increased sensitivity towards angiotensin II. It has been shown that angiotensin II activity is increased in vessels of EH patients [22]. Cell cultures of vascular smooth muscle and endothelial cells suggest that a rise in sodium concentration within culture medium increases the number of AT_1 receptors, transforming growth factor *β* (TGF*β*) production, and the functional response of the cells to stimulation with angiotensin II (Fig. 16.3) [20]. TGF*β*1 was demonstrated to be stimulated by angiotensin II in different cultured cells (mesangium, vascular smooth muscle, and rat glomerular mesangial cells) [23]. TGF*β*1 can modulate nitric oxide (NO) production through controlling the transcriptional rate of endothelial isoform of nitric oxide synthase (NOS3). The BP response to an increase in salt intake is dependent on NO production in rat. Thus, dietary salt intake regulates TGF*β*1, which in turn regulates expression of NOS3 in potential targets of hypertension-induced damage. Increased production of NO produces feedback inhibition of TGF*β*1 production and further serves as a vasodilatory function, which decreases shear forces [23]. The system appears to become dysfunctional when NO production is impaired (Fig. 16.3). The studies indicate that structural changes in hypertension are due in part to an increase in tissue angiotensin II activity in vessels and the heart, independent of BP, and these may be due to an increase in plasma sodium (Fig. 16.2).

16.2.2
Sodium Handling, Genetics Variants, and Hypertension

Genetic, environmental, and demographic factors together contribute to hypertension. Among the environmental factors, high salt intake contributes to the most common form of hypertension, known as salt-sensitive hypertension. The evolution of humans in a salt-poor environment might have provided selective pressure favoring alleles promoting salt retention. The presence of such alleles in modern times is bound to make humans susceptible to hypertension. Therefore, the genes that coordinately participate in the maintenance of sodium/water homeostasis are the potential candidate genes (Fig. 16.4).

Here we discuss some of the results of some studies designed to define the molecular basis of hypertension with an emphasis on those molecular variants that are implicated in salt/water homeostasis.

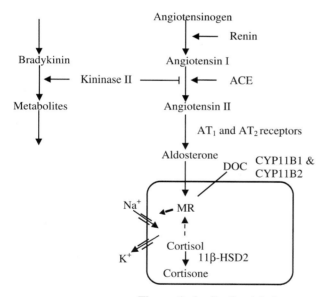

The cortical collecting tubule

Fig. 16.4. Candidate genes of human hypertension along the pathway of renin–angiotensin–aldosterone system. ACE, angiotensin-converting enzyme; AT_1 and AT_2 receptor, angiotensin II receptor type I and II; DOC, deoxycorticosterone; CYP11B1, 11β-hydroxylase; CYP11B2, aldosterone synthase; MR, mineralocorticoid receptor; 11βHSD2, type 2 11β-hydroxysteroid dehydrogenase.

16.3
Genetic Analysis of Hypertension

16.3.1
Mendelian Forms of Hypertension (Secondary Hypertension)

Molecular genetic studies have identified mutations in eight nuclear genes and one mitochondrial gene that cause Mendelian forms of hypertension (Table 16.1). They include the 11β-hydroxylase/aldosterone synthase (CYP11B1/CYP11B2, Chr8p) in glucocorticoid remediable aldosteronism [24], the β- and γ-subunits of the epithelial sodium channel (ENaC, Chr16p) in Liddle's syndrome [25–28], type 2 11β-hydroxysteroid dehydrogenase (11β-HSD2, Chr16q) in the syndrome of apparent mineralocorticoid excess [29, 30], the mineralocorticoid receptor (MR, Chr4q) in

Table 16.1. Causative mutations for Mendelian forms of hypertension (update).

Monogenic syndrome	Causative gene	Characteristics of mutations	Enzyme function	MOI	Chromosome
Glucocorticoid-remediable aldosteronism (GRA) [24]	*CYP11B1* and *CYP11B2*	Fusion gene arising from unequal crossover	Gain	AD	8p
Pseudo-aldosteronism (Liddle's syndrome) [25–28]	β and γ-subunits of ENaC	Truncation mutations in C-terminal region and missense mutation	Gain	AD	16p
Pseudohypoaldosteronism type II (Gordon's syndrome) [32]	WNK1 and WNK4	Deletion and missense mutations	Gain	AD	12p and 17q
Apparent mineralocorticoid excess (AME) [29, 30]	11β-HSD2	Missense and deletion mutations	Loss	AR	16q
Hypertension exacerbated in pregnancy [31]	MR	Missense	Gain	AD	4q
Hypertension plus brachydactyly (HTNB) [36]	Unknown	Unknown	Unknown	AD	12p
Hypertension, hypercholesterolemia, and hypomagnesemia [10]	M. tRNA^{Ile}	Missense mutation	M. dysfunction	M	M

MOI, mode of inheritance; AD, autosomal dominant; AR, autosomal recessive; CYP11B1, 11β-hydroxylase; CYP11B2, aldosterone synthase; ENaC, epithelial sodium channel; WNK1 and WNK4, Serine/threonine protein kinases, lysine-deficient 1 and 4; 11β-HSD2, type 2 11β-hydroxysteroid dehydrogenase; MR, mineralocorticoid receptor; tRNA, transfer RNA; M, mitochondria.

hypertension exacerbated in pregnancy [31], and mutated genes of the serine-threonine protein kinases, WNK1 (Chr12p) and WNK4 (Chr17q) in pseudohypoaldosteronism type II (FHH) [32]. The molecular basis of the former four syndromes was reviewed in detail by Lifton *et al.* [9]. Extensive studies of pseudohypoaldosteronism type II by Mayan *et al.* found that affected subjects with WNK4 Q565E mutations have hypercalciuria accompanied by lower serum calcium levels supporting a mechanism of renal calcium wasting. Together with the result that WNK4 regulates the renal outer medullary potassium channel as well as epithelial Cl^-/base exchange and the $Na^+/K^+/2Cl^-$ cotransporter, the interaction between WNK4 and a calcium channel or transporter was suggested [33]. Most of these disorders are due to defective genes acting in the same physiological pathway in the kidney, altering net renal salt reabsorption [9].

Hypertension and dyslipidemia cluster more often than expected for the risk of many common cardiovascular diseases (i.e. myocardial infarction, congestive heart failure, and stroke) [34]. A cluster of metabolic defects caused by mutation in a mitochondrial transfer RNA (mtRNA) was identified in one large Caucasian kindred by Wilson *et al.* [10]. The kindred features a cluster of hypertension, hypercholesterolemia, and hypomagnesemia.

Direct sequencing and single-strand conformational polymorphism analysis of the entire mitochondrial genome have been performed and a novel mutation at nucleotide 4291 uridine-to-cytidine transition was identified, which lies within the mitochondrial tRNA^{Ile} gene. The mutation occurs immediately 5' to the tRNA^{Ile} anticodon. Uridine at this position is one of the most conserved bases. Biochemical studies with anticodon stem–loop analogs of tRNA have been performed and indicate that substitution of cytidine for uridine at this position markedly impairs ribosome binding [35]. Thus, the authors speculated that the complexity can arise from a single mutation because of the combined effects of reduced penetrance and pleiotropy. This underscores the value of studying very large kindreds [10].

One further Mendelian form of hypertension – hypertension and brachydactyly – has been mapped to a defined chromosomal region [36], but the molecular basis of the underlying gene still awaits identification. This genetic region nearly overlaps with a later whole genome scan linkage analysis for EH in a large Chinese pedigree [37], which indicates that the susceptibility gene(s) for EH may reside on chromosome 12p.

Even though these rare syndromes with Mendelian inheritance only account for a small fraction of the pathological blood pressure variation in the general population, they provide new insight into the pathophysiology of hypertension.

16.3.2
Essential Hypertension

16.3.2.1 Association Studies in Hypertension

Association between sodium intake, genetics variants, and hypertension A large population-based cross-sectional study was conducted in 2823 Japanese men and women aged 30–74 to examine the association between angiotensinogen (AGT) T174M polymorphism and BP levels stratified by age, body mass index (BMI), and salt intake (median) estimated by 24-h urine collection and dietary questionnaire. There was no difference in systolic or diastolic BP between the TT and TM+MM genotypes overall. However, among young subjects aged 30–64, mean diastolic BP tended to be 1 mmHg higher in the TM+MM group than in the TT group ($P = 0.06$), but not among the older groups. Similarly, mean diastolic BP was 1.7 mmHg higher in the TM+MM group than in the TT group ($P = 0.01$) in subjects with lower BMI (≤ 23.4 kg/m^2), but not in those with higher BMI (>23.5 kg/m^2). Furthermore, among younger and non-overweight subjects, diastolic BP differences were larger for those with higher urinary sodium excretion (3.1 mmHg, $P = 0.03$), those with a higher sodium/potassium excretion ratio (4.1 mmHg, $P = 0.007$), those with higher present sodium intake score (3 mmHg, $P = 0.003$), and those with higher past sodium intake score (3.4 mmHg, $P = 0.001$). The study shows that there was a significant association between AGT T174M polymorphism and diastolic BP levels among younger non-overweight individuals with high sodium intake, but not among individuals who were older [38].

The relationship between the 174M allele and BP level was more obvious among younger subjects than older subjects. It may be the younger subjects are more likely to be affected by genetic factors rather than by environmental factors, and high salt intake strengthens the association. Key enzymes in the RAAS cascade are the angiotensin-converting enzyme (ACE) and aldosterone synthase (CYP11B2). Thus, the relationship between these two genes and different hypertension complications were carried out. Investigators from the European Projects On Genes in Hypertension (EPOGH) [39] studied the relationship between left ventricular mass (LVM) and the ACE D/I, CYP11B2 −344C/T polymorphisms according to sodium intake. In 219 nuclear families (382 parents and 436 offspring) from Cracow, Novosibirsk and Mirano, no association was found between left ventricular mass index (LVMI) and CYP11B2 −344C/T polymorphism. LVMI increased with higher sodium excretion in ACE II homologous offspring of both Slavic and Italian extraction ($+4.2 \pm 2.1$ g/m^2 per mmol; $P = 0.04$). In ACE D allele carriers, LVMI also increased with higher sodium excretion in Slavic offspring, but not in Italians. The interaction between ACE D/I genotype and urinary sodium excretion was significant in Italian offspring ($P = 0.03$). The relationship between LVMI and the ACE D/I polymorphism differs across populations, the authors speculated it is possibly as consequence of varying levels of salt intake [39].

The same group studied the relationship between LVM and angiotensin II receptors (AT1R and AT2R), while accounting for possible gene–gene interactions with *AGT* −532C/T and *ACE* I/D [40]. LVMI was unrelated to the *AT1R* A1166C polymorphism. In population- and family-based analysis in men, the allelic effects of the *AT2R* G1675A on LVMI differed ($P = 0.01$) according to sodium excretion. In women, this gene–environment interaction was not significantly different. In untreated men, LVMI (4.2 g/m^2 per 100 mmol) and left ventricular internal diameter (0.73 mm/100 mmol) increased ($P < 0.02$) with higher sodium excretion in the presence of the G allele, with an opposite tendency in A allele carriers. The *ACE*

D/I polymorphism, together with the *ACE* genotype-by-sodium interaction term, significantly and independently improved the models relating LVMI to the *AT2R* polymorphism and the *AT2R* genotype-by-sodium interaction [40]. Similarly, the EPOGH investigators studied the relationship between heart rate and *CYPB11B2* −344T/C and *AT1R* A1166C at varying levels of salt intake in 1797 participants from six European populations [41]. The low-frequency (LF) and high-frequency (HF) components of heart rate variability and their ratio (LF:HF) in the supine and standing positions was measured. In subjects with sodium excretion < 190 mmol/day (median), supine heart rate, LF, and LF:HF increased and HF decreased with the number of *CYP11B2* −344T alleles ($0.01 < P \le 0.04$), and the orthostatic changes in LF, HF, and LF:HF were blunted in carriers of the *AT1R* 1166C allele ($0.02 < P \le 0.03$). In subjects with sodium excretion > 190 mmHg/day, these associations with the two gene polymorphisms were non-significant or in the opposite direction, respectively. The above results demonstrated the genetic effects of LVM and heart rate depend on sodium excretion.

Overall, it seems clear that the body's handing of salt has an important effect on cardiovascular status, and recent evidence strongly supports the use of saluretics, in the shape of thiazide diuretics, as first-line therapy for hypertension and coronary heart disease prevention. Low-sodium and low-fat foods are strongly recommended for the daily diet to prevent hypertension and cardiovascular diseases.

The renin–angiotensin–aldosterone system Due to the important role of the renin–angiotensin–aldosterone system (RAAS) in the regulation of water and sodium balance as well as BP [42], numerous studies have investigated the relationship between the RAAS and EH [43] (Table 16.2). Two genes of the renin–angiotensin system have provided evidence for association. The angiotensinogen gene variant *AGT* M235T is associated with higher circulating angiotensinogen levels and EH in several but not all populations [44–47]. A second variant in the angiotensin-converting enzyme gene (*ACE*) is strongly associated with increased blood pressure in men [48–50]. However, negative associations were also detected in some linkage and association studies [51, 52].

Bozec *et al.* [53] studied the mechanical properties of the carotid artery according to *AGT* M235/T genotype in 98 never-treated hypertensive patients (aged 24–80) and in *Agt* mutant mice. Few studies had investigated the effect of candidate genes of RAAS on large artery stiffness in hypertensive patients and this was the first study to include never-treated hypertensive patients only. It is important to exclude previously treated patients because antihypertensive drugs may affect arterial wall components and thus arterial stiffness. The study found that patients homozygous for the T allele had reduced carotid distensibility and increased stiffness of the carotid wall independent of blood pressure, compared with patients homozygous for the M allele. The carotid distensibility in *Agt1/2* mice, however, was not significantly different from that of *Agt1/1* (wild type) mice. The arterial wall was less stiff in the *Agt1/2* mice than in wild-type mice.

A mutant mouse model for the angiotensinogen gene (*Agt*) has been developed by Smithies *et al.* [54, 55]. The genotypes are *Agt1/1* (wild-type with two singleton

Table 16.2. Summary of candidate genes studies acting on pathways for blood pressure regulation.

Candidate genes	Molecular basis in EH	Polymorphisms	Studied population	Cases and controls	Association with EH and other EH-dependent damage
AGT [53, 59]	RAAS pathway, cleaved by renin to form angiotensin I (AngI)	M235T	Chinese Tibetan	173 EH and 193 NT	Yes with EH in women
			Caucasian	98 never treated EH	Yes with artery stiffness and arterial wall hypertrophy
		G−6A	Chinese Tibetan	173 EH and 193 NT	Yes with EH
ACE [56, 59, 63]	RAAS pathway, cleavage of AngI into AngII	I/D	Caucasian	311 untreated EH and 128 NT	Yes with blood wave velocity
			Chinese Tibetan	173 EH and 193 NT	No with EH
			Chinese Han	132 EH and 345 NT from 125 NF	Yes with EH
AT1R [56, 59]	Receptor of angiotensin II involved in vasoconstriction, vascular hypertrophy, aldosterone release, left ventricular contractility and renal perfusion	A1166C	Caucasian	311 untreated EH and 128 NT	Yes with blood wave velocity
			Chinese Tibetan	173 EH and 193 NT	No with EH
ACE2 [58]	Negative regulator of ACE in the heart	I1 A/G1075, I3 G/A 8790, I11 C/G 28330, I16 G/C36787	Anglo-Celtic Australian	152 EH and 193 NT	No with EH
CYPB11B2 [63]	Regulation of sodium and water homeostasis; involved in vascular smooth muscle hypertrophy; vascular matrix impairment and endothelial dysfunction	−344T/C	Chinese Han	132 EH and 345 NT from 125 NF	Yes with EH

Gene	Function	Polymorphism	Population	Sample size	Association
Alpha-adducin [63]	Cytoskeletal protein potentially involved in cellular signal transduction; interacts with other membrane-skeleton proteins that affect ion transport across the cell membrane	Gly460Trp	Chinese Han	132 EH and 345 NT from 125 NF	Yes with EH
GRK4 [65]	Desensitization of G protein-coupled receptors including the D1 receptor in proximal tubules	R65L A142V A486V	Caucasian	168 EH and 312 controls	No with EH No with EH Yes with EH
HSD3B1 [65]	The biosynthesis of steroid hormones including aldosterone	Leu338 T → C	Caucasian	168 EH and 312 controls	No with EH
PTP1B [65]	Regulating insulin signaling via receptor dephosphorylation	1484insG	Caucasian	168 EH and 312	No with EH
NHE3 [71]	Regulation of sodium reabsorption in the proximal tubule	G1579A, G1709A, G1867A, C1945T, A2041G, C2405T	African and Afro-Caribbean and Caucasian origin	399 (68% with EH) and 292 (50% with EH)	No with EH
EnaC [75]	Key determinant of sodium homeostasis	βENaC G589S βENaC i12–17CT γENaC V546I	Caucasian	347 EH, 175 NT and 301 randomly NT	Yes with EH Yes with EH No with EH
GREB1 [76]	Depressor effect through the improvement of endothelial dysfunction and modulation of sympathetic nerves activation	−13945A > T and 45718A > G	Japanese	796 EH and 1084 NT	Yes with EH in men
HPGAL1 [76]	Protection of neurons against calcium-induced death stimuli in cooperation with neuronal apoptosis inhibitory protein	IMS-JST126186 A > C	Japanese	796 EH and 1084 NT	Yes with EH in women

Table 16.2 (continued)

Candidate genes	Molecular basis in EH	Polymorphisms	Studied population	Cases and controls	Association with EH and other EH-dependent damage
BDKRB1 [90]	Activates arachidonic acid–nitric oxide cascade	SNPs in NT 026437 76646507, 76623594, and 76647595	AC	121 HT and 99 NT	Yes with EH in AC
BDKRB2 [90]	Activates arachidonic acid–nitric oxide cascade and affects insulin-dependent glucose transport/utilization	−58C/T, SNP in NT 026437 76648043	AA / AC	120 HT and 98 NT / 121 HT and 99 NT	Yes with EH in AA / Yes with EH in AC
Catalase [85]	Reduction of smooth muscle cell contraction and proliferation induced by endothelia, angiotensin II and α adrenoreceptor agonists	−844 A/G / −262 T/C	GC / AA	100 EH and 93 NT / 129 EH and 98 NT	Yes with EH only in GC
UCP2 [97]	The inner mitochondrial membrane-associated proteins and act as proton channels or transporters, marching electrochemical gradient across the mitochondrial inner membrane	−866 G/A	Japanese	342 DM2 (158 with HT) and 156 EH and 134 NT	Yes with EH

AGT, angiotensinogen; ACE, angiotensin-converting enzyme; CYP11B2, aldosterone synthase; AT1R, angiotensin II receptor type I; GRK4, G-protein-coupled receptor kinases 4; HSD3B1, 3β-hydroxysteroid dehydrogenase/delta isomerase, type 1 gene; PTP1B, protein phosphatase 1B gene; NHE3, sodium/hydrogen exchanger type 3 gene; ENaC, epithelial sodium channel; GREB1, estrogen in breast cancer 1; HPCAL1, hippocalcin-like 1; UCP2, uncoupling protein; IMS-JST, JSNP ID from JSNP database (Japan); BDKRB1, bradykinin receptor B1; BDKRB2, bradykinin receptor B2. I, intron; EH, essential hypertension; NT, Normotensive; NF, nuclear families AC, American Caucasian; AA, African American; GC, Greek Caucasian; DM2 diabetes mellitus type 2; HT, hypertension; Yes, associated; No, no association.

copies of the gene) and *Agt1/2* (one wild-type and one duplicated copy). These mice are characterized by increasing BP and plasma angiotensinogen concentrations as the number of *Agt* gene copies increases (plasma concentrations of angiotensinogen are 24% greater in *Agt1/2* than in wild-type mice) [54, 55]. Hence, *Agt1/2* mice have a genetically determined increase in plasma angiotensinogen concentrations similar to that of TT-homologous patients. The higher blood pressure was not associated with arterial hypertrophy, resulting in greater circumferential wall stress in *Agt1/2* mice [53]. The *in vivo* and *in vitro* pressure responses to angiotensin II were reduced in *Agt1/2* mice, whereas the contractile response to phenylephrine was not significantly different between *Agt1/1* and *Agt1/2* mice, indicating the integrity of the contractile apparatus and suggesting a dysfunction of the angiotensin II type 1 (AT$_1$) receptor signaling pathways in *Agt1/2* mice.

In hypertensive patients in whom treatment was stopped at least three weeks before investigation, Benetos *et al.* [56] found a positive association between the wave velocity, a marker of aortic stiffness, and both the A1166C polymorphism of the AT$_1$ receptor gene and the insertion/deletion polymorphism of angiotensin-converting enzyme 1 gene (*ACE* II/DD). These results suggest that the angiotensinogen 235TT genotype could be a genetic marker for arterial stiffness in never-treated hypertensive patients; the arterial wall hypertrophy and stiffening in 235TT patients are probably mediated by an increased stimulation of angiotensin II type 1 receptor, whereas the opposite carotid phenotype of *Agt1/2* mice is probably the result of a dysfunction on the angiotensin II type 1 receptor pathways [53].

Angiotensin-converting enzyme 2 gene (*ACE2*), a homolog of *ACE* has been recently discovered. *ACE2* appears to be a negative regulator of *ACE* in the heart [57]. A case–control study investigating four single-nucleotide polymorphisms (SNPs) of *ACE2* and EH provided no evidence for association in an Anglo-Celtic Australian population [58]. In this study, the 152 studied hypertensive subjects were the offsprings of parents who both had hypertension, and similarly the 193 normotensive subjects were from both normotensive parents over 50 years, which could have inherently high biological power. However, the data indicate little support for *ACE2* in genetic predisposition to EH [58].

The relationship between AGT, ACE, as well as AT$_1$ receptor genes and EH was studied in 173 hypertensive individuals and 193 normotensive individuals of Chinese Tibetans. The *AGT* M235T and the promoter G-6A polymorphisms showed association with EH in Tibetan women. No association could be detected for polymorphisms in ACE and AT$_1$ receptor and EH [59]. The *AGT* M235T allele was demonstrated to be in linkage disequilibrium with allelic variants in the AGT promoter region (G-6A and A-20C), which may affect the basal rate of angiotensinogen transcription and could account for phenotypic variation in plasma angiotensinogen concentrations [60, 61]. The relationship between ACE and environmental factors predisposing to EH has been investigated in 1099 subjects from one Mongolian population. The study claimed evidence for an interaction between the *ACE* DD and ID polymorphism and cigarette smoking, alcohol drinking, and body mass index (BMI) [62]. Another study in a Chinese Han population (479 subjects from 125 nuclear families) revealed that *ACE* I/D, α-adducin Gly460Trp, and aldosterone synthase -344C/T polymorphisms interact to influence systolic BP ($P < 0.05$), suggesting these genes might indeed predispose to hypertension, especially in an ecogenetic context characterized by high salt intake [63].

Association studies of genes acting on pathways for blood pressure regulation outside the renin–angiotensin–aldosterone system   An ever-expanding repertoire of genes outside of the RAAS has been tested for involvement in the genetic basis of EH (Table 16.2). A number of studies showed a correlation between hyperinsulinemia, insulin resistance, and hypertension [64]. Speirs *et al.* tested several novel potential candidates, including the G protein-coupled receptor kinases 4 (*GRK4*), 3-β hydroxysteroid dehydrogenase/delta isomerase, type 1 gene (*HSD3B1*), and protein phosphatase 1B gene (*PTP1B*) in 168 Caucasian EH patients and 312 normotensive controls [65]. The regulation of sodium excretion by the kidney is of paramount importance for homeostasis of the extracellular fluid volume and thereby of arterial blood pressure. *GRK4* was implicated in human hypertension by desensitization of G protein-coupled receptors including the dopamine 1 (D1) receptor [66]. In humans with EH, there is a decrease of the responsiveness of the D1 receptor in proximal tubules due to the uncoupling of the D1 receptor from its G protein/effector enzyme complex [67]. *HSD3B1* plays a role in the biosynthesis of steroid hormones including aldosterone [68]. It has been proposed that allelic variations of *HSD3B1* could lead to elevated plasma aldosterone, resulting in an increased intravascular volume and hypertension [69]. *PTP1B* negatively regulates insulin signaling via receptor dephosphorylation [70]. No association between *HSD3B1* and *PTP1B* variants and hypertension could be detected. In contrast, the V allele of the A486V variant of GRK4γ showed association with elevated blood pressure ($P = 0.02$ for EH).

Zhu *et al.* [71] studied the relationship between sodium/hydrogen exchanger type 3 gene (*NHE3*) and EH in 399 subjects African or Afro-Caribbean origin (68% with EH) and 292 subjects Caucasian origin (50% with EH), trying to examine the relationship with hypertension and biochemical indices of sodium balance. Six variants were identified in total. *NHE3* is a member of an increasing number of sodium/hydrogen exchangers responsible for transport of sodium and hydrogen ions across the proximal tubule [72]. Moreover, animal studies point out that this class of genes has potential importance in the control of blood pressure [73, 74]. However, no association between the variants and EH was detected in EH patients of either African and Afro-Caribbean origin or Caucasian origin [71].

Gain-of-function mutations in the β and γ-subunits of the epithelial sodium channel (ENaC) cause the monogenic form of hypertension, Liddle's syndrome [25–28]. One recent investigation in a Finnish population showed higher prevalence of three ENaC variants (βENaC G589S, βENaC i12–17CT, γENaC V546I) in 347 hypertensive patients compared with 175 normotensive individuals and 301 randomly chosen blood donors ($P < 0.01$). When frequencies of the individual gene variants in the hypertensive patients were compared to those in the other two groups combined, only the frequency of the βENaC i12–17CT variant was significantly higher among the hypertensive patients than in the other two groups

($P = 0.001$), whereas there is no significant difference in the prevalence of βENaC G589S and γENaC V546I variants between the hypertensive and control groups. Patients carrying the three variant alleles also showed an increased urinary potassium excretion rate in relation to their renin levels ($P = 0.034$). However, no activity change of the two ENaC amino acid variants could be detected when they were expressed in *Xenopus* oocytes compared with ENaC wild-type [75].

A large study was carried out in a Japanese general population investigating the association of polymorphisms on chromosome 2p24–p25 with BP. Forty-seven polymorphisms in 14 genes in the region between D2S2278 and D2S168 and in the region just outside of these two markers (between nucleotide 8845292 and nucleotide 11946689) were genotyped in 1880 individuals, 796 of whom were hypertensive and 1084 normotensive [76], and this region was showed to be linked with hypertension in several studies [77–79]. The study revealed that an SNP in the hippocalcin-like 1 (*HPCAL1*) gene was significantly associated with both the prevalence of hypertension and increased blood pressure levels in women ($P = 0.003$) [76]. *HPCAL1* shares 94% amino acid identity with hippocalcin, which functions as a neuronal calcium sensor, and possesses a Ca^{2+}/myristoyl switch, allowing it to translocate to the membrane [80].

Gene regulated by estrogen in breast cancer 1 (*GREB1*) was identified as a direct target gene of estrogen receptor α and is evolutionarily conserved compared with mouse genome [81]. Estrogen has depression effects through the improvement of endothelial dysfunction [82] and modulation of sympathetic nerve activation [83] in animal experiments. Estrogen insufficiency may be related to postmenopausal hypertension [84]. Two SNPs in *GREB1* were associated with hypertension in men ($P = 0.008$ for both SNPs) in this study. The authors thus concluded that *GREB1* might play a role in blood pressure regulation [76].

Genetic heterogeneity may exist in different populations for the genesis of hypertension. One association was assessed between the SNPs in the promoter region of the catalase gene (*CAT*) and EH in Greek Caucasians and African Americans. An association was found with specific genotype combination of *CAT* −844 homozygous AA together with *CAT* −262CT or TT in Caucasians only (100 hypertensive and 93 normotensive subjects, $P = 0.0339$), but no association was observed in African Americans (129 hypertensive and 98 normotensive subjects) [85]. The role of oxidative stress in hypertension has been tested in a number of studies [86]. Catalase, a protein converting H_2O_2 to water and oxygen, has been shown to reduce smooth muscle cell contraction and proliferation induced by endothelia, angiotensin II, and α-adrenoreceptor agonists [87]. Experimental studies showed a protective role of higher catalase expression levels in hypertensive animal models [88, 89]. Similarly, one SNP in the bradykinin receptor B2 (*BDKRB2*) and three SNPs in bradykinin receptor B1 (*BDKRB1*) were associated with hypertension in American Caucasians ($n = 220$, *P*-values between 0.026 and 0.0004). One SNP in the promoter region of *BDKRB2* was associated with hypertension in African Americans ($n = 218$, $P = 0.044$) [90].

Bradykinin (BK) has a variety of vasoactive and metabolic effects, including vasodilatation via interaction with components of the arachidonic acid cascade and enhancing insulin-independent glucose transport through B1 and B2 receptors [91]. Genetic variations of the receptors may alter the function capacity of bradykinin, which may thus alter an individual susceptibility to hypertension.

The above results suggest that individual SNP may not be as important as the interaction among several SNPs. The genetic factors that contribute to hypertension are likely to be different among different ethnic populations. Further studies of association in a large number of genes in different pathways will be required to identify the possible interaction among genes and the full array of genetics factors causing hypertension.

Mitochondrial genome mutations and essential hypertension DeStefano *et al.* studied maternal and paternal effects in the development of human EH in American Caucasians, Greek Caucasians, and African Americans. They found that among the parents with known hypertensive status, the proportion of affected mothers was significantly higher than the proportion of affected fathers in all three ethnic groups [92]. The fraction of patients with EH potentially due to mtDNA mutation involvement is estimated at 55% (95% CI 45–65%) [93]. A complete sequencing of the mitochondrial genome from 20 hypertensive probands in Africa American ($n = 10$) and Caucasian families ($n = 10$) was carried out. A total of 297 base exchanges were identified, including 24 in the ribosomal RNA (rRNA) genes, 15 in the tRNA genes, and 46 amino acid substitutions, with the remainder involving the non-coding regions or synonymous changes [94]. Several of these were associated with cardiovascular and renal pathologies in the earlier studies. Among them, an A10398G mutation in the NADH dehydrogenase subunit 3 gene, identified in 12 hypertensive individuals of both ethnicities, had been shown to occur with increased frequency in African Americans with the EH associated with end-stage renal disease [95]. These mitochondrial mutations data can thus serve as a starting point for case–control association studies.

Uncoupling proteins (UCPs) are inner mitochondrial membrane-associated proteins and act as proton channels or transporters. They are key to marching electrochemical gradients across the mitochondrial inner membrane. A functional polymorphism ($-866G/A$) in the UCP2 promoter has been reported to be associated with obesity in an analysis of 340 obese and 256 never-obese middle-aged Caucasian subjects ($P = 0.007$) [96] (Table 16.2). Another association study between this polymorphism and obesity, hypertension, as well as type 2 diabetes mellitus was carried out in one Japanese population with 342 type 2 diabetic patients (among them, 158 patients complicated with hypertension), 156 hypertensive patients without diabetes mellitus, and 134 control subjects. The polymorphism was significantly associated with hypertension (frequency of A allele: 51.8% in hypertensives versus 46.6% in normotensives, $P < 0.05$), but not associated with obesity in the Japanese population, which is in contrast to the significant association with obesity in Caucasian populations [97].

Mitochondrial coupling factor 6 is an essential component of mitochondrial adenosine triphosphate (ATP) synthase, suppressing the synthesis of prostacyclin in vascular endothelial cells [98]. The role of the gene was studied in spontaneously hypertension rats (SHRs) [99]. *In vivo*, the peptide circulates in the rat vascular system, and its gene expression and plasma concentration are higher in SHRs than in normotensive controls. Functional analysis suggests that it acts as a potent endogenous vasoconstrictor in the fashion of a circulating hormone [99]. Circulating coupling factor 6 is elevated in human hypertensive patients ($n = 30$) compared with normotensive subjects ($n = 27$, $P < 0.01$) and was increased after salt loading in hypertensive patients. The percentage changes in plasma coupling factor 6 level after salt restriction and loading were positively correlated with those in mean blood pressure ($r = 0.57$, $P < 0.01$), and negatively correlated with those in plasma nitric oxide level ($r = -0.51$, $P < 0.05$) [99, 100]. The elevated circulating coupling factor 6 in SHRs rats and human hypertension patients indicates that it is involved in the regulation of arterial blood pressure in physiological and pathological conditions [99, 100]. All the above studies suggest that EH may be not only polygenic, but also a "polygenomic" disorder.

Further investigation of the genetic causes of hypertension should consider dysfunction not only in the nuclear genome but also in the mitochondrial genome. Comprehensive analysis of the genetic cause for EH in both genomes will be appreciated in the future.

16.3.2.2 Genome-wide Linkage Analysis in Hypertension

Genome-wide linkage analysis suggests that multiple chromosomal regions may play a role in the development of human EH. However, lack of consistency across studies makes it difficult to draw any general conclusion for the genetic cause of human EH [101] (Table 16.3). An investigation focusing on only systolic and diastolic BP in 1109 Caucasian female dizygotic twin pairs was carried out. No significant linkage to BP could be detected in this study, but several suggestive linkage regions were replicated and one novel suggestive linkage for systolic BP on chromosome 11p was detected [102].

Significant linkage for longitudinal systolic BP from the Framingham Heart study was detected on chromosome 1q. In this study, the systolic BP for each individual was modeled as a function of age using a mixed modeling methodology. It was thus the best linear unbiased predictor of the individual's deviation from the population rate of change in systolic BP for each year of age while controlling for gender, BMI, and hypertension treatment [103]. Two previous linkage studies of a hypertension phenotype had found peak logarithm of the odds (LOD) scores in the same region [104, 105]. Similarly, linkage on chromosomes 12q, 15q, and 17q for mean systolic BP and linkage for both systolic BP slope and curvature on chromosome 20q were detected in the other study of Framingham heart data [106].

In the linkage analysis for age at diagnosis of hypertension and early-onset hypertension in the HyperGen cohort of different populations several suggestive linkage loci were detected, some of which had been reported to be linked to hypertension and BP in the former studies [107]. It is encouraging to note that linkage can be replicated from other studies, which suggests that new genetic factors with moderate to large effect sizes may be discovered. In view of the power of individual studies, two genome scan meta-analyses for hypertension were carried out individ-

Table 16.3. Genome-wide scans of human essential hypertension and blood pressure.

Study design	Population	Chromosomal location	Results
Linear mixed model for longitudinal SBP [103]	Framingham cohort	1q32–44	2×10^{-5} for SBP
Sib pair analysis for longitudinal SBP covariates [106]	Framingham cohort	12q13	2.9×10^{-7} for mean SBP
		15q26	2.1×10^{-7} for mean SBP
		17q24	4.8×10^{-6} for mean SBP
		20q13	4.2×10^{-5} for slope SBP
		20q13	2.8×10^{-6} for curvature SBP
Family-based linkage analysis [37]	Chinese	12p11	LOD score of 3.44 for EH
Dizygotic twin pairs [102]	Caucasian	11p12	LOD score of 2.28 for SBP
Family based linkage analysis [107]	Caucasian and African American	4q25	No linkage for Caucasians LOD score of 2.44 for EH in AA
		4q32	LOD score of 2.05 for early onset EH in AA
		15q21–22	LOD score of 2.31 for EH in AA
Meta-analysis [108]	Mixed populations	5q11–14	$P_{weighted} = 0.0288$ for EH
		5q23–34	$P_{weighted} = 0.0251$ for EH
		6q25–qter	$P_{weighted} = 0.0315$ for EH
		11q22–24	$P_{weighted} = 0.0084$ for EH
Meta-analysis [109]	Caucasian	2p12–q22.1	$P_{weighted} < 10^{-4}$ for EH
		3p14.1–q12.3	$P_{weighted} < 10^{-4}$ for EH
Admixture mapping [110]	African American	6q24	$P < 0.05$ for EH
		21q21	$P < 0.05$ for EH

SBP, systolic blood pressure; EH, essential hypertension; LOD, logarithm of the odds; $P_{weighted}$, P-values with a study sample size-weighting factor.

ually [108, 109]. Interestingly, the first of these, with different populations, failed to detect significant linkage to hypertension, and only several regions with suggestive linkage were identified, including 2p, 5q, 6q, 8p, 9p, 9q, and 11q. Of these, only regions on 5q, 6q, and 11q showed P-values lower than 0.05 [108]. Controversially, meta-analysis of genome-wide scans for hypertension and BP in Caucasians showed significant linkage on chromosomes 2p12–q22 and 3p14–q12 [109]. The results strongly suggest the population difference of the common phenotype. The mixed populations probably have a considerable degree of genetic heterogeneity, which is one of the main reasons why pooling of the results in different populations in the meta-analysis did not enhance the signals. However, pooling of the results in

Caucasians possesses a smaller degree of genetic heterogeneity. One admixture mapping for hypertension loci with genome-scan markers was carried out in African Americans [110], using individuals from Nigeria as the African ancestral population and European Americans for the estimates of allele frequencies for European ancestors. The distribution of marker location-specific African ancestry was shifted upward in hypertensive cases versus normotensive controls, and the markers are located on chromosome 6q24 and 21q21.

Even though numerous whole genome screens have been carried out in different populations; no positionally cloned genes that are associated with EH have been identified to date. This confirms the complex polygenic nature of the disorder. Different hypertension genes might play a role in different ethnic groups or even different subsets of large families; thus consistent linkage could be difficult to detect through different studies.

16.4
Perspective

Future analysis of complex diseases like essential hypertension will benefit from the development and application of analytical methods that have the ability to systematically evaluate the contribution of genes operating in heterogeneous environments. Comprehensive analysis of the genes lying in different pathway(s) that are necessary for the development of hypertension will be an essential tool, since it seems very likely from our current knowledge that many molecular variants acting in concert may be required to alter blood pressure homeostasis.

References

1 World Health Organization/ International Society of Hypertension Guidelines Subcommittee (**1999**) Guidelines for the Management of Hypertension. *J. Hypertens.* 17, 151–183.

2 ZINNER, S.H., P.S. LEVY, E.H. KASS (**1971**) Familial aggregation of blood pressure in childhood. *N. Engl. J. Med.* 284, 401–404.

3 HAVLIK, R.J., M. FEINLEIB (**1982**) Epidemiology and genetics of hypertension. *Hypertension* 4, III121–III127.

4 LEVINE, R.S., C.H. HENNEKENS, A. PERRY, J. CASSADY, H. GELBAND, M.J. JESSE (**1982**) Genetic variance of blood pressure levels in infant twins. *Am. J. Epidemiol.* 116, 759–764.

5 BIRON, P., J.G. MONGEAU, D. BERTRAND (**1976**) Familial aggregation of blood pressure in 558 adopted children. *Can. Med. Assoc. J.* 115, 773–774.

6 RICE, T., G.P. VOGLER, L. PERUSSE, C. BOUCHARD, D.C. RAO (**1989**) Cardiovascular risk factors in a French Canadian population: resolution of genetic and familial environmental effects on blood pressure using twins, adoptees, and extensive information on environmental correlates. *Genet. Epidemiol.* 6, 571–588.

7 LONGINI, I.M., JR., M.W. HIGGINS, P.C. HINTON, P.P. MOLL, J.B. KELLER (**1984**) Environmental and genetic sources of familial aggregation of blood pressure in Tecumseh,

Michigan. *Am. J. Epidemiol.* 120, 131–144.

8 HUNT, S.C., S.J. HASSTEDT, H. KUIDA, B.M. STULTS, P.N. HOPKINS, R.R. WILLIAMS (**1989**) Genetic heritability and common environmental components of resting and stressed blood pressures, lipids, and body mass index in Utah pedigrees and twins. *Am. J. Epidemiol.* 129, 625–638.

9 LIFTON, R.P., A.G. GHARAVI, D.S. GELLER (**2001**) Molecular mechanisms of human hypertension. *Cell* 104, 545–556.

10 WILSON, F.H., A. HARIRI, A. FARHI, H. ZHAO, K.F. PETERSEN, H.R. TOKA, C. NELSON-WILLIAMS, K.M. RAJA, M. KASHGARIAN, G.I. SHULMAN, S.J. SCHEINMAN, R.P. LIFTON (**2004**) A cluster of metabolic defects caused by mutation in a mitochondrial tRNA. *Science* 306, 1190–1194.

11 SAMANI, N.J. (**2003**) Genome scans for hypertension and blood pressure regulation. *Am. J. Hypertens.* 16, 167–171.

12 Intersalt Cooperative Research Group (**1988**) Intersalt: an international study of electrolyte excretion and blood pressure. Results for 24 hour urinary sodium and potassium excretion. *BMJ* 297, 319–328.

13 ELLIOTT, P., S. ROGERS, G. SCALLY, D.G. BEEVERS, M.J. LICHTENSTEIN, G. KEENAN, R. HORNBY, A. EVANS, M.J. SHIPLEY, P.C. ELWOOD (**1990**) Sodium, potassium, body mass, alcohol and blood pressure in three United Kingdom centres (the INTERSALT study). *Eur. J Clin. Nutr.* 44, 637–645.

14 ELLIOTT, P., J. STAMLER, R. NICHOLS, A.R. DYER, R. STAMLER, H. KESTELOOT, M. MARMOT (**1996**) Intersalt revisited: further analyses of 24 hour sodium excretion and blood pressure within and across populations. Intersalt Cooperative Research Group. *BMJ* 312, 1249–1253.

15 CARVALHO, J.J., R.G. BARUZZI, P.F. HOWARD, N. POULTER, M.P. ALPERS, L.J. FRANCO, L.F. MARCOPITO, V.J. SPOONER, A.R. DYER, P. ELLIOTT (**1989**) Blood pressure in four remote populations in the INTERSALT Study. *Hypertension* 14, 238–246.

16 LAW, M.R., C.D. FROST, N.J. WALD (**1991**) By how much does dietary salt reduction lower blood pressure? III – Analysis of data from trials of salt reduction. *BMJ* 302, 819–824.

17 WHELTON, P.K., J. HE (**1999**) Potassium in preventing and treating high blood pressure. *Semin. Nephrol.* 19, 494–499.

18 SACKS, F.M., L.P. SVETKEY, W.M. VOLLMER, L.J. APPEL, G.A. BRAY, D. HARSHA, E. OBARZANEK, P.R. CONLIN, E.R. MILLER, III, D.G. SIMONS-MORTON, N. KARANJA, P.H. LIN (**2001**) Effects on blood pressure of reduced dietary sodium and the Dietary Approaches to Stop Hypertension (DASH) diet. DASH-Sodium Collaborative Research Group. *N. Engl. J Med.* 344, 3–10.

19 APPEL, L.J., T.J. MOORE, E. OBARZANEK, W.M. VOLLMER, L.P. SVETKEY, F.M. SACKS, G.A. BRAY, T.M. VOGT, J.A. CUTLER, M.M. WINDHAUSER, P.H. LIN, N. KARANJA (**1997**) A clinical trial of the effects of dietary patterns on blood pressure. DASH Collaborative Research Group. *N. Engl. J Med.* 336, 1117–1124.

20 DE WARDENER, H.E., F.J. HE, G.A. MacGREGOR (**2004**) Plasma sodium and hypertension. *Kidney Int.* 66, 2454–2466.

21 HE, F.J., N.D. MARKANDU, G.A. SAGNELLA, H.E. DE WARDENER, G.A. MacGREGOR (**2005**) Plasma sodium: ignored and underestimated. *Hypertension* 45, 98–102.

22 SCHIFFRIN, E.L., J.B. PARK, H.D. INTENGAN, R.M. TOUYZ (**2000**) Correction of arterial structure and endothelial dysfunction in human essential hypertension by the angiotensin receptor antagonist losartan. *Circulation* 101, 1653–1659.

23 SANDERS, P.W. (**2004**) Salt intake, endothelial cell signaling, and progression of kidney disease. *Hypertension* 43, 142–146.

24 LIFTON, R.P., R.G. DLUHY, M. POWERS, G.M. RICH, S. COOK, S. ULICK, J.M. LALOUEL (**1992**) A

chimaeric 11 beta-hydroxylase/
aldosterone synthase gene causes
glucocorticoid-remediable
aldosteronism and human
hypertension. *Nature* 355, 262–265.

25 SHIMKETS, R.A., D.G. WARNOCK, C.M.
BOSITIS, C. NELSON-WILLIAMS, J.H.
HANSSON, M. SCHAMBELAN, J.R. GILL,
JR., S. ULICK, R.V. MILORA, J.W.
FINDLING (**1994**) Liddle's syndrome:
heritable human hypertension caused
by mutations in the beta subunit of
the epithelial sodium channel. *Cell* 79,
407–414.

26 HANSSON, J.H., L. SCHILD, Y. LU, T.A.
WILSON, I. GAUTSCHI, R. SHIMKETS,
C. NELSON-WILLIAMS, B.C. ROSSIER,
R.P. LIFTON (**1995**) A de novo
missense mutation of the beta subunit
of the epithelial sodium channel
causes hypertension and Liddle
syndrome, identifying a proline-rich
segment critical for regulation of
channel activity. *Proc. Natl Acad. Sci.
USA* 92, 11495–11499.

27 HANSSON, J.H., C. NELSON-WILLIAMS,
H. SUZUKI, L. SCHILD, R. SHIMKETS,
Y. LU, C. CANESSA, T. IWASAKI,
B. ROSSIER, R.P. LIFTON (**1995**)
Hypertension caused by a truncated
epithelial sodium channel gamma
subunit: genetic heterogeneity of
Liddle syndrome. *Nat. Genet.* 11, 76–
82.

28 TAMURA, H., L. SCHILD, N. ENOMOTO,
N. MATSUI, F. MARUMO, B.C. ROSSIER
(**1996**) Liddle disease caused by a
missense mutation of beta subunit of
the epithelial sodium channel gene.
J. Clin. Invest. 97, 1780–1784.

29 MUNE, T., F.M. ROGERSON, H.
NIKKILA, A.K. AGARWAL, P.C. WHITE
(**1995**) Human hypertension caused by
mutations in the kidney isozyme of 11
beta-hydroxysteroid dehydrogenase.
Nat. Genet. 10, 394–399.

30 STEWART, P.M., Z.S. KROZOWSKI, A.
GUPTA, D.V. MILFORD, A.J. HOWIE,
M.C. SHEPPARD, C.B. WHORWOOD
(**1996**) Hypertension in the syndrome
of apparent mineralocorticoid excess
due to mutation of the 11 beta-
hydroxysteroid dehydrogenase type 2
gene. *Lancet* 347, 88–91.

31 GELLER, D.S., A. FARHI, N.
PINKERTON, M. FRADLEY, M. MORITZ,
A. SPITZER, G. MEINKE, F.T. TSAI,
P.B. SIGLER, R.P. LIFTON (**2000**)
Activating mineralocorticoid receptor
mutation in hypertension exacerbated
by pregnancy. *Science* 289, 119–123.

32 WILSON, F.H., S. DISSE-NICODEME,
K.A. CHOATE, K. ISHIKAWA, C.
NELSON-WILLIAMS, I. DESITTER, M.
GUNEL, D.V. MILFORD, G.W. LIPKIN,
J.M. ACHARD, M.P. FEELY, B. DUSSOL,
Y. BERLAND, R.J. UNWIN, H. MAYAN,
D.B. SIMON, Z. FARFEL, X.
JEUNEMAITRE, R.P. LIFTON (**2001**)
Human hypertension caused by
mutations in WNK kinases. *Science*
293, 1107–1112.

33 MAYAN, H., G. MUNTER, M.
SHAHARABANY, M. MOUALLEM, R.
PAUZNER, E.J. HOLTZMAN, Z. FARFEL
(**2004**) Hypercalciuria in familial
hyperkalemia and hypertension
accompanies hyperkalemia and
precedes hypertension: description of
a large family with the Q565E WNK4
mutation. *J. Clin. Endocrinol. Metab.*
89, 4025–4030.

34 STERGIOU, G.S., E.V. SALGAMI (**2004**)
New European, American and
International guidelines for
hypertension management: agreement
and disagreement. *Expert. Rev.
Cardiovasc. Ther.* 2, 359–368.

35 ASHRAF, S.S., R. GUENTHER, P.F.
AGRIS (**1999**) Orientation of the tRNA
anticodon in the ribosomal P-site:
quantitative footprinting with U33-
modified, anticodon stem and loop
domains. *RNA* 5, 1191–1199.

36 SCHUSTER, H., T.E. WIENKER, S.
BAHRING, N. BILGINTURAN, H.R.
TOKA, H. NEITZEL, E. JESCHKE, O.
TOKA, D. GILBERT, A. LOWE, J. OTT,
H. HALLER, F.C. LUFT (**1996**) Severe
autosomal dominant hypertension and
brachydactyly in a unique Turkish
kindred maps to human chromosome
12. *Nat. Genet.* 13, 98–100.

37 GONG, M., H. ZHANG, H. SCHULZ,
Y.A. LEE, K. SUN, S. BAHRING, F.C.
LUFT, P. NURNBERG, A. REIS, K.
ROHDE, D. GANTEN, R. HUI, N.
HUBNER (**2003**) Genome-wide linkage

reveals a locus for human essential (primary) hypertension on chromosome 12p. *Hum. Mol. Genet.* 12, 1273–1277.

38 YAMAGISHI, K., H. ISO, T. TANIGAWA, R. CUI, M. KUDO, T. SHIMAMOTO (**2004**) High sodium intake strengthens the association between angiotensinogen T174M polymorphism and blood pressure levels among lean men and women: a community-based study. *Hypertens. Res.* 27, 53–60.

39 KUZNETSOVA, T., J.A. STAESSEN, K. STOLARZ, A. RYABIKOV, V. TIKHONOFF, A. OLSZANECKA, G. BIANCHI, E. BRAND, E. CASIGLIA, A. DOMINICZAK, R. FAGARD, S. MALYUTINA, Y. NIKITIN, K. KAWECKA-JASZCZ (**2004**) Relationship between left ventricular mass and the ACE D/I polymorphism varies according to sodium intake. *J. Hypertens.* 22, 287–295.

40 KUZNETSOVA, T., J.A. STAESSEN, L. THIJS, C. KUNATH, A. OLSZANECKA, A. RYABIKOV, V. TIKHONOFF, K. STOLARZ, G. BIANCHI, E. CASIGLIA, R. FAGARD, S.M. BRAND-HERRMANN, K. KAWECKA-JASZCZ, S. MALYUTINA, Y. NIKITIN, E. BRAND (**2004**) Left ventricular mass in relation to genetic variation in angiotensin II receptors, renin system genes, and sodium excretion. *Circulation* 110, 2644–2650.

41 STOLARZ, K., J.A. STAESSEN, K. KAWECKA-JASZCZ, E. BRAND, G. BIANCHI, T. KUZNETSOVA, V. TIKHONOFF, L. THIJS, T. REINEKE, S. BABEANU, E. CASIGLIA, R. FAGARD, J. FILIPOVSKY, J. PELESKA, Y. NIKITIN, H. STRUIJKER-BOUDIER, T. GRODZICKI (**2004**) Genetic variation in CYP11B2 and AT1R influences heart rate variability conditional on sodium excretion. *Hypertension* 44, 156–162.

42 NABER, C.K., W. SIFFERT (**2004**) Genetics of human arterial hypertension. *Minerva Med.* 95, 347–356.

43 TIMBERLAKE, D.S., D.T. O'CONNOR, R.J. PARMER (**2001**) Molecular genetics of essential hypertension: recent results and emerging strategies. *Curr. Opin. Nephrol. Hypertens.* 10, 71–79.

44 JEUNEMAITRE, X., F. SOUBRIER, Y.V. KOTELEVTSEV, R.P. LIFTON, C.S. WILLIAMS, A. CHARRU, S.C. HUNT, P.N. HOPKINS, R.R. WILLIAMS, J.M. LALOUEL (**1992**) Molecular basis of human hypertension: role of angiotensinogen. *Cell* 71, 169–180.

45 ATWOOD, L.D., C.M. KAMMERER, P.B. SAMOLLOW, J.E. HIXSON, R.E. SHADE, J.W. MACCLUER (**1997**) Linkage of essential hypertension to the angiotensinogen locus in Mexican Americans. *Hypertension* 30, 326–330.

46 BRAND, E., N. CHATELAIN, B. KEAVNEY, M. CAULFIELD, L. CITTERIO, J. CONNELL, D. GROBBEE, S. SCHMIDT, H. SCHUNKERT, H. SCHUSTER, A.M. SHARMA, F. SOUBRIER (**1998**) Evaluation of the angiotensinogen locus in human essential hypertension: a European study. *Hypertension* 31, 725–729.

47 VASKU, A., M. SOUCEK, S. TSCHOPLOVA, A. STEJSKALOVA (**2002**) An association of BMI with A (-6) G, M235T and T174M polymorphisms in angiotensinogen gene in essential hypertension. *J. Hum. Hypertens.* 16, 427–430.

48 O'DONNELL, C.J., K. LINDPAINTNER, M.G. LARSON, V.S. RAO, J.M. ORDOVAS, E.J. SCHAEFER, R.H. MYERS, D. LEVY (**1998**) Evidence for association and genetic linkage of the angiotensin-converting enzyme locus with hypertension and blood pressure in men but not women in the Framingham Heart Study. *Circulation* 97, 1766–1772.

49 FORNAGE, M., C.I. AMOS, S. KARDIA, C.F. SING, S.T. TURNER, E. BOERWINKLE (**1998**) Variation in the region of the angiotensin-converting enzyme gene influences inter-individual differences in blood pressure levels in young white males. *Circulation* 97, 1773–1779.

50 SUNDER-PLASSMANN, G., H. KITTLER, C. EBERLE, M.M. HIRSCHL, C. WOISETSCHLAGER, U. DERHASCHNIG, A.N. LAGGNER, W.H. HORL, M. FODINGER (**2002**) Angiotensin converting enzyme DD genotype is associated with hypertensive crisis. *Crit. Care Med.* 30, 2236–2241.

51 JEUNEMAITRE, X., R.P. LIFTON, S.C. HUNT, R.R. WILLIAMS, J.M. LALOUEL (**1992**) Absence of linkage between the angiotensin converting enzyme locus and human essential hypertension. *Nat. Genet.* 1, 72–75.

52 SUGIYAMA, T., H. MORITA, N. KATO, H. KURIHARA, Y. YAMORI, Y. YAZAKI (**1999**) Lack of sex-specific effects on the association between angiotensin-converting enzyme gene polymorphism and hypertension in Japanese. *Hypertens. Res.* 22, 55–59.

53 BOZEC, E., P. LACOLLEY, S. BERGAYA, P. BOUTOUYRIE, P. MENETON, M. HERISSE-LEGRAND, C.M. BOULANGER, F. ALHENC-GELAS, H.S. KIM, S. LAURENT, H. DABIRE (**2004**) Arterial stiffness and angiotensinogen gene in hypertensive patients and mutant mice. *J. Hypertens.* 22, 1299–1307.

54 SMITHIES, O., H.S. KIM (**1994**) Targeted gene duplication and disruption for analyzing quantitative genetic traits in mice. *Proc. Natl Acad. Sci. USA* 91, 3612–3615.

55 KIM, H.S., J.H. KREGE, K.D. KLUCKMAN, J.R. HAGAMAN, J.B. HODGIN, C.F. BEST, J.C. JENNETTE, T.M. COFFMAN, N. MAEDA, O. SMITHIES (**1995**) Genetic control of blood pressure and the angiotensinogen locus. *Proc. Natl Acad. Sci. USA* 92, 2735–2739.

56 BENETOS, A., S. GAUTIER, S. RICARD, J. TOPOUCHIAN, R. ASMAR, O. POIRIER, E. LAROSA, L. GUIZE, M. SAFAR, F. SOUBRIER, F. CAMBIEN (**1996**) Influence of angiotensin-converting enzyme and angiotensin II type 1 receptor gene polymorphisms on aortic stiffness in normotensive and hypertensive patients. *Circulation* 94, 698–703.

57 ERIKSSON, U., U. DANILCZYK, J.M. PENNINGER (**2002**) Just the beginning: novel functions for angiotensin-converting enzymes. *Curr. Biol.* 12, R745–R752.

58 BENJAFIELD, A.V., W.Y. WANG, B.J. MORRIS (**2004**) No association of angiotensin-converting enzyme 2 gene (ACE2) polymorphisms with essential hypertension. *Am. J. Hypertens.* 17, 624–628.

59 SUN, B., T. DRONMA, W.J. QIN, C.Y. CUI, D. TSE, T. PINGTSO, Y. LIU, C.C. QIU (**2004**) Polymorphisms of renin-angiotensin system in essential hypertension in Chinese Tibetans. *Biomed. Environ. Sci.* 17, 209–216.

60 JEUNEMAITRE, X., I. INOUE, C. WILLIAMS, A. CHARRU, J. TICHET, M. POWERS, A.M. SHARMA, A.P. GIMENEZ-ROQUEPLO, A. HATA, P. CORVOL, J.M. LALOUEL (**1997**) Haplotypes of angiotensinogen in essential hypertension. *Am. J. Hum. Genet.* 60, 1448–1460.

61 ZHAO, Y.Y., J. ZHOU, C.S. NARAYANAN, Y. CUI, A. KUMAR (**1999**) Role of C/A polymorphism at −20 on the expression of human angiotensinogen gene. *Hypertension* 33, 108–115.

62 XU, Q., Y.H. WANG, W.J. TONG, M.L. GU, G. WU, B. BUREN, Y.Y. LIU, J. WANG, Y.S. LI, H. FENG, S.L. BAI, H.H. PANG, G.R. HUANG, M.W. FANG, Y.H. ZHANG, Z.L. WU, C.C. QIU (**2004**) Interaction and relationship between angiotensin converting enzyme gene and environmental factors predisposing to essential hypertension in Mongolian population of China. *Biomed. Environ. Sci.* 17, 177–186.

63 WANG, J.G., L. LIU, L. ZAGATO, J. XIE, R. FAGARD, K. JIN, J. WANG, Y. LI, G. BIANCHI, J.A. STAESSEN, L. LIU (**2004**) Blood pressure in relation to three candidate genes in a Chinese population. *J. Hypertens.* 22, 937–944.

64 WANG, C.C., M.L. GOALSTONE, B. DRAZNIN (**2004**) Molecular mechanisms of insulin resistance that impact cardiovascular biology. *Diabetes* 53, 2735–2740.

65 SPEIRS, H.J., K. KATYK, N.N. KUMAR, A.V. BENJAFIELD, W.Y. WANG, B.J. MORRIS (**2004**) Association of G-protein-coupled receptor kinase 4 haplotypes, but not HSD3B1 or PTP1B polymorphisms, with essential hypertension. *J. Hypertens.* 22, 931–936.

66 JOSE, P.A., G.M. EISNER, R.A. FELDER

(**2003**) Dopamine and the kidney: a role in hypertension? *Curr. Opin. Nephrol. Hypertens.* 12, 189–194.

67 SANADA, H., P.A. JOSE, D. HAZEN-MARTIN, P.Y. YU, J. XU, D.E. BRUNS, J. PHIPPS, R.M. CAREY, R.A. FELDER (**1999**) Dopamine-1 receptor coupling defect in renal proximal tubule cells in hypertension. *Hypertension* 33, 1036–1042.

68 SIMARD, J., F. DUROCHER, F. MEBARKI, C. TURGEON, R. SANCHEZ, Y. LABRIE, J. COUET, C. TRUDEL, E. RHEAUME, Y. MOREL, V. LUU-THE, F. LABRIE (**1996**) Molecular biology and genetics of the 3 beta-hydroxysteroid dehydrogenase/delta5-delta4 isomerase gene family. *J. Endocrinol.* 150 (Suppl), S189–S207.

69 AZIZI, M., T.T. GUYENE, G. CHATELLIER, M. WARGON, J. MENARD (**1997**) Additive effects of losartan and enalapril on blood pressure and plasma active renin. *Hypertension* 29, 634–640.

70 HASHIMOTO, N., B.J. GOLDSTEIN (**1992**) Differential regulation of mRNAs encoding three protein-tyrosine phosphatases by insulin and activation of protein kinase C. *Biochem. Biophys. Res. Commun.* 188, 1305–1311.

71 ZHU, H., G.A. SAGNELLA, Y. DONG, M.A. MILLER, A. ONIPINLA, N.D. MARKANDU, G.A. MACGREGOR (**2004**) Molecular variants of the sodium/hydrogen exchanger type 3 gene and essential hypertension. *J. Hypertens.* 22, 1269–1275.

72 ORLOWSKI, J., S. GRINSTEIN (**1997**) Na$^+$/H$^+$ exchangers of mammalian cells. *J. Biol. Chem.* 272, 22373–22376.

73 SCHULTHEIS, P.J., J.N. LORENZ, P. MENETON, M.L. NIEMAN, T.M. RIDDLE, M. FLAGELLA, J.J. DUFFY, T. DOETSCHMAN, M.L. MILLER, G.E. SHULL (**1998**) Phenotype resembling Gitelman's syndrome in mice lacking the apical Na$^+$–Cl$^-$ cotransporter of the distal convoluted tubule. *J. Biol. Chem.* 273, 29150–29155.

74 ALDRED, K.L., P.J. HARRIS, E. EITLE (**2000**) Increased proximal tubule NHE-3 and H+-ATPase activities in spontaneously hypertensive rats. *J. Hypertens.* 18, 623–628.

75 HANNILA-HANDELBERG, T., K. KONTULA, I. TIKKANEN, T. TIKKANEN, F. FYHRQUIST, K. HELIN, H. FODSTAD, K. PIIPPO, H.E. MIETTINEN, J. VIRTAMO, J. KRUSIUS, S. SARNA, I. GAUTSCHI, L. SCHILD, T.P. HILTUNEN (**2005**) Common variants of the beta and gamma subunits of the epithelial sodium channel and their relation to plasma renin and aldosterone levels in essential hypertension. *BMC Med. Genet.* 6, 4.

76 KAMIDE, K., Y. KOKUBO, J. YANG, C. TANAKA, H. HANADA, S. TAKIUCHI, N. INAMOTO, M. BANNO, Y. KAWANO, A. OKAYAMA, H. TOMOIKE, T. MIYATA (**2005**) Hypertension susceptibility genes on chromosome 2p24-p25 in a general Japanese population. *J. Hypertens.* 23, 955–960.

77 ZHU, D.L., H.Y. WANG, M.M. XIONG, X. HE, S.L. CHU, L. JIN, G.L. WANG, W.T. YUAN, G.S. ZHAO, E. BOERWINKLE, W. HUANG (**2001**) Linkage of hypertension to chromosome 2q14-q23 in Chinese families. *J. Hypertens.* 19, 55–61.

78 ANGIUS, A., E. PETRETTO, G.B. MAESTRALE, P. FORABOSCO, G. CASU, D. PIRAS, M. FANCIULLI, M. FALCHI, P.M. MELIS, M. PALERMO, M. PIRASTU (**2002**) A new essential hypertension susceptibility locus on chromosome 2p24-p25, detected by genomewide search. *Am. J. Hum. Genet.* 71, 893–905.

79 LAIVUORI, H., P. LAHERMO, V. OLLIKAINEN, L. WIDEN, L. HAIVA-MALLINEN, H. SUNDSTROM, T. LAITINEN, R. KAAJA, O. YLIKORKALA, J. KERE (**2003**) Susceptibility loci for preeclampsia on chromosomes 2p25 and 9p13 in Finnish families. *Am. J. Hum. Genet.* 72, 168–177.

80 MERCER, E.A., L. KORHONEN, Y. SKOGLOSA, P.A. OLSSON, J.P. KUKKONEN, D. LINDHOLM (**2000**) NAIP interacts with hippocalcin and protects neurons against calcium-induced cell death through caspase-3-dependent and -independent pathways. *EMBO J.* 19, 3597–607.

81 RAE, J.M., M.D. JOHNSON, J.O. SCHEYS, K.E. CORDERO, J.M. LARIOS, M.E. LIPPMAN (2005) GREB1 is a critical regulator of hormone dependent breast cancer growth. *Breast Cancer Res. Treat.* 92, 141–149.

82 YEN, C.H., Y.T. LAU (2004) 17beta-Oestradiol enhances aortic endothelium function and smooth muscle contraction in male spontaneously hypertensive rats. *Clin. Sci. (Lond.)* 106, 541–546.

83 BRANDIN, L., G. BERGSTROM, K. MANHEM, H. GUSTAFSSON (2003) Oestrogen modulates vascular adrenergic reactivity of the spontaneously hypertensive rat. *J. Hypertens.* 21, 1695–702.

84 DUBEY, R.K., S. OPARIL, B. IMTHURN, E.K. JACKSON (2002) Sex hormones and hypertension. *Cardiovasc. Res.* 53, 688–708.

85 ZHOU, X.F., J. CUI, A.L. DeSTEFANO, I. CHAZARO, L.A. FARRER, A.J. MANOLIS, H. GAVRAS, C.T. BALDWIN (2005) Polymorphisms in the promoter region of catalase gene and essential hypertension. *Dis. Markers* 21, 3–7.

86 TOUYZ, R.M., E.L. SCHIFFRIN (2004) Reactive oxygen species in vascular biology: implications in hypertension. *Histochem. Cell Biol.* 122, 339–352.

87 WASSMANN, S., K. WASSMANN, G. NICKENIG (2004) Modulation of oxidant and antioxidant enzyme expression and function in vascular cells. *Hypertension* 44, 381–386.

88 UDDIN, M., H. YANG, M. SHI, M. POLLEY-MANDAL, Z. GUO, H. YANG, M. SHI, H. VANREMMEN, X. CHEN, J. VIJG, A. RICHARDSON, Z. GUO (2003) Elevation of oxidative stress in the aorta of genetically hypertensive mice. *Mech. Ageing Dev.* 124, 811–817.

89 YANG, H., M. SHI, H. VANREMMEN, X. CHEN, J. VIJG, A. RICHARDSON, Z. GUO (2003) Reduction of pressor response to vasoconstrictor agents by overexpression of catalase in mice. *Am. J. Hypertens.* 16, 1–5.

90 CUI, J., E. MELISTA, I. CHAZARO, Y. ZHANG, X. ZHOU, A.J. MANOLIS, C.T. BALDWIN, A.L. DeSTEFANO, H. GAVRAS (2005) Sequence variation of bradykinin receptors B1 and B2 and association with hypertension. *J. Hypertens.* 23, 55–62.

91 MERKUS, D., D.J. DUNCKER, W.M. CHILIAN (2002) Metabolic regulation of coronary vascular tone: role of endothelin-1. *Am. J. Physiol. Heart Circ. Physiol.* 283, H1915–H1921.

92 DeSTEFANO, A.L., H. GAVRAS, N. HEARD-COSTA, M. BURSZTYN, A. MANOLIS, L.A. FARRER, C.T. BALDWIN, I. GAVRAS, F. SCHWARTZ (2001) Maternal component in the familial aggregation of hypertension. *Clin. Genet.* 60, 13–21.

93 SUN, F., J. CUI, H. GAVRAS, F. SCHWARTZ (2003) A novel class of tests for the detection of mitochondrial DNA-mutation involvement in diseases. *Am. J. Hum. Genet.* 72, 1515–1526.

94 SCHWARTZ, F., A. DUKA, F. SUN, J. CUI, A. MANOLIS, H. GAVRAS (2004) Mitochondrial genome mutations in hypertensive individuals. *Am. J. Hypertens.* 17, 629–635.

95 WATSON, B., JR., M.A. KHAN, R.A. DESMOND, S. BERGMAN (2001) Mitochondrial DNA mutations in black Americans with hypertension-associated end-stage renal disease. *Am. J. Kidney Dis.* 38, 529–536.

96 ESTERBAUER, H., C. SCHNEITLER, H. OBERKOFLER, C. EBENBICHLER, B. PAULWEBER, F. SANDHOFER, G. LADURNER, E. HELL, A.D. STROSBERG, J.R. PATSCH, F. KREMPLER, W. PATSCH (2001) A common polymorphism in the promoter of UCP2 is associated with decreased risk of obesity in middle-aged humans. *Nat. Genet.* 28, 178–183.

97 JI, Q., H. IKEGAMI, T. FUJISAWA, Y. KAWABATA, M. ONO, M. NISHINO, M. OHISHI, T. KATSUYA, H. RAKUGI, T. OGIHARA (2004) A common polymorphism of uncoupling protein 2 gene is associated with hypertension. *J. Hypertens.* 22, 97–102.

98 KNOWLES, A.F., R.J. GUILLORY, E. RACKER (1971) Partial resolution of the enzymes catalyzing oxidative phosphorylation. XXIV. A factor

required for the binding of mitochondrial adenosine triphosphatase to the inner mitochondrial membrane. *J. Biol. Chem.* 246, 2672–2679.

99 OSANAI, T., M. TANAKA, T. KAMADA, T. NAKANO, K. TAKAHASHI, S. OKADA, K. SIRATO, K. MAGOTA, S. KODAMA, K. OKUMURA (**2001**) Mitochondrial coupling factor 6 as a potent endogenous vasoconstrictor. *J. Clin. Invest.* 108, 1023–1030.

100 OSANAI, T., S. SASAKI, T. KAMADA, N. FUJIWARA, T. NAKANO, H. TOMITA, T. MATSUNAGA, K. MAGOTA, K. OKUMURA (**2003**) Circulating coupling factor 6 in human hypertension: role of reactive oxygen species. *J. Hypertens.* 21, 2323–2328.

101 HARRAP, S.B. (**2003**) Where are all the blood-pressure genes? *Lancet* 361, 2149–2151.

102 DE LANGE, M., T.D. SPECTOR, T. ANDREW (**2004**) Genome-wide scan for blood pressure suggests linkage to chromosome 11, and replication of loci on 16, 17, and 22. *Hypertension* 44, 872–877.

103 JAMES, K., L.R. WEITZEL, C.D. ENGELMAN, G. ZERBE, J.M. NORRIS (**2003**) Genome scan linkage results for longitudinal systolic blood pressure phenotypes in subjects from the Framingham Heart Study. *BMC. Genet.* 4 (Suppl 1), S83.

104 PEROLA, M., K. KAINULAINEN, P. PAJUKANTA, J.D. TERWILLIGER, T. HIEKKALINNA, P. ELLONEN, J. KAPRIO, M. KOSKENVUO, K. KONTULA, L. PELTONEN (**2000**) Genome-wide scan of predisposing loci for increased diastolic blood pressure in Finnish siblings. *J. Hypertens.* 18, 1579–1585.

105 HUNT, S.C., R.C. ELLISON, L.D. ATWOOD, J.S. PANKOW, M.A. PROVINCE, M.F. LEPPERT (**2002**) Genome scans for blood pressure and hypertension: the National Heart, Lung, Blood Institute Family Heart Study. *Hypertension* 40, 1–6.

106 JACOBS, K.B., C. GRAY-MCGUIRE, K.C. CARTIER, R.C. ELSTON (**2003**) Genome-wide linkage scan for genes affecting longitudinal trends in systolic blood pressure. *BMC Genet.* 4 (Suppl 1), S82.

107 WILK, J.B., L. DJOUSSE, D.K. ARNETT, S.C. HUNT, M.A. PROVINCE, G. HEISS, R.H. MYERS (**2004**) Genome-wide linkage analyses for age at diagnosis of hypertension and early-onset hypertension in the HyperGEN study. *Am. J. Hypertens.* 17, 839–844.

108 LIU, W., W. ZHAO, G.A. CHASE (**2004**) Genome scan meta-analysis for hypertension. *Am. J. Hypertens.* 17, 1100–1106.

109 KOIVUKOSKI, L., S.A. FISHER, T. KANNINEN, C.M. LEWIS, F. VON WOWERN, S. HUNT, S.L. KARDIA, D. LEVY, M. PEROLA, T. RANKINEN, D.C. RAO, T. RICE, B.A. THIEL, O. MELANDER (**2004**) Meta-analysis of genome-wide scans for hypertension and blood pressure in Caucasians shows evidence of susceptibility regions on chromosomes 2 and 3. *Hum. Mol. Genet.* 13, 2325–2332.

110 ZHU, X., A. LUKE, R.S. COOPER, T. QUERTERMOUS, C. HANIS, T. MOSLEY, C.C. GU, H. TANG, D.C. RAO, N. RISCH, A. WEDER (**2005**) Admixture mapping for hypertension loci with genome-scan markers. *Nat. Genet.* 37, 177–181.

17
Gene Variants, Nutrition, and Cancer

Julian Little and Linda Sharp

17.1
Introduction

In this chapter we focus on the investigation of the effects of genetic variations in the host that are thought to influence metabolism of nutrients and food groups that have been found to be associated with cancer risk in observational studies. Much of the research so far has concentrated on polymorphisms influencing metabolism of folate, heterocyclic amines formed by cooking meats, insulin status, and alcohol. We review evidence for specific cancers within each of these areas, and discuss issues in the interpretation of the evidence on gene variants, nutrition and cancer.

17.1.1
Goals of Research on Genetic Variation in the Metabolism of Dietary Factors in Relation to Cancer

The goals of research on the relations between genetic variation in nutrient and food group metabolism and cancer (and other diseases) allow more precise public health advice about dietary intake, use of supplements, and genetic testing to be given. This, in turn, ultimately contributes to the prevention of cancer in the population.

These goals are important for several reasons. First, although diet appears to be important in the etiology of many different types of cancer [1], few definite relationships between diet and cancer risk have been established [2]. In consequence, it has been hard to provide clear, understandable, precise (and realistic) dietary advice for the population for the purposes of cancer prevention (discussed further below). Second, the extent of change in diet that has been achieved by population-based interventions has in general not been substantial, and it is possible that strategies tailored to specific subgroups of the population, defined on the basis of their genetic makeup, might be more successful [3–6]. Third, behavioral interventions may be inadequate to produce preventive effects in groups at high risk of disease (e.g. cardiovascular complications in persons with type 2 diabetes), and other

Nutritional Genomics. Edited by Regina Brigelius-Flohé and Hans-Georg Joost
Copyright © 2006 WILEY-VCH Verlag GmbH & Co. KGaA, Weinheim
ISBN: 3-527-31294-3

approaches, such as chemoprevention (delivered via supplements, pharmaceutical agents or food fortification), might be needed. Fourth, a number of companies are offering tests for genotypic and/or phenotypic markers of gene variants influencing nutrient metabolism to the consumer [7–9] or the physician [10] and it will be important to evaluate these tests.

17.1.2
Limitations of Studying Main Effects of Dietary Factors Only and the Potential for Advances Offered by Investigating Gene–Disease Relations

It has proved difficult to develop clear guidance about the role of diet in cancer prevention on the basis of observational data. For example, leads from observational studies suggesting that β-carotene supplementation might prevent cancer of the lung and other sites have not been supported by evidence from randomized intervention trials [11–13]. Moreover, while the evidence on the possible protective effects of fruit and vegetables against many types of cancer in 1997 was considered to be "convincing" [1], in more recent cohort studies the association between cancer and consumption of vegetables and fruit is less clear [14–18]. In aggregate, the evidence is considered to be less strong than was previously the case [19].

It would be hoped that the investigation of genetic variation in metabolism of diet will advance understanding of the etiology of cancer helping, for example, to clarify issues such as those raised above. For example, studies of associations between a disease and variants of genes coding for enzymes known to be involved in the metabolism of specific nutrients or food components could corroborate other evidence of the relation between nutrients or food groups and the disease [20]. The investigation of gene–disease associations differs from the investigation of exposure–disease associations in two important respects. First, the assessment of genotypes by DNA assays (polymerase chain reaction (PCR) methods) is more accurate (although it still suffers from some degree of misclassification) than is generally the case for exposure assessment, and is less heavily dependent on study design. Second, because of "Mendelian randomization" [21, 22], an association between a disease and a genotype is unlikely to be due to confounding, provided that the study is designed according to principles of population-based studies [23].

17.2
Mendelian Randomization and Associations between Genetic Polymorphisms Influencing Nutrient Metabolism and Cancer

In a population-based study of a genotype–disease association, the random assortment of alleles at the time of gamete formation (Mendel's second law) results in a random association between loci in a population and is independent of environmental factors [22]. In theory, this random assortment brings about a similar distribution of variants at unlinked genetic loci between individuals with and without disease. This situation is analogous to a randomized controlled trial (provided the trial is of adequate size) in which the random assignment to the intervention or placebo results in similar distributions of confounders (both measured and unmeasured) between the trial arms. For genes known to modulate the effects of environmental exposure, genetic variants with known functional effects can be considered as markers of altered exposure to an environmental factor of putative causal importance, such as intake of a specific nutrient. Therefore, the investigation of associations between genes affecting the metabolism of nutrients and other food components potentially enables the effect of the nutrients and food components themselves to be determined, excluding confounding as an explanation for the association.

17.3
Issues in the Interpretation of Studies of Gene–Disease Associations and Gene–Nutrient Interactions

Concerns about the interpretation and synthesis of evidence in studies of gene–disease associations include study size and quality, linkage disequilibrium, population stratification, gene–gene and gene–environment interaction that may mask gene association effects, and incomplete understanding of gene functions and biologic pathways important in the pathogenesis of common diseases [20, 24, 25]. There are also issues that relate particularly to studies of interactions (or effect modification) between gene variants and dietary components. These include specification of the model of interaction, summarizing the data and testing for that interaction statistically.

17.3.1
Study Quality and Size

Factors that impact on study quality include selection bias, in relation to the representativeness of study subjects, and information bias, specifically the analytical validity of genotyping.

In several studies of genotype and cancer, prevalent cases have been included to varying extents [26]. This can introduce bias (systematic error) if the genotype affected survival or if genotypes were assayed by a phenotypic test that was influenced by disease progression or treatment. In a number of case–control studies of gene–disease associations with unrelated controls, controls were not selected from the same source population as the case subjects [27], which can also result in bias.

Misclassification of genotype can bias genotype–disease associations, usually towards the null [28, 29]. The most marked bias occurs when genotyping sensitivity is poor and genotype prevalence is high (>85%), or genotyping specificity is poor and genotype prevalence genotype is low (<15%) [28]. The extent of genotyping error has been reported to vary between about 1% and 30% [30].

Sample size and genotype frequencies in the population are the major determinants of the statistical power of a study. The danger inherent in small studies, with limited statistical power, is that a true underlying gene–disease association could fail to be detected. Variation in statistical power may in part explain non-replication of gene–disease association studies [22, 31, 32].

17.3.2
Linkage Disequilibrium

Linkage disequilibrium is the tendency for the alleles of two separate but already linked loci on the same chromosome to be found together more than would be expected by chance in the general population. In consequence, when an allele at a specific locus appears to be associated with a disease, an issue is whether the allele is causal, or whether the association exists only because the allele is associated with a truly causal allele at another locus. Linkage disequilibrium depends on population history and on the genetic makeup of the founders of that population [33–35]. Linkage disequilibrium varies between populations [33] and therefore potentially could be a source of heterogeneity between studies of gene–disease associations and related interactions.

17.3.3
Population Stratification

Concern has been raised about the possible effects of population stratification on the results of population-based case–control studies. Population stratification includes differences between groups in ethnic origin and can arise because of differences between groups of similar ethnic origin but between which there has been limited admixture, such as in isolated populations. For example, a population might comprise the descendants of waves of immigrants from the same source but differ generally because of founder effects. The differences may then be apparent because insufficient time has elapsed for mixture between the groups.

The most discussed examples [36–38] have generated controversy as to whether population stratification represents a fundamental problem for association studies, or whether it is part of more general issues about rigorous application of epidemiologic study design principles [39–41]. In an exploration of possible population stratification in US studies of cancer among non-Hispanic Americans of European descent, the effect was considered unlikely to be substantial when epidemiologic principles of study design, conduct, and analysis were rigorously applied [42]. The possibility that the potential problem of population stratification may be greater in other ethnic groups has been considered [43, 44]. Meta-analyses of 43 gene–disease associations comprising 697 individual studies show consistent associations across groups of different ethnic origin [45], and so provides evidence against population stratification, hidden or otherwise. Nevertheless, there remains controversy about the potential importance of population stratification for population-based studies of gene–disease association and for studies of gene–environment interaction [46–48].

17.3.4
Gene Function and Biologic Pathways

Information on the effects of gene variants on enzyme activity and structure may be relatively limited or determined in contexts (biochemical assays in cell lines or animal models) that do not reflect human biology. Investigation of the C677T and A1298C variants in the methylenetetrahydrofolate reductase (*MTHFR*) gene illustrates this. Even for these relatively intensively studied variants, there are few papers on enzyme activity *in vitro* [49–51], and enzyme activity in compound heterozygotes is unclear [52]. Moreover, data on the relations between the two common *MTHFR* variants and other markers of function effects, including plasma and red cell folate, plasma homocysteine, and measures of methylation or DNA stability are limited and not entirely consistent [53].

17.3.5
Gene–Environment Interaction

The absence of a gene–disease association in an epidemiologic study may not exclude the possibility of different effects of genotype (or the nutrient) in subgroups (i.e. gene–environment interaction) [54, 55]. For example, there appears to be no overall association between homozygosity for the C677T variant of the *MTHFR* gene colorectal adenoma (see below) but in studies in which joint effects of *MTHFR* genotype and folate intake or status were assessed, the highest risk for adenomas was for individuals homozygous for the variant with the lowest folate intake or status [20].

17.3.6
Gene–Gene Interaction

It is unlikely that a single polymorphism will determine the response to a nutrient [56]. The metabolism of any exposure is likely to depend on the balance between the relative activities of all the enzymes active within the metabolic pathway [57]. For example, there are common functional variants of several genes coding enzymes in the absorption, metabolism and transport of folate [58], and it is plausible that they interact to influence disease risk. Moreover, it is theoretically possible that combinations of a small number of such variants may account for a high proportion of disease [59].

17.3.7
Specific Issues in Studies of Gene–Nutrient Interaction

The main issues in the specific investigation of joint effects of genes and specific nutrients or food groups/components additional to those for gene–disease associations are that statistical power is a bigger problem, misclassification of exposure has to be considered, and prior specification of the model for interaction requires special consideration.

In most of the earliest studies of associations between genetic polymorphisms influencing nutrient metabolism and cancer, detection of gene–environment interaction was a secondary aim. Most were based on a candidate gene approach, with strong biologic evidence of the importance of the genes and some evidence about the functional effects of variants of the genes [60]. Most were of modest size, and while their statistical power was adequate for the detection of gene–disease associations, it was inadequate to detect gene–environment interactions. To test for departures from multiplicative effects, it has been noted that study size should be at least four times larger than needed to detect only the main effects of the individual factors [61]. When non-differential misclassification of nutrient intake, genotype or both is taken into account, this in turn increases the required study size [62].

Many hypotheses of interaction can potentially be tested. In the analysis of data on nutrient intake, in which multiple categories of intake may be defined, many different dose–response models can be tested in the data. Supplement use can be classified as "ever" versus "never", "use of supplement A" versus "use of reference supplement B," as a continuous variable, and further defined on the basis of period of use. In a two-allele system, heterozygotes potentially could be considered separately, included in the reference category with homozygotes for the common variant, or grouped with homozygotes for the rarer variant. This is more complex for multi-allelic systems. The use of different approaches to summarizing data for the analysis of interactions has caused considerable problems in the synthesis of evidence from different studies [53]. A further issue is differences in statistical approaches to test for interaction [63].

17.4
Polymorphisms Influencing Folate Metabolism and Cancer

Vegetables, particularly green leafy vegetables, are a major source of folate. Folate is involved in the synthesis and methylation of DNA and mechanisms have been postulated by which low folate status might increase the risk of malignancy [64]. This has prompted considerable investigation of the role of folate, and its synthetic form, folic acid, in several types of cancer. Higher levels of folate intake or blood folate have been associated with lower risks of colon, and perhaps other, cancers [65].

Many of the genes involved in the absorption, metabolism, and transport of folate contain polymorphisms [58]. *MTHFR*, the gene encoding 5,10-methylenetetrahydrofolate reductase, functions at a key branch point in folate metabolism, directing the folate pool towards DNA methylation or synthesis. *MTHFR* contains two common variants – C677T and A1298C – both of which appear to have functional effects, including lowered enzyme activity associated with the variant alleles [53]. These polymorphisms have been investigated in relation to several types of cancer [53].

17.4.1
Colorectal Cancer

The studies of colon cancer and measures of folate intake or status are not wholly consistent, which in part reflects the well-known problems of investigating the association between nutrients and disease in observational studies [66]. Most, however, suggest that there is an inverse association between folate status and both colon cancer and adenomatous polyps of the colorectum [67], from which most invasive colorectal cancers are thought to arise.

Homozygosity for the *MTHFR* C677T and A1298C variants has been associated with a moderate reduction in the risk for colorectal cancer in most studies [53, 68–71], and this at first sight appears the opposite of what would be expected on the basis of the associations with folate intake or status. The findings of six studies of the *MTHFR* C677T variant and adenomatous polyps are inconsistent [53, 72, 73].

Interactions between genes and folate and related nutrients have been reported more for colorectal cancer than for other types of cancer. In studies in which joint effects of *MTHFR* genotype and diet have been investigated, those homozygous for the C677T variant who had higher folate levels (or a high-methyl diet) had the lowest risk for colorectal cancer [53]. The data suggest that the effect of a low folate diet overrides the effect of genotype, but this is based on a limited number of observational studies and is not entirely consistent. Moreover, two studies of adenomas suggested the opposite [74, 75].

Variation of a specific gene may impact on more than one pathway and therefore on more than one causal process. For instance, methylenetetrahydrofolate reductase (MTHFR) may affect DNA synthesis and repair in addition to its more established role in the conversion of homocysteine to methionine and Chen *et al.* [76] proposed this as an explanation for the inverse association between homozygosity for the *MTHFR* C677T variant and colon cancer. However, the evidence on the role of MTHFR in DNA synthesis and repair is not strong and, specifically, the relations between the *MTHFR* variants and incorporation of uracil into DNA, DNA strand breakage, and methylation are not clear [20, 77]. Altogether this suggests that the role of folate and folate pathway genes in colorectal neoplasia are complex.

17.4.2
Breast Cancer

Several observational studies have reported an inverse association between folate intake and blood status and breast cancer [65]. This is particularly apparent amongst women with high alcohol intake [78], presumably reflecting the adverse effects of alcohol on folate [79].

There have been at least 15 studies of breast cancer and *MTHFR* C667T genotype and, while several suggest that the TT genotype may be associated with increased risk, the evidence overall is inconsistent [80–94]. In a large study in China, a significant interaction between folate intake and C677T genotype was observed, such that presence of the T allele enhanced the increased risk associated with low folate intake [93]. Compared with those with the CC genotype and high folate in-take, the odds ratios for low folate were 1.94 (95% confidence interval (CI) 1.15–3.26), 2.17 (95% CI 1.34–3.51) and 2.51 (95% CI 1.37–4.60) for those with the CC, CT, and TT genotypes respectively. Chen *et al.*, in a large study in the USA, also observed that the group at highest risk were those with the TT genotype and low folate intake [94]. The relatively few studies that have considered A1298C are in-consistent [82, 83, 90, 93, 94].

Therefore, the data on breast cancer and *MTHFR* do not appear to entirely cor-roborate the data on breast cancer and folate intake or status but a detailed system-atic review and meta-analysis is required to clarify this.

17.4.3
Cervical Cancer and Precancerous Lesions

A small randomized controlled trial (RCT) suggested that folate supplementation would prevent progression of cervical dysplasia in women taking oral contracep-tives [95], while a later RCT from the same group was interpreted as suggesting that folate status was most likely involved in the earliest stages of cervical neoplasia [96]. There is conflicting evidence from observational studies about the relation-ship between folate intake or status and high-grade cervical precancer or invasive cervical cancer [97–103]. Higher levels of plasma homocysteine (which is inversely related to plasma folate) have been associated with an increased risk of cervical cancer in the few studies in which this has been investigated [104]. Studies of the *MTHFR* C677T variant and both cervical precancer [105–110] and invasive cervical cancer [109–111] have been inconsistent.

17.4.4
Esophageal Cancer

All four available studies of folate intake and esophageal cancers observed an in-verse relation, with an up to 50% reduction in risk in the highest versus lowest intake groups [112–115]. Song *et al.* reported a 6-fold increased risk associated with the *MTHFR* 677 TT genotype [116]. More modest, but none-the-less positive, effects have been reported in two other studies [117, 118]. As regards A1298C, in the single available study, risk of esophageal cancer was raised in those who either smoked tobacco or drank alcohol and carried the C allele [119].

17.4.5
Hematopoietic Malignancies

One study suggests that folate supplementation in pregnancy reduces the risk for common acute lymphocytic leukemia in the child [120]. Homozygosity for the C677T and A1298C variants has been associated with a reduced risk for leukemia and lymphomas in most [121–124] but not all [125] studies.

17.4.6
Other Polymorphisms Influencing Folate Metabolism

Other polymorphisms in genes of the folate pathway – such as variants of the genes coding for methionine synthase (*MTR*) [126–130], methionine synthase reductase (*MTRR*) [128], cystathione *β*-synthase (*CBS*) [128, 131] or thymidylate synthase (*TS*) [132–136] – have been investigated primarily in relation to colorectal cancer, but as yet the results are inconclusive. Three studies of *TS* and one of *MTRR* suggest associations with cancer of the esophagus [117, 118, 137, 138]. Associations between *MTR* and *TS* genotypes and different types of hematopoietic malignancies have been reported in a small number of studies [123, 139–141]. Replication of these findings is required.

While the metabolism of any exposure is likely to depend on the balance between the relative activities of all the enzymes active within the metabolic pathway [57], to date joint effects of folate pathway genes have only been little investigated [123, 128, 133, 139].

17.5
Polymorphisms Influencing Metabolism of Heterocyclic Amines Formed by Cooking Meats

Heterocyclic amines are generated during the cooking of red meat and chicken at high temperatures until well done [142]. For the heterocyclic amines to be carcinogenic they must be metabolized by enzymes including glutathione *S*-transferase (GST), *N*-acetyltransferase 1 (NAT1), and *N*-acetyltransferase 2 (NAT2) (see also chapter 19). This has prompted investigation of associations between variants in phase I and phase II metabolism genes, and interactions between gene variants and meat intake, with regard to risk of several types of cancer, most notably colorectal and breast cancer.

The glutathione *S*-transferases (GSTs) play a central role in the detoxification of carcinogens by catalyzing the conjugation of glutathione to potentially genotoxic compounds, including polycyclic aromatic hydrocarbons (PAHs) [26]. In humans, seven classes of cytosolic GSTs have been described [143]. For two of these, mu and theta, no enzyme activity has been detected in substantial proportions in several populations [144, 145], and this variation in activity may be important in modifying cancer risk. Homozygosity for deletion variants that occur at the *GSTM1* and *GSTT1* loci are associated with reduced, or no conjugation activity [26]. Evidence is lacking as to whether heterozygosity for either deletion variant affects gene function. NAT1 and NAT2 function as phase II conjugating enzymes [146]. Both enzymes are capable of *N*-acetylation and *O*-acetylation and *N*,*O*-acetylation and are implicated in the activation and detoxification of known carcinogens. Variants of the genes coding for NAT2 have been associated with altered rapidity of acetylation, and this has been postulated to affect the risk for cancer [147]. However, clarification of functional effects has proved difficult as most nucleotide polymorphisms occur in combination. Compared with the rapid acetylation phenotype associated with presence of the *NAT2*4* allele, variant alleles are thought to influence protein expression or stability, and the effects of combinations of these are heterogeneous and not fully understood [146, 147]. Similarly, the relationship between *NAT1* genotype and phenotype is unclear (see also chapter 19).

It is worth noting that this area of research is particularly challenging because of the difficulty in adequately assessing exposure to carcinogens in cooked meats in epidemiologic studies. Moreover, and in common with other areas discussed in this chapter, the comparison of studies and overall interpretation of the evidence is complicated by differences in the genes and polymorphisms that have been investigated, different approaches to statistical analysis and low statistical power.

17.5.1
Colorectal Cancer

Two meta-analyses of meat consumption and colorectal cancer have been reported in the last few years, both of which showed no statistically significant overall association between total meat consumption and colorectal cancer [148, 149]. However, the data on red meat and processed meat suggest positive associations, although it is possible that in part this reflects publication bias. In addition, increased consumption of well-done red meat has been associated with increased risk of colorectal neoplasia in some studies [150, 151].

Ishibe *et al.* observed a six-fold increased risk of adenomas among rapid NAT1 acetylators (defined as those carrying the *NAT1*10* allele), who were estimated, on the basis of reported meat intake, cooking methods, and doneness level, to consume more than 27 ng/day of the heterocyclic amine MeIQx, whereas among slow acetylators the increase in risk was two-fold [152]. While other investigators have also reported patterns in risk suggestive of interactions between particular genetic variants and meat intake (e.g. Welfare *et al.* for *NAT2*; Gertig *et al.* for *GSTM1* and *GSTT1*; Turner *et al.* for *GSTT1* and *GSTP1*; Cortessis *et al.* for microsomal epoxide hydrolase (*mEH*)) [153–156], the direction of the associations have not always been consistent with the underlying hypotheses [26]. Other studies have failed to find any evidence that the relationship between red meat intake and colorectal neoplasia is modified by genotype [151, 157–159].

17.5.2
Breast Cancer

While cohort studies of breast cancer and total meat, red meat, and white meat consumption have had conflicting results [160], in several studies a positive association has been observed between consumption of well-done meat and breast cancer [161–164]. Exposure to the most abundant heterocyclic amine in cooked meats, PhIP, causes mammary tumors in rats [165]. A significant association between higher levels of PhIP/DNA adducts in breast tissue and increased risk of breast cancer has been shown in humans [166].

The results of studies of associations between *GSTM1* and *GSTT1* genotypes and breast cancer risk have been inconsistent [167, 168]. In a study of post-menopausal breast cancer in Iowa, USA, women with both *GSTM1* and *GSTT1* deleted had a 60% increased disease risk, compared with those who had both genes present [167]. There was some evidence of effect modification by meat doneness; the increased risk associated with consuming meat well-done or very well-done was mainly confined to women with the null genotype (OR *GSTM1*-null and/or *GSTT1*-null and ate meat well-done versus *GSTM1* and *GSTT1* present and ate meat rare/medium = 3.4, 95% CI 1.6–7.1). In a case–control study in the Netherlands, of pre- and post-menopausal disease, van der Hel *et al.* reported risk estimates consistent with an interaction between *GSTM1* genotype and red meat intake, but the results did not reach statistical significance [168].

Further analyses from the Iowa study found a significant interaction between sulfotransferase (*SULT*) *1A1* genotype and well-done meat intake [169]. In the same study, *NAT2* genotype also modified associations between well-done meat intake and breast cancer, such that a preference for increased meat doneness raised breast cancer risk among women with genotypes associated with rapid/intermediate acetylation, but was not associated with risk among women with slow acetylation genotypes [170]. Four further studies did not find any significant interactions between *NAT2* genotype and red meat consumption or estimated intake of heterocyclic amines [168, 171–173]. The relatively few studies of *NAT1* genotype and breast cancer are inconclusive [168, 174–176]. A study in Korea suggested that, while *NAT1* and *NAT2* genotype may not have an independent association with breast cancer, they might act together with *GSTM1* and *GSTT1* to influence risk [176]. This is plausible since metabolism of heterocyclic amines involves both phase I and phase II enzymes. Further studies of the effects of combinations of variants in different genes, and interaction with markers of heterocylic amine exposure, are required to resolve this issue.

17.6
Polymorphisms Affecting Insulin Status

Hyperinsulinemia and insulin resistance have been implicated in several types of cancer, with suggestions that elevated levels of insulin, glucose or triglycerides could have tumor growth-promoting effects [177–179]. This has led, in recent years, to investigation of whether dietary glycemic load and/or glycemic index is related to cancer risk [65].

One mechanism by which raised insulin levels could affect cancer risk is by increasing the bioactivity of insulin-like growth factor I (IGF-I) and inhibiting production of two of the main binding proteins, IGFBP-1 and IGFBP-2 [180]. IGF-I has mitogenic effects on normal and neoplastic cells, inhibiting apoptosis and stimulating cell proliferation [180]. More than 75% of serum IGF-I circulates as complexes with IGFBP-3. IGFBP-3 thus modulates IGF-I bioavailability and, in addition, independently of IGF-I, inhibits replication and promotes apoptosis [181]. Serum levels of IGF-I and IGFBP-3 are determined by a combination of genetic and environmental effects. A recent review concluded that while there was no convincing evidence that a CA repeat polymorphism in the IGF-I gene influenced serum IGF-I levels, an A–C polymorphism at nucleotide −202 in the *IGFBP-3* gene did appear to have a modest effect on circulating IGFBP-3 levels [182].

17.6.1
Colorectal Cancer

The similarity of risk factors for colon cancer and diabetes, and the observation that insulin promotes the growth of colon cells *in vitro* and colon tumors *in vivo* [183, 184], prompted suggestions that hyperinsulinemia and insulin resistance may lead to colorectal cancer [177, 178]. While several strands of epidemiologic evidence support the hypothesis, inconsistencies remain and a number of areas require clarification [66, 185]. Glycemic control is likely to be important, as an increased risk of colorectal cancer has been found to be associated with a high level of dietary glycemic load in two large studies [186, 187] and changes in glycated hemoglobin concentrations seem to account for the increase in risk associated with type 2 diabetes [188]. Higher levels of IGF-I have been associated with an increased risk of colorectal cancer, but there is no dose–response relationship [189].

A genetic variant at position 1663 in the human growth hormone 1 gene (*GH1*) is thought to be associated with lower IGF-I levels. In a single study, the variant A allele was related, in a dose–response fashion, to a reduced risk of both colorectal cancer and adenomas [190]. In a study in Singapore, the *IGF1* (CA)21 repeat allele, but not the more common (CA)19 repeat allele, was associated with reduced colorectal cancer risk [191]. In the same study a novel polymorphism in *IGF1*, −533T/C, was also related to disease risk; the effect was apparent for colon, but not rectal, cancer but the numbers of homozygous variant cases was small. There was no association between *IGFBP-3* genotype and disease risk in this study. Also in a single study, polymorphisms in the genes encoding the insulin receptor substrates (*IRS1*, *IRS2*) were associated with risk of colon, but not rectal, cancer [192]. In the same study, variants in the *IGF1* and *IGFBP3* genes were not independently related to cancer but did appear to act together with *IRS1* to influence risk. Moreover, specific sources of energy appeared to be more strongly related to colon cancer risk in the presence of specific *IRS2* and *IGF1* genotypes [193]; a high sucrose-to-fiber ratio increased risk in those with the *IRS2* DD genotype and those without the *IGF1* (CA)19/(CA)19 genotype.

17.6.2
Prostate Cancer

Deregulation of the IGF system may be implicated in the etiology of prostate cancer [194]. IGFBP-3 protein levels have been found to be decreased in malignant prostatic tissue as compared to benign epithelium [195]. In some but not all studies, both cross-sectional and prospective, high plasma IGF-I and low IGFBP-3 levels have been associated with increased risk of prostate cancer overall [196–201], while a high IGF-I:IGFBP-3 ratio has been related to advanced stage prostate cancer [202]. Men with higher fasting serum insulin levels have been found to have increased prostate cancer risk [203]. Dietary glycemic load and glycemic index have also been positively associated with disease risk [204].

The four studies of the *IGF1* repeat allele and prostate cancer have had inconsistent results. While in two studies risk was raised in men carrying (CA)19 [205, 206], this was not observed in the other two studies [201, 207]. There was no association between *IGFBP-3* genotype and prostate cancer in two studies [194, 205], although a third suggested there may be a relation in Caucasian, but not African-American, men [201]. In a single study, possession of the *IRS1* variant R allele was associated with an almost three-fold increased prostate cancer risk overall, and with more advanced disease stage [207], but there was no association with *IRS2* or *INS* genotype. Combinations of polymorphisms and gene–diet interactions do not yet appear to have been investigated.

17.6.3
Breast Cancer

There is limited evidence for an association between dietary glycemic load and glycemic index and breast cancer [208]. Recent studies suggest that the effect, if any, is probably confined to particular subgroups, defined by lifestyle and hormonal factors such as menopausal status, BMI, level of physical activity and/or hormone replacement therapy (HRT) use [209–211], although the groups at highest and lowest risk are not yet clear. Studies of breast cancer risk and plasma insulin levels have been inconsistent [212, 213]. As regards circulating levels of IGF-I and IGFBP-3, while there appears to be no consistent relation with risk of post-menopausal breast cancer, the majority of studies of pre-menopausal disease have found risk to be at least doubled in women in the highest compared to the lowest quantile of IGF-I or IGFBP-3 [182].

While the first reported study of *IGF1* genotype and breast cancer, which included 53 cases and 53 controls, observed an significant association with the (CA)19 allele [214], this has not been confirmed in subsequent studies [182]. The single study of *IGFBP3* genotype and breast cancer was null [215].

17.7
Polymorphisms Influencing Alcohol Metabolism

Drinking alcohol has been associated with raised risk of several different types of cancer [216]. Strong trends of increasing risk with increasing consumption are evident for head and neck cancers (including the oral cavity, pharynx, and larynx) and cancers of the esophagus, while less strong relations have been reported for colorectal, stomach, liver, breast, and ovarian cancer.

Alcohol is metabolized to the carcinogen acetaldehyde by oxidation by the enzyme alcohol dehydrogenase (ADH) and is subsequently detoxified into acetate by aldehyde dehydrogenase (ALDH). The ADH isoenzymes involved in these reactions include subunits encoded by the *ADH2* (also known as *ADH1B*) and *ADH3* (*ADH1C*) genes, both of which are polymorphic. The alleles *ADH3*1* and *ADH2*2* code for "fast" metabolism of ethanol [217]. Several *ALDH* genes have been identified, one of which, the mitochondrial *ALDH2*, contains an inactive variant (the *ALDH2*2* allele). Those who are homozygous for the variant are unable to oxidize acetaldehyde, while heterozygotes have reduced oxidation capacity [217]. The *ADH3*2* allele is common in Caucasian populations, while *ADH2*2* and *ALDH2*2* are thought to be specific to Asian populations [217].

17.7.1
Colorectal Neoplasia

Two studies of adenomas have reported patterns consistent with an interaction between *ADH3* genotype and alcohol intake [72, 218]. Among subjects in the Male Health Professional Follow-up Study (HPFS), compared with those who consumed low levels of alcohol per day and carried the fast alcohol catabolism genotype (*ADH3*1/*1*), high consumers of alcohol with the slow catabolism genotype (*2/*2) had a substantially increased risk of disease (odds ratio (OR) > 30 g/day and *2/*2 versus ≤ 5 g/day and *1/*1 = 2.94, 95% CI 1.24–6.92); those who consumed high quantities of alcohol but had the fast catabolism genotype had only minimally increased risk (OR > 30 g/day and *1/*1 versus ≤ 5 g/day and *1/*1 = 1.27, 95% CI 0.63–2.53) [72]. The pattern of interaction described in the other study, from the Netherlands [218], was very similar to the HPFS result. In addition, the relation was apparent in both male and female subjects.

Since alcohol adversely affects folate metabolism, Giovannucci *et al.* investigated, in the HPFS, whether *ADH3* acted together with alcohol and folate intake to influence disease risk. Individuals with high alcohol and low folate and the slow catabolism genotype were at particularly high risk compared with fast catabolizers with low alcohol and high folate intake (OR 17.1, 95% CI 2.13–137.0, *P* interaction = 0.006), although the result was based on small numbers in the high alcohol/low folate/slow catabolism group [72]. This analysis illustrates some of the complexity of investigating the joint effects of genes and diet on disease – as well as there being multiple genes in pathways, several dietary factors may also be relevant. In addition, if verified, it is an example of the possibility that a small number of gene variants of moderate effect which have a modest interaction with exposure can have a high predictive value [59].

Again with the rationale that alcohol affects the metabolism of folate, interactions between alcohol intake and *MTHFR* genotype have been investigated in several studies [53, 69–73]. While some have documented patterns consistent with effect modification [71, 72, 76, 219], others found no evidence of any interaction [69, 71, 73, 220].

Two studies failed to find any significant associations between *ALDH2* genotype and colorectal cancer or adenomas, but both studies were small [221, 222].

17.7.2
Head and Neck Cancers

In a pooled analysis of seven case–control studies, there was no evidence of a significantly increased risk of head and neck cancers in individuals who were homozygous or heterozygous for the *ADH3* fast metabolism allele *1 [217]. There was some suggestion of an interaction between genotype and amount of alcohol consumed, with risk increased among heavy drinkers (>60 drinks per week) with the *ADH3*1/*1* genotype, but this was driven by the results of two studies. The three available studies of *ALDH2* genotype and head and neck cancer point to a possible increased risk associated with the inactive 2 allele [223–225], but since the effect appears to be limited to those possessing one, but not two, inactive alleles [217], further investigation is needed.

17.7.3
Esophageal Cancer

Several studies, mainly in Chinese and Japanese populations and mostly involving alcoholics or habitual drinkers, have investigated risk of esophageal cancer according to *ALDH2* and *ADH2* genotypes. Disease risk has consistently been found to be raised (by between 4- and 16-fold) in those carrying the inactive form of *ALDH2* [226]. Similarly, possession of the less active *ADH2* allele has been associated with raised risk in most studies [226]. A single study, in a Japanese population, found that *ADH3* genotype did not influence risk of esophageal cancer [227].

17.7.4
Breast Cancer

Freudenheim *et al.* observed an increased risk of premenopausal breast cancer among women with the *ADH3* fast metabolism genotype (OR *1/*1 versus (*1/*2 or *2/*2) = 2.3, 95% CI 1.2–4.3) [228]. In addition, there was evidence of an interaction with alcohol intake in that disease risk was raised 3.6-fold among women who had the *ADH3*1/*1* genotype and above median alcohol levels compared with those with lower levels of alcohol intake and the *ADH*2* allele. There was no significant relation between *ADH3* and postmenopausal disease, an observation confirmed by Hines *et al.* [229]. In a case-only analysis, Sturmer and colleagues reported a statistically significant interaction between *ADH2* genotype and alcohol intake in relation to breast cancer [230]. The single study of *ALDH2* and breast cancer reported no significant association [231].

17.8
Conclusion

Most of the current evidence on associations between specific types of cancer and genetic polymorphisms thought to influence nutrient metabolism is inconsistent.

This phenomenon has been noted more generally for studies of gene–disease associations, and has prompted considerable concern [34, 232–238]. A number of potential explanations for these inconsistencies have been proposed, including differences between studies in statistical power, different opportunities for bias to occur, and differences in the frequency of genetic variants, patterns of linkage disequilibrium and patterns of dietary exposure between populations. Investigation of the reasons for non-replication will be facilitated by efforts to improve the reporting of such studies [27]. It is critically important to integrate evidence across studies, and to minimize publication bias, as the combination of high-throughput genotyping applied to studies of moderate size, exploratory statistical analysis, and selective reporting has the potential to generate an enormous number of false-positive results [60, 239]. Collaborative networks such as the Human Genome Epidemiology Network (HuGENet™: http://www.cdc.gov/genomics/hugenet/default.htm) should facilitate this. Application of the concept of Mendelian randomization has considerable theoretical appeal in corroborating evidence on the relation between diet and chronic disease. However, to realize its value will require the careful application of epidemiologic principles in study design and integration of evidence across studies. This also applies to data on the functional effects of gene variants, which tend to be limited. As yet, fewer studies have investigated gene–nutrient interaction than have assessed associations with variants postulated to affect nutrient metabolism. Moreover, the methods used to test for the same putative interaction have differed between studies, making it difficult to integrate evidence across studies. There is increasing appreciation of the complexity of the effects of gene variants and their interaction with dietary and other exposures, and the analytical challenges of moving from consideration of one gene and one exposure at a time to the investigation of complex pathways involving multiple genes and multiple exposures [25, 240]. However, it is also possible that a relative small number of common gene variants and even one exposure could account for a substantial proportion of the population-attributable risk of the commoner types of cancer [59].

Although there remains a long way to go before nutrigenomic research can have practical value in disease prevention, the increasing emphasis on collaborative research and the importance of applying rigorous methods for systematic review in integrating evidence should avoid wasteful investment in false leads.

References

1 World Cancer Research Fund in Association with American Institute for Cancer Research (**1997**) *Food, Nutrition and the Prevention of Cancer: a global perspective.* Menasha: American Institute for Cancer Research.

2 World Health Organization (**2003**) *Diet, Nutrition and the Prevention of Chronic Diseases. Report of a Joint WHO/FAO Expert Consultation.* 916, 1–160.

3 Ames, B.N. (**1999**) Cancer prevention and diet: help from single nucleotide polymorphisms. *Proc. Natl Acad. Sci. USA* 96, 12216–12218.

4 Gillies, P.J. (**2003**) Nutrigenomics: the Rubicon of molecular nutrition. *J. Am. Diet. Assoc.* 103, S50–5.

5 MULLER, M., S. KERSTEN (**2003**) Nutrigenomics: goals and strategies. *Nat. Rev. Genet.* 4, 315–322.

6 KAPUT, J., R.L. RODRIGUEZ (**2004**) Nutritional genomics: the next frontier in the postgenomic era. *Physiol. Genomics* 16, 166–177.

7 Genovations. Genovations: Predictive genomics for personalized medicine. (http://www.genovations.com) Accessed 13 May **2005**.

8 Sciona. (http://www.sciona.com) Accessed 13 May **2005**.

9 GeneLink, Inc. (http://www.bankdna. com) Accessed 13 May **2005**.

10 GeneLex. Health & DNA. (http:// www.healthanddna.com) Accessed 13 May **2005**.

11 IARC Working Group (**1998**) *IARC Handbooks of Cancer Prevention*, Volume 2: *Carotenoids*. Lyon: IARC.

12 The Alpha-Tocopherol Beta Carotene Cancer Prevention Study Group (**1994**) The effect of vitamin E and beta carotene on the incidence of lung cancer and other cancers in male smokers. *N. Engl J. Med.* 330, 1029–1035.

13 DE GAETANO, G. and the Collaborative Group of the Primary Prevention Project (**2001**) Low-dose aspirin and vitamin E in people at cardiovascular risk: a randomised trial in general practice. Collaborative Group of the Primary Prevention Project. *Lancet* 357, 89–95.

14 FESKANICH, D., R.G. ZIEGLER, D.S. MICHAUD, E.L. GIOVANNUCCI, F.E. SPEIZER, W.C. WILLETT, G.A. COLDITZ (**2000**) Prospective study of fruit and vegetable consumption and risk of lung cancer among men and women. *J. Natl Cancer Inst.* 92, 1812–1823.

15 MICHELS, K.B., E. GIOVANNUCCI, K.J. JOSHIPURA, B.A. ROSNER, M.J. STAMPFER, C.S. FUCHS, G.A. COLDITZ, F.E. SPEIZER, W.C. WILLETT (**2000**) Prospective study of fruit and vegetable consumption and incidence of colon and rectal cancers. *J. Natl Cancer Inst.* 92, 1740–1752.

16 VOORRIPS, L.E., R.A. GOLDBOHM, D.T. VERHOEVEN, G.A. VAN POPPEL, F. STURMANS, R.J. HERMUS, P.A. VAN DEN BRANDT (**2000**) Vegetable and fruit consumption and lung cancer risk in the Netherlands Cohort Study on diet and cancer. *Cancer Causes Control* 11, 101–115.

17 SMITH-WARNER, S.A., D. SPIEGELMAN, S.S. YAUN, H.O. ADAMI, W.L. BEESON, P.A. VAN DEN BRANDT, A.R. FOLSOM, G.E. FRASER, J.L. FREUDENHEIM, R.A. GOLDBOHM, S. GRAHAM, A.B. MILLER, J.D. POTTER, T.E. ROHAN, F.E. SPEIZER, P. TONIOLO, W.C. WILLETT, A. WOLK, A. ZELENIUCH-JACQUOTTE, D.J. HUNTER (**2001**) Intake of fruits and vegetables and risk of breast cancer. *JAMA* 285, 769–776.

18 FLOOD, A., E.M. VELIE, N. CHATERJEE, A.F. SUBAR, F.E. THOMPSON, J.V.j. LACEY, *et al.* (**2002**) Fruit and vegetable intakes and the risk of colorectal cancer in the Breast Cancer Detection Demonstration Project follow-up cohort. *Am. J. Clin. Nutr.* 75, 936–943.

19 IARC Working Group (**2003**) *IARC Handbook on Cancer Prevention*, Volume 8: *Fruit and vegetables*. Lyon: IARC.

20 LITTLE, J., L. SHARP, S. DUTHIE, S. NARAYANAN (**2003**) Colon cancer and genetic variation in folate metabolism: the clinical bottom line. *J. Nutr.* 133, 3758S–3766S.

21 YOUNGMAN, L.D., B.D. KEAVNEY, A. PALMER, *et al.* (**2000**) Plasma fibrinogen and fibrinogen genotypes in 4685 cases of myocardial infarction and in 6002 controls: test of causality by "Mendelian randomisation". *Circulation* 102, 31–32.

22 DAVEY SMITH, G., S. EBRAHIM (**2003**) "Mendelian randomization": can genetic epidemiology contribute to understanding environmental determinants of disease? *Int. J. Epidemiol.* 32, 1–22.

23 CLAYTON, D., P.M. MCKEIGUE (**2001**) Epidemiological methods for studying genes and environmental factors in complex diseases. *Lancet* 358, 1356–1360.

24 LITTLE, J., M.J. KHOURY (**2003**) Mendelian randomisation: a new spin or real progress? *Lancet* 362, 930–931.

25 KHOURY, M.J., R. MILLIKAN, J. LITTLE,

M. Gwinn (**2004**) The emergence of epidemiology in the genomics age. *Int. J. Epidemiol.* 33, 936–944.

26 Cotton, S.C., L. Sharp, J. Little, N. Brockton (**2000**) Glutathione S-transferase polymorphisms and colorectal cancer. *Am. J. Epidemiol.* 151, 7–32.

27 Little, J., L. Bradley, M.S. Bray, M. Clyne, J. Dorman, D.L. Ellsworth, J. Hanson, M. Khoury, J. Lau, T.R. O'Brien, N. Rothman, D. Stroup, E. Taioli, D. Thomas, H. Vainio, S. Wacholder, C. Weinberg (**2002**) Reporting, appraising, and integrating data on genotype prevalence and gene-disease associations. *Am. J. Epidemiol.* 156, 300–310.

28 Rothman, N., W.F. Stewart, N.E. Caporaso, R.B. Hayes (**1993**) Misclassification of genetic susceptibility biomarkers: implications for case-control studies and cross-population comparisons. *Cancer Epidemiol. Biomarkers Prev.* 2, 299–303.

29 Garcia-Closas, M., S. Wacholder, N. Caporaso, N. Rothman (**2004**) Inference issues in cohort and case-control studies of genetic effects and gene-environment interactions. In: *Human Genome Epidemiology: a Scientific Foundation for Using Genetic Information to Improve Health and Prevent Disease.* Khoury, M.J., J. Little, W. Burke, editors. New York: Oxford University Press, pp. 127–144.

30 Akey, J.M., K. Zhang, M. Xiong, P. Doris, L. Jin (**2001**) The effect that genotyping errors have on the robustness of common linkage-disequilibrium measures. *Am. J. Hum. Genet.* 68, 1447–1456.

31 Colhoun, H.M., P.M. McKeigue, G. Davey Smith (**2003**) Problems of reporting genetic associations with complex outcomes. *Lancet* 361, 865–872.

32 Ioannidis, J.P. (**2003**) Genetic associations: false or true? *Trends Mol. Med.* 9, 135–138.

33 Ardlie, K.G., L. Kruglyak, M. Seielstad (**2002**) Patterns of linkage disequilibrium in the human genome. *Nat. Rev. Genet.* 3, 299–309.

34 Hirschhorn, J.N., K. Lohmueller, E. Byrne, K. Hirschhorn (**2002**) A comprehesive review of genetic association studies. *Genet. Med.* 4, 45–61.

35 Salanti, G., S. Sanderson, J.P. Higgins (**2005**) Obstacles and opportunities in meta-analysis of genetic association studies. *Genet. Med.* 7, 13–20.

36 Knowler, W.C., R.C. Williams, D.J. Pettitt, A.G. Steinberg (**1988**) Gm3;5,13,14 and type 2 diabetes mellitus: an association in American Indians with genetic admixture. *Am. J. Hum. Genet.* 43, 520–526.

37 Gelernter, J., D. Goldman, N. Risch (**1993**) The A1 allele at the D2 dopamine receptor gene and alcoholism: a reappraisal. *JAMA* 269, 1673–1677.

38 Kittles, R.A., W. Chen, R.K. Panguluri, C. Ahaghotu, A. Jackson, C.A. Adebamowo, R. Griffin, T. Williams, F. Ukoli, L. Adams-Campbell, J. Kwagyan, W. Isaacs, V. Freeman, G.M. Dunston (**2002**) CYP3A4-V and prostate cancer in African Americans: causal or confounding association because of population stratification? *Hum. Genet.* 110, 553–560.

39 Thomas, D.C., J.S. Witte (**2002**) Point: population stratification: a problem for case control studies of candidate-gene associations? *Cancer Epidemiol. Biomarkers Prev.* 11, 505–512.

40 Cardon, L.R., L.J. Palmer (**2003**) Population stratification and spurious allelic association. *Lancet* 361, 598–604.

41 Wacholder, S., N. Rothman, N. Caporaso (**2002**) Counterpoint: bias from population stratification is not a major threat to the validity of conclusions from epidemiology studies of common polymorphisms and cancer. *Cancer Epidemiol. Biomarkers Prev.* 11, 513–520.

42 Wacholder, S., N. Rothman, N. Caporaso (**2000**) Population stratification in epidemiologic studies of common genetic variants and

cancer: quantification of bias. *J. Natl Cancer Inst.* 92, 1151–1158.

43 GARTE, S. (**1998**) The role of ethnicity in cancer susceptibility gene polymorphisms: the example of CYP1A1. *Carcinogenesis* 19, 1329–1332.

44 EDLAND, S.D., S. SLAGER, M. FARRER (**2004**) Genetic association studies in Alzheimer's disease research: challenges and opportunities. *Stat. Med.* 23, 169–178.

45 IOANNIDIS, J.P., E.E. NTZANI, T.A. TRIKALINOS (**2004**) "Racial" differences in genetic effects for complex diseases. *Nat. Genet.* 36, 1312–1318.

46 FREEDMAN, M.L., D. REICH, K.L. PENNEY, G.J. McDONALD, A.A. MIGNAULT, N. PATTERSON, S.B. GABRIEL, E.J. TOPOL, J.W. SMOLLER, C.N. PATO, M.T. PATO, T.L. PETRYSHEN, L.N. KOLONEL, E.S. LANDER, P. SKLAR, B. HENDERSON, J.N. HIRSCHHORN, D. ALTSHULER (**2004**) Assessing the impact of population stratification on genetic association studies. *Nat. Genet.* 36, 388–393.

47 KHLAT, M., M.H. CAZES, E. GENIN, M. GUIGUET (**2004**) Robustness of case-control studies of genetic factors to population stratification: magnitude of bias and type I error. *Cancer Epidemiol. Biomarkers Prev.* 13, 1660–1664.

48 MARCHINI, J., L.R. CARDON, M.S. PHILLIPS, P. DONNELLY (**2004**) The effects of human population structure on large genetic association studies. *Nat. Genet.* 36, 512–517.

49 ROZEN, R. (**1997**) Genetic predisposition to hyperhomocysteinemia: deficiency of methylenetetrahydrofolate reductase (MTHFR). *Thromb. Haemost.* 78, 523–526.

50 VAN DER PUT, N.M.J., F. GABREËLS, E.M.B. STEVENS, J.A.M. SMEITINK, F.J.M. TRIJBELS, T.K.A.B. ESKES, L.P. VAN DEN HEUVEL, H.J. BLOM (**1998**) A second common mutation in the methylenetetrahydrofolate reductase gene: an additional risk factor for neural-tube defects? *Am. J. Hum. Genet.* 62, 1044–1051.

51 WEISBERG, I., P. TRAN, B. CHRISTEN-SEN, S. SIBANI, R. ROZEN (**1998**) A second genetic polymorphism in methylenetetrahydrofolate reductase (MTHFR) associated with decreased enzyme activity. *Mol. Genet. Metab.* 64, 169–172.

52 LIEVERS, K.J.A., G.H.J. BOERS, P. VERHOEF, M. DEN HEIJER, L.A.J. KLUIJTMANS, N.M.J. VAN DER PUT, F.J.M. TRIJBELS, H.J. BLOM (**2001**) A second common variant in the methylenetetrahydrofolate reductase (MTHFR) gene and its relationship to MTHFR enzyme activity, homocysteine, and cardiovascular disease risk. *J. Mol. Med.* 79, 522–528.

53 SHARP, L., J. LITTLE (**2004**) Polymorphisms in genes involved in folate metabolism and colorectal neoplasia: a HuGE review. *Am. J. Epidemiol.* 159, 423–443.

54 KHOURY, M.J., W. STEWART, T.H. BEATY (**1987**) The effect of genetic susceptibility on causal inference in epidemiologic studies. *Am. J. Epidemiol.* 126, 561–567.

55 KHOURY, M.J., M.J. ADAMS, JR., W.D. FLANDERS (**1988**) An epidemiologic approach to ecogenetics. *Am. J. Hum. Genet.* 42, 89–95.

56 SING, C.F., J.H. STENGARD, S.L.R. KARDIA (**2003**) Genes, environment, and cardiovascular disease. *Arterioscler. Thromb. Vasc. Biol.* 23, 1190–1196.

57 WOLF, C.R., G. SMITH (**1999**) *Cytochrome P450 CYP2D6.* Lyon: IARC Scientific Publications, 148, pp. 209–229.

58 JOHNSON, W.G. (**1999**) DNA polymorphism-diet-cofactor-development hypothesis and the gene-teratogen model for schizophrenia and other developmental disorders. *Am. J. Med. Genet.* 88, 311–323.

59 KHOURY, M.J., Q. YANG, M. GWINN, J. LITTLE, W.D. FLANDERS (**2004**) An epidemiologic assessment of genomic profiling for measuring susceptibility to common diseases and targeting interventions. *Genet. Med.* 6, 38–47.

60 WACHOLDER, S., S. CHANOCK, M. GARCIA-CLOSAS, L. EL GHORMLI, N. ROTHMAN (**2004**) Assessing the probability that a positive report is

false: an approach for molecular epidemiology studies. *J. Natl Cancer Inst.* 96, 434–442.

61 SMITH, P.G., N.E. DAY (**1984**) The design of case–control studies: the influence of confounding and interaction effects. *Int. J. Epidemiol.* 13, 356–365.

62 GARCIA-CLOSAS, M., J.H. LUBIN (**1999**) Power and sample size calculations in case-control studies of gene-environment interactions: comments on different approaches. *Am. J. Epidemiol.* 149, 689–692.

63 LITTLE, J. (**2004**) Reporting and review of human genome epidemiology studies. In: *Human Genome Epidemiology: a Scientific Foundation for Using Genetic Information to Improve Health and Prevent Disease.* KHOURY, M.J., J. LITTLE, W. BURKE, editors. New York: Oxford University Press, pp. 168–192.

64 DUTHIE, S.J. (**1999**) Folic acid instability and cancer: mechanisms of DNA instability. *Br. Med. Bull.* 55, 578–592.

65 MCCULLOUGH, M.L., E.L. GIOVANNUCCI (**2004**) Diet and cancer prevention. *Oncogene* 23, 6349–6364.

66 SHARP, L., J. LITTLE. in press. Epidemiology of colorectal cancer. In: *Colon Cancer.* CASSIDY, J., P. JOHNSTON, E. VAN CUSTEM, editors. New York: Marcel Dekker.

67 SHARP, L., J. LITTLE (**2004**) Methylenetetrahydrofolate reductase gene (*MTHFR*), folate, and colorectal neoplasia. In: *Human Genome.* KHOURY, M.J., J. LITTLE, W. BURKE, editors. New York: Oxford University Press, pp. 333–364.

68 CURTIN, K., J. BIGLER, M.L. SLATTERY, B. CAAN, J.D. POTTER, C.M. ULRICH (**2004**) MTHFR *C677T* and *A1298C* polymorphisms: diet, estrogen and risk of colon cancer. *Cancer Epidemiol. Biomarkers Prev.* 13, 285–292.

69 JIANG, Q.T., K. CHEN, X.Y. MA, X.P. MIAO, K.Y. YAO, W.P. YU, L.Y. LI, Y.M. ZHU, H.G. ZHOU (**2004**) A case-control study on the polymorphisms of methylenetetrahydrofolate reductases, drinking interaction and susceptibility in colorectal cancer.

Zhonghua Liu Xing Bing Xue Za Zhi 25, 612–616.

70 KIM, D., Y. AHN, B. LEE, E. TSUJI, C. KIYOHARA, S. KONO (**2004**) Methylenetetrahydrofolate reductase polymorphism, alcohol intake, and risks of colon and rectal cancers in Korea. *Cancer Lett.* 216, 199–205.

71 YIN, G., S. KONO, K. TOYOMURA, T. HAGIWARA, J. NAGANO, T. MIZOUE, R. MIBU, M. TANAKA, Y. KAKEJI, Y. MAEHARA, T. OKAMURA, K. IKEJIRI, K. FUTAMI, Y. YASUNAMI, T. MAEKAWA, K. TAKENAKA, H. ICHIMIYA, N. IMAIZUMI (**2004**) *Methylenetetrahydrofolate reductase C677T and A1298C polymorphisms and colorectal cancer: the Fukuoka Colorectal Cancer Study.* *Cancer Sci.* 95, 908–913.

72 GIOVANNUCCI, E., J. CHEN, S.A. SMITH-WARNER, E.B. RIMM, C.S. FUCHS, C. PALOMEQUE, W.C. WILLETT, D.J. HUNTER (**2003**) Methylenetetrahydrofolate reductase, alcohol dehydrogenase, diet and risk of colorectal adenomas. *Cancer Epidemiol. Biomarkers Prev.* 12, 970–979.

73 BOYAPATI, S.M., R.M. BOSTICK, K.A. MCGLYNN, M.F. FINA, W.M. ROUFAIL, K.R. GEISINGER, J.R. HEBERT, A. COKER, M. WARGOVICH (**2004**) Folate intake, MTHFR C677T polymorphism, alcohol consumption, and risk for sporadic colorectal adenoma (United States). *Cancer Causes Control* 15, 493–501.

74 ULRICH, C.M., E. KAMPMAN, J. BIGLER, S.M. SCHWARTZ, C. CHEN, R. BOSTICK, L. FOSDICK, S.A.A. BERESFORD, Y. YASUI, J.D. POTTER (**1999**) Colorectal adenomas and the C677T MTHFR polymorphism: evidence for gene-environment interaction? *Cancer Epidemiol. Biomarkers Prev.* 8, 659–668.

75 LEVINE, A.J., K.D. SIEGMUND, C.M. ERVIN, A. DIEP, E.R. LEE, H.D. FRANKL, R.W. HAILE (**2000**) The methylenetetrahydrofolate reductase 677CrT polymorphism and distal colorectal adenoma risk. *Cancer Epidemiol. Biomarkers Prev.* 9, 657–663.

76 CHEN, J., E. GIOVANNUCCI, K. KELSY, E.B. RIMM, M.J. STAMPFER, G.A.

Colditz, D. Spiegelman, W.C. Willett, D.J. Hunter (**1996**) A methylenetetrahydrofolate reductase polymorphism and the risk of colorectal cancer. *Cancer Res.* 56, 4862–4864.

77 Narayanan, S., J. McConnell, J. Little, L. Sharp, C.J. Piyathilake, H. Powers, G. Basten, S.J. Duthie (**2004**) Associations between two common variants C677T and A1298C in the methylenetetrahydrofolate reductase gene and measures of folate metabolism and DNA stability (strand breaks, misincorporated uracil and DNA methylation status) in human lymphocytes *in vivo. Cancer Epidemiol. Biomarkers Prev.* 13, 1436–1443.

78 Zhang, S.M. (**2004**) Role of vitamins in the risk, prevention, and treatment of breast cancer. *Curr. Opin. Obstet. Gynecol.* 16, 19–25.

79 Herbert, V. (**1987**) Recommended dietary intakes (RDI) of folate in humans. *Am. J. Clin. Nutr.* 45, 661–670.

80 Gershoni-Baruch, R., E. Dagan, D. Israeli, L. Kasinetz, E. Kadouri, E. Friedman (**2000**) Association of the C677T polymorphism in the MTHFR gene with breast and/or ovarian cancer risk in Jewish women. *Eur. J. Cancer* 36, 2313–2316.

81 Campbell, I.G., S.W. Baxter, D.M. Eccles, D.Y.H. Choong (**2002**) Methylenetetrahydrofolate reductase polymorphism and susceptibility to breast cancer. *Breast Cancer Res.* 4, 1–4.

82 Sharp, L., J. Little, A.C. Schofield, E. Pavlidou, S.C. Cotton, Z. Miedzybrodzka, J.O.C. Baird, N.E. Haites, S.D. Heys, D.A. Grubb (**2002**) Folate and breast cancer: the role of polymorphisms in methylenetetra-hydrofolate reductase (MTHFR). *Cancer Lett.* 181, 65–71.

83 Ergul, E., A. Sazci, Z. Utkan, N.Z. Canturk (**2003**) Polymorphisms in the MTHFR gene are associated with breast cancer. *Tumor Biol.* 24, 286–290.

84 Langsenlehner, U., P. Krippl, W. Renner, B. Yazdani-Biuki, G. Wolf, T.C. Wascher, B. Paulweber, W.

Weitzer, H. Samonigg (**2003**) The common 677C > T gene polymorphism of methylenetetrahydrfolate reductase gene is not associated with breast cancer risk. *Breast Cancer Res. Treat.* 81, 169–172.

85 Semenza, J.C., R.J. Delfino, A. Ziogas, H. Anton-Culver (**2003**) Breast cancer risk and methylenetetra-hydrofoilate reductase polymorphism. *Breast Cancer Res. Treat.* 77, 217–223.

86 Beilby, J., D. Ingram, R. Hahnel, E. Rossi (**2004**) Reduced breast cancer risk with increasing serum folate in a case-control study of the C677T genotype of the methylenetetrahydrofolate reductase gene. *Eur. J. Cancer* 40, 1250–1254.

87 Forsti, A., S. Angelini, F. Festa, S. Sanyal, Z. Zhang, E. Grzybowska, J. Pamula, W. Pekala, H. Zientek, K. Hemminki, R. Kumar (**2004**) Single nucleotide polymorphisms in breast cancer. *Oncol. Rep.* 11, 917–922.

88 Grieu, F., B. Powell, J. Beilby, B. Iacopetta (**2004**) Methylenetetra-hydrofolate reductase and thymidylate synthase polymorphisms are not associated with breast cancer risk of phenotype. *Anticancer Res.* 24, 3215–3219.

89 Lee, S., D. Kang, H. Nishio, M.J. Lee, D. Kim, W. Han, K. Yoo, S. Ahn, K. Choe, A. Hirvonen, D. Noh (**2004**) Methylenetetrahydrofolate reductase polymorphism, diet and breast cancer in Korean women. *Exp. Mol. Med.* 36, 116–121.

90 Le Marchand, L., C.A. Haiman, L.R. Wilkens, L.N. Kolonel, B.E. Henderson (**2004**) MTHFR polymorphism, diet, HRT and breast cancer risk: the multiethnic cohort study. *Cancer Epidemiol. Biomarkers Prev.* 13, 2071–2077.

91 Lin, W., Y.-C. Chou, M.-H. Wu, H. Huang, Y. Jeng, C. Wu, C. Yu, J. Yu, S. You, T. Chu, C. Chen, C.-A. Sun (**2004**) The MTHFR C677T polymorphism, estrogen exposure and breast cancer risk: a nested case-control study in Taiwan. *Anticancer Res.* 24, 3863–3868.

92 Qi, J., X.P. Miao, W. Tan, C.Y. Yu, G. Liang, W.F. Lu, D.X. Lin (**2004**)

Association between genetic polymorphisms in methylenetetrahydrofolate reductase and risk of breast cancer. *Zhonghua Zhong Liu Za Zhi* 26, 287–289.

93 SHRUBSOLE, M.J., Y. GAO, Q. CAI, X.O. SHU, Q. DAI, J.R. HEBERT, F. JIN, W. ZHENG (2004) MTHFR polymorphisms, dietary folate intake, and breast cancer risk: results from the Shanghai Breast Cancer Study. *Cancer Epidemiol. Biomarkers Prev.* 13, 190–196.

94 CHEN, J., M.D. GAMMON, W. CHAN, C. PALOMEQUE, J.G. WETMUR, G.C. KABAT, S.L. TEITELBAUM, J.A. BRITTON, M.B. TERRY, A.I. NEUGUT, R.M. SANTELLA (2005) One-carbon metabolism, MTHFR polymorphisms, and risk of breast cancer. *Cancer Res.* 65, 1606–1614.

95 BUTTERWORTH, C.E., JR., K.D. HATCH, H. GORE, H. MUELLER, C.L. KRUMDIECK (1982) Improvement in cervical dysplasia associated with folic acid therapy in users of oral contraceptives. *Am. J. Clin. Nutr.* 35, 73–82.

96 BUTTERWORTH, C.E., JR., K.D. HATCH, S.J. SOONG, P. COLE, T. TAMURA, H.E. SAUBERLICH, M. BORST, M. MACALUSO, V. BAKER (1992) Oral folic acid supplementation for cervical dysplasia: a clinical intervention trial. *Am. J. Obstet. Gynecol.* 166, 803–809.

97 LITTLE, J. (1995) Is folic acid pluripotent? A review of the associations with congenital anomalies, cancer and other diseases. In: *Drugs, Diet and Disease*, Volume 1: *Mechanistic Approaches to Cancer.* IOANNIDES, C., D.F.V. LEWIS, editors. New York: Ellis Horwood, pp. 259–308.

98 EICHHOLZER, M., J. LUTHY, U. MOSER, B. FOWLER (2001) Folate and the risk of colorectal, breast and cervix cancer: the epidemiological evidence. *Swiss Med. Weekly* 131, 539–549.

99 RAMPERSAUD, G.C., L.B. BAILEY, G.P. KAUWELL (2002) Relationship of folate to colorectal and cervical cancer; review and recommendations for practitioners. *J. Am. Diet. Assoc.* 102, 1273–1282.

100 SHANNON, J., D.B. THOMAS, R.M. RAY, M. KESTIN, A. KOETSAWANG, S. KOETSAWANG, K. CHITNARONG, N. KIVIAT, J. KUYPERS (2002) Dietary risk factors for invasive and in-situ cervical carcinomas in Bangkok, Thailand. *Cancer Causes Control* 13, 691–699.

101 ZIEGLER, R.G., S.J. WEISTEIN, T.R. FEARS (2002) Nutritional and genetic inefficiencies in one-carbon metabolism and cervical cancer risk. *J. Nutr.* 132, 2345s–2349s.

102 HERNANDEZ, B.Y., K. MCDUFFIE, L.R. WILKENS, L. KAMEMOTO, M.T. GOODMAN (2003) Diet and premalignant lesions of the cervix: evidence of a protective role for folate, riboflavin, thiamin, and vitamin B12. *Cancer Causes Control* 14, 859–870.

103 KWANBUNJAN, K., P. SAENGKAR, C. CHEERAMAKARA, W. THANOMSAK, W. BENJACHAI, S. TANGJITGAMOL, P. LAISUPASIN, K. SONGMUAENG (2004) Folate status of Thai women cervical dysplasia. *Asia Pacif. J. Clin. Nutr.* 13, S171.

104 WEINSTEIN, S.J., R.G. ZIEGLER, J. SELHUB, T.R. FEARS, H.D. STRICKLER, L.A. BRINTON, R.F. HAMMAN, R.S. LEVINE, K. MALLIN, P.D. STOLLEY (2001) Elevated serum homocysteine levels and increased risk of invasive cervical cancer in US women. *Cancer Causes Control* 12, 317–324.

105 PIYATHILAKE, C.J., M. MACALUSO, G.L. JOHANNING, M. WHITESIDE, D.C. HEIMBURGER, A. GIULIANO (2000) Methylenetetrahydrofolate reductase (MTHFR) polymorphism increases the risk of cervical intraepithelial neoplasia. *Anticancer Res.* 20, 1751–1757.

106 GOODMAN, M.T., K. MCDUFFIE, B. HERNANDEZ, L.R. WILKENS, C.C. BERTRAM, J. KILLEEN, L. LE MARCHAND, J. SELHUB, S. MURPHY, T.A. DONLON (2001) Association of methylenetetrahydrofolate reductase polymorphism C677T and dietary folate with the risk of cervical dysplasia. *Cancer Epidemiol. Biomarkers Prev.* 10, 1275–1280.

107 LAMBROPOULOS, A.F., T. AGORASTOS, Z.J. FOKA, S. CHRISAFI, T.C.

Constantinidis, J. Bontis, A. Kotsis (**2003**) Methylenetetrahydrofolate reductase polymorphism C677T is not associated to the risk of cervial dysplasia. *Cancer Lett.* 191, 187–191.

108 Henao, O.L., C.J. Piyathilake, J.W. Waterbor, E. Funkhouser, G.L. Johanning, D.C. Heimburger, E.E. Partridge (**2005**) Women with polymorphisms of methylenetrahydrofolate reductase (MTHFR) and methionine synthase (MS) and less likely to have cervical intrepithelial neoplasia (CIN) 2 or 3. *Int. J. Cancer* 113, 991–997.

109 Sull, J.W., S.H. Jee, S. Yi, J.E. Lee, J.S. Park, S. Kim, H. Ohrr (**2004**) The effect of methylenetetrahydrofolate reductase polymorphism C677T on cervical cancer in Korean women. *Gynecol. Oncol.* 95, 557–563.

110 Zoodsma, M., I.M. Nolte, M. Schipper, E. Oosterom, G. van der Steege, E.G. de Vries, G.J. te Meerman, A.G. van der Zee (**2005**) Methylenetetrahydrofolate reductase (MTHFR) and susceptibility for (pre)neoplastic cervical disease. *Hum. Genet.* 116, 247–254.

111 Gerhard, D.S., L.T. Nguyen, Z.Y. Zhang, I.B. Borecki, B.I. Coleman, J.S. Rader (**2003**) A relationship between methylenetetrahydrofolate reductase variants and the development of invasive cervical cancer. *Gynecol. Oncol.* 90, 560–565.

112 Zhang, Z.F., R.C. Kurtz, G.P. Yu, M. Sun, N. Gargon, M. Karpeh, Jr., J.S. Fein, S. Harlap (**1997**) Adenocarcnomas of the esophagus and gastric cardia: the role of diet. *Nutr. Cancer* 27, 298–309.

113 Bollschweiler, W., E. Wolfgarten, T. Nowroth, U. Rosendahl, S.P. Monig, A.H. Holscher (**2001**) Vitamin intake and risk of subtypes of esophageal cancer in Germany. *J. Cancer Res. Clin. Oncol.* 128, 575–580.

114 Mayne, S.T., H.A. Risch, R. Dubrow, W.H. Chow, M.D. Gammon, T.L. Vaughan, D.C. Farrow, J.B. Schoenberg, J.L. Stanford, H. Ahsan, A.B. West, H. Rotterdam, W.J. Blot, J.F. Fraumeni, Jr. (**2001**) Nutrient intake and risk of sub-types of esophageal and gastric cancer. *Cancer Epidemiol. Biomarkers Prev.* 10, 1055–1062.

115 Chen, H., K.L. Tucker, B.I. Graubard, E.F. Heineman, R.S. Markin, N.A. Potischman, R.M. Russell, D.D. Weisenburger, M.H. Ward (**2002**) Nutrient intakes and adenocarcinoma of the esophagus and distal stomach. *Nutr. Cancer* 42, 33–40.

116 Song, C., D. Xing, W. Tan, Q. Wei, D. Lin (**2001**) Methylenetetrahydrofolate reductase polymorphisms increase risk of esophageal squamous cell carcinoma in a Chinese population. *Cancer Res.* 15, 3272–3275.

117 Stolzenberg-Solomon, R.Z., Y.L. Qiao, C.C. Abnet, D.L. Ratnasinghe, S.M. Dawsey, Z.W. Dong, P.R. Taylor, S.D. Mark (**2003**) Esophageal and gastric cardia cancer risk and folate- and vitamin B(12)-related polymorphisms in Linxian, China. *Cancer Epidemiol. Biomarkers Prev.* 12, 1222–1226.

118 Zhang, J., Y. Cui, G. Kuang, Y. Li, N. Wang, R. Wang, R. Guo, D. Wen, L. Wei, F. Yu, S. Wang (**2004**) Association of the thymidylate synthase polymorphisms with esophageal squamous cell carcinoma and gastric cardiac adenocarcinoma. *Carcinogen* 25, 2479–2485.

119 Gao, C.M., T. Toshiro, J.Z. Wu, H.X. Cao, Y.T. Liu, J.H. Ding, S.P. Li, P. Su, X. Hu, H.T. Kai, T. Kazuo (**2004**) A case-control study on the polymorphisms of methylenetetrahydrofolate reductase 1298A → C and susceptibility of esophageal cancer. *Zhonghua Liu Xing Bing Xue Za Zhi* 25, 341–345.

120 Thompson, J.R., P.F. Gerald, M.L.N. Willoughby, B.K. Armstrong (**2001**) Maternal folate supplementation in pregnancy and protection against acute lymphoblastic leukaemia in childhood: a case-control study. *Lancet* 358, 1935–1940.

121 Robien, K., C.M. Ulrich (**2003**) 5,10-methylenetetrahydrofolate reductase polymorphisms and leukemia risk. A HuGE mini-review. *Am. J. Epidemiol.* 157, 571–582.

122 Matsuo, K., R. Suzuki, N. Hamajima, M. Ogura, Y. Kagami, H. Taji, E. Kondoh, S. Maeda, S. Asakura, S. Kaba, S. Nakamura, M. Seto, Y. Morishima, K. Tajima (**2001**) Association between polymorphisms of folate- and methionine-metabolizing enzymes and susceptibility to malignant lymphoma. *Blood* 97, 3205–3209.

123 Gemmati, D., A. Ongaro, G.L. Scapoli, M. Della Porta, S. Tognazzo, M.L. Serino, E. Di Bona, F. Rodeghiero, G. Gilli, R. Reverberi, A. Caruso, M. Pasello, A. Pellati, M. De Mattei (**2004**) Common gene polymorphisms in the metabolic folate and methylation pathway and the risk of acute lymphoblastic leukemia and non-Hodgkin's lymphoma in adults. *Cancer Epidemiol. Biomarkers Prev.* 13, 787–794.

124 Krajinovic, M., S. Lamothe, D. Labuda, E. Lemieux-Blanchard, Y. Theoret, A. Moghrabi, D. Sinnett (**2004**) Role of MTHFR genetic polymorphisms in the susceptibility to childhood acute lymphoblastic leukemia. *Blood* 103, 252–257.

125 Chiusolo, P., G. Reddiconto, G. Cimino, S. Sica, A. Fiorini, G. Farina, A. Vitale, F. Sora, L. Laurenti, F. Bartolozzi, P. Fazi, F. Mandelli, G. Leone (**2004**) Methylenetetrahydrofolate reductase genotypes do not play a role in acute lymphoblastic leukemia pathogenesis in the Italian population. *Haematologica* 89, 139–144.

126 Chen, J., E. Giovannucci, S.E. Hankinson, J. Ma, W.C. Willett, D. Spiegelman, K.T. Kelsey, D.J. Hunter (**1998**) A prospective study of methylenetetrahydrofolate reductase and methionine synthase gene polymorphisms, and risk of colorectal adenoma. *Carcinogenesis* 19, 2129–2132.

127 Ma, J., M.J. Stampfer, B. Christensen, E. Giovannucci, D.J. Hunter, J. Chen, W.C. Willett, J. Selhub, C.H. Hennekens, R. Gravel, R. Rozen (**1999**) A polymorphism of the methionine synthase gene: assocation with plasma folate, vitamin B12, homocyst(e)ine, and colorectal cancer risk. *Cancer Epidemiol. Biomarkers Prev.* 8, 825–829.

128 Le Marchand, L., T. Donlon, J.H. Hankin, L.N. Kolonel, L.R. Wilkens, A. Seifried (**2002**) B-vitamin intake, metabolic genes, and colorectal cancer risk (United States). *Cancer Causes Control* 13, 239–248.

129 Goode, E.L., C.M. Ulrich, J. Bigler, R.M. Bostick, L. Fosdick, J.D. Potter (**2002**) Methionine synthase D919G polymorphism: possible interaction with dietary nutrients and association with colorectal adenoma risk. *Proc. Am. Assoc. Cancer Res.* 43, 659–660.

130 Ulvik, A., S.E. Vollset, S. Hansen, R. Gislefoss, E. Jellum, P.M. Ueland (**2004**) Colorectal cancer and the methylenetetrahydrofolate reductase 677C → T and methionine synthase 2756A → G polymorphisms: a study of 2,168 case-control pairs from the JANUS cohort. *Cancer Epidemiol. Biomarkers Prev.* 13, 2175–2180.

131 Shannon, B., S. Gnanasampanthan, J. Beilby, B. Iacopetta (**2002**) A polymorphism in the methylene-tetrahydrofolate reductase gene predisposes to colorectal cancers with mircrosatellite instability. *Gut* 50, 520–524.

132 Lenz, H.J., W. Zhang, S. Zahedy, J. Gil, M. Yu, J. Stoehlmacher (**2002**) A 6 base-pair deletion in the 3 UTR of the thymidylate synthase (TS) gene predicts TS mRNA expression in colorectal tumours: a possible candidate gene for colorectal cancer risk. *Proc. Am. Assoc. Cancer Res.* 43, 660.

133 Ulrich, C.M., J. Bigler, R. Bostick, L. Fosdick, J.D. Potter (**2002**) *Thymidylate synthase* promoter polymorphism, interaction with folate intake, and risk of colorectal adenomas. *Cancer Res.* 62, 3361–3364.

134 Chen, J., D.J. Hunter, M.J. Stampfer, J.G. Wetmur, C. Kyte, J. Selhub, J. Ma (**2002**) A novel polymorphism in the thymidylate synthase gene promoter influences plasma folate level and may modify the risk of colorectal cancer in a prospective study. *Proc. Am. Assoc. Cancer Res.* 43, 659–659.

135 CHEN, J., D.J. HUNTER, M.J. STAMPFER, C. KYTE, W. CHAN, J.G. WETMUR, R. MOSIG, J. SELHUB, J. MA (2003) Polymorphism in the thymidylate synthase promoter enhancer region modifies the risk and survival of colorectal cancer. *Cancer Epidemiol. Biomarkers Prev.* 12, 958–962.

136 ADLEFF, V., E. HITRE, I. KOVES, Z. OROSZ, A. HAJNAL, J. KRALOVANSZKY (2004) Heterozygote deficiency in thymidylate snthase enhancer region polymorphism genotype distribution in Hungarian colorectal cancer patients. *Int. J. Cancer* 108, 852–856.

137 GAO, C.M., T. TAKEZAKI, J.Z. WU, Y.T. LIU, J.H. DING, S.P. LI, P. SU, X. HU, H.T. KAI, Z.Y. LI, K. MATSUO, N. HAMAJIMA, H. SUGIMURA, K. TAJIMA (2004) Polymorphisms in thymidylate synthase and methylenetetrahydrofolate reductase genes and the susceptibility to esophageal and stomach cancer with smoking. *Asian Pacif. J. Cancer Prev.* 5, 133–138.

138 TAN, W., X. MIAO, L. WANG, C. YU, P. XIONG, G. LIANG, T. SUN, Y. ZHOU, X. ZHANG, H. LI, D. LIN (2005) Significant increase in risk of gastroesophageal cancer is associated with interaction between promoter polymorphisms in thymidylate synthase and serum folate status. *Carcinogen* 26, 1430–1435.

139 SKIBOLA, C.F., M.T. SMITH, A. HUBBARD, B. SHANE, A.C. ROBERTS, G.R. LAW, S. ROLLINSON, E. ROMAN, R.A. CARTWRIGHT, G.J. MORGAN (2002) Polymorphisms in the thymidylate synthase and serine hydroxymethyltransferase genes and risk of adult acute lymphocytic leukemia. *Blood* 99, 3786–3791.

140 MATSUO, K., N. HAMAJIMA, R. SUZUKI, M. OGURA, Y. KAGAMI, H. TAJI, T. YASUE, N.E. MUELLER, S. NAKAMURA, M. SETO, Y. MORISHIMA, K. TAJIMA (2004) Methylenetetrahydrofolate reductase gene (MTHFR) polymor-

phisms and reduced risk of malignant lymphoma. *Am. J. Hematol.* 77, 351–357.

141 SKIBOLA, C.F., M.S. FORREST, F. COPPEDE, L. AGANA, A. HUBBARD, M.T. SMITH, P.M. BRACCI, E.A. HOLLY (2004) Polymorphisms and haplotypes in folate-metabolizing genes and risk of non-Hodgkin lymphoma. *Blood* 104, 2155–2162.

142 SINHA, R. (2002) An epidemiologic approcah to studying heterocyclic amines. *Mutat. Res.* 506–7, 197–204.

143 YIN, Z.L., J.E. DAHLSTROM, D.G. LE COUTEUR, P.G. BOARD (2001) Immunohistochemistry of omega class glutathione S-transferase in human tissues. *J. Histochem. Cytochem.* 49, 983–987.

144 BOARD, P., M. COGGAN, P. JOHNSTON, V. ROSS, T. SUZUKI, G. WEBB (1990) Genetic heterogeneity of the human glutathione transferases: a complex of gene families. *Pharmacol. Ther.* 48, 357–369.

145 PEMBLE, S., K.R. SCHROEDER, S.R. SPENCER, D.J. MEYER, E. HALLIER, H.M. BOLT, B. KETTERER, J.B. TAYLOR (1994) Human glutathione S-transferase theta (GSTT1): cDNA cloning and the characterization of a genetic polymorphism. *Biochem. J.* 300, 271–276.

146 BROCKTON, N., J. LITTLE, L. SHARP, S.C. COTTON (2000) N-acetyltransferase polymorphisms and colorectal cancer: a review. *Am. J. Epidemiol.* 151, 846–861.

147 HEIN, D. (2002) Molecular genetics and function of NAT1 and NAT2, role in aromatic amine metabolism and carcinogenesis. *Mutat. Res.* 506–507, 65–77.

148 SANDHU, M.S., I.R. WHITE, K. MCPHERSON (2001) Systematic review of the prospective cohort studies on meat consumption and colorectal cancer risk: a meta-analytical approach. *Cancer Epidemiol. Biomarkers Prev.* 10, 439–446.

149 NORAT, T., A. LUKANOVA, P. FERRARI, E. RIBOLI (2002) Meat consumption and colorectal cancer risk: dose-response meta-analysis of epidemio-

logical studies. *Int. J. Cancer* 98, 241–256.

150 NOWELL, S., B. COLES, R. SINHA, S. MACLEOD, D.L. RATNASINGHE, C. STOTTS, F.F. KADLUBAR, C.B. AMBROSONE, N.P. LANG (**2002**) Analysis of total meat intake and exposure to individual heterocyclic amines in a case-control study of colorectal cancer: contribution of metabolic varaition to risk. *Mutat. Res. Fund. Mol. Mech. Mutagen.* 506–507, 175–185.

151 MURTAGH, M.A., K. MA, C. SWEENEY, B.J. CAAN, M.L. SLATTERY (**2004**) Meat consumption patterns and prepartion, genetic variants of metabolic enzymes, and their association with rectal cancer in men and women. *J. Nutr.* 134, 776–784.

152 ISHIBE, N., R. SINHA, D.W. HEIN, M. KULLDORFF, P. STRICKLAND, A.J. FRETLAND, W. CHOW, F.F. KADLUBAR, N.P. LANG, N. ROTHMAN (**2002**) Genetic polymorphisms in hetero-cyclic amine metabolism and risk of adenomas. *Pharmacogenetics* 12, 145–150.

153 WELFARE, M.R., J. COOPER, M.F. BASSENDINE, A.K. DALY (**1997**) Relationship between acetylator status, smoking, diet and colorectal cancer risk in the north-east of England. *Carcinogenesis* 18, 1351–1354.

154 GERTIG, D.M., M. STAMPFER, C.H. HAIMAN, C.H. HENNEKENS, K. KELSEY, D.J. HUNTER (**1998**) Glutathione S-Transferase GSTM1 and GSTT1 polymorphisms and colorectal cancer risk: a prospective study. *Cancer Epidemiol. Biomarkers Prev.* 7, 1001–1005.

155 TURNER, F., G. SMITH, C. SACHSE, T. LIGHTFOOT, R.C. GARNER, C.R. WOLF, D. FORMAN, D.T. BISHOP, J.H. BARRETT (**2004**) Vegetable, fruit and meat consumption and potential risk modifying genes in relation to colorectal cancer. *Int. J. Cancer* 112, 259–264.

156 CORTESSIS, V., K. SIEGMUND, Q. CHEN, N. ZHOU, A. DIEP, H. FRANKL, E. LEE, Q.S. ZHU, R. HAILE, D. LEVY (**2001**) A case-control study of

microsomal epoxide hydrolase, smoking, meat consumption, glutathione S-transferase M3, and risk of colorectal adenomas. *Cancer Res.* 61, 2381–2385.

157 KAMPMAN, E., M.L. SLATTERY, J. BIGLER, M. LEPPERT, W. SAMOWITZ, B.J. CAAN, J.D. POTTER (**1999**) Meat consumption, genetic susceptibility, and colon cancer risk: A United States multicenter case-control study. *Cancer Epidemiol. Biomarkers Prev.* 8, 15–24.

158 TIEMERSMA, E.W., E. KAMPMAN, H.B. BUENO DE MESQUITA, A. BUNSCHOTEN, E.M. VAN SCHOTHORST, F.J. KOK, D. KROMHOUT (**2002**) Meat consumption, cigarette smoking, and genetic susceptibility in the etiology of colorectal cancer: results from a Dutch prospective study. *Cancer Causes Control* 13, 383–393.

159 BARRETT, J.H., G. SMITH, R. WAXMAN, N. GOODERHAM, T. LIGHTFOOT, R.C. GARNER, K. AUGUSTSSON, C.R. WOLF, D.T. BISHOP, D. FORMAN, A.R. BOOBIS, R. COLIN, C. SACHSE (**2003**) Investigation of intercation between N-acetyltransferase 2 and heterocylic amines as potential risk factors for colorectal cancer. *Carcinogenesis* 24, 275–282.

160 MISSMER, S.A., S.A. SMITH-WARNER, D. SPIEGELMAN, S.S. YAUN, H.O. ADAMI, W.L. BEESON, P.A. VAN DEN BRANDT, G.E. FRASER, J.L. FREUDEN-HEIM, R.A. GOLDBOHM, S. GRAHAM, L.H. KUSHI, A.B. MILLER, J.D. POTTER, T.E. ROHAN, F.E. SPEIZER, P. TONIOLO, W.C. WILLETT, A. WOLK, A. ZELENIUCH-JACQUOTTE, D.J. HUNTER (**2002**) Meat and dairy food consump-tion and breast cancer: a pooled analy-sis of cohort studies. *Int. J. Epidemiol.* 31, 78–85.

161 KNEKT, P., G. STENECK, R. JARVINEN, T. HAKULINNEN, A. AROMAA (**1994**) Intake of fried meat and risk of cancer: a follow-up study in Finland. *Int. J. Cancer* 59, 756–760.

162 DE STEFANI, E., A. RONCO, M. MENDILAHARSU, M. GUIDOBONO, H. DENEO-PELLEGRINI (**1997**) Meat intake, heterocyclic amines, and risk of breast cancer: a case-control study in Uruguay. *Cancer Epidemiol. Biomarkers Prev.* 6, 573–581.

163 ZHENG, W., D.R. GUSTAFSON, R. SINHA, J.R. CERHAN, D. MOORE, C. HONG, K.E. ANDERSON, L.H. KUSHI, T.A. SELLERS, A.R. FOLSOM (1998) Well-done meat intake and the risk of breast cancer. *J. Natl Cancer Inst.* 90, 1724–1729.

164 DAI, Q., X. XHU, F. JIN, Y. GAO, Z. RUAN, W. ZHENG (2002) Consumption of animal foods, cooking methods, and risk of breast cancer. *Cancer Epidemiol. Biomarkers Prev.* 11, 801–808.

165 IMAIDA, K., A. HAGIWARA, H. YADA, T. MASUI, R. HASEGAWA, M. HIROSE, T. SUGIMURA, N. ITO, T. SHIRAI (1996) Dose-dependent induction of mammary carcinomas in female Sprague-Dawley rats with 2-amino-1-methyl-6-phenylimidazol[4,5-b]pyridine. *Jap. J. Cancer Res.* 87, 1116–1120.

166 ZHU, J., P. CHANG, M.L. BONDY, A.A. SAHIN, S.E. SINGLETARY, S. TAKAHASHI, T. SHIRAI, D. LI (2003) Detection of 2-amino-1-methyl-6-phenylimidazo[4,5-b]-pyridine-DNA adducts in normal breast tissues and risk of breast cancer. *Cancer Epidemiol. Biomarkers Prev.* 12, 830–837.

167 ZHENG, W., W. WEN, D.R. GUSTAFSON, M. GROSS, J.R. CERHAN, A.R. FOLSOM (2002) GSTM1 and GSTT1 polymorphisms and postmenopausal breast cancer risk. *Breast Cancer Res. Treat.* 74, 9–16.

168 VAN DER HEL, O.L., P.H.M. PEETERS, D.W. HEIN, M.A. DOLL, D.E. GROBBEE, M. OCKE, H. BAS BUENO DE MESQUITA (2004) GSTM1 null genotype, red meat consumption and breast cancer risk. *Cancer Causes Control* 13, 295–303.

169 ZHENG, W., D. XIE, J.R. CERHAN, T.A. SELLERS, W. WEN, A.R. FOLSOM (2001) Sulfotransferase 1A1 polmorphsims, endogenous estrogen exposure, well-done meat intake, and breast cancer risk. *Cancer Epidemiol. Biomarkers Prev.* 10, 89–94.

170 DIETZ, A.C., W. ZHENG, M.A. LEFF, M. GROSS, W. WEN, M.A. DOLL, G.H. XIAO, A.R. FOLSOM, D.W. HEIN (2000) N-Acetyltransferase genetic polymorphism, well-done meat intake, and breast cancer risk among postmenopausal women. *Cancer Epidemiol. Biomarkers Prev.* 9, 905–910.

171 AMBROSONE, C.B., J.L. FREUDENHEIM, R. SINHA, S. GRAHAM, J.R. MARSHALL, J.E. VENA, R. LAUGHLIN, T. NEMOTO, P.G. SHIELDS (1998) Breast cancer risk, meat consumption and N-acetyltransferase (NAT2) genetic polymorphisms. *Int. J. Cancer* 75, 825–830.

172 GERTIG, D.M., S.E. HANKINSON, H. HOUGH, D. SPIEGELMAN, G.A. COLDITZ, W.C. WILLETT, K.T. KELSEY, D.J. HUNTER (1999) N-Acetyl transferase 2 genotypes, meat intake and breast cancer risk. *Int. J. Cancer* 80, 13–17.

173 DELFINO, R.J., R. SINHA, C. SMITH, J. WEST, E. WHITE, H.J. LIN, S.Y. LIAO, J.S.Y. GIM, H.L. MA, J. BUTLER, H. ANTON-CULVER (2000) Breast cancer, heterocyclic aromatic amines from meat and N-acetyltransferase 2 genotype. *Carcinogenesis* 21, 607–615.

174 ZHENG, W., A.C. DEITZ, D.R. CAMPBELL, W. WEN, J.R. CERHAN, T.A. SELLERS, A.R. FOLSOM, D.W. HEIN (1999) N-Acetyltransferase 1 genetic polymorphism, cigarette smoking, well-done meat intake, and breast cancer risk. *Cancer Epidemiol. Biomarkers Prev.* 8, 233–239.

175 MILLIKAN, R.C. (2000) NAT1*10 and NAT1*11 polymorphisms and breast cancer risk. *Cancer Epidemiol. Biomarkers Prev.* 9, 217–220.

176 LEE, K., S. PARK, S. KIM, M.A. DOLL, K. YOO, S. AHN, D. NOH, A. HIRVONEN, D.W. HEIN, D. KANG (2003) N-Acetyltransferase (NAT1, NAT2) and glutathione S-transferase (GSTM1, GSTT1) polymorphisms and breast cancer. *Cancer Lett.* 196, 179–186.

177 MCKEOWN-EYSSEN, G. (1994) Epidemiology of colorectal cancer revisited: are serum triglycerides and/ or plasma glucose associated witk risk? *Cancer Epidemiol. Biomarkers Prev.* 3, 687–695.

178 Giovannucci, E. (**1995**) Insulin and colon cancer. *Cancer Causes Control* 6, 164–179.

179 Cooney, K.A., S.B. Gruber (**2005**) Hyperglycemia, obesity, and cancer risks on the horizon. *JAMA* 293, 235–236.

180 Kaaks, R. (**2004**) Nutrition, insulin, IGF-1 metabolism and cancer risk: a summary of epidemiological evidence. *Novartis Foundation Symp.* 262, 247–260.

181 Collard, T.J., M. Guy, A.J. Butt, C.M. Perks, J.M. Holly, C. Paraskeva, A.C. Williams (**2003**) Transcriptional upregulation of the insulin-like growth factor binding protein IGFBP-3 by sodium butyrate increases IGF-independent apoptosis in human colonic adenoma-derived epithelial cells. *Carcinogenesis* 24, 393–401.

182 Fletcher, O., L. Gibson, N. Johnson, D.R. Altmann, J.M. Holly, A. Ashworth, J. Peto, S. Silva Idos (**2005**) Polymorphisms and circulating levels in the insulin-like growth factor system and risk of breast cancer: a systematic review. *Cancer Epidemiol. Biomarkers Prev.* 14, 2–19.

183 Koenuma, M., T. Yamori, T. Tsuruo (**1989**) Insulin and insulin-like growth factor 1 stimulate proliferation of metastatic varaints of colon carcinoma 26. *Jpn J. Cancer Res.* 80, 51–58.

184 Tran, T.T., A. Medline, W.R. Bruce (**1996**) Insulin promotion of colon tumors in rats. *Cancer Epidemiol. Biomarkers Prev.* 5, 1013–1015.

185 Renehan, A.G., S.M. Shalet (**2005**) Diabetes, insulin therapy, and colorectal cancer. *BMJ* 330, 551–552.

186 Franceschi, S., L. Dal Maso, L. Augustin, E. Negri, M. Parpinel, P. Boyle, D.J.A. Jenkins, C. La Vecchia (**2001**) Dietary glycemic load and colorectal cancer risk. *Ann. Oncol.* 12, 173–178.

187 Higginbotham, S., Z. Zhang, I. Lee, N.R. Cook, E. Giovannucci, J.E. Buring, S. Liu (**2004**) Dietary glycemic load and risk of colorectal cancer in the women's health study. *J. Natl Cancer Inst.* 96, 229–233.

188 Khaw, K.T., N. Wareham, S. Bingham, R. Luben, A. Welch, N. Day (**2004**) Preliminary communication: glycated hemoglobin, diabetes, and incident colorectal cancer in men and women: a prospective analysis from the European prospective investigation into cancer-Norfolk study. *Cancer Epidemiol. Biomarkers Prev.* 13, 915–919.

189 Renehan, A.G., M. Zwahlen, C. Minder, S.T. O'Dwyer, S.M. Shalet, M. Egger (**2004**) Insulin-like growth factor (IGF)-I, IGF binding protein-3, and cancer risk: systematic review and meta-regression analysis. *Lancet* 363, 1346–1353.

190 Le Marchand, L., T. Donlon, A. Seifried, R. Kaaks, S. Rinaldi, L.R. Wilkens (**2002**) Association of a common polymorphism in the human GH1 gene with colorectal neoplasia. *J. Natl Cancer Inst.* 94, 454–460.

191 Wong, H.L., K. Delellis, N. Probst-Hensch, W.P. Koh, D. Van Den Berg, H.P. Lee, M.C. Yu, S.A. Ingles (**2005**) A new single nucleotide polymorphism in the insulin-like growth factor I regulatory region associates with colorectal cancer risk in Singapore Chinese. *Cancer Epidemiol. Biomarkers Prev.* 14, 144–151.

192 Slattery, M.L., W. Samowitz, K. Curtin, K.N. Ma, M. Hoffman, B. Caan, S. Neuhausen (**2004**) Associations among *IRS1*, *IRS2*, *IGF1*, and *IGFBP3* genetic polymorphisms and colorectal cancer. *Cancer Epidemiol. Biomarkers Prev.* 13, 1206–1214.

193 Slattery, M.L., M. Murtaugh, B. Cann, K.N. Ma, S. Neuhausen, W. Samowitz (**2005**) Energy balance, insulin-related genes and risk of colon and rectal cancer. *Int. J. Cancer* 115, 148–154.

194 Wang, L., T. Habuchi, N. Tsuchiya, K. Mitsumori, C. Ohyama, K. Sato, H. Kinoshita, T. Kamoto, A. Nakamura, O. Ogawa, T. Kato (**2003**) Insulin-like growth factor-binding protein-3 gene −202 A/C polymorphism is correlated with advanced stage disease in prostate cancer. *Cancer Res.* 63, 4407–4411.

195 TENNANT, M.K., J.B. TRASHER, P.A. TWOMEY, R.S. BIRNBAUM, S.R. PLYMATE (**1996**) Insulin-like growth factor-binding protein-2 and -3 expresssion in benign human prostate epithelium, prostate intraepithelial neoplasia and adenocarcinoma of the prostate. *J. Clin. Endocrinol. Metab.* 81, 411–420.

196 KANETY, H., Y. MAGJAR, Y. DAGAN, J. LEVI, M.Z. PAPA, C. PARIENTE, B. GOLDWASSER, A. KARASIK (**1993**) Serum insulin-like growth factor-binding protein-2 (IGFBP-2) is increased and IGFBP-3 is decreased in patients with prostate cancer: correlation with serum prostate-specific antigen. *J. Clin. Endocrinol. Metab.* 77, 229–233.

197 CHAN, J.M., M.J. STAMPFER, E. GIOVANNUCCI, P.H. GANN, J. MA, P. WILKINSON, C.H. HENNEKENS, M. POLLAK (**1998**) Plasma insulin-like growth factor-I and prostate cancer risk: a prospective study. *Science* 279, 563–566.

198 HARMAN, S.M., E.J. METTER, M.R. BLACKMAN, P.K. LANDIS, H.B. CARTER, Baltimore Longitudinal Study on Aging (**2000**) Serum levels of insulin-like growth factor I (IGF-I), IGF-II, IGF-binding protein-3, and prostate-specific antigen as predictors of clinical prostate cancer. *J. Clin. Endocrinol. Metab.* 85, 4258–4265.

199 STATTIN, P., A. BYLUND, S. RINALDI, C. BIESSY, H. DECHAUD, U. STENMAN, L. EGEVAD, E. RIBOLI, G. HALLMANS, R. KAAKS (**2000**) Plasma insulin-like growth factor-I, insulin-like growth factor-binding proteins, and prostate cancer risk: a prospective study. *J. Natl Cancer Inst.* 92, 1910–1917.

200 CHOKKALINGAM, A.P., M. POLLAK, C. FILLMORE, Y. GAO, F.Z. STANCZYK, J. DENG, I.A. SESTERHENN, F.K. MOSTOFI, T.R. FEARS, M.P. MADIGAN, R.G. ZIEGLER, J.F. FRAUMENI, JR., A.W. HSING (**2001**) Insulin-like growth factors and prostate cancer: a population-based case-control study in China. *Cancer Epidemiol. Biomarkers Prev.* 421–427.

201 SCHILDKRAUT, J.M., W. DEMARK-WAHNEFRIED, R.M. WENHAM, J. GRUBBER, A.S. JEFFREYS, S.C. GRAMBOW, J.R. MERKS, P.G. MOORMAN, C. HOYO, S. ALI, P.J. WALTHER (**2005**) IGF1 (CA)19 repeat and IGFBP3 −202 A/C genotypes and the risk of prostate cancer in black and white men. *Cancer Epidemiol. Biomarkers Prev.* 14, 403–408.

202 CHAN, J.M., M.J. STAMPFER, J. MA, P. GANN, J.M. GAZIANO, M. POLLAK, E. GIOVANNUCCI (**2002**) Insulin-like growth factor-I (IGF-I) and IGF binding protein-3 as predictors of advanced-stage prostate cancer. *J. Natl Cancer Inst.* 94, 1099–1106.

203 HSING, A.W., S. CHUA, JR., Y. GAO, E. GENTZSCHEIN, L. CHANG, J. DENG, F.Z. STANCZYK (**2001**) Prostate canncer risk and serum levesl of insulin and leptin: a population-based study. *J. Natl Cancer Inst.* 93, 783–789.

204 AUGUSTIN, L.S., C. GALEONE, L. DEL MASO, C. PELUCCHI, V. RAMAZZOTTI, D.J. JENKINS, M. MONTELLA, R. TALAMINI, E. NEGRI, S. FRENCESCHI, C. LA VECCHIA (**2004**) Glycemic index, glycemic load and risk of prostate cancer. *Int. J. Cancer* 112, 446–450.

205 NAM, R.K., W.W. ZHANG, J. TRACHTENBERG, M.A.S. JEWETT, M. EMAMI, D. VESPRINI, W. CHU, M. HO, J. SWEET, A. EVANS, A. TOI, M. POLLAK, S.A. NAROD (**2003**) Comprehensive assessment of candidate genes and serological markers for the detection of prostate cancer. *Cancer Epidemiol. Biomarkers Prev.* 12, 1429–1437.

206 TSUCHIYA, N., L. WANG, Y. HORIKAWA, T. INOUE, H. KAKINUMA, S. MATSUURA, K. SATO, O. OGAWA, T. KATO, T. HABUCHI (**2005**) CA repeat polymorphism in the insulin-like growth factor-I gene is associated with increased risk of prostate cancer and benign prostatic hyperplasia. *Int. J. Oncol.* 26, 225–231.

207 NEUHAUSEN, S.L., M.L. SLATTERY, C.P. GARNER, Y.C. DING, M. HOFFMAN, A.R. BROTHMAN (**2005**) Prostate cancer risk and IRSI, IRS2, IGF1, and INS polymorphisms: strong associations of IRS1 G972R variant and cancer risk. *Prostate* 64, 168–174.

208 Brand-Miller, J.C. (2003) Glycemic load and chronic disease. *Nutr. Rev.* 61, S49–55.

209 Cho, E., D. Spiegelman, D.J. Hunter, W.Y. Chen, G.A. Colditz, W.C. Willett (2003) Premenopausal dietary carbohydrate, glycemic index, glycemic load and fiber in relation to breast cancer risk. *Cancer Epidemiol. Biomarkers Prev.* 12, 1153–1158.

210 Higginbotham, S., Z. Zhang, I. Lee, N.R. Cook, J.E. Buring, S. Liu (2004) Dietary glycemic load a breast cancer risk in the Iowa Women's Health Study. *Cancer Epidemiol. Biomarkers Prev.* 13, 65–70.

211 Silvera, S.A., M. Jain, G.R. Howe, A.B. Miller, T.E. Rohan (2005) Dietary carbohydrates and breast cancer risk: a prospective study of the roles of overall glycemic index and glycemic load. *Int. J. Cancer* 114, 653–658.

212 Kaaks, R., E. Lundin, S. Rinaldi, J. Manjer, C. Biessy, S. Soderberg, P. Lenner, L. Janzon, E. Riboli, G. Berglund, G. Hallmans (2002) Prospective study of IGF-I, IGF-binding proteins, and breast cancer risk, in northern and southern Sweden. *Cancer Causes Control* 13, 307–316.

213 Hirose, K., T. Toyama, H. Iwata, T. Takezaki, N. Hamajima, K. Tajima (2003) Insulin, insulin-like growth factor-I and breast cancer risk in Japanese women. *Asian Pacif. J. Cancer Prev.* 4, 239–246.

214 Yu, H., B.D. Li, M. Smith, R. Shi, H.J. Berkel, I. Kato (2001) Polymorphic CA repeats in the IGF-I gene and breast cancer. *Breast Cancer Res. Treat.* 70, 117–122.

215 Schernhammer, E.S., S.E. Hankinson, D.J. Hunter, M.J. Blouin, M.N. Pollak (2003) Polymorphic variation at the −202 locus in IGFBP3, influence on serum levels of insulin-like growth factors, intercation with plasma retinol and vitamin D and breast cancer risk. *Int. J. Cancer* 107, 60–64.

216 Bagnardi, V., M. Blangiardo, C. La Vecchia, G. Corrao (2001) A meta-analysis of alcohol drinking and cancer risk. *Br. J. Cancer* 85, 1700–1705.

217 Brennan, P., S. Lewis, M. Hashibe, D.A. Bell, P. Boffetta, C. Bouchardy, N. Caporaso, C. Chen, C. Coutelle, S.R. Diehl, R.B. Hayes, A.F. Olshan, S.M. Schwartz, E.M. Sturgis, Q. Wei, A.I. Zavras, S. Benhamou (2004) Pooled analysis of alcohol dehydrogenase genotypes and head and neck cancer: a HuGE review. *Am. J. Epidemiol.* 159, 1–16.

218 Tiemersma, E.W., P.A. Wark, M.C. Ock, A. Bunschoten, M.H. Otten, F.J. Kok, E. Kampman (2003) Alcohol consumption, alcohol dehydrogenase 3 polymorphism, and colorectal adenomas. *Cancer Epidemiol. Biomarkers Prev.* 12, 419–425.

219 Ma, J., M.J. Stampfer, E. Giovan-nucci, C. Artigas, D.J. Hunter, C. Fuchs, W.C. Willett, J. Selhub, C.H. Hennekens, R. Rozen (1997) Methylenetetrahydrofolate reductase polymorphism, dietary interactions and risk of colorectal cancer. *Cancer Res.* 57, 1098–1102.

220 Keku, T., R. Millikan, K. Worley, S. Winkel, A. Eaton, L. Bisocho, C. Martin, R. Sandler (2002) 5,10-Methylenetetrahydrofolate reductase codon 677 and 1298 polymorphisms and colon cancer in African Americans and whites. *Cancer Epidemiol. Biomarkers Prev.* 11, 1611–1621.

221 Takeshita, T., K. Morimoto, N. Yamaguchi, S. Watanabe, I. Todoroki, S. Honjo, K. Nakagawa, S. Kono (2000) Relationships between cigarette smoking, alcohol drinking, the ALDH2 genotype and adenoma-tous types of colorectal polyps in male self-defense force officials. *J. Epidemiol.* 10, 366–371.

222 Matsuo, K., N. Hamajima, T. Hirai, T. Kato, K. Koike, M. Inoue, T. Takezaki, K. Tajima (2002) Aldehyde dehydrogenase 2 (ALDH2) genotype affects rectal cancer susceptibility due to alcohol consumption. *J. Epidemiol.* 12, 70–76.

223 Ktaoh, T., S. Kaneko, K. Kohshi, M. Munaka, K. Kitagawa, N. Kunugita,

K. Ikemura, T. Kawamoto (**1999**)
Genetic polymorphisms of tabacco-
and alcohol-related metabolizing
enzymes and oral cavity cancer. *Int.
J. Cancer* 83, 606–609.

224 Nomura, T., H. Noma, T. Shibahara,
A. Yokoyama, T. Muramatuse,
T. Ohmori (**2000**) Aldehyde
dehydrogenase 2 and glutathione S-
transferase M1 polymorphisms in
relation to the risk for oral cancer in
Japanese drinkers. *Oral Oncol.* 36, 42–
46.

225 Yokoyama, A., T. Muramatsu, T.
Omori, T. Yokoyama, S. Matsushita,
S. Higuchi, K. Maruyama, H. Ishii
(**2001**) Alcohol and aldehyde
deydrogenase gene polymorphisms
and oropharyngolaryngeal, esophageal
and stomach cancers in Japanese
alcoholics. *Carcinogen* 22, 433–439.

226 Yokoyama, A., T. Omori (**2003**)
Genetic polymorphisms of alcohol and
aldehyde dehydrogenases and risk for
esophageal and head and neck cancers.
Jpn. J. Clin. Oncol. 33, 111–121.

227 Yokoyama, A., H. Kato, T. Yokoyama,
T. Tsujinaka, M. Muto, T. Omori, T.
Haneda, Y. Kamagai, H. Igaki, M.
Yokoyama, H. Watanabe, H. Fukada,
H. Yoshimizu (**2002**) Genetic
polymorphisms of alcohol and
aldehyde dehydrogenases and
glutathione S-transferase M1 and
drinking, smoking, and diet in
Japanese men with esophageal
squamous cell carcinoma. *Carcinogen*
11, 1851–1859.

228 Freudenheim, J.L., C.B. Ambrosone,
K.B. Moysich, J.E. Vena, S. Graham,
J.R. Marshall, P. Muti, R. Laughlin, T. Nemoto, L.C. Harty, G.A.
Crits, A.W.K. Chan, P.G. Shields
(**1999**) Alcohol dehydrogenase 3 genotype modification of the association
of alcohol consumption with breast
cancer risk. *Cancer Causes Control* 10,
369–377.

229 Hines, L.M., S.E. Hankinson, S.A.
Smith-Warner, D. Spiegelman, K.T.
Kelsey, G.A. Colditz, W.C. Willett,
D.J. Hunter (**2000**) A prospective
study of the effect of alcohol
consumption and ADH3 genotype on

plasma steriod hormone levels and
breast cancer risk. *Cancer Epidemiol.
Biomarkers Prev.* 9, 1099–1105.

230 Sturmer, T., S. Wang-Gohrke, V.
Arndt, H. Boeing, X. Kong, R.
Kreienberg, H. Brenner (**2002**)
Interaction between alcohol
dehydrogenase II gene, alcohol
consumption, and risk of breast
cancer. *Br. J. Cancer* 27, 519–523.

231 Choi, J.Y., J. Abel, T. Neuhaus, Y.
Ko, V. Harth, N. Hamajima, K.
Tajima, K.Y. Yoo, S.K. Park, D.Y.
Noh, W. Han, K.J. Choe, S.H. Ahn,
S.U. Kim, A. Hirvonen, D. Kang
(**2003**) Role of alcohol and genetic
polymorphisms of CYP2E1 and
ALDH2 in breast cancer development.
Pharmacogenetics 13, 67–72.

232 Nature Genetics (**2001**) Challenges for
the 21st century. *Nat. Genet.* 29, 353–
354.

233 Cardon, L.R., J.I. Bell (**2001**)
Association study designs for complex
diseases. *Nat. Rev. Genet.* 2, 91–99.

234 Ioannidis, J.P.A., E.E. Ntzani, T.A.
Trikalinos, D.G. Contopoulos-
Ionnadis (**2001**) Replication validity of
genetic association studies. *Nat. Genet.*
29, 306–309.

235 Gambaro, G., F. Anglani, A.
D'Angelo (**2000**) Association studies
of genetic polymorphisms and
complex disease. *Lancet* 355, 308–311.

236 Tabor, H.K., N.J. Risch, R.M. Myers
(**2002**) Opinion: candidate-gene
approaches for studying complex
genetic traits: practical considerations.
Nat. Rev. Genet. 3, 391–397.

237 Lancet (**2003**) In search of genetic
precision. *Lancet* 361, 357–357.

238 Lohmueller, K.E., C.L. Pearce, M.
Pike, E.S. Lander, J.N. Hirschhorn
(**2003**) Meta-analysis of genetic
association studies supports a
contribution of common variants to
susceptibility to common disease.
Nat. Genet. 33, 177–182.

239 Thomas, D.C., D.G. Clayton (**2004**)
Betting odds and genetic associations.
J. Natl Cancer Inst. 96, 421–423.

240 Hunter, D.J. (**2005**) Gene–
environment interactions in human
diseases. *Nat. Rev. Genet.* 6, 287–298.

18
Taste Receptors and Their Variants

Bernd Bufe and Wolfgang Meyerhof

18.1
Introduction

The prevalence of obesity is increasing dramatically in most Western countries [1]. Long-term overweight drastically enhances the risk for many ailments including diabetes, cardiovascular diseases, and cancer [2] causing enormous costs for our public health care systems now and even more in coming decades [3]. An important prerequisite to develop efficient strategies that counteract obesity is the detailed understanding of its etiology. A key factor for weight gain is a positive energy balance usually due to the overconsumption of energy-rich food [1]. Therefore, a change in dietary habits would be an appropriate preventive measure or even a therapeutic option. Unfortunately, the development of our eating habits is not well understood, yet involves the taste or the flavor of food as a major determinant, as well as many other factors including genetics, age, and education [4–6]. Taste, smell, and touch are the key sensations in flavor perception. Thus, these senses crucially influence our decision to ingest food or not [7]. Interestingly, the sense of taste is more systematically correlated with caloric and metabolic effects than odor and texture [8]. In humans and in a wide range of animal species an inherent connection exists between the taste of a substance, the hedonic tone it elicits, and the consequences of its ingestion [8]. Sweet and umami tastes for example, indicate carbohydrates and protein, respectively, and are therefore perceived as attractive. In contrast, bitter and sour tastes induce repulsive behaviors thereby preventing intoxication and intake of spoiled food [9]. Salt taste is involved in the regulation of the ion homeostasis and osmoregulation and consequently is perceived attractive in low and repulsive in high concentrations [9].

Significant progress in the molecular neurobiology of taste perception has been made over the last years. We now know specific receptors that mediate sweet, bitter, and umami taste, and we are beginning to understand some of their functional properties. These discoveries opened new opportunities to deepen our understanding of taste biology. For example, interspecies comparison and functional analysis of taste receptors revealed a surprisingly high degree of divergence between species

Nutritional Genomics. Edited by Regina Brigelius-Flohé and Hans-Georg Joost
Copyright © 2006 WILEY-VCH Verlag GmbH & Co. KGaA, Weinheim
ISBN: 3-527-31294-3

[10, 11]. The elucidation of the underlying evolutionary processes will improve our understanding of the relation of taste perception, food preferences, and environmental influences. Beyond that, the functional characterization of taste receptors should facilitate the identification of new taste-active compounds, such as taste enhancers or bitter blockers, which could influence human eating habits [12]. In addition, newly developed cell-based assays [13, 14] might allow the development of novel tools for the quality control in food processing. It has been recently shown that variations in taste receptor genes alter receptor function and thereby taste perception in humans [15, 16]. Thus, it is now possible to study the influence of variations in taste receptor genes on taste perception and food choice. Moreover, a link has been established between the metabolism and taste perception through the action of hormones such as leptin and cholecystokinin that control food intake and modulate taste perception [17–19]. This review summarizes the most recent findings in the molecular biology of taste with special emphasis on taste receptor gene polymorphisms and assesses their possible relevance for eating behavior.

18.2
Salt Taste

Salt taste is involved in the maintenance of electrolyte homeostasis. It is not only elicited by sodium chloride but also by other salts including $LiCl$, KCl, NH_4Cl, and $CaCl_2$ [20–23]. Several lines of evidence suggest that Na^+ salt taste involves the epithelial sodium channel ENaC [21]. First, amiloride, a well-known blocker of the ENaC channel, largely inhibits the salt taste in rodents [24]. Second, reverse transcriptase polymerase chain reaction (RT-PCR) experiments detected mRNA for the α, β, and γ subunits of the ENaC channel in rat taste buds [25]. Third, immunohistochemistry showed the presence of the α, β, and γ ENaC subunits in rat taste cells, although the majority of the protein was located inside the cell instead on the apical microvilli extending into the taste porus [25, 26]. Fourth, electrophysiological recordings identified amiloride-blockable sodium currents in the apical membranes of taste receptor cells [27].

The ENaC channel is permeable for Na^+ and Li^+ ions but nearly impermeable for other ions such as K^+, Ca^{2+}, and NH_4^+ [28, 29]. Thus, an additional, yet unidentified, receptor has to mediate the salt taste of these ions. This unknown receptor seems to be especially important in humans in whom just 20% of the salt taste is amiloride-blockable [30]. Based on electrophysiological evidence the amiloride-insensitive salt taste in rodents is mediated by a non-selective cation channel that is permeable to Na^+, K^+, Ca^{2+}, and NH_4^+ [31, 32]. The pharmacological characterization of currents recorded from taste receptor cells in wild-type and vanilloid receptor 1-knockout mice suggested that a variant of the vanilloid receptor 1 channel mediates the amiloride-insensitive salt taste [31, 33, 34]. Further studies will have to show if this holds true.

18.3
Sour Taste

Acids evoke sour taste. From all taste qualities the molecular mechanisms of sour taste perception are least well understood. Depending on the species, various molecular mechanisms have been proposed, including the direct blockade of potassium channels in mudpuppies and the induction of amiloride-sensitive proton currents in hamsters and rats [9, 35–37]. Currently, the acid-sensing ion channels (ASIC) and hyperpolarization and cyclic nucleotide-gated cation channels of the HCN family are being discussed as sour taste receptors in rodents [38–41]. In particular, ASIC2a (synonyms: MDEG1, BNC1a, BNaC1α) appears to be a promising candidate sour taste receptor [41]. A number of observations support the role of ASIC2 in sour taste perception. First, ASIC2a is expressed in a large subset of rat taste receptor cells [39]. Second, a proton-gated sodium current can be recorded in taste receptor cells upon acid stimulation [42]. Third, when expressed in *Xenopus-laevis* oocytes ASIC2a responded at equal pH stronger to acetic acid than to HCl, a phenomenon that has also been observed in taste experiments *in vivo* [39, 43]. Although these data seem to be convincing, it is quite problematic that sour taste is not diminished in ASIC2a-knockout mice [44, 45]. It remains to be seen if the proposed coexpression and heteromerization of ASIC2 and ASIC2b [41] fully explains the persisting sour taste in ASIC2a-knockout mice.

The HCN channels HCN1 and HCN4 can be detected in a subset of rat taste receptor cells by RT-PCR, *in situ* hybridization, and immunohistochemsitry [38]. Moreover, electrophysiological recordings of taste receptor cells in intact taste buds demonstrated that acid stimulation of taste buds at the taste pore elicited an I_h current in a subset of taste receptor cells, the hallmark of members of the HCN family [38]. Furthermore, functional expression of HCN1 and HCN4 in HEK293 cells showed that extracellular protons activated both channels [38]. Unfortunately, it is not known if these channels show a stronger response to acetic acid than to HCl and studies of tissue-specific HCN knockout are still missing. Thus, our knowledge about the molecular mechanisms of sour taste perception is still insufficient.

18.4
Sweet Taste

Sweet taste is elicited by various natural occurring compounds such as sugars, some amino acids and proteins of certain tropical fruits, but also by structurally diverse artificial sweeteners including saccharin, cyclamate, and aspartame [46–48]. Sugars and sweet amino acids, which are the most relevant natural sweeteners, have a relatively low potency [47]. This may serve as a quantity check because only high concentrations of these compounds indicate energy-rich food sources. In contrast, artificial sweeteners usually have a high potency [47].

Interestingly, the perception of sweet compounds varies across species. Unlike humans, rodents do not perceive the sweetness of the sweet proteins monellin, thaumatin, and the artificial sweeteners aspartame and cyclamate [49–52]. Moreover, even within a single species strain-specific differences exist. It has been known for many decades that mice strains can differ in their sensitivity towards sweet compounds [53]. Studies of congenic mice strains revealed that these differences are due to a single chromosomal locus called "*sac*" [53] at the distal end of chromosome 4 [54, 55]. Assuming that the sweet taste receptor is located in the *sac* locus, six groups independently analyzed the human genome data of the distal end of chromosome 1, which is syntenic to the distal end of mouse chromosome 4 [14, 56–60]. They discovered a gene encoding a G protein-coupled receptor named TAS1R3 (synonym T1R3) as a new member of the taste receptor family TAS1R (synonym T1R) [61]. The *TAS1R3* gene is directly located in the *sac* locus while the two other family members *TAS1R1* (synonyms T1R1, TR1) and *TAS1R2* (synonyms T1R2, TR2) are located 12 centimorgan more centomeric [14]. Consistent with its proposed role as a sweet receptor, *in situ* hybridizations showed the presence of *Tas1r3* mRNA in a subset of mouse and human taste receptor cells [14, 56–60, 62]. Further studies revealed that *Tas1r3* is coexpressed with its two other known family members *Tas1r2* or *Tas1r1* in non-overlapping subsets of taste receptor cells [14, 56, 58], suggesting that the functional sweet receptor could be a heteromer. Consistent with this assumption, expression studies in HEK293 cells showed that cells co-transfected with the human TAS1R2/TAS1R3 or rodent Tas1r2/Tas1r3 cDNAs responded to sweeteners of multiple chemical classes [10, 14, 63]. In line with the observed species differences the human receptor TAS1R2/TAS1R3 but not the mouse counterpart could be activated by monellin, thaumatin, cyclamate, and aspartame [10, 50]. In accordance with human taste perception relative low concentrations of the artificial sweeteners activated the recombinant sweet taste receptor, while carbohydrates and sweet amino acids acted at much higher concentrations [10, 14, 50]. Notably, all tested compounds that are sweet to humans activated the human TAS1R2/TAS1R3 receptor [10]. These results show that the TAS1R2/TAS1R3 heteromer mediates human sweet taste perception.

Mouse models provided further evidence for the role of *Tas1r3* in sweet taste perception. The transgenic expression of the *Tas1r3* taster gene in a non-taster mouse strain rescued the taster phenotype [14]. Moreover, behavioral experiments and nerve recordings of *Tas1r2*- and *Tas1r3*-knockout mice confirmed that the deletion of either gene strongly reduced the nerve responses and attraction to various sweeteners [64, 65]. In contrast, the knockout of the third family member, the *Tas1r1* gene, did not influence the perception of sweet compounds [65]. Taken together there is overwhelming evidence that the rodent Tas1r2/Tas1r3 heteromer is a genuine sweet taste receptor. Interestingly, both the *Tas1r2*- and the *Tas1r3*-knockout mice kept a residual behavioral and nerve response to high concentrations of carbohydrate sweeteners [64, 65]. Thus the absence of the Tas1r2/Tas1r3 heteromer reveals the existence of an additional low-affinity detection mechanism for natural sweeteners [64]. It remains to be seen if this residual taste response can be explained by activation of Tas1r monomers or homomers [65]. It is also possible that pathways independent of Tas1r contribute to the residual response [64], such as direct activation of G proteins by sweet-tasting compounds [66–68].

18.5
Umami Taste

In humans, umami taste is mainly triggered by sodium glutamate and enhanced by ribonucleotides such as inosine-5'-monophosphate (IMP) and guanosine-5'-monophosphate (GMP) [69, 70]. Umami taste indicates protein-rich food and is therefore perceived as an attractive stimulus that strongly enhances the palatability of food [69]. Besides glutamate, other metabotropic glutamate receptor agonists such as ibotenate and L-AP4 also elicit umami taste [71]. In addition, various physiological and molecular studies showed that the metabotropic glutamate receptors (mGluRs) are expressed in taste receptor cells in mice [72–76]. This initially led to the hypothesis that mGluRs are involved in umami taste [75]. Indeed, the cDNA of an N-terminally truncated variant (mGluR4t) of the mGluR4 was isolated from rat tongue tissue [77]. Subsequently, heterologous expression and functional analysis showed that it responded to L-AP4 and glutamate at high concentrations, which are typically used to elicit umami taste in humans [77]. Thus, the truncated mGluR4 variant seemed to be an attractive candidate umami taste receptor, although a number of inconsistencies exist [65]. First, mGluR4-knockout mice show a ~20% increased preference for glutamate [78] instead of a reduced response as one would expect. Second, so far, there is no evidence for the translation of the mGluR4t mRNA into a functional protein in taste receptor cells. This is quite an important issue as this receptor variant lacks the normal cell surface targeting sequence of mGluR4 [65]. Third, the responses of the recombinant mGluR4t are not enhanced by IMP [77].

Recently, it has been shown that receptors of the TAS1R family mediate umami taste. *In vitro* expression studies showed that cells co-transfected with human *TAS1R1* and *TAS1R3* responded to glutamate and that the response is strongly enhanced in the presence of 5'-ribonucleotides such as IMP [10, 63]. Moreover, cells co-transfected with cDNAs for the human receptor *TAS1R1/TAS1R3* not only responded to glutamate and but also to other known umami compounds such as L-AP4 [10]. Thus, this receptor heteromer sufficiently fulfils all functional properties that are known for umami taste. In addition, *in situ* hybridization and immunohistochemisty clearly showed that these receptors are expressed in a subset of taste receptor cells in humans and rodents [58, 61, 62]. Moreover, in *Tas1r1*- and *Tas1r3*-knockout mice behavioral and nerve responses to umami compounds were greatly abolished [64, 65]. Taken together, anatomical, genetic, functional, and physiological evidence convincingly demonstrates that *Tas1r1* and *Tas1r3* are umami taste receptors.

18.6
Bitter Taste

Bitter tastants comprise thousands of compounds, belonging to many chemical classes ranging from inorganic salts such as KCl and $MgSO_4$ to complex organic molecules such as peptides, amino acids, fatty acids, amines, amides, ureas, alkaloids, lactones, terpenes, polyphenols, or flavonoides [22, 48, 79]. Although bitter taste is generally aversive, the consumption of foods and beverages such as bitter chocolate, beer, or coffee shows that humans accept moderate bitter taste as tolerable or even attractive. In humans and rodents, heritable variations in the perception of bitter compounds have been observed [80–83]. Genetic studies mapped bitter taste to loci on human chromosomes 5 and 7 and mouse chromosome 6 [84]. As in the case of the sweet taste receptor, database mining of the human genome by two independent groups also led to the discovery of a novel family of GPCRs, the TAS2Rs (synonyms: T2R, TRB), [84, 85]. In humans, this family of G protein-coupled receptors comprises 25 receptor genes and 11 pseudogenes that are located on chromosomes 5, 7, and 12 [86, 87]. The corresponding mouse Tas2r family consists of 35 genes and six pseudogenes located on the chromosomes 2, 6, and 15 [86]. The chromosomal distribution of *TAS2R* genes correlates well with the gene loci that determine variations in bitter perception in humans and mice [84]. Moreover, *in situ* hybridizations using several mouse, rat, and human *TAS2R* genes as probes showed their expression in a subset of human and mouse taste receptor cells [16, 84, 85, 87, 88, 89]. Interestingly, *in situ* hybridization in rodents with mixtures of two and 10 different *Tas2*r probes showed that *TAS2R*s are coexpressed in the same subset of taste receptor cells [84]. Although these experiments do not formally prove the presence of all TAS2Rs in the same subset of taste receptor cells, they deliver a molecular explanation for the observation that different compounds elicit the same uniform bitter taste. Moreover, expression studies so far identified for three rodent and eight human TAS2Rs bitter tastants as ligands [13, 16, 87–90]. For example, the mouse receptor Tas2r105 and its rat counterpart rT2R9 responded to cycloheximide, while its closet human relative TAS2R10 responded to strychnine [13, 87]. It will be interesting to understand which evolutionary processes led to different tunings of the human and rodent receptors. Notably, TAS2R14, TAS2R16, TAS2R43, TAS2R44, and TAS2R38 were activated by many structurally divergent bitter tastants [16, 87–89]. If this holds true for all TAS2Rs it would explain how thousands of chemically diverse bitter compounds can be detected by just 25 receptors.

Sequence polymorphisms in *TAS2R* genes have been linked to differences in bitter taste sensitivity in mice and humans. Variations in the *Tas2r105* gene are linked to the sensitivity of mice for cycloheximide [13]. Moreover, in humans the inherited loss of sensitivity for the bitter compound phenylthiocarbamide (PTC) is caused by polymorphisms in the *TAS2R38* gene [15, 16]. Taken together, there is strong molecular, genetic, anatomical, and functional evidence that the TAS2R receptors mediate bitter taste in mammals. Recently, it has been shown that TAS2Rs are present in solitary chemoreceptor cells in the nasal cavity which serve as senti-

nels of respiration [91]. This finding and the fact that the *TAS2R* genes are expressed in other peripheral tissues suggest additional roles for TAS2Rs [91, 92]. It will be interesting to elucidate the biological role of the TAS2Rs in other tissues.

18.7
Signal Transduction of Sweet, Bitter, and Umami Taste

The first molecule discovered in taste signal transduction was the G protein alpha-subunit gustducin by the laboratory of Robert F. Margolskee [93]. *In situ* hybridization and immunocytochemisty showed, that α-gustducin is coexpressed in a subset of taste receptor cells together with the Tas2r receptors [84, 93]. Sequence comparisons predicted that gustducin, like its closest relative transducin, decreases cyclic nucleotide levels by activation of phosphodiesterase activity [94]. Later on, quench flow assays showed that the exposure of taste tissue extracts to bitter compounds decreased cAMP levels, an effect that was blocked by an α-gustducin antibody [95]. Stimulation of membrane preparations of bovine taste tissue and of cells expressing Tas2r105 with bitter compounds directly activated α-gustducin, further proofing the importance of this G protein subunit in bitter taste transduction [13, 96].

Surprisingly, α-gustducin-knockout mice showed not only a reduced bitter taste sensitivity but also an impaired sweet taste [97]. This led to the hypothesis that α-gustducin, apart from its role in bitter taste, is also involved in sweet taste transduction [97]. Recent single-cell RT-PCR and *in situ* hybridization experiments showed the coexpression of Tas1r3 and α-gustducin in taste receptor cells, further supporting the dual role of gustducin in sweet and bitter taste [56, 58]. Since in the knockout mice sweet and bitter taste was impaired but not completely abolished [97], α-gustducin appears to be not the only G protein alpha-subunit transducing sweet and bitter taste. A good additional candidate is α-transducin, which is also present in taste tissue and can partially rescue the phenotype of α-gustducin-knockout mice [98, 99]. Various studies showed rising inositoltrisphosphate (IP$_3$) levels in taste receptor cells upon stimulation with bitter compounds [95, 100], which were suppressed by antibodies directed against phospholipase C beta 2 (PLCβ2) or the G protein γ-subunit 13 [95]. Moreover, single-cell RT-PCR and *in situ* hybridizations localized the G protein subunits β1, β3, γ13 in α-gustducin-positive taste receptor cells [99, 101]. Beta 1 and β3-subunits interact with γ13 to activate PLCβ2 [102], suggesting that β1 or β3/γ13/α-gustducin complexes activate PLCβ2 in taste receptor cells. Behavioral studies and nerve recordings of PLCβ2-knockout animals, which lost their responses to bitter, sweet, and umami stimuli [103], further supported this assumption. Interestingly, animals gene-targeted for the ion channel TRPM5 also lost their responses to sweet, bitter, and umami stimuli [103]. Taken together, these findings show that sweet, bitter, and umami tastes share key molecules of their signal transduction pathways. Sweet, bitter and amino acid taste receptors couple through G$_i$ type G protein alpha-subunits to the cAMP and simultaneously via β/γ complexes to the IP$_3$ pathways, respectively. The IP$_3$ production triggers a release of calcium from intracellular stores, which then activates TRPM5. The resulting ion fluxes finally lead to depolarization of the taste receptor cells and to the secretion of a yet unknown neurotransmitter.

18.8
Taste Quality Coding

Humans can perceive all taste qualities on any area of the tongue that contains papillae [37, 104]. Only the perceived intensities of the taste stimuli differ depending on the tongue region and papilla type. The sweet taste of saccharin for instance is highest at the tip of the tongue, whereas the bitter taste of quinine is best perceived at the back of the tongue [104].

Taste sensations are mediated by specialized epithelial cells, the taste receptor cells that are located within the taste buds of the papillae on the surface of the tongue. These elongated taste receptors cells are embedded by the surrounding epithelium. Only at the gustatory porus of the taste buds do the receptor cells have access to the outside world. Thus, interactions between the taste receptor molecules and their tastants only occur at the apical part of the taste receptor cells in the porus. The anterior part of the tongue that contains the fungiform papillae is innervated by the VIIth cranial nerve, whereas the posterior part of the tongue that contains the foliate and vallate papillae is innervated by the IXth cranial nerve [105]. This nerve also innervates isolated taste buds in the palate and epiglottis, while the Xth cranial nerve innervates the pharyngeal taste buds. Each of these three cranial nerves also carries somatosensory afferents that innervate regions of the tongue neighboring lingual taste buds. This type of innervation makes it difficult to distinguish gustatory from somatosensory information.

The coding of taste qualities is a fiercely debated field. In principle, two competing models exist. The labeled line model favors a separate coding of the five basic taste qualities [106]. This model predicts specialized taste receptor cells for each taste quality, which are innervated by dedicated fibers [106]. Consequently, in this model, the taste quality is encoded at the level of the taste receptor cells. In the competing across-fiber pattern model, the cells and the innervating neurons are not strictly specialized and respond to all taste stimuli, albeit with different strength [107]. Each taste stimulus is thought to generate discriminative patterns of excited and non-excited neurons that are decoded by the brain [107]. Consequently, in this model the information about the taste qualities is encoded at higher levels. Both models are supported by experimental data. Various electrophysiological recordings of taste receptor cells, and afferent nerves, obtained from rodents and amphibians showed that nerve fibers frequently responded to stimuli of more than one taste quality [108–110]. In addition, *in situ* calcium imaging of rat taste buds also suggested a broad tuning of taste receptor cells across modalities [73].

In marked contrast, nerve recordings in primates and hamster revealed that fibers for sweet, bitter and salt transduction are relatively narrowly tuned [111–113]. While a subset of ~30% of the taste receptor cells expresses the Tas1r receptors, another subset of ~20% expresses the Tas2r receptors [14, 84]. Moreover, although *Tas1r1* and *Tas1r2* are almost always coexpressed with *Tas1r3* they do not colocalize with each other or the Tas2r receptors [14, 61]. These findings strongly argue for the existence of independent subsets of taste receptor cells that mediate bitter, sweet and umami taste and are thus consistent with a labeled line model. Also behavioral and physiological studies of various transgenic animals which showed that PLCβ2- and TRPM5-knockout animals lost their ability to taste sweet, bitter, and amino acid but not salt and sour stimuli support the labeled line theory [93, 103]. Transgenic expression of PLCβ2 under control of *Tas2r* promoters rescued only bitter but not sweet and umami taste in PLCβ2-knockout animals [103]. Transgenic animals expressing an artificial opiate receptor under control of the *Tas1r2* promoter acquired a high preference for the synthetic opiate spiralidone, while wild-type mice were indifferent to it [65]. If, in contrast, this receptor was expressed under control of a *Tas2r* gene promoter, the transgenic animals showed a pronounced aversion to spiralidone. Together, these results convincingly demonstrate that sweet and bitter tastes are encoded by separate subpopulations of taste receptor cells [114].

Thus, a considerable number of independent observations point to a labeled line coding of taste information in the periphery. The final proof of the model depends on the elucidation of the mode of innervations of taste receptor cells, taste buds, and papillae. In the coming years *trans*-synaptic tracing studies will likely clarify how excitation of taste receptor cells is conveyed to the brain.

18.9
Peripheral Modulation of Taste

Taste perception varies depending on the nutritional status and specific needs of the animal [8]. One of the most well-documented examples for such a sensory alteration is the specific hunger for salt. Severe salt depletion elicits salt craving in animals and humans. Remarkably, salt depletion not only increases the attractiveness of salt but also lowers salt taste threshold levels [115, 116]. Salt depletion leads to an increase in plasma aldosterone and angiotensin II levels [117]. Intriguingly, increasing aldosterone levels have been shown to upregulate the expression of β and γ ENaC subunits in taste receptor cells [26], providing a molecular explanation for the increase in salt sensitivity. This example clearly documents that specific physiological needs influence taste sensitivity.

Notably, the modulation of taste perception is not limited to salt taste. Leptin, an adipocyte hormone that signals the size of the body's fat stores to the brain and hence plays a central role in the regulation of energy metabolism, influences sweet taste [19]. For example, db/db mice that do not express a functional leptin receptor display enhanced neuronal and behavioral responses to sweet, but not to sour, salty, or bitter stimuli [19]. Intraperitoneal leptin injections into ob/ob mice that lack leptin but not into db/db mice, reduced the behavioral responses to sweet but not to bitter, sour, and salt stimuli [118]. RT-PCR, *in situ* hybridization and imunohistochemistry showed that the signaling form of the leptin receptor and its effector, the signal transducer and activator of transcription 3, is expressed in mouse taste receptor cells [18, 19]. Patch clamp recordings of isolated taste receptor cells revealed that leptin induced the hyperpolarization of taste receptor cells through closure of K^+ currents [19]. Taken together, molecular, genetic, neurological, and behavioral evidence suggests that the anorexigenic hormone leptin directly influences sweet taste perception. Intriguingly, in humans high serum-leptin levels have been correlated with an altered palatability [119, 120]. Although this might be solely attributed to leptin's action in the brain, it is tempting to speculate that the leptin-dependent modulation of sweet taste contributes to altered intake behavior.

Apart from leptin, also other hormones that are involved in the regulation of eating behavior interact with taste receptor cells. Cholecystokinin (CCK), a hormone released by the gut, is expressed in a subset of rat taste receptor cells [17]. *In situ* hybridization revealed that the CCK-expressing cells partially overlap with those expressing the sweet receptor subunit Tas1r2 and α-gustducin. Another subpopulation of *Tas1r2*-expressing cells coexpressed vasoactive intestinal peptide (VIP) [121]. Moreover, the functional analysis of receptor cells by calcium imaging and patch clamp recordings showed that a subset of taste receptor cells responded to CCK with elevated calcium levels and an inhibition of K^+ currents [17]. Pharmacological evidence suggested that this response was mediated by the CCK-A receptor [17]. Subsequently, calcium-imaging studies demonstrated that the majority of CCK responsive cells responded to bitter stimuli, suggesting that these cells might also express bitter taste receptors [122]. In summary, these data indicate that sweet receptor-expressing cells express CCK, while bitter responsive cells express the CCK-A receptor. Moreover, behavioral tests in humans and mice suggest a CCK level-dependent modulation of palatability and food preferences, although these data are still somewhat inconsistent [123–125]. It is therefore tempting to speculate that CCK and the CCK-A receptor play a role in the cross-talk between sweet and bitter taste perception.

Recently, the expression of neuropeptide Y (NPY), in a subset of taste receptor cells also has been reported [126], making it another potential regulator of taste sensitivity. The observation that orexigenic and anorexigenic hormones influence taste perception clearly argues for a specific role of taste in feeding/eating behavior, perhaps through mechanisms such as sensory specific satiety. However, many more studies are needed to elucidate its importance.

18.10
Evolution of Taste Receptor Genes

Analysis of *TAS2R* bitter taste receptor genes across species revealed a surprisingly high degree of variability between humans and rodents [11]. In contrast to many other gene families, the majority of *TAS2R* genes lack a single clearly identifiable counterpart in the other species (Fig. 18.1). Instead, each species developed separate *TAS2R* subfamilies from ancestral genes [11]. The different diets of humans

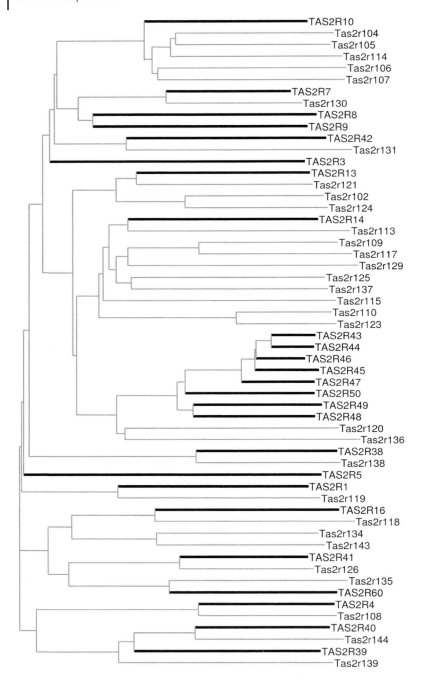

and rodents [127] likely expose both types of organisms to different poisons. As the failure to detect toxins can be lethal, natural selection should favor optimized bitter taste receptors for the detection of "species-relevant" poisons. Thus, the observed divergence of bitter receptors between species could be an adaptation to different dietary habits. Notably, the comparison of TAS2R genes in 13 monkey species showed that frequently the same genes became pseudogenes, although the pseudo-genizations were caused by independent mutations [86]. This convergent evolution of the bitter taste receptor genes in different primate species suggests that environmental factors such as the adaptation to new food sources were more important for this process than genetic causes [86].

The faster pseudogenization of and the ratio of non-synonymous to synonymous substitutions in the human *TAS2R* genes [86, 128–130] suggests a relaxation of selective constraints in humans which probably has been caused by the increased proportion of animal food and the controlled use of fire during hominid evolution [129, 131, 132]. Notably, the general relaxation of constraints for the *TAS2R* genes as a group does not exclude the possibility for selective constraint or positive selection of individual *TAS2Rs*. Indeed, further studies analyzing the distribution of the *TAS2R38* gene polymorphisms in four different world populations suggested a balancing natural selection for the two predominant *TAS2R38* haplotypes, suggesting that heterozygotes might be preferred by natural selection [133]. In another study, the genetic analysis of *TAS2R16* variants revealed two frequent non-synonymous polymorphisms within the *TAS2R16* gene leading to an N172K and an R222H exchange in the second extracellular and third intracellular loop of the encoded receptor, respectively [134]. The subsequent analysis showed that the mutation leading from the ancestral K172 to N172 occurred before the migration of hominids out of Africa. The allele frequency in 60 different world populations revealed that the evolutionary-derived N172 allele is predominant in all major human populations, while the ancestral K172 variant is virtually absent outside Africa [134]. Population models strongly suggested that the high frequency of the N172 allele was fixed in the populations by positive selection, while no selective pressure was observed for the R222 and H222 alleles [134].

In line with these results the N172 receptor variant has a ~two-fold higher sensitivity for toxic cyanogenic glycosides than the K172 variant, while there was no functional difference between the R222 and H222 variants [134]. Thus, a better protection of humans carrying the *TAS2R16*–N172 allele against intoxication by cyanogenic bitter compounds might explain its high frequency in most modern human world populations. Intriguingly, the K172 variant is still moderately abun-

---

Fig. 18.1. Sequence relationship of human and mouse TAS2R receptors. The sequence comparison is based on the derived amino acid sequence of TAS2R genes in the human genome database (version 35.1) and the mouse genome database (version 34.1). Human TAS2Rs are indicated by bold lines, while the mouse Tas2rs are indicated by thin lines. Note that homologous pairs of genes are rare. Usually, a single human TAS2R clusters with several mouse receptors and vice versa.

dant throughout Central Africa with a distribution that resembles that of some malaria-resistance alleles [134]. The lower sensitivity of *TAS2R16*–K172 carriers likely promoted the chronic consumption of sublethal doses of cyanogenic glyco-sides which may protect against malaria through various mechanisms including cyanide-induced sickle cell anemia, oxidative stress, and inhibition of *Plasmodium*'s respiratory chain [135–137]. It is reasonable to assume that this health benefit accounts for the relatively high frequency of the *hTAS2R16*–K172 allele in Sub-Saharan Africa.

18.11
Genetic Variation of Taste Perception

Humans and rodents show substantial differences in their sensitivity to sweet, bitter, and umami stimuli [54, 55, 80, 83, 138, 139]. In rodents, the strain-specific variability of sweet taste perception is one of the best-studied genetic traits. Ini-tially, it was noticed that C57BL/6 mice showed a higher preference for the artifi-cial sweetener saccharin than DBA/2J mice [53]. Breeding experiments revealed that the five-fold increased sensitivity of the C57BL/6 to saccharin is inherited and controlled by a single locus called "*sac*" [53]. Further analysis of the *sac* locus led to the discovery that polymorphisms in the *Tas1r3* gene are associated with the taster phenotype [58]. Subsequently, functional expression studies revealed that Tas1r3 is involved in sweet and umami taste [10, 14, 63] which is consistent with the obser-vation that saccharin taster and non-taster mice also differ in their preference for amino acids [140]. Moreover, the transgenic expression of the *Tas1r3* gene of the taster strain C57BL/6 completely rescued saccharin tasting in a non-taster mouse strain [14], formally proving that variations in the *Tas1r3* gene define the saccharin taster status. A subsequent extensive genotypic and phenotypic analysis of 30 mice strains revealed that eight out of 89 observed polymorphisms within the *Tas1r3* gene strongly correlate with saccharin preference [141]. Thus, the taster status of different mice strains is determined by several alleles within the *Tas1r3* gene.

In mice, heritable variations for tasting bitter compounds such as sucrose octa-acetate, raffinose undedecaacetate, cycloheximide, and quinine have also been ob-served [54, 55, 81–83, 142]. Although the chromosomal loci have been mapped, the responsible genes have, with one exception, not yet been identified. So far, only variations in cyclohexmide tasting of several strains of mice have been correlated with polymorphisms in the bitter taste receptor gene *Tas2r105* [13]. Moreover heterologous expression studies revealed different receptor responses to cyclohexi-mide of the taster and non taster receptor variants [13].

In humans the most prominent example of a taste variation is the perception of the bitter compound phenylthiocarbamide (PTC). More than 70 years ago, it was shown that strong individual variations for the perception of this compound exist [143]. PTC tasters can detect micromolar concentrations of this compound, whereas the non-tasters are nearly taste-blind even at several hundred-fold higher concentrations [16]. Family studies subsequently revealed that the ability to taste

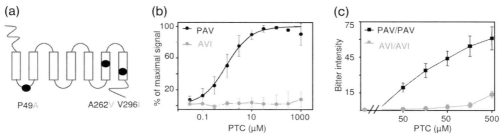

Fig. 18.2. Genetics of PTC tasting in humans. (a) Scheme of the TAS2R38 receptor and localization of its polymorphic sites. (b) Concentration response curves of the effects of PTC on cells expressing the PAV or AVI variants of *TAS2R38*. Note that the AVI variant does not respond to PTC even in 100-fold higher concentrations. (c) Averaged responses to various concentrations of PTC in subjects homozygous for the PAV and AVI *TAS2R38* haplotypes. In line with the receptor function determined *in vitro*, the subjects homozygous for PAV are very sensitive to PTC, while subjects homozygous for AVI are nearly taste-blind.

PTC is inherited [138, 144]. Genetic mapping studies in humans correlated the ability to taste PTC with the KEL blood group antigen locus [145] on chromosome 7. Recently, a genetic study in humans extended these findings. It showed that single-nucleotide polymorphisms in the *TAS2R38* gene, which is near the KEL locus, predict the PTC taster status [15, 146]. There are three common non-synonymous variations in the *TAS2R38* gene, leading to the amino acid exchanges P49A, A262V, V296I (Fig. 18.2). The variations define two frequent haplotypes encoding the two predominant receptor variants *TAS2R38*-PAV and *TAS2R38*-AVI, as well as some additional rare variants [15, 133]. Homozygotes for *TAS2R38*-AVI are PTC non-tasters, whereas humans that are homo- or heterozygous for *TAS2R38*-PAV are PTC tasters [15] (Fig. 18.2). Recently, heterologous expression studies demonstrated that the PAV form but not the AVI form of the receptor responded to PTC [16] (Fig. 18.2). Thus, there is strong molecular, genetic, and functional evidence that polymorphisms in the *TAS2R38* gene define the PTC taster status. This findings provide the first evidence that variations in taste receptors are the molecular basis for heritable variations in taste perception.

Whole family analysis revealed that the occurrence of polymorphisms within the coding regions of *TAS2R* genes is the rule, not the exception [147, 148]. Many of the identified polymorphisms lead to amino acid exchanges or introduce translational stop codons or frameshift mutations. At least 104 different *TAS2R* haplotypes are known (Table 18.1). In light of this variability it appears likely that PTC tasting is just the first example of a heritable variation in bitter taste perception; likely many more will be discovered in the next future. This assumption is further supported by the observation that individual variations in the perception of other bitter compounds such as chloramphenicol, strychnine, quinine, and cascara have been reported [80, 149].

Table 18.1. Single-nucleotide polymorphisms (SNPs) that alter the primary structure of TAS2Rs.

TAS2R gene	Coding SNPs	Number of haplotypes
1	2	3
3	1	2
4	7	8
5	6	7
7	5 + 1 stop	5
8	5	6
9	7	8
10	3	4
13	1	2
14	2	3
16	4	5
38	7	7
39	2	2
40	n.d.	n.d.
41	2	3
42	4	n.d.
43	5	n.d.
44	11 + 1 stop	7
45	6	n.d.
46	3 + 2 stop	6
47	3	4
48	9	9
49	9	7
50	3	4
60	1	2

Adapted from Ref. [147].
n.d., no data.

Interestingly, taste variations are not limited to bitter, but have also been described for human sweet and umami taste [139]. So far, 47 different polymorphisms in the human *TAS1R* genes have been described, 29 of them cause amino acid substitutions. Therefore, variations in human sweet and umami taste might be attributed to polymorphisms in the responsible taste receptor genes [147]. In contrast to sweet, bitter, and umami taste, so far little variations in the perception of sour and salt taste have been observed [147]. Only in African populations does a bimodal distribution of salt taste threshold levels and twin studies suggest a possible genetic cause [150–152]. Given that the ENaC and HCN channels are likely involved in the transduction of these taste qualities the lack of variability is easy to understand. These channels have important functions elsewhere in the body. Therefore, severe functional changes would likely be lethal.

18.12
Taste Variations and Eating Behavior

Our dietary habits depend on a complex combination of social, environmental, and genetic factors. Therefore, not only flavor perception but also many other factors such as age, education, social status, hormone levels, and metabolic rate influence our food choice [5, 7]. Food perception is also strongly associated with hedonic responses and learned experiences [153]. Moreover, the postingestive analysis by chemoreceptors in the lumen of the intestine also influences eating behavior [6]. Therefore, it is difficult to clearly discriminate between sensoric effects mediated by taste and smell and postingestive chemosensation or central actions of hormones in the brain. Nonetheless, multiple independent lines of evidence strongly suggest an impact of taste variations on eating behavior. First, various behavioral studies indicate that flavor is an important factor of food selection [154, 155]. Second, the ingestion of food modulates its perception and palatability [156, 157]. Third, there is strong evidence that several hormones that are involved in the regulation of hunger and satiety modulate taste perception at the periphery [17, 19]. Fourth, it has been shown that specific needs such as salt depletion lead to an increased sensitivity of salt taste [115, 116]. Fifth, several reports associate a loss of sensory perception during aging or medical problems with a reduced food consumption and body weight, although the causal relationship has not yet been established [5, 158–160]. Sixth, comparisons of the taste receptor genes from different species as well as physiological studies clearly show a species-specific evolution of taste perception, which is most likely caused by adaptation to different dietary habits [11]. Thus, inherited differences in taste perception could potentially lead to different dietary habits and contribute to nutrition-related diseases.

In humans, PTC tasting is one of the best-studied genetic traits. PTC tasters and non-tasters are quite frequent. In Europe ~30% of the population are non-tasters while the other ~70% are PTC tasters [161]. PTC tasters and non-tasters differ several hundred-fold in their bitter perception of PTC and also many other N–C=S group-containing compounds [16, 162, 163]. Some of these chemicals are natural food ingredients that occur in many cruciferous vegetables such as Brussels sprouts and cabbage [153]. Due to the strong phenotype, its high frequency within the population, and the observation that N–C=S compounds are food ingredients, several research groups tried to correlate the taster phenotype with differences in food choice and dietary behavior (for review see Ref. [164]). Indeed, the PTC taster status has been associated with various differences in food preferences for vegetables, fruits, but also fats, alcohol, and sweeteners [165–167]. Moreover, several studies found a correlation of PTC tasting and the risk for goiter, alcoholism, cancer, and depression [168–172].

While these studies showed that PTC tasting can be used as a risk marker, the causal link between PTC tasting and disease risks is still elusive. The PTC gene may be in close proximity to other genes and therefore serve as a marker without a crucial involvement in the etiology of a disease. On the other hand, technical issues may account for the difficulty. One limitation of the studies carried out so far is that they used PTC tasting as a phenotypic marker, although PTC taster status is determined by various *TAS2R38* genotypes (for review see Ref. [173]). In addition, the content of N–C=S compounds, the bitter compounds activating *hTAS2R38*, frequently varies by more then one order of magnitude in a given vegetable under study [174]. Moreover, not all N–C=S compounds correlate equally well with PTC tasting [162].

The recent discovery that PTC tasting is mediated by variations in the bitter receptor gene *TAS2R38* [15] now allows researchers to design improved studies by using genetic instead of phenotypic markers. A first pilot study using *TAS2R38* genotypes as markers rather than PTC tasting indeed associated alcohol consumption with the AVI haplotype [175]. The functional characterization of TAS2R38 will establish a precise ligand profile for this receptor. This will help future study designers to better identify the relevant bitter compounds in foodstuffs to be tested. Moreover, reasonable estimations about the relevant concentrations of TAS2R38's bitter compounds can be made. This will permit a better selection of foods that contain appropriate amounts of TAS2R38 agonists for taste preference experiments. The combined use of genetic taste markers and improved methods to select relevant foods will provide powerful new tools to link taster status with different dietary habits.

The high number of polymorphisms already described for the *TAS1R* and *TAS2R* genes implies that in addition to PTC tasting many other heritable variations in sweet, bitter, and umami taste perception exist that need to be discovered in the near future. Given that 25 *TAS2R* genes are involved in bitter perception, genetic variations of a single bitter taste receptor gene will only have a limited influence on bitter taste in general and on the liking or disliking of foods and thereby on nutritional status and disease risk. We therefore need to know about variations in the whole *TAS2R* gene family. Functional variations in the three *TAS1R* genes will potentially alter sweet and umami taste in general and could therefore cause more global alteration in food choice. Therefore, the elucidation of genetic traits in sweet and umami taste and their impact on food consumption remains one of the important challenges in future taste research.

References

1 McCrory, M.A., V.M. Suen, S.B. Roberts (2002) Biobehavioral influences on energy intake and adult weight gain. *J Nutr* 132, 3830S–3834S.

2 Formiguera, X., A. Canton (2004) Obesity: epidemiology and clinical aspects. *Best Pract Res Clin Gastroenterol* 18, 1125–1146.

3 Finkelstein, E.A., C.J. Ruhm, K.M. Kosa (2005) Economic causes and consequences of obesity. *Annu Rev Public Health* 26, 239–257.

4 Drewnowski, A., N. Darmon (2005) Food choices and diet costs: an economic analysis. *J Nutr* 135, 900–904.

5 Drewnowski, A. (1997) Taste preferences and food intake. *Annu Rev Nutr* 17, 237–253.

6 Sclafani, A. (2001) Psychobiology of food preferences. *Int J Obes Relat Metab Disord* 25 (Suppl 5), S13–16.

7 Nasser, J. (2001) Taste, food intake and obesity. *Obes Rev* 2, 213–218.

8 NACHMAN, M., L.P. COLE (**1971**) Role of taste in specific hungers. In: *Handbook of Sensory Physiology* IV: *Chemical Senses*, Part 2. H. ATRUM, R. JUNG, W.R. LOEWENSTEIN, D.M. MacKAY, H.L. TEUBNER, editors. Berlin, Heidelberg, New York: Springer-Verlag, pp. 338–358.

9 LINDEMANN, B. (**1996**) Taste reception. *Physiol Rev* 76, 718–766.

10 LI, X., L. STASZEWSKI, H. XU, K. DURICK, M. ZOLLER, E. ADLER (**2002**) Human receptors for sweet and umami taste. *Proc Natl Acad Sci USA* 99, 4692–4696.

11 SHI, P., J. ZHANG, H. YANG, Y.P. ZHANG (**2003**) Adaptive diversification of bitter taste receptor genes in Mammalian evolution. *Mol Biol Evol* 20, 805–814.

12 GUESRY, P.R. (**2005**) Impact of "functional food". *Forum Nutr* 73–83.

13 CHANDRASHEKAR, J., K.L. MUELLER, M.A. HOON, E. ADLER, L. FENG, W. GUO, C.S. ZUKER, N.J. RYBA (**2000**) T2Rs function as bitter taste receptors. *Cell* 100, 703–711.

14 NELSON, G., M.A. HOON, J. CHANDRASHEKAR, Y. ZHANG, N.J. RYBA, C.S. ZUKER (**2001**) Mammalian sweet taste receptors. *Cell* 106, 381–390.

15 KIM, U.K., E. JORGENSON, H. COON, M. LEPPERT, N. RISCH, D. DRAYNA (**2003**) Positional cloning of the human quantitative trait locus underlying taste sensitivity to phenylthiocarbamide. *Science* 299, 1221–1225.

16 BUFE, B., P.A. BRESLIN, C. KUHN, D.R. REED, C.D. THARP, J.P. SLACK, U.K. KIM, D. DRAYNA, W. MEYERHOF (**2005**) The molecular basis of individual differences in phenylthiocarbamide and propylthiouracil bitterness perception. *Curr Biol* 15, 322–327.

17 HERNESS, S., F.L. ZHAO, S.G. LU, N. KAYA, T. SHEN (**2002**) Expression and physiological actions of cholecystokinin in rat taste receptor cells. *J Neurosci* 22, 10018–10029.

18 SHIGEMURA, N., H. MIURA, Y. KUSAKABE, A. HINO, Y. NINOMIYA (**2003**) Expression of leptin receptor (Ob-R) isoforms and signal transducers and activators of transcription (STATs) mRNAs in the mouse taste buds. *Arch Histol Cytol* 66, 253–260.

19 KAWAI, K., K. SUGIMOTO, K. NAKASHIMA, H. MIURA, Y. NINOMIYA (**2000**) Leptin as a modulator of sweet taste sensitivities in mice. *Proc Natl Acad Sci USA* 97, 11044–11049.

20 DANIELS, D., S.J. FLUHARTY (**2004**) Salt appetite: a neurohormonal viewpoint. *Physiol Behav* 81, 319–337.

21 LINDEMANN, B. (**1997**) Sodium taste. *Curr Opin Nephrol Hypertens* 6, 425–429.

22 SKRAMLIK, E. (**1926**) *Handbuch der Physiologie der Niederen Sinne*. Leipzig: Georg Thieme Verlag.

23 MIYAMOTO, T., R. FUJIYAMA, Y. OKADA, T. SATO (**2000**) Acid and salt responses in mouse taste cells. *Prog Neurobiol* 62, 135–157.

24 HECK, G.L., S. MIERSON, J.A. DeSIMONE (**1984**) Salt taste transduction occurs through an amiloride-sensitive sodium transport pathway. *Science* 223, 403–405.

25 KRETZ, O., P. BARBRY, R. BOCK, B. LINDEMANN (**1999**) Differential expression of RNA and protein of the three pore-forming subunits of the amiloride-sensitive epithelial sodium channel in taste buds of the rat. *J Histochem Cytochem* 47, 51–64.

26 LIN, W., T.E. FINGER, B.C. ROSSIER, S.C. KINNAMON (**1999**) Epithelial Na^+ channel subunits in rat taste cells: localization and regulation by aldosterone. *J Comp Neurol* 405, 406–420.

27 AVENET, P., B. LINDEMANN (**1991**) Noninvasive recording of receptor cell action potentials and sustained currents from single taste buds maintained in the tongue: the response to mucosal NaCl and amiloride. *J Membr Biol* 124, 33–41.

28 KELLENBERGER, S., I. GAUTSCHI, L. SCHILD (**1999**) A single point mutation in the pore region of the epithelial Na+ channel changes ion selectivity by modifying molecular sieving. *Proc Natl Acad Sci USA* 96, 4170–4175.

29 KELLENBERGER, S., M. AUBERSON, I. GAUTSCHI, E. SCHNEEBERGER, L. SCHILD (2001) Permeability properties of ENaC selectivity filter mutants. *J Gen Physiol* 118, 679–692.

30 OSSEBAARD, C.A., D.V. SMITH (1995) Effect of amiloride on the taste of NaCl, Na-gluconate and KCl in humans: implications for Na$^+$ receptor mechanisms. *Chem Senses* 20, 37–46.

31 LYALL, V., G.L. HECK, A.K. VINNIKOVA, S. GHOSH, T.H. PHAN, R.I. ALAM, O.F. RUSSELL, S.A. MALIK, J.W. BIGBEE, J.A. DESIMONE (2004) The mammalian amiloride-insensitive non-specific salt taste receptor is a vanilloid receptor-1 variant. *J Physiol* 558, 147–159.

32 DESIMONE, J.A., V. LYALL, G.L. HECK, T.H. PHAN, R.I. ALAM, G.M. FELDMAN, R.M. BUCH (2001) A novel pharmacological probe links the amiloride-insensitive NaCl, KCl, and NH$_4$Cl chorda tympani taste responses. *J Neurophysiol* 86, 2638–2641.

33 LYALL, V., G.L. HECK, T.H. PHAN, S. MUMMALANENI, S.A. MALIK, A.K. VINNIKOVA, J.A. DESIMONE (2005) Ethanol modulates the VR-1 variant amiloride-insensitive salt taste receptor. II. Effect on chorda tympani salt responses. *J Gen Physiol* 125, 587–600.

34 LYALL, V., G.L. HECK, T.H. PHAN, S. MUMMALANENI, S.A. MALIK, A.K. VINNIKOVA, J.A. DESIMONE (2005) Ethanol modulates the VR-1 variant amiloride-insensitive salt taste receptor. I. Effect on TRC volume and Na$^+$ flux. *J Gen Physiol* 125, 569–585.

35 HERNESS, M.S., T.A. GILBERTSON (1999) Cellular mechanisms of taste transduction. *Annu Rev Physiol* 61, 873–900.

36 KINNAMON, S.C., V.E. DIONNE, K.G. BEAM (1988) Apical localization of K+ channels in taste cells provides the basis for sour taste transduction. *Proc Natl Acad Sci USA* 85, 7023–7027.

37 LINDEMANN, B. (2001) Receptors and transduction in taste. *Nature* 413, 219–225.

38 STEVENS, D.R., R. SEIFERT, B. BUFE, F. MULLER, E. KREMMER, R. GAUSS, W. MEYERHOF, U.B. KAUPP, B. LINDEMANN (2001) Hyperpolarization-activated channels HCN1 and HCN4 mediate responses to sour stimuli. *Nature* 413, 631–635.

39 UGAWA, S., Y. MINAMI, W. GUO, Y. SAISHIN, K. TAKATSUJI, T. YAMAMOTO, M. TOHYAMA, S. SHIMADA (1998) Receptor that leaves a sour taste in the mouth. *Nature* 395, 555–556.

40 UGAWA, S. (2003) Identification of sour-taste receptor genes. *Anat Sci Int* 78, 205–210.

41 UGAWA, S., T. YAMAMOTO, T. UEDA, Y. ISHIDA, A. INAGAKI, M. NISHIGAKI, S. SHIMADA (2003) Amiloride-insensitive currents of the acid-sensing ion channel-2a (ASIC2a)/ASIC2b heteromeric sour-taste receptor channel. *J Neurosci* 23, 3616–3622.

42 LIN, W., T. OGURA, S.C. KINNAMON (2002) Acid-activated cation currents in rat vallate taste receptor cells. *J Neurophysiol* 88, 133–141.

43 GANZEVLES, P.G., J.H. KROEZE (1987) Effects of adaptation and cross-adaptation to common ions on sourness intensity. *Physiol Behav* 40, 641–646.

44 KINAMON, S.C.P., M.P. STONE, L.M. LIN, W. WELSH, M.J. (2000) The acid sensing ion channel BNC1 is not required for sour taste transduction. Paper presented at the XII International Symposium on Olfaction and Taste (Brighton, UK), p. 104.

45 RICHTER, T.A., G.A. DVORYANCHIKOV, S.D. ROPER, N. CHAUDHARI (2004) Acid-sensing ion channel-2 is not necessary for sour taste in mice. *J Neurosci* 24, 4088–4091.

46 DREWNOWSKI, A. (1995) Energy intake and sensory properties of food. *Am J Clin Nutr* 62, 1081S–1085S.

47 SCHIFFMAN, S.S., C.A. GATLIN (1993) Sweeteners: state of knowledge review. *Neurosci Biobehav Rev* 17, 313–345.

48 CHON, G. (1914) Die Organischen Geschmacksstoffe. Franz Seimenroth, Berlin.

49 Brouwer, J.N., G. Hellekant, Y. Kasahara, H. van der Wel, Y. Zotterman (**1973**) Electrophysiological study of the gustatory effects of the sweet proteins monellin and thaumatin in monkey, guinea pig and rat. *Acta Physiol Scand* 89, 550–557.

50 Sclafani, A., M. Abrams (**1986**) Rats show only a weak preference for the artificial sweetener aspartame. *Physiol Behav* 37, 253–256.

51 Sclafani, A., C. Perez (**1997**) Cypha [propionic acid, 2-(4-methoxyphenol) salt] inhibits sweet taste in humans, but not in rats. *Physiol Behav* 61, 25–29.

52 Tonosaki, K., K. Miwa, F. Kanemura (**1997**) Gustatory receptor cell responses to the sweeteners, monellin and thaumatin. *Brain Res* 748, 234–236.

53 Fuller, J.L. 1974. Single-locus control of saccharin preference in mice. *J Hered* 65, 33–36.

54 Lush, I.E., N. Hornigold, P. King, J.P. Stoye (**1995**) The genetics of tasting in mice. VII. Glycine revisited, and the chromosomal location of Sac and Soa. *Genet Res* 66, 167–174.

55 Lush, I.E. (**1989**) The genetics of tasting in mice. VI. Saccharin, acesulfame, dulcin and sucrose. *Genet Res* 53, 95–99.

56 Montmayeur, J.P., S.D. Liberles, H. Matsunami, L.B. Buck (**2001**) A candidate taste receptor gene near a sweet taste locus. *Nat Neurosci* 4, 492–498.

57 Sainz, E., J.N. Korley, J.F. Battey, S.L. Sullivan (**2001**) Identification of a novel member of the T1R family of putative taste receptors. *J Neurochem* 77, 896–903.

58 Max, M., Y.G. Shanker, L. Huang, M. Rong, Z. Liu, F. Campagne, H. Weinstein, S. Damak, R.F. Margolskee (**2001**) Tas1r3, encoding a new candidate taste receptor, is allelic to the sweet responsiveness locus Sac. *Nat Genet* 28, 58–63.

59 Kitagawa, M., Y. Kusakabe, H. Miura, Y. Ninomiya, A. Hino (**2001**) Molecular genetic identification of a candidate receptor gene for sweet

taste. *Biochem Biophys Res Commun* 283, 236–242.

60 Bachmanov, A.A., X. Li, D.R. Reed, J.D. Ohmen, S. Li, Z. Chen, M.G. Tordoff, P.J. de Jong, C. Wu, D.B. West, A. Chatterjee, D.A. Ross, G.K. Beauchamp (**2001**) Positional cloning of the mouse saccharin preference (Sac) locus. *Chem Senses* 26, 925–933.

61 Hoon, M.A., E. Adler, J. Lindemeier, J.F. Battey, N.J. Ryba, C.S. Zuker (**1999**) Putative mammalian taste receptors: a class of taste-specific GPCRs with distinct topographic selectivity. *Cell* 96, 541–551.

62 Liao, J., P.G. Schultz (**2003**) Three sweet receptor genes are clustered in human chromosome 1. *Mamm Genome* 14, 291–301.

63 Nelson, G., J. Chandrashekar, M.A. Hoon, L. Feng, G. Zhao, N.J. Ryba, C.S. Zuker (**2002**) An amino-acid taste receptor. *Nature* 416, 199–202.

64 Damak, S., M. Rong, K. Yasumatsu, Z. Kokrashvili, V. Varadarajan, S. Zou, P. Jiang, Y. Ninomiya, R.F. Margolskee (**2003**) Detection of sweet and umami taste in the absence of taste receptor T1r3. *Science* 301, 850–853.

65 Zhao, G.Q., Y. Zhang, M.A. Hoon, J. Chandrashekar, I. Erlenbach, N.J. Ryba, C.S. Zuker (**2003**) The receptors for mammalian sweet and umami taste. *Cell* 115, 255–266.

66 Peri, I., H. Mamrud-Brains, S. Rodin, V. Krizhanovsky, Y. Shai, S. Nir, M. Naim (**2000**) Rapid entry of bitter and sweet tastants into liposomes and taste cells: implications for signal transduction. *Am J Physiol Cell Physiol* 278, C17–25.

67 Naim, M., R. Seifert, B. Nurnberg, L. Grunbaum, G. Schultz (**1994**) Some taste substances are direct activators of G-proteins. *Biochem J* 297 (Pt 3), 451–454.

68 Naim, M., B.J. Striem, M. Tal (**1998**) Cellular signal transduction of sweetener-induced taste. *Adv Food Nutr Res* 42, 211–243.

69 Yamaguchi, S., K. Ninomiya (**2000**)

Umami and food palatability. *J Nutr* 130, 921S–926S.

70 Bellisle, F. (**1999**) Glutamate and the UMAMI taste: sensory, metabolic, nutritional and behavioural considerations. A review of the literature published in the last 10 years. *Neurosci Biobehav Rev* 23, 423–438.

71 Kurihara, K., M. Kashiwayanagi (**2000**) Physiological studies on umami taste. *J Nutr* 130, 931S–934S.

72 Caicedo, A., M.S. Jafri, S.D. Roper (**2000**) In situ Ca^{2+} imaging reveals neurotransmitter receptors for glutamate in taste receptor cells. *J Neurosci* 20, 7978–7985.

73 Caicedo, A., K.N. Kim, S.D. Roper (**2002**) Individual mouse taste cells respond to multiple chemical stimuli. *J Physiol* 544, 501–509.

74 Caicedo, A., K.N. Kim, S.D. Roper (**2000**) Glutamate-induced cobalt uptake reveals non-NMDA receptors in rat taste cells. *J Comp Neurol* 417, 315–324.

75 Chaudhari, N., H. Yang, C. Lamp, E. Delay, C. Cartford, T. Than, S. Roper (**1996**) The taste of monosodium glutamate: membrane receptors in taste buds. *J Neurosci* 16, 3817–3826.

76 Toyono, T., Y. Seta, S. Kataoka, S. Kawano, R. Shigemoto, K. Toyoshima (**2003**) Expression of metabotropic glutamate receptor group I in rat gustatory papillae. *Cell Tissue Res* 313, 29–35.

77 Chaudhari, N., A.M. Landin, S.D. Roper (**2000**) A metabotropic glutamate receptor variant functions as a taste receptor. *Nat Neurosci* 3, 113–119.

78 Chaudhari, N., S.D. Roper (**1998**) Molecular and physiological evidence for glutamate (umami) taste transduction via a G protein-coupled receptor. *Ann NY Acad Sci* 855, 398–406.

79 Delwiche, J.F., Z. Buletic, P.A. Breslin (**2001**) Covariation in individuals' sensitivities to bitter compounds: evidence supporting multiple receptor/transduction mechanisms. *Percept Psychophys* 63, 761–776.

80 Blakeslee, A. (**1935**) A dinner demonstration of threshold differences in taste and smell. *Science* 81, 504–507.

81 Lush, I.E. (**1981**) The genetics of tasting in mice. I. Sucrose octaacetate. *Genet Res* 38, 93–95.

82 Lush, I.E. (**1984**) The genetics of tasting in mice. III. Quinine. *Genet Res* 44, 151–160.

83 Lush, I.E. (**1986**) The genetics of tasting in mice. IV. The acetates of raffinose, galactose and beta-lactose. *Genet Res* 47, 117–123.

84 Adler, E., M.A. Hoon, K.L. Mueller, J. Chandrashekar, N.J. Ryba, C.S. Zuker (**2000**) A novel family of mammalian taste receptors. *Cell* 100, 693–702.

85 Matsunami, H., J.P. Montmayeur, L.B. Buck (**2000**) A family of candidate taste receptors in human and mouse. *Nature* 404, 601–604.

86 Go, Y., Y. Satta, O. Takenaka, N. Takahata (**2005**) Lineage-specific loss of function of bitter taste receptor genes in humans and nonhuman primates. *Genetics* 170, 313–326.

87 Bufe, B., T. Hofmann, D. Krautwurst, J.D. Raguse, W. Meyerhof (**2002**) The human TAS2R16 receptor mediates bitter taste in response to beta-glucopyranosides. *Nat Genet* 32, 397–401.

88 Behrens, M., A. Brockhoff, C. Kuhn, B. Bufe, M. Winnig, W. Meyerhof (**2004**) The human taste receptor hTAS2R14 responds to a variety of different bitter compounds. *Biochem Biophys Res Commun* 319, 479–485.

89 Kuhn, C., B. Bufe, M. Winnig, T. Hofmann, O. Frank, M. Behrens, T. Lewtschenko, J.P. Slack, C.D. Ward, W. Meyerhof (**2004**) Bitter taste receptors for saccharin and acesulfame K. *J Neurosci* 24, 10260–10265.

90 Pronin, A.N., H. Tang, J. Connor, W. Keung (**2004**) Identification of ligands for two human bitter T2R receptors. *Chem Senses* 29, 583–593.

91 Finger, T.E., B. Bottger, A. Hansen,

K.T. Anderson, H. Alimohammadi, W.L. Silver (**2003**) Solitary chemo-receptor cells in the nasal cavity serve as sentinels of respiration. *Proc Natl Acad Sci USA* 100, 8981–8986.

92 Wu, S.V., N. Rozengurt, M. Yang, S.H. Young, J. Sinnett-Smith, E. Rozengurt (**2002**) Expression of bitter taste receptors of the T2R family in the gastrointestinal tract and enteroendocrine STC-1 cells. *Proc Natl Acad Sci USA* 99, 2392–2397.

93 McLaughlin, S.K., P.J. McKinnon, R.F. Margolskee (**1992**) Gustducin is a taste-cell-specific G protein closely related to the transducins. *Nature* 357, 563–569.

94 McLaughlin, S.K., P.J. McKinnon, N. Spickofsky, W. Danho, R.F. Margolskee (**1994**) Molecular cloning of G proteins and phosphodiesterases from rat taste cells. *Physiol Behav* 56, 1157–1164.

95 Yan, W., G. Sunavala, S. Rosenzweig, M. Dasso, J.G. Brand, A.I. Spielman (**2001**) Bitter taste transduced by PLC-beta$_2$-dependent rise in IP$_3$ and alpha-gustducin-dependent fall in cyclic nucleotides. *Am J Physiol Cell Physiol* 280, C742–751.

96 Ming, D., L. Ruiz-Avila, R.F. Margolskee (**1998**) Characterization and solubilization of bitter-responsive receptors that couple to gustducin. *Proc Natl Acad Sci USA* 95, 8933–8938.

97 Wong, G.T., K.S. Gannon, R.F. Margolskee (**1996**) Transduction of bitter and sweet taste by gustducin. *Nature* 381, 796–800.

98 Ruiz-Avila, L., S.K. McLaughlin, D. Wildman, P.J. McKinnon, A. Robichon, N. Spickofsky, R.F. Margolskee (**1995**) Coupling of bitter receptor to phosphodiesterase through transducin in taste receptor cells. *Nature* 376, 80–85.

99 Perez, C.A., L. Huang, M. Rong, J.A. Kozak, A.K. Preuss, H. Zhang, M. Max, R.F. Margolskee (**2002**) A transient receptor potential channel expressed in taste receptor cells. *Nat Neurosci* 5, 1169–1176.

100 Spielman, A.I., T. Huque, H. Nagai, G. Whitney, J.G. Brand (**1994**) Generation of inositol phosphates in bitter taste transduction. *Physiol Behav* 56, 1149–1155.

101 Huang, L., Y.G. Shanker, J. Dubauskaite, J.Z. Zheng, W. Yan, S. Rosenzweig, A.I. Spielman, M. Max, R.F. Margolskee (**1999**) Ggamma13 colocalizes with gustducin in taste receptor cells and mediates IP3 responses to bitter denatonium. *Nat Neurosci* 2, 1055–1062.

102 Blake, B.L., M.R. Wing, J.Y. Zhou, Q. Lei, J.R. Hillmann, C.I. Behe, R.A. Morris, T.K. Harden, D.A. Bayliss, R.J. Miller, D.P. Siderovski (**2001**) G beta association and effector interaction selectivities of the divergent G gamma subunit G gamma(13). *J Biol Chem* 276, 49267–49274.

103 Zhang, Y., M.A. Hoon, J. Chandrashekar, K.L. Mueller, B. Cook, D. Wu, C.S. Zuker, N.J. Ryba (**2003**) Coding of sweet, bitter, and umami tastes: different receptor cells sharing similar signaling pathways. *Cell* 112, 293–301.

104 Häning, D.P. (**1901**) Zur Psycho-physik des Geschmackssinnes. *Philosophische Studien* 17, 576–623.

105 Smith, D.V., S.J. St John (**1999**) Neural coding of gustatory information. *Curr Opin Neurobiol* 9, 427–435.

106 Hellekant, G., Y. Ninomiya, V. Danilova (**1998**) Taste in chimpanzees. III. Labeled-line coding in sweet taste. *Physiol Behav* 65, 191–200.

107 Smith, D.V., S.J. John, J.D. Boughter (**2000**) Neuronal cell types and taste quality coding. *Physiol Behav* 69, 77–85.

108 Yamamoto, T., N. Yuyama (**1987**) On a neural mechanism for cortical processing of taste quality in the rat. *Brain Res* 400, 312–320.

109 Woolston, D.C., R.P. Erickson (**1979**) Concept of neuron types in gustation in the rat. *J Neurophysiol* 42, 1390–1409.

110 Dahl, M., R.P. Erickson, S.A. Simon

(**1997**) Neural responses to bitter compounds in rats. *Brain Res* 756, 22–34.

111 SATO, M., H. OGAWA, S. YAMASHITA (**1994**) Gustatory responsiveness of chorda tympani fibers in the cynomolgus monkey. *Chem Senses* 19, 381–400.

112 DANILOVA, V., G. HELLEKANT, J.M. TINTI, C. NOFRE (**1998**) Gustatory responses of the hamster Mesocricetus auratus to various compounds considered sweet by humans. *J Neurophysiol* 80, 2102–2112.

113 DANILOVA, V., Y. DANILOV, T. ROBERTS, J.M. TINTI, C. NOFRE, G. HELLEKANT (**2002**) Sense of taste in a new world monkey, the common marmoset: recordings from the chorda tympani and glossopharyngeal nerves. *J Neurophysiol* 88, 579–594.

114 MUELLER, K.L., M.A. HOON, I. ERLENBACH, J. CHANDRASHEKAR, C.S. ZUKER, N.J. RYBA (**2005**) The receptors and coding logic for bitter taste. *Nature* 434, 225–229.

115 OKADA, Y., T. MIYAMOTO, T. SATO (**1990**) Aldosterone increases gustatory neural response to NaCl in frog. *Comp Biochem Physiol A* 97, 535–536.

116 HERNESS, M.S. (**1992**) Aldosterone increases the amiloride-sensitivity of the rat gustatory neural response to NaCl. *Comp Biochem Physiol Comp Physiol* 103, 269–273.

117 EPSTEIN, A.N. (**1991**) Neurohormonal control of salt intake in the rat. *Brain Res Bull* 27, 315–320.

118 SHIGEMURA, N., R. OHTA, Y. KUSAKABE, H. MIURA, A. HINO, K. KOYANO, K. NAKASHIMA, Y. NINOMIYA (**2004**) Leptin modulates behavioral responses to sweet substances by influencing peripheral taste structures. *Endocrinology* 145, 839–847.

119 RAYNAUD, E., J.F. BRUN, A. PEREZ-MARTIN, C. SAGNES, A.M. BOULARAN, C. FEDOU, J. MERCIER (**1999**) Serum leptin is associated with the perception of palatability during a standardized high-carbohydrate breakfast test. *Clin Sci (Lond)* 96, 343–348.

120 KARHUNEN, L.J., R.I. LAPPALAINEN, S.M. HAFFNER, R.H. VALVE, H. TUORILA, H. MIETTINEN, M.I. UUSITUPA (**1998**) Serum leptin, food intake and preferences for sugar and fat in obese women. *Int J Obes Relat Metab Disord* 22, 819–821.

121 SHEN, T., N. KAYA, F.L. ZHAO, S.G. LU, Y. CAO, S. HERNESS (**2005**) Co-expression patterns of the neuropeptides vasoactive intestinal peptide and cholecystokinin with the transduction molecules alpha-gustducin and T1R2 in rat taste receptor cells. *Neuroscience* 130, 229–238.

122 LU, S.G., F.L. ZHAO, S. HERNESS (**2003**) Physiological phenotyping of cholecystokinin-responsive rat taste receptor cells. *Neurosci Lett* 351, 157–160.

123 DE JONGHE, B.C., A. HAJNAL, M. COVASA (**2005**) Increased oral and decreased intestinal sensitivity to sucrose in obese, prediabetic CCK-A receptor-deficient OLETF rats. *Am J Physiol Regul Integr Comp Physiol* 288, R292–300.

124 SIMON, S.A., L. LIU, R.P. ERICKSON (**2003**) Neuropeptides modulate rat chorda tympani responses. *Am J Physiol Regul Integr Comp Physiol* 284, R1494–1505.

125 CROSS-MELLOR, S.K., W.D. KENT, K.P. OSSENKOPP, M. KAVALIERS (**1999**) Differential effects of lipopolysaccharide and cholecystokinin on sucrose intake and palatability. *Am J Physiol* 277, R705–715.

126 ZHAO, F.L., T. SHEN, N. KAYA, S.G. LU, Y. CAO, S. HERNESS (**2005**) Expression, physiological action, and coexpression patterns of neuropeptide Y in rat taste-bud cells. *Proc Natl Acad Sci USA* 102, 11100–11105.

127 LASKA, M. (**2002**) Gustatory responsiveness to food-associated saccharides in European rabbits, Oryctolagus cuniculus. *Physiol Behav* 76, 335–341.

128 FISCHER, A., Y. GILAD, O. MAN, S. PAABO (**2005**) Evolution of bitter taste receptors in humans and apes. *Mol Biol Evol* 22, 432–436.

129 WANG, X., S.D. THOMAS, J. ZHANG (**2004**) Relaxation of selective

constraint and loss of function in the evolution of human bitter taste receptor genes. *Hum Mol Genet* 13, 2671–2678.

130 CONTE, C., M. EBELING, A. MARCUZ, P. NEF, P.J. ANDRES-BARQUIN (**2003**) Evolutionary relationships of the Tas2r receptor gene families in mouse and human. *Physiol Genomics* 14, 73–82.

131 MILTON, K. (**2003**) The critical role played by animal source foods in human (Homo) evolution. *J Nutr* 133, 3886S–3892S.

132 LEONARD, W.R. (**2002**) Food for thought. Dietary change was a driving force in human evolution. *Sci Am* 287, 106–115.

133 WOODING, S., U.K. KIM, M.J. BAMSHAD, J. LARSEN, L.B. JORDE, D. DRAYNA (**2004**) Natural selection and molecular evolution in PTC, a bitter-taste receptor gene. *Am J Hum Genet* 74, 637–646.

134 SORANZO, N., B. BUFE, P.C. SABETI, J.F. WILSON, M.E. WEALE, R. MARGUERIE, W. MEYERHOF, D.B. GOLDSTEIN. Positive selection on a functional allele of the hT2R16 bitter taste receptor associated with higher sensitivity to b-glucopyranosides. *Curr Biol* 15, 1257–1265.

135 NAGEL, R.L., C. RAVENTOS, H.B. TANOWITZ, M. WITTNER (**1980**) Effect of sodium cyanate on Plasmodium falciparum in vitro. *J Parasitol* 66, 483–487.

136 JACKSON, F. (**1990**) Two evolutionary models for the interactions of dietary cyanogens, hemoglobin S, and falciparum malaria. *Am J Hum Biol* 2, 521–532.

137 MADUAGWU, E.N. (**1989**) Metabolism of linamarin in rats. *Food Chem Toxicol* 27, 451–454.

138 BLAKESLEE, A.F. (**1932**) Genetics of sensory thresholds: taste for pheny thio carbamide. *Proc Natl Acad Sci USA* 18, 120–130.

139 LUGAZ, O., A.M. PILLIAS, A. FAURION (**2002**) A new specific ageusia: some humans cannot taste L-glutamate. *Chem Senses* 27, 105–115.

140 BACHMANOV, A.A., M.G. TORDOFF, G.K. BEAUCHAMP (**2001**) Sweetener preference of C57BL/6ByJ and 129P3/J mice. *Chem Senses* 26, 905–913.

141 REED, D.R., S. LI, X. LI, L. HUANG, M.G. TORDOFF, R. STARLING-RONEY, K. TANIGUCHI, D.B. WEST, J.D. OHMEN, G.K. BEAUCHAMP, A.A. BACHMANOV (**2004**) Polymorphisms in the taste receptor gene (Tas1r3) region are associated with saccharin preference in 30 mouse strains. *J Neurosci* 24, 938–946.

142 LUSH, I.E., G. HOLLAND (**1988**) The genetics of tasting in mice. V. Glycine and cycloheximide. *Genet Res* 52, 207–212.

143 FOX, A.L. (**1932**) The relationship between chemical constitution and taste. *Proc. Natl Acad. Sci. USA* 18, 115–120.

144 BLAKESLEE, A.F. (**1931**) Genetics of sensory thresholds: taste for pheny thio carbamide. *Science* 74, 607.

145 CONNEALLY, P.M., M. DUMONT-DRISCOLL, R.S. HUNTZINGER, W.E. NANCE, C.E. JACKSON (**1976**) Linkage relations of the loci for Kell and phenylthiocarbamide taste sensitivity. *Hum Hered* 26, 267–271.

146 DRAYNA, D., H. COON, U.K. KIM, T. ELSNER, K. CROMER, B. OTTERUD, L. BAIRD, A.P. PEIFFER, M. LEPPERT (**2003**) Genetic analysis of a complex trait in the Utah Genetic Reference Project: a major locus for PTC taste ability on chromosome 7q and a secondary locus on chromosome 16p. *Hum Genet* 112, 567–572.

147 DRAYNA, D. (**2005**) Human taste genetics. *Annu Rev Genomics Hum Genet* 6, 217–235.

148 UEDA, T., S. UGAWA, Y. ISHIDA, Y. SHIBATA, S. MURAKAMI, S. SHIMADA (**2001**) Identification of coding single-nucleotide polymorphisms in human taste receptor genes involving bitter tasting. *Biochem Biophys Res Commun* 285, 147–151.

149 SUGINO, Y., A. UMEMOTO, S. MIZUTANI (**2002**) Insensitivity to the bitter taste of chloramphenicol: an autosomal recessive trait. *Genes Genet Syst* 77, 59–62.

150 GREENE, L.S., J.A. DESOR, O. MALLER (**1975**) Heredity and experience: their

relative importance in the development of taste preference in man. *J Comp Physiol Psychol* 89, 279–284.

151 OKORO, E.O., G.E. UROGHIDE, E.T. JOLAYEMI (**1998**) Salt taste sensitivity and blood pressure in adolescent school children in southern Nigeria. *East Afr Med J* 75, 199–203.

152 OKORO, E.O., F. BRISIBE, E.T. JOLAYEMI, G. HADIZATH TAIMAGARI (**2000**) Taste sensitivity to sodium chloride and sucrose in a group of adolescent children in Northern Nigeria. *Ethn Dis* 10, 53–59.

153 DREWNOWSKI, A. (**2001**) The science and complexity of bitter taste. *Nutr Rev* 59, 163–169.

154 HETHERINGTON, M.M. (**1996**) Sensory-specific satiety and its importance in meal termination. *Neurosci Biobehav Rev* 20, 113–117.

155 SORENSEN, L.B., P. MOLLER, A. FLINT, M. MARTENS, A. RABEN (**2003**) Effect of sensory perception of foods on appetite and food intake: a review of studies on humans. *Int J Obes Relat Metab Disord* 27, 1152–1166.

156 BERRIDGE, K.C. (**1991**) Modulation of taste affect by hunger, caloric satiety, and sensory-specific satiety in the rat. *Appetite* 16, 103–120.

157 ZVEREV, Y.P. (**2004**) Effects of caloric deprivation and satiety on sensitivity of the gustatory system. *BMC Neurosci* 5, 5.

158 MATTES, R.D., B.J. COWART (**1994**) Dietary assessment of patients with chemosensory disorders. *J Am Diet Assoc* 94, 50–56.

159 MATTES, R.D., B.J. COWART, M.A. SCHIAVO, C. ARNOLD, B. GARRISON, M.R. KARE, L.D. LOWRY (**1990**) Dietary evaluation of patients with smell and/or taste disorders. *Am J Clin Nutr* 51, 233–240.

160 SCHIFFMAN, S.S., B.G. GRAHAM (**2000**) Taste and smell perception affect appetite and immunity in the elderly. *Eur J Clin Nutr* 54 (Suppl 3), S54–63.

161 GUO, S.W., D.R. REED (**2001**) The genetics of phenylthiocarbamide perception. *Ann Hum Biol* 28, 111–142.

162 BARNICOT, N.A., H. HARRIS, H. KALMUS (**1951**) Taste thresholds of further eighteen compounds and their correlation with PTC thresholds. *Ann Eugen* 16, 119–128.

163 HARRIS, H., H. KALMUS (**1949**) Chemical specificity in genetical differences of taste sensitivity. *Ann Eugen* 15, 32–45.

164 PRESCOTT, J., J.T. BEVERLY, editors (**2004**) *Genetic Variations in Taste Sensitivity.* New York: Marcel Dekker.

165 BARTOSHUK, L.M. (**1979**) Bitter taste of saccharin related to the genetic ability to taste the bitter substance 6-n-propylthiouracil. *Science* 205, 934–935.

166 TEPPER, B.J., R.J. NURSE (**1997**) Fat perception is related to PROP taster status. *Physiol Behav* 61, 949–954.

167 DREWNOWSKI, A., C.L. ROCK (**1995**) The influence of genetic taste markers on food acceptance. *Am J Clin Nutr* 62, 506–511.

168 FACCHINI, F., A. ABBATI, S. CAMPAG-NONI (**1990**) Possible relations between sensitivity to phenylthio-carbamide and goiter. *Hum Biol* 62, 545–552.

169 HARRIS, H., H. KALMUS, W.R. TROTTER (**1949**) Taste sensitivity to phenylthiourea in goitre and diabetes. *Lancet* 2, 1038.

170 PELCHAT, M.L., S. DANOWSKI (**1992**) A possible genetic association between PROP-tasting and alcoholism. *Physiol Behav* 51, 1261–1266.

171 JOINER, T.E., JR., M. PEREZ (**2004**) Phenylthiocarbamide tasting and family history of depression, revisited: low rates of depression in families of supertasters. *Psychiatry Res* 126, 83–87.

172 DREWNOWSKI, A., S.A. HENDERSON, C.S. HANN, W.A. BERG, M.T. RUFFIN (**2000**) Genetic taste markers and preferences for vegetables and fruit of female breast care patients. *J Am Diet Assoc* 100, 191–197.

173 BARTOSHUK, L.M. (**2000**) Comparing sensory experiences across individuals: recent psychophysical advances illuminate genetic variation in taste perception. *Chem Senses* 25, 447–460.

174 Drewnowski, A., C. Gomez-Carneros (**2000**) Bitter taste, phytonutrients, and the consumer: a review. *Am J Clin Nutr* 72, 1424–1435.

175 Duffy, V.B., A.C. Davidson, J.R. Kidd, K.K. Kidd, W.C. Speed, A.J. Pakstis, D.R. Reed, D.J. Snyder, L.M. Bartoshuk (**2004**) Bitter receptor gene (TAS2R38), 6-n-propylthiouracil (PROP) bitterness and alcohol intake. *Alcohol Clin Exp Res* 28, 1629–1637.

19
Cancer and Gene Variants in Enzymes Metabolizing Dietary Xenobiotics

Susan Nowell and Fred F. Kadlubar

19.1
Introduction

Cancer etiology is complex, and many factors, both environmental and genetic, contribute to susceptibility to this disease. Humans are exposed to compounds from dietary sources that can be either protective against cancer or can elevate the risk of cancer occurrence. Indeed, the link between diet and cancer risk has long been recognized. Many epidemiologic investigations of these associations arose from the observation that there are significant differences in the incidence of specific cancers by geographical regions. In 1981, Doll and Peto attempted to quantify the contributions of several environmental exposures such as tobacco and alcohol consumption, occupation, and diet to cancer risk [1]. They estimated that approximately 35% of all cancers (with a range of 10–70% depending on the cancer site) could be prevented by changes in dietary habits. They went on to postulate that the contribution of diet to cancer risk was plausibly comparable to risk associated with tobacco use.

Migration studies have lent support to the hypothesis that diet contributes to cancer risk; migrant populations who move from low to high cancer incidence areas experience an increase in cancer rates, with risk approaching that of the newer region within a few generations. Nutritional epidemiologic studies have identified numerous dietary components that are associated with increased/decreased cancer risk. But identified nutritional risk factors are not consistent across all studies and only explain a portion of disease variability in populations. Thus, the question arises, "Why can we not explain more of cancer etiology in nutritional epidemiologic studies, and what is the basis of inconsistencies in observations?" Several different factors could account for discrepancies in findings from nutritional epidemiologic studies. These include biases inherent to epidemiologic and, particularly, case–control studies, such as recall, selection, and misclassification bias which could influence associations. Another contributing factor to the inability to discern risk/protective relationships may be the heterogeneity of study populations and the fact that most nutritional epidemiologic studies of diet and cancer have not taken into account the contribution of individual biochemistry to risk.

Nutritional Genomics. Edited by Regina Brigelius-Flohé and Hans-Georg Joost
Copyright © 2006 WILEY-VCH Verlag GmbH & Co. KGaA, Weinheim
ISBN: 3-527-31294-3

Humans have a myriad of enzymes that function to maintain cellular homeostasis. This includes enzymes that metabolize exogenous environmental compounds and nutrients ingested in food. Metabolism allows the utilization of nutrients and the subsequent detoxification and excretion of potentially harmful compounds and metabolites. The genes encoding metabolic enzymes are polymorphically expressed in humans; molecular biology and enzymology studies have shown that there are many polymorphisms that have a functional consequence for the expressed protein. Therefore, the interaction of genetic polymorphisms with consumed nutrients or with food-borne promutagens could serve to modulate diet-influenced cancer etiology. Table 19.1 lists identified genetic polymorphisms in genes involved in the metabolism of dietary xenobiotics.

19.2
Food Mutagens and Antimutagens

Food mutagens are known to cause DNA damage that can include both nucleotide alterations and chromosomal aberrations which are initiated by the formation of

Table 19.1. Genetic polymorphisms in dietary xenobiotic metabolizing enzymes

Gene	Polymorphic alleles	Effect on enzyme function
CYP1A1	MspI (intron 7)	Unknown
	Val/Ile (exon 7)	Unknown
CYP1B1	M1 (Val432Leu)	Unknown
	M2 (Asn453Ser)	Unknown
NAT1	*4	Low activity (wild type)
	*10	High activity
NAT2	*4	High activity
	*5A, *5B, *6A, *6B, *7A, *7B, *12A, *13, *13B	Low activity alleles
GSTA1	A1*1	High transcriptional activity
	A1*B	Low transcriptional activity
GSTM1	Null allele	No activity
GSTT1	Null allele	No activity
GSTP1	Val105	High activity (wild type)
	Ile105	Low activity
SULT1A1	Arg213	High activity (wild type)
	His213	Low activity
UGT1A1	*1	High transcriptional activity
	*28	Lower transcriptional activity
	*33	Lower transcriptional activity
	*34	Lower transcriptional activity
UGT1A6	T181 + R184	High activity
	A181 + S184	Low activity

carcinogen/DNA adducts. Adduct formation is the result of covalent binding of a carcinogen to a DNA base. However, adduct formation can be modified by host factors that affect the ultimate exposure of DNA to a food mutagen, such as variations in low-penetrance genes responsible for metabolic activation/detoxification of the putative mutagen or in genes responsible for cellular response to damage (i.e. genes controlling DNA repair or apoptosis). The most characterized food mutagens include heterocyclic amines, nitrosamines, polycyclic aromatic hydrocarbons, and mycotoxins.

Once consumed, most food mutagens undergo metabolic activation catalyzed by endogenous enzymes such as cytochromes P450 (CYPs) whose normal function is to eliminate compounds from the organism. For most food mutagens, however, phase I metabolism by CYPs results in the generation of electrophilic species capable of binding to cellular macromolecules, including DNA.

In addition to potentially harmful promutagens, there are several bioactive compounds from dietary sources that may be protective against carcinogenesis. These components include essential nutrients such as calcium, zinc, selenium, and folate in addition to non-essential food components such as carotenoids, flavonoids, and indoles. Genetic polymorphisms in enzymes that metabolize these components may influence the ultimate effect of the potentially protective agent. And some dietary components have also been shown to either induce or inhibit the expression of enzymes that are responsible for metabolism.

19.3
Meat Consumption, Genetics, and Cancer

19.3.1
Heterocyclic Amines and Colon Cancer

Epidemiologic studies have indicated that components of the Western-type diet, particularly high fat and meat consumption, are closely associated with the risk of colorectal cancer; prospective studies addressing meat consumption, particularly red meat, and colorectal cancer have shown a fairly consistent association [2]. In the early 1980s, researchers demonstrated that specific heterocyclic amines (HCAs) produced by pyrolysis of meats cooked at high temperatures were highly mutagenic [3]. In animal studies, exposure to HCAs results in the development of colon tumors in rodents. Recently, in its 11th Report on Carcinogens, the National Toxicology Program listed three HCAs (2-amino-3,4-dimethylimidazo[4,5-*f*]quinoline (MeIQ), 2-amino-3,8-dimethylimidazo[4,5-*f*]quinoxaline (MeIQx), and 2-amino-1-methyl-6-phenylimidazo[4,5-*b*]pyridine (PhIP)) as "reasonably anticipated to be human carcinogens." However, HCAs are considered to be promutagens in that they require metabolic activation to realize their full mutagenic potency. The major pathways of HCA metabolism are shown in Fig. 19.1. Bioactivation of HCAs to carcinogenic species *in vivo* is initiated by *N*-oxidation of the compound (reviewed in Ref. [4]). This reaction occurs primarily in the liver and,

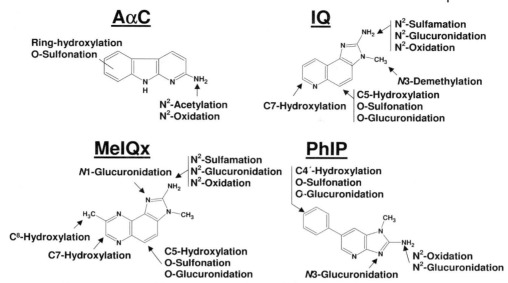

Fig. 19.1. Chemical structures of four common heterocyclic amines (HCAs) and the major pathways of their metabolism that have been reported in humans and animals. IQ, imidazo[4,5-*f*]quinoline; MeIQx, 2-amino-3,8-dimethylimidazo[4,5-*f*]quinoxaline; PhIP, 2-amino-1-methyl-6-phenylimidazo[4,5-*b*]pyridine.

in humans, is catalyzed by cytochrome P4501A2 (CYP1A2). Subsequent acetylation or sulfation of the *N*-hydroxy-HCA, catalyzed by acetyltransferases (NAT) and sulfotransferases (SULT), generates *N*-acetoxy and *N*-sulfonyloxy esters, electrophiles that are much more reactive with DNA. Figure 19.2 demonstrates the proposed pathway of HCA carcinogenesis in extrahepatic tissues.

Case–control studies have examined the relationship between HCA exposure, genetic variation in metabolic enzymes, and colorectal cancer risk and these studies were the subject of a recent review article by Pisani and Mitton [5]. Briefly, in one study, metabolic phenotypes for CYP1A2 and NAT2, in combination with lifestyle variables collected in a patient questionnaire were evaluated. Rapid phenotypes for these enzymes in conjunction with a preference for well-done meat conferred an increased risk of colorectal cancer (odds ratio (OR) = 6.45). Subsequent studies examining meat intake and metabolic genotypes and phenotypes in relation to both colorectal cancer and the occurrence of colorectal adenomas have not provided consistent results, but this discrepancy could be due to differences in study design, and the fact that dietary content of individual HCAs was not addressed in most instances. A majority of studies have not addressed the issue of the doneness of meat, which can substantially affect the amount of carcinogens in the meat [6–9]. In two studies that provided study participants with photographs of meat cooked to varying degrees of doneness, there was a significant association between doneness and the occurrence of colorectal adenomas and of colon cancer

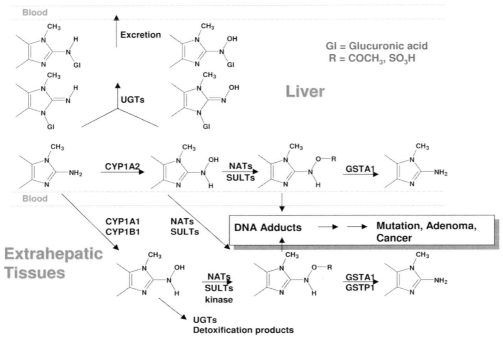

Fig. 19.2. Summary of the key pathways of 2-amino-1-methyl-6-phenylimidazo[4,5-*b*]pyridine (PhIP) metabolism in human liver and transport of genotoxic metabolites to extrahepatic tissues.

risk, presumably due to increased exposure to HCAs. In a large molecular epide-miologic study of colorectal cancer, meat consumption, and NAT2 genotype and CYP1A2 phenotype, LeMarchand and colleagues [10] showed that risk was great-est for those who were rapid for both NAT2 and CYP1A2, and were consumers of well-done meat (OR = 8.8, 95% confidence interval (CI) 1.7–44.9). SULT1A1 and SULT1E1 have been demonstrated to be involved in the activation of HCAs, but studies have not been able to demonstrate a convincing relationship between ex-posure to HCAs, genetic polymorphisms in these enzymes, and risk of colorectal cancer.

While oxidation, followed by acetylation or sulfation are activation pathways for HCAs, several detoxification pathways exist, including reduction of reactive electrophilic HCA metabolites by glutathione. PhIP is the most mass-abundant HCA and human GSTA1 has been shown to be the most effective member of the glutathione-*S*-transferase (GST) family in detoxifying the reactive PhIP metabolite *N*-acetoxy-PhIP [11]. There is a polymorphism in the promoter region of the *hGSTA1* gene that predicts the levels of hepatic expression of both hGSTA1 and hGSTA2 and, perhaps more importantly, the ratio of hGSTA1/GSTA2 expression [11]. One study examined the effect of *hGSTA1* genotype on susceptibility to colorectal cancer [11], and found that the allele associated with the lowest amount of hGSTA1 expression was significantly associated with risk of colorectal cancer.

In contrast to sulfation, glucuronidation of HCAs by the UDP-glucuronosyl-transferase (UGT) family of phase II enzymes can represent a detoxification pathway. Studies have identified several UGTs that are capable of catalyzing the glucuronidation of N-OH-PhIP. Research with recombinant enzymes has indicated that UGT1A1 plays a major role in the detoxification of N-OH-PhIP [12], although no studies of dietary exposures, UGT genotype, and cancer risk have been performed to date. But taken together, these studies indicate a role for exposure to a dietary carcinogen, metabolic variation, and risk of colorectal cancer.

19.3.2
Heterocyclic Amines and Breast Cancer

High consumption of meat has also been associated with an increased risk of breast cancer. Several prospective studies have also identified a link between meat consumption and breast cancer risk (reviewed in Ref. [2]). Case–control studies examining NAT2 genotype, meat consumption, and breast cancer risk have not been entirely in agreement in their conclusions. A case–control study by Ambrosone *et al.* found no association, while a later study by Dietz *et al.* demonstrated that NAT2 rapid acetylators who consumed well-done meat were at increased risk of breast cancer. A subsequent study in a Turkish population, however, did not find an association between NAT2 polymorphisms and breast cancer risk. Recent studies by Sinha *et al.* demonstrated an association between PhIP exposure and breast cancer risk [13]. Using an HCA database, the authors were able to estimate exposure to specific HCAs from food-frequency questionnaires, and found that PhIP was more strongly associated with breast cancer than intake of red meat alone.

Since most of the initial metabolism of HCAs from the diet occurs in the liver, the question of exposure of the breast target tissue to HCAs arises. A recent study by Zhu *et al.* [14] demonstrated the presence of PhIP/DNA adducts in the normal breast tissue obtained from breast cancer patients and from normal tissue of women undergoing reduction mammoplasty. These investigators found that significantly more adducts were detected in the tissue of breast cancer patients compared with tissue from healthy women. A potential interaction between *NAT2* genotype, well-done meat consumption and levels of PhIP/DNA adducts was also identified. PhIP has also been detected in epithelial cells isolated from the breast milk of healthy women [15] and PhIP/DNA adducts have been detected in human breast tissues by accelerator mass spectrometry after administration of ^{14}C-labeled PhIP in patients undergoing breast surgery [16], and in exfoliated epithelial cells obtained from human breast milk [17]. Hence, it is biologically plausible that HCA exposure and genetic variability could be involved in the etiology of breast cancer.

A genetic polymorphism in *SULT1A1* results in an amino acid change (Arg to His, designated *SULT1A1*1* and *SULT1A1*2*) at the conserved residue 213. In a predominantly White population, the frequency of the polymorphic allele has been reported to be 0.674 and 0.313 for *SULT1A1*1* and *SULT1A1*2*, respectively. The *SULT1A1*2* allele is associated with reduced sulfotransferase activity and thermostability in platelets, although the relationship to activity in human liver cytosol is not clear [18]. This polymorphism has been investigated in relation to breast cancer in several studies, with conflicting results, but only one study thus far has examined SULT1A1 polymorphism, consumption of well-done cooked meat and breast cancer risk. In this study, individuals who possessed the high-activity *SULT1A1* allele were at increased risk of breast cancer when they consumed large amounts of well-done cooked meat [19].

19.3.3
Heterocyclic Amines and Prostate Cancer

Prostate cancer incidence varies substantially by ethnicity and geographical location. Chinese men have the lowest rates of prostate cancer in the world, while African-American men have the highest. As with breast cancer, consumption of meat has been suggested to be involved in prostate cancer risk. PhIP also produces prostate tumors in rodents and human prostate tissue has been shown to metabolically activate *N*-hydroxy-PhIP to DNA-binding species [20, 21]. Moreover, transplantation of human prostate into athymic mice followed by exposure to PhIP resulted in PhIP/DNA adducts in approximately 95% of samples [22]. Finally, evidence has been presented that human prostate DNA contains the same PhIP/DNA adduct formed by metabolic activation of *N*-hydroxy-PhIP [21].

Some studies have examined the potential interaction of meat consumption and heterocyclic amine exposure in relation to prostate cancer. A large study in New Zealand did not find a clear association between HCA intake and prostate cancer risk, but did find an association between well-done beefsteak and the HCA 2-amino-1,6-dimethylfuro[3,2-*e*]imidazo[4,5-*b*]pyridine and risk of prostate cancer. Another study in the United States demonstrated that genotypes for both rapid NAT1 and rapid NAT2 were associated with increased risk of prostate cancer; authors speculated that metabolism of HCAs might be involved in prostate cancer etiology. While this study examined the NAT pathway of HCA activation, heterologous expression of human NAT and SULT enzymes in *Salmonella typhimurium* and mutagenicity testing of HCAs indicated that *N*-OH-PhIP was activated specifically by SULT1A1 in humans. *SULT1A1* transcript has been detected in human prostate, so a role for SULT1A1 in PhIP-induced prostate carcinogenesis is biologically plausible. A coding region polymorphism in *SULT1A1* was investigated in relation to prostate cancer risk in two studies. One report found no association of *SULT1A1* genotype with prostate cancer risk [23], while another study found that both *SULT1A1* genotype and SULT1A1 enzymatic activity was significantly associated with risk of this disease [24]. This difference is likely due to the size of the respective study populations. The study population from the latter study consisted of 403 prostate cancer patients and 450 control individuals, while that of the former consisted of 134 patients and 184 control individuals, all of White origin. Although their results did not achieve statistical significance, Steiner also reported a lower frequency of the *SULT1A1*2/*2* genotype in prostate cancer patients compared with control individuals, which is consistent with the study by Nowell *et al.* [24].

19.3.4
Nitrosamines and Colon Cancer

In addition to HCAs, recent results from large prospective studies have indicated that both exposures to nitrosamines in the diet and smoking cigarettes are positively associated with the risk of developing colorectal cancer. Several studies (reviewed in Ref. [25]) have reported increased colorectal cancer risk associated with consumption of processed meats, which are important sources of a variety of nitrosamines. One cohort study examined exposure to nitrate, nitrite, and *N*-nitroso compounds in relation to cancer risk in Finland and found that of all cancers examined, *N*-nitroso intake was associated with colorectal cancer risk. Another study examined the role of *CYP2E1* polymorphisms in relation to red meat intake and colorectal cancer risk. CYP2E1 catalyzes the biotransformation of relatively short-chain nitrosamines, while CYP2A6 metabolizes higher molecular weight nitrosamine compounds. The study involving the *CYP2E1* polymorphism found a significant association between variant genotype, consumption of salted fish and risk of rectal cancer [26]. While *CYP2A6* is polymorphic in humans, these polymorphisms are rare in the White population, at least in North America. Therefore, another study examined CYP2A6 activity, measured by analysis of caffeine metabolites in urine, and found that CYP2A6 activity was positively associated with colorectal cancer incidence [27].

19.3.5
Polycyclic Aromatic Hydrocarbons and Lung Cancer

Polycyclic aromatic hydrocarbons (PAH) are substances formed by the incomplete combustion of organic matter. PAHs have been demonstrated to be carcinogenic in animal models and in humans as well. PAHs are abundant in cigarette smoke, but PAH exposures from diet, particularly charbroiled meat, are also significant. Diet is estimated to result in PAH exposures of 3 µg/day, while cigarette smokers receive an additional 2–5 µg PAH/day [28]. Therefore, the exposures in diet are comparable to what smokers experience, although dietary routes of PAH exposure and cancer risk have received less attention than the risk associated with PAHs from smoking.

Although dietary exposures are significant, PAHs are ubiquitous in the environment, making it difficult to assess the contribution of dietary PAHs to cancer risk. Benzo[a]pyrene (BaP) is the most-characterized PAH found in dietary sources. It exerts its mutagenic effects by the binding of the bay region diol epoxide metabolites to DNA, generating an N^2-deoxyguanosine adduct [29]. Metabolic activation of BaP to mutagenic species is catalyzed by CYP1A1, CYP1B1, and CYP3A4.

19.3.6
Aflatoxin B$_1$ and Liver Cancer

Aflatoxin B$_1$ (AFB$_1$) is a mycotoxin found in improperly stored foodstuffs such as edible nuts, oil seeds, and grains. It is an extremely potent hepatocarcinogen in rodent studies and is acutely toxic in all species studied; experimental animals who succumb to AFB$_1$ exposure inevitably die of hepatotoxicity. In humans, exposure comes from consumption of contaminated corn and peanuts and also from animal feed. AFB$_1$ can be transmitted transplacentally and in breast milk to newborns. AFB$_1$ exposure in the United States is low, but it is a significant health problem in areas of Africa and Asia, particularly parts of China.

Epidemiologic studies have linked AFB$_1$ exposure and the occurrence of liver cancer with an observed synergistic effect of the hepatitis B virus [30]. The mechanistic basis of cancer risk related to dietary exposure to AFB$_1$ is the oxidation of this promutagen to the AFB-8,9-exo-epoxide, a species capable of binding to cellular macromolecules like proteins and DNA. This activation step is catalyzed primarily by CYP3A4 and CYP3A5. Metabolic defense against AFB-8,9-exo-epoxide is provided by conjugation of the reactive intermediate with glutathione, catalyzed by GSTM1, GSTT1, and GSTA4. Studies of CYP3A5 genotype and the presence of AFB$_1$/albumin adducts showed that CYP3A5 haplotypes that conferred high CYP3A5 activity were significantly related to increased levels of AFB$_1$/albumin adducts. The magnitude of the effect was found to be higher in individuals who possessed low CYP3A4 activity [31]. Another study found that the incidence of hepatocellular carcinoma was significantly greater in individuals with *GSTM1*- and *GSTT1*-null genotypes [32] and with an intron 7 polymorphism in *GSTA4* [33]. Therefore, genetic variation in CYP3A4/CYP3A5 and GSTM1/T1/A4 might modulate susceptibility to aflatoxin-associated hepatocellular carcinoma.

19.4
Fruit and Vegetable Consumption and Cancer

High consumption of fruits and vegetables has been thought to provide protection against many types of cancers. One major review of the literature published in 1997 concluded that evidence of the role of fruits and vegetables in protecting against cancer is convincing [34]. However, the majority of studies have been case–control studies, and in cohort studies the associations identified are considerably weaker. But in 2005, analysis of the European Prospective Investigation into Cancer and Nutrition (EPIC) cohort found no significant associations between breast cancer risk and fruit and vegetable consumption [35]. EPIC is a prospective cohort designed to examine the relationship between food, nutritional status and incidence of various types of cancers. It recruited almost 300 000 women between the years 1992 and 1998, and of those, 3659 cases of invasive breast cancer were reported. While no significant associations were found, it is noteworthy that at the time of analysis, follow-up was relatively short, exposures at younger ages (when cancer initiation is thought to begin) were not measured, and the influence of gene–environment interactions were not assessed. Nonetheless, consumption of fruits and vegetables could decrease cancer risk through several mechanisms. For example, vegetable consumption could reduce cancer risk due to the abundance of antioxidants they contain and also because they contain compounds as isoflavones, lignans, indoles, and isothiocyanates, that have been shown to have anticarcinogenic properties.

19.4.1
Cruciferous Vegetables and Cancer Risk

Cruciferous vegetables, which include broccoli, cauliflower, Brussels sprouts, cabbage, mustard, and watercress, represent more than 10% of vegetable intake. Levels of consumption vary geographically, with the highest consumption reported in China; Americans consume approximately one-quarter the amount of cruciferous vegetables consumed by people in China. There is also significant variability in cruciferous vegetable consumption in Europe, with Northern and Central Europeans consuming more than Southern Europeans. Epidemiological studies that employ food-frequency questionnaires to assess the relationship between cruciferous vegetable consumption and cancer risk have indicated that these vegetables may be important in protecting against stomach and lung cancer [36].

The putative anticarcinogenic effects of cruciferous vegetables may derive from the consumption of glucosinolates, which are degraded into indoles and isothiocyanates. Indoles have been studied in relation to breast cancer partly because of their effects on estrogen metabolism. Michnovicz and Bradlow [37] and others hypothesized that indole-3-carbinol has anti-estrogenic effects, primarily through induction of 2-hydroxylation of estradiol, which gives rise to a non-estrogenic metabolite. Indole-3-carbinol can bind to the estrogen receptor with low affinity, therefore it may repress 17β-estradiol-activated ER-α signaling, thus downregulating the expression of estrogen-responsive genes. In addition to indole-3-carbinol, cruciferous vegetables also contain glucosinolates, which are hydrolyzed to isothiocyanates (ITCs) by the action of myrosinase, an enzyme present in the plant itself. ITCs have been demonstrated to possess chemoprotective properties in animal models. They are sulfur-containing compounds that are responsible for the flavor of cruciferous vegetables. ITCs could also be responsible for the observed inverse associations between cruciferous vegetable consumption and cancer risk. The chemopreventive effect of ITCs observed in animal models is due, in part, to their indirect effects on metabolism of xenobiotics. ITCs are known to inhibit phase I activating enzymes (e.g. cytochrome P450s), and also to induce the expression of phase II detoxifying enzymes such as the glutathione-S-transferase family (GSTs).

19.4.2
Fruits, Vegetables, and Glutathione-S-transferases

The GSTs are a family of phase II enzymes that are involved in the detoxification of carcinogen metabolites and reactive oxidative products. They comprise seven classes: alpha, mu, pi, omega, kappa, theta, and zeta. Of these, the alpha class appears to possess the greatest peroxidase activity but enzymes of this class are expressed at low levels in both normal and breast tumor tissue. A substantial proportion of the White population have a homozygous deletion of the *GSTM1* and *GSTT1* genes, which results in lack of enzyme activity, with the *GSTM1*-null polymorphism present in approximately 50% of the population [38].

As reviewed in Ref. [36], studies have shown that individuals who possess the homozygous null allele are at increased risk of lung and bladder cancer, both of which are associated with exposure to chemical carcinogens. However, studies of possible associations between *GSTM1* and breast cancer risk have yielded inconsistent results. While a western New York study and another study performed in Australia did not show an association between GSTM1 and breast cancer, Helzlsouer noted a more than two-fold increased risk with the null allele of these genes. Zheng *et al.* have also reported an association between *GSTM1*- and *GSTP1*-null genotype with a 60% increased risk of breast cancer.

Because ITCs are substrates for GSTs, and also induce the expression of GSTs, studies were conducted to investigate interactions between GST genotypes, cruciferous vegetable intake, and risk of cancer. Associations were also investigated in relation to lung cancer risk for which carcinogen exposure is a known risk factor, and the detoxifying effects of the GSTs may play a more important role. In a nested case–control study from a cohort of men in Shanghai, China, investigators found that individuals with detectable urinary isothiocyanates were at decreased risk of lung cancer, and risk was lowest among those who carried deletions in *GSTM1* and *GSTT1*. This observation also held true when the analysis was restricted to smokers. Others found that lung cancer risk was greatest among those with GST-null genotypes who were low consumers of ITCs. Among higher consumers of ITCs, risk was greater for those with null alleles than those with present genotypes. These data indicate that the inducing effects of ITCs on GSTs may be more important than their role in the metabolism and excretion of the chemopreventive agents.

Among non-smoking women in China, however, the strongest inverse associations with ITC intake on lung cancer risk were among those with *GSTM1*- and *GSTT1*-null genotypes. Another study of breast cancer and urinary isothiocyanates in China showed decreased risk of breast cancer in women with either the null or non-null genotype, but the effect of isothiocyanates was statistically significant only for those with null genotypes [36]. Lin *et al.* evaluated the hypothesis that GST genotype and ITCs could modulate the occurrence of colorectal adenomas, and found inverse associations between broccoli consumption and risk of colorectal adenomas in patients with *GSTM1*-null genotypes; however, this association was not evident when colon cancer was examined [39].

Ambrosone *et al.* [40] investigated associations between consumption of cruciferous vegetables and breast cancer risk, and the potential modifying effects of *GSTM1* and *GSTT1* genotypes. Consumption of cruciferous vegetables, particularly broccoli, was marginally inversely associated with breast cancer risk in premenopausal women, although no significant interaction effects of GST genotype on risk were observed.

19.4.3
Fruits, Vegetables, and Sulfotransferases

Sulfotransferases are classified as phase II detoxification enzymes, and are responsible for conjugating xeno- and endobiotics with 5'-phosphoadenosine 3'-phosphosulfate, thus rendering the substrate more water-soluble. In some instances, as earlier discussed concerning HCAs, sulfation of some xenobiotics results in their metabolic activation. In this instance, sulfation is a double-edged sword. Several SULT isoforms are involved in the metabolism of compounds found in fruits and vegetables, but it appears that SULT1A1 plays a major role in catalyzing the conjugation of dietary phytoestrogens, including diadzein and genistein. SULT1A1 is also potently inhibited by a variety of dietary chemicals (reviewed in Ref. [41]). For example, curcumin, which is under investigation as a chemopreventive agent for colorectal cancer, inhibits SULT1A1 activity with an IC_{50} of 12.8 nmol/l, a level easily achievable *in vivo* [41].

SULT1A1 activity, as measured in human platelet preparations, has also been shown to be inhibited by as much as 99% by red wine extracts [41]. In 1996, Harris and Waring reported the inhibitory effect of more than 30 dietary constituents found in vegetables on SULT1A1 activity [42]. Using vegetable cytosols, they found that, of all cytosols tested, constituents found in radishes, spinach, broccoli, bananas, and leeks were the most potent inhibitors of both SULT1A1 and SULT1A3 activity, although the identity of the inhibitory substance was not identified. Later studies examined the identity of these compounds and found that quercetin, an abundant flavonoid found in fruit, vegetables (and in wine) inhibited SULT1A1 activity with an IC_{50} of 0.1 μmol/l, and was also an inhibitor of SULT1A3, SULT1E1, and SULT2A1 (DHEA sulfotransferase) [41]. This is well within range of the peak plasma levels of quercetin, which have been reported to be between 0.3 and 0.7 μmol/l. As with quercetin, other dietary compounds inhibit SULT1A1, such as 5-OH-flavone and 3-OH-flavone, and both epicatechin gallate and epigallocatechin gallate. For this reason, association studies of SULT1A1 genotype and cancer risk may very well be confounded by the effect of fruit and vegetable components on SULT1A1 activity.

Thus far, no studies have examined the interaction between fruit and vegetable intake, SULT1A1 genotype, and cancer risk.

19.4.4
Fruits, Vegetables and UDP-Glucuronosyltransferases

The UDP-glucuronosyltransferases (UGTs) (EC 2.4.1.17) are a gene superfamily whose principal role is to convert endo- and xenobiotics into water-soluble derivatives. UGTs are divided into two families based on evolutionary divergence. The UGT1A locus is on chromosome 2 and can potentially encode nine functional isoforms and three pseudogenes. The UGT1A gene complex is composed of multiple tandem first exons that encode the variable N-terminal part of the enzyme and are linked by differential splicing to common exons that encode the C-terminal region.

The first exons have unique TATA elements approximately 30 bp upstream, allowing for independent regulation of the isoforms. Members of the UGT2B family are unique gene products and preferentially glucuronidate steroids and bile acids, in addition to xenobiotics. To date, at least eight 2B isoforms have been identified. UGTs are expressed primarily in the liver but recent findings indicate that extrahepatic glucuronidation contributes significantly to detoxification of endo- and xenobiotics, particularly in the gastrointestinal tract [43].

Studies in animal models have shown that phenethyl isothiocyanate, a constituent of cruciferous vegetables, leads to a slight increase in hepatic UGT activity in F344 rats. Piperine, which is a major component of black pepper, has been shown to potently inhibit UGT activity in intestinal epithelial cells of guinea-pigs [44]. This inhibition was observed to be time- and concentration-dependent. Induction of UGT activity has been reported in human-derived HepG2 cells by exposure to extracts of garden cress and white mustard, sprouts which vary in their glucosinolate content [45]. Exposure of HepG2 cells to the extracts resulted in a 1.4- and 1.8-fold induction of UGT activity for garden cress and white mustard, respectively. While these investigations showed induction of UGT activity toward 4-methylumbelliferone, the specific isoform(s) induced were not identified. Studies using the human cell lines HepG2 and Caco-2 have demonstrated that the flavonoid chrysin is a potent and fairly selective inducer of UGT1A1 [46]. Treatment of cells with 25 μmol/l chrysin resulted in a 20-fold increase in the activity of UGT1A1 for bilirubin, its primary endogenous substrate.

Induction of UGT1A1 may be particularly important given the role of UGT1A1 in the detoxification of the heterocyclic amine N-OH-PhIP, as discussed previously. One study in human subjects has examined the effect of cruciferous vegetable consumption on the metabolism of PhIP in 20 non-smoking male White volunteers [47]. This study was divided into three phases: one where the participants avoided cruciferous vegetables for 12 days, one where the participants ingested 250 g each of Brussels sprouts and broccoli for 12 days and another 12-day avoidance period. At the end of each phase of the study, participants ingested a cooked meat meal containing relatively high levels of PhIP and urine samples were collected. Urinary levels of N^2-hydroxy-N^2-PhIP-glucuronide (the major urinary metabolite of PhIP in humans) were increased 127 and 136% in the cruciferous vegetable-consuming phase compared with the first and last phases, respectively. Since UGT1A1 has a major role in the production of this metabolite and has been shown to be inducible by dietary components of fruits and vegetables, it may play a crucial role in the balance between mutagenic and antimutagenic outcomes of exposures to dietary PhIP and to cruciferous vegetables.

Numerous mutations in *UGT1A1* (reviewed in Ref. [43]) have been identified but only a few are of sufficient frequency in the general population to be classified as polymorphisms. One particular genetic variation that has been investigated in several studies of cancer risk is a dinucleotide repeat $(A(TA)_n TAA)$ in the atypical TATA-box region of the *UGT1A1* promoter. Four variant alleles are the result of variation in the number of dinucleotide repeats. Five repeats generates *UGT1A1*33*; six repeats generates *UGT1A1*1* ("wild type"); seven repeats generates *UGT1A1*28*; and eight repeats generates *UGT1A1*34*. Functional studies have shown that increasing numbers of repeats lead to decreased transcription of *UGT1A1*. *UGT1A1*28* is the most common variant allele and has been associated with the occurrence of Gilbert's syndrome, a mild form of unconjugated hyperbilirubinemia. The distribution of *UGT1A1* alleles varies by ethnicity, with *UGT1A1*34* much more common in African-American than in White populations.

UGT1A1 genotype has been investigated in relation to breast cancer risk and breast density; thus far no one has investigated the effect of this polymorphism on risk of cancers that are influenced by either consumption of well-done cooked meat or by vegetable consumption.

19.5
Tea Polyphenols and Cancer Risk

In recent years, epidemiologic studies have suggested that consumption of green tea is associated with reduced risk of cancer at several sites, including breast, colon, ovary, and lung. The benefits conferred by consumption of tea, particularly green tea, are attributed to the presence of polyphenolic compounds. Tea has many different polyphenols present, but the most abundant polyphenols are the catechins, which include (−)-epigallocatechin gallate (EGCG), the most pharmacologically active tea polyphenol. The exact mechanisms by which EGCG exerts its protective effects are not well-defined. These studies are complicated by the fact that EGCG has low absorption and bioavailability, such that humans would need to drink more than one liter of green tea per day to achieve concentrations that are associated with anticarcinogenic effects [48]. In addition, while the peak concentration of EGCG in human plasma occurs between 1 and 2 h after oral administration with a half-life of 3–5 h, EGCG is not detected in urine as unchanged compound, suggesting that it is a target of extensive biotransformation [48]. This biotransformation appears to involve methylation, glucuronidation, and sulfation reactions, with methylated metabolites excreted in the urine and glucuronidated/sulfated metabolites excreted in bile. At present, it is not clear if any of these metabolites is responsible for the pharmacological actions of EGCG.

Several studies have indicated that methylation is an important biotransformation mechanism for EGCG. Lu *et al.* recently demonstrated that EGCG is a substrate for liver cytosolic catechol-*O*-methyltransferase (COMT) [49]. There is wide interindividual variation in the excretion of methylated EGCG, which could be due to polymorphisms in COMT. One polymorphism in COMT has been the target of several investigations in relation to cancer risk. This polymorphism consists of a G to A transversion that results in an amino acid change from methionine to valine at amino acid 158 of the translated protein (*COMT*2*) [50]. *COMT*2* generates an enzyme with greatly decreased catalytic activity compared with the *COMT*1* common allele. The allele distribution varies by ethnicity, with approximately 25–30% of White and Asian populations homozygous for the variant allele, compared with 7–9% of individuals with African heritage [48]. To date, studies of EGCG consumption, COMT genotype, and cancer risk are lacking.

EGCG is also subject to conjugation with glucuronic acid, and recent studies have indicated that human UGT1A1, UGT1A3, UGT1A8, and UGT1A9 are capable of catalyzing this reaction [51]. Interestingly, these investigators found that when the major glucuronidation site of EGCG is methylated, this serves to enhance UGT1A9 activity. Since both COMT and UGT1A9 are highly expressed in the kidney, this could contribute to the renal excretion of EGCG. As noted previously in this chapter, all of the aforementioned UGT isoforms are polymorphic in humans, although studies of the interaction between UGT polymorphisms, EGCG consumption, and cancer risk have not been performed to date.

Catechins, particularly EGCG, have been shown to inhibit the sulfation of 1-naphthol in Caco-2 cells [52]. Further work demonstrated that EGCG inhibits SULT1A1 activity with a K_i value of 0.04 µmol/l, concentrations that are easily achievable *in vivo* [53]. SULT1A3 is also inhibited, but at almost six-fold higher concentrations. Since bioactivation of HCAs by SULT1A1 is thought to contribute to HCA-induced carcinogenesis, inhibition of this enzyme by EGCG could account for part of the chemoprotective effects associated with green tea consumption. However, as with COMT and UGT isoforms, investigations into the interaction between SULT1A1 genotype and tea consumption in relation to cancer risk are lacking.

Cancer is a multifactorial disease, with both genetic and environmental factors contributing to its occurrence. Understanding of the interaction between exposures to xenobiotics and genetic polymorphisms in enzymes involved in their biotransformation is critically important to the individual in making informed lifestyle choices. Furthermore, it is quite likely that positive or negative associations found by traditional epidemiologic methods will be modified when genetic variation is taken into account.

References

1 DOLL, R., R. PETO (**1981**) The causes of cancer: quantitative estimates of avoidable risks of cancer in the United States today. *J Natl Cancer Inst* 66, 1191–1308.

2 NOWELL, S.A., J. AHN, C.B. AMBROSONE (**2004**) Gene–nutrient interactions in cancer etiology. *Nutr Rev* 62, 427–438.

3 NAGAO, M., Y. FUJITA, K. WAKABAYASHI, T. SUGIMURA (**1983**) Ultimate forms of mutagenic and carcinogenic heterocyclic amines produced by pyrolysis. *Biochem Biophys Res Commun* 114, 626–631.

4 KIM, D., F.P. GUENGERICH (**2004**) Cytochrome P450 activation of arylamines and heterocyclic amines. *Annu Rev Pharmacol Toxicol.*

5 PISANI, P., N. MITTON (**2002**) Cooking methods, metabolic polymorphisms and colorectal cancer. *Eur J Cancer Prev* 11, 75–84.

6 SINHA, R., N. ROTHMAN, C.P. SALMON, M.G. KNIZE, E.D. BROWN, C.A. SWANSON, D. RHODES, S. ROSSI, J.S. FELTON, O.A. LEVANDER (**1998**) Heterocyclic amine content in beef cooked by different methods to varying degrees of doneness and gravy

made from meat drippings. *Food Chem Toxicol* 36, 279–287.

7 KNIZE, M.G., C.P. SALMON, P. PAIS, J.S. FELTON (1999) Food heating and the formation of heterocyclic aromatic amine and polycyclic aromatic hydrocarbon mutagens/carcinogens. *Adv Exp Med Biol* 459, 179–193.

8 KNIZE, M.G., K.S. KULP, C.P. SALMON, G.A. KEATING, J.S. FELTON (2002) Factors affecting human heterocyclic amine intake and the metabolism of PhIP. *Mutat Res* 506–507, 153–162.

9 SINHA, R., N. ROTHMAN (1997) Exposure assessment of heterocyclic amines (HCAs) in epidemiologic studies. *Mutat Res* 376, 195–202.

10 LE MARCHAND, L., J.H. HANKIN, L.R. WILKENS, L.M. PIERCE, A. FRANKE, L.N. KOLONEL, A. SEIFRIED, L.J. CUSTER, W. CHANG, A. LUM-JONES, T. DONLON (2001) Combined effects of well-done red meat, smoking, and rapid N-acetyltransferase 2 and CYP1A2 phenotypes in increasing colorectal cancer risk. *Cancer Epidemiol Biomarkers Prev* 10, 1259–1266.

11 COLES, B., S.A. NOWELL, S.L. MACLEOD, C. SWEENEY, N.P. LANG, F.F. KADLUBAR (2001) The role of human glutathione S-transferases (hGSTs) in the detoxification of the food-derived carcinogen metabolite N-acetoxy-PhIP, and the effect of a polymorphism in hGSTA1 on colorectal cancer risk. *Mutat Res* 482, 3–10.

12 MALFATTI, M.A., J.S. FELTON (2001) N-glucuronidation of 2-amino-1-methyl-6-phenylimidazo[4,5-b]pyridine (PhIP) and N-hydroxy-PhIP by specific human UDP-glucuronosyltransferases. *Carcinogenesis* 22, 1087–1093.

13 SINHA, R., D.R. GUSTAFSON, M. KULLDORFF, W.Q. WEN, J.R. CERHAN, W. ZHENG (2000) 2-amino-1-methyl-6-phenylimidazo[4,5-b]pyridine, a carcinogen in high-temperature-cooked meat, and breast cancer risk. *J Natl Cancer Inst* 92, 1352–1354.

14 ZHU, J., P. CHANG, M.L. BONDY, A.A. SAHIN, S.E. SINGLETARY, S. TAKAHASHI, T. SHIRAI, D. LI (2003) Detection of 2-amino-1-methyl-6-phenylimidazo[4,5-b]-pyridine-DNA adducts in normal breast tissues and risk of breast cancer. *Cancer Epidemiol Biomarkers Prev* 12, 830–837.

15 DEBRUIN, L.S., P.A. MARTOS, P.D. JOSEPHY (2001) Detection of PhIP (2-amino-1-methyl-6-phenylimidazo[4,5-b]pyridine) in the milk of healthy women. *Chem Res Toxicol* 14, 1523–1528.

16 LIGHTFOOT, T.J., J.M. COXHEAD, B.C. CUPID, S. NICHOLSON, R.C. GARNER (2000) Analysis of DNA adducts by accelerator mass spectrometry in human breast tissue after administration of 2-amino-1-methyl-6-phenylimidazo[4,5-b]pyridine and benzo[a]pyrene. *Mutat Res* 472, 119–127.

17 GORLEWSKA-ROBERTS, K., B. GREEN, M. FARES, C.B. AMBROSONE, F.F. KADLUBAR (2002) Carcinogen-DNA adducts in human breast epithelial cells. *Environ Mol Mutagen* 39, 184–192.

18 MACLEOD, S.L., S. NOWELL, N.P. LANG (2000) Genetic polymorphisms. In: *Food Borne Carcinogens: Heterocyclic Amines.* M. NAGAO, T. SIGUMURA, editors. Chichester: John Wiley & Sons, pp. 112–130.

19 ZHENG, W., D. XIE, J.R. CERHAN, T.A. SELLERS, W. WEN, A.R. FOLSOM (2001) Sulfotransferase 1A1 polymorphism, endogenous estrogen exposure, well-done meat intake, and breast cancer risk. *Cancer Epidemiol Biomarkers Prev* 10, 89–94.

20 WILLIAMS, J., E. STONE, G. FAKIS, N. JOHNSON, W. MEINL, H. GLATT, E. SIM, D. PHILLIPS (2000) Human mammary NAT and SULT enzymes metabolically activate N-hydroxylated heterocyclic amines, but NAT enzyme activity is not influenced by NAT genotype. *Proc Am Assoc Cancer Res* 41, 551.

21 DIPAOLO, O., C. TEITEL, S. NOWELL, B. COLES, F. KADLUBAR (2005) Expression of cytochromes P450 and glutathione S-transferases in human prostate, and the potential for activation of heterocyclic amine carcinogens via acetyltransferase-, PAPS- and ATP-

dependent pathways. *Int J Cancer*, in press.

22 CUI, L., S. TAKAHASHI, M. TADA, K. KATO, Y. YAMADA, K. KOHRI, T. SHIRAI (2000) Immunohistochemical detection of carcinogen-DNA adducts in normal human prostate tissues transplanted into the subcutis of athymic nude mice: results with 2-amino-1-methyl-6-phenylimidazo[4,5-b]pyridine (PhIP) and 3,2'-dimethyl-4-aminobiphenyl (DMAB) and relation to cytochrome P450s and N-acetyltransferase activity. *Jpn J Cancer Res* 91, 52–58.

23 STEINER, M., M. BASTIAN, W.A. SCHULZ, T. PULTE, K.H. FRANKE, A. ROHRING, J.M. WOLFF, H. SEITER, P. SCHUFF-WERNER (2000) Phenol sulphotransferase SULT1A1 polymorphism in prostate cancer: lack of association. *Arch Toxicol* 74, 222–225.

24 NOWELL, S., D.L. RATNASINGHE, C.B. AMBROSONE, S. WILLIAMS, T. TEAGUE-ROSS, L. TRIMBLE, G. RUNNELS, A. CARROL, B. GREEN, A. STONE, D. JOHNSON, G. GREENE, F.F. KADLUBAR, N.P. LANG (2004) Association of SULT1A1 phenotype and genotype with prostate cancer risk in African-Americans and Caucasians. *Cancer Epidemiol Biomarkers Prev* 13, 270–276.

25 NORAT, T., E. RIBOLI (2001) Meat consumption and colorectal cancer: a review of epidemiologic evidence. *Nutr RevRev* 59, 37–47.

26 LE MARCHAND, L., T. DONLON, A. SEIFRIED, L.R. WILKENS (2002) Red meat intake, CYP2E1 genetic polymorphisms, and colorectal cancer risk. *Cancer Epidemiol Biomarkers Prev* 11, 1019–1024.

27 NOWELL, S., C. SWEENEY, G. HAMMONS, F.F. KADLUBAR, N.P. LANG (2002) CYP2A6 activity determined by caffeine phenotyping: association with colorectal cancer risk. *Cancer Epidemiol Biomarkers Prev* 11, 377–383.

28 GOLDMAN, R., P.G. SHIELDS (2003) Food mutagens. *J Nutr* 133 (Suppl 3), 965S–973S.

29 CHENG, S.C., B.D. HILTON, J.M.

ROMAN, A. DIPPLE (1989) DNA adducts from carcinogenic and noncarcinogenic enantiomers of benzo[a]pyrene dihydrodiol epoxide. *Chem Res Toxicol* 2, 334–340.

30 QIAN, G.S., R.K. ROSS, M.C. YU, J.M. YUAN, Y.T. GAO, B.E. HENDERSON, G.N. WOGAN, J.D. GROOPMAN (1994) A follow-up study of urinary markers of aflatoxin exposure and liver cancer risk in Shanghai, People's Republic of China. *Cancer Epidemiol Biomarkers Prev* 3, 3–10.

31 WOJNOWSKI, L., P.C. TURNER, B. PEDERSEN, E. HUSTERT, J. BROCKMOLLER, M. MENDY, H.C. WHITTLE, G. KIRK, C.P. WILD (2004) Increased levels of aflatoxin-albumin adducts are associated with CYP3A5 polymorphisms in The Gambia, West Africa. *Pharmacogenetics* 14, 691–700.

32 DENG, Z.L., Y.P. WEI, Y. MA (2005) Polymorphism of glutathione S-transferase mu 1 and theta 1 genes and hepatocellular carcinoma in southern Guangxi, China. *World J Gastroenterol* 11, 272–274.

33 McGLYNN, K.A., K. HUNTER, T. LeVOYER, J. ROUSH, P. WISE, R.A. MICHIELLI, F.M. SHEN, A.A. EVANS, W.T. LONDON, K.H. BUETOW (2003) Susceptibility to aflatoxin B1-related primary hepatocellular carcinoma in mice and humans. *Cancer Res* 63, 4594–4601.

34 FUND, W.C.R., A.I.f.C. RESEARCH (1997) Food, Nutrition and the Prevention of Cancer: a Global Perspective. American Institute for Cancer Research, Washington.

35 VAN GILS, C.H., P.H. PEETERS, H.B. BUENO-DE-MESQUITA, H.C. BOSHUIZEN, P.H. LAHMANN, F. CLAVEL-CHAPELON, A. THIEBAUT, E. KESSE, S. SIERI, D. PALLI, R. TUMINO, S. PANICO, P. VINEIS, C.A. GONZALEZ, E. ARDANAZ, M.J. SANCHEZ, P. AMIANO, C. NAVARRO, J.R. QUIROS, T.J. KEY, N. ALLEN, K.T. KHAW, S.A. BINGHAM, T. PSALTOPOULOU, M. KOLIVA, A. TRICHOPOULOU, G. NAGEL, J. LINSEISEN, H. BOEING, G. BERGLUND, E. WIRFALT, G. HALLMANS, P. LENNER, K. OVERVAD, A.

Tjonneland, A. Olsen, E. Lund, D. Engeset, E. Alsaker, T. Norat, R. Kaaks, N. Slimani, E. Riboli (**2005**) Consumption of vegetables and fruits and risk of breast cancer. *JAMA* 293, 183–193.

36 Bianchini, F., H. Vainio (**2004**) Isothiocyanates in cancer prevention. *Drug Metab Rev* 36, 655–667.

37 Michnovicz, J.J., H.L. Bradlow (**1990**) Dietary and pharmacological control of estradiol metabolism in humans. *Ann NY Acad Sci* 595, 291–299.

38 Brockmoller, J., D. Gross, R. Kerb, N. Drakoulis, I. Roots (**1992**) Correlation between trans-stilbene oxide-glutathione conjugation activity and the deletion mutation in the glutathione S-transferase class mu gene detected by polymerase chain reaction. *Biochem Pharmacol* 43, 647–650.

39 Slattery, M.L., E. Kampman, W. Samowitz, B.J. Caan, J.D. Potter (**2000**) Interplay between dietary inducers of GST and the GSTM-1 genotype in colon cancer. *Int J Cancer* 87, 728–733.

40 Ambrosone, C.B., S.E. McCann, J.L. Freudenheim, J.R. Marshall, Y. Zhang, P.G. Shields (**2004**) Breast cancer risk in premenopausal women is inversely associated with consumption of broccoli, a source of isothiocyanates, but is not modified by GST genotype. *J Nutr* 134, 1134–1138.

41 Pacifici, G.M. (**2004**) Inhibition of human liver and duodenum sulfotransferases by drugs and dietary chemicals: a review of the literature. *Int J Clin Pharmacol Ther* 42, 488–495.

42 Harris, R.M., R.H. Waring (**1996**) Dietary modulation of human platelet phenolsulphotransferase activity. *Xenobiotica* 26, 1241–1247.

43 Guillemette, C. (**2003**) Pharmacogenomics of human UDP-glucuronosyltransferase enzymes. *Pharmacogenomics J* 3, 136–158.

44 Grancharov, K., Z. Naydenova, S. Lozeva, E. Golovinsky (**2001**) Natural and synthetic inhibitors of UDP-glucuronosyltransferase. *Pharmacol Ther* 89, 171–186.

45 Lhoste, E.F., K. Gloux, I. De Waziers, S. Garrido, S. Lory, C. Philippe, S. Rabot, S. Knasmuller (**2004**) The activities of several detoxication enzymes are differentially induced by juices of garden cress, water cress and mustard in human HepG2 cells. *Chem Biol Interact* 150, 211–219.

46 Walle, T., Y. Otake, A. Galijatovic, J.K. Ritter, U.K. Walle (**2000**) Induction of UDP-glucuronosyltransferase UGT1A1 by the flavonoid chrysin in the human hepatoma cell line hep G2. *Drug Metab Dispos* 28, 1077–1082.

47 Walters, D.G., P.J. Young, C. Agus, M.G. Knize, A.R. Boobis, N.J. Gooderham, B.G. Lake (**2004**) Cruciferous vegetable consumption alters the metabolism of the dietary carcinogen 2-amino-1-methyl-6-phenylimidazo[4,5-b]pyridine (PhIP) in humans. *Carcinogenesis* 25, 1659–1669.

48 Moyers, S.B., N.B. Kumar (**2004**) Green tea polyphenols and cancer chemoprevention: multiple mechanisms and endpoints for phase II trials. *Nutr Rev* 62, 204–211.

49 Lu, H., X. Meng, C.S. Yang (**2003**) Enzymology of methylation of tea catechins and inhibition of catechol-O-methyltransferase by (–)-epigallocatechin gallate. *Drug Metab Dispos* 31, 572–579.

50 Shield, A.J., B.A. Thomae, B.W. Eckloff, E.D. Wieben, R.M. Weinshilboum (**2004**) Human catechol O-methyltransferase genetic variation: gene resequencing and functional characterization of variant allozymes. *Mol Psychiatry* 9, 151–160.

51 Lu, H., X. Meng, C. Li, S. Sang, C. Patten, S. Sheng, J. Hong, N. Bai, B. Winnik, C.T. Ho, C.S. Yang (**2003**) Glucuronides of tea catechins: enzymology of biosynthesis and biological activities. *Drug Metab Dispos* 31, 452–461.

52 Isozaki, T., H. Tamura (**2001**) Epigallocatechin gallate (EGCG) inhibits the sulfation of 1-naphthol in a human colon carcinoma cell line, Caco-2. *Biol Pharm Bull* 24, 1076–1078.

53 COUGHTRIE, M.W., L.E. JOHNSTON (**2001**) Interactions between dietary chemicals and human sulfotransferases-molecular mechanisms and clinical significance. *Drug Metab Dispos* 29, 522–528.

Index

Nutritional Genomics. Edited by Regina Brigelius-Flohé and Hans-Georg Joost
Copyright © 2006 WILEY-VCH Verlag GmbH & Co. KGaA, Weinheim
ISBN: 3-527-31294-3